Geospatial Analysis of Environmental Health

Geotechnologies and the Environment

Volume 4

Series Editors:

Jay D. Gatrell, *School of Graduate Studies and Department of Geography, Geology, and Anthropology, Indiana State University, Terre Haute, IN, USA*
Ryan R. Jensen, *Department of Geography, Brigham Young University, Provo, UT, USA*

The "Geotechnologies and the Environment" series is intended to provide specialists in the geotechnologies and academics who utilize these technologies, with an opportunity to share novel approaches, present interesting (sometimes counter-intuitive) case studies, and most importantly to situate GIS, remote sensing, GPS, the internet, new technologies, and methodological advances in a real world context. In doing so, the books in the series will be inherently applied and reflect the rich variety of research performed by geographers and allied professionals.

Beyond the applied nature of many of the papers and individual contributions, the series interrogates the dynamic relationship between nature and society. For this reason, many contributors focus on human-environment interactions. The series are not limited to an interpretation of the environment as nature per se. Rather, the series "places" people and social forces in context and thus explore the many socio-spatial environments humans construct for themselves as they settle the landscape. Consequently, contributions will use geotechnologies to examine both urban and rural landscapes.

For further volumes:
http://www.springer.com/series/8088

Juliana A. Maantay · Sara McLafferty
Editors

Geospatial Analysis of Environmental Health

Springer

Editors
Prof. Juliana A. Maantay
Earth, Environmental, and Geospatial Sciences Department
City University of New York Lehman College
Bronx, NY 10468
USA
Juliana.maantay@lehman.cuny.edu

Prof. Sara McLafferty
Department of Geography
University of Illinois
Urbana, IL 61801
USA
smclaff@uiuc.edu

Cover artwork by Kristen Grady, Urban GISc Lab, Lehman College, City University of New York, adapted from Figure 2.19, showing a continuous surface created by a Geographically Weighted Regression (GWR) analysis.

ISBN 978-94-007-0328-5 e-ISBN 978-94-007-0329-2
DOI 10.1007/978-94-007-0329-2
Springer Dordrecht Heidelberg London New York

Library of Congress Control Number: 2011921119

Printed on acid-free paper

Springer is part of Springer Science+Business Media (www.springer.com)

Foreword

Recently, two disciplines – geospatial analysis and environmental health began to interact with each other. Each of them emerged from a decade of growth in their knowledge base, numbers of engaged professionals, and validated methodologies. Out of these separate growths, a new interdisciplinary area of study, geospatial analysis and environmental health, has emerged and this book captures this moment. It is an area of practice as well as an area of knowledge, because many studies in this area lead to policy changes and actions directly associated with improved health outcomes for affected populations. Studies in this area directly contribute to accelerating the health transition in those areas and populations where society is disposed to go from knowledge to action when environmental causes of ill-health are identified.

"Environment" in "environmental health" is broadly defined. From physical elements such as water, air, toxic materials; to the built environment with its access to elements that promote well-being or that promote poor diets and little physical exercise; to societal decisions that promote health or more negatively increase health risks. All these elements have spatial patterns and populations differ in their interactions with them. Advances in methods of measuring, storing, and accessing such data have brought the concept of geospatial data into prominence and an increasing recognition that exposures to these elements by individuals or groups can be measured using geospatial methods. Along with developments in measuring human exposure to environmental factors comes the search for relationships, if any, between measures of environmental exposure and health outcomes. After relationships are established, questions arise about placing responsibilities for reducing the risks of exposures, particularly those that are experienced by vulnerable population groups.

This useful and timely book provides examples of the principles described above as well as descriptions of recently developed methods such as participatory mapping, spatial regression and distance decay methods that are often used in studies of the environment and health. This book brings together the kind of studies that are more often widely distributed among specialist journals that are difficult to find and access. It will be valuable especially for students and the general public whose widespread interest in population health is coupled with their commitment to developing a more healthy environment for the future.

Iowa City, IA Gerard Rushton

Contents

Contributors

Tunde S. Agbola Department of Urban and Regional Planning, University of Ibadan, Ibadan, Nigeria, babatundeagbola@yahoo.com

Tim E. Aldrich College of Public Health, East Tennessee State University, Johnson City, TN 37614-1709, USA, Aldrich@etsu.edu

Ryan Allen Faculty of Health Sciences, Simon Fraser University, Burnaby, BC, Canada V5A 1S6, allenr@sfu.ca

Sam Atkinson Institute of Applied Science, University of North Texas, Denton, TX, USA, sam.atkinson@unt.edu

Aldo Aviña Department of Geography, University of North Texas, Denton, TX, USA, aldo.avina@unt.edu

Fatoumata B. Barry School of International Service, The American University, Washington, DC, USA, fbbarry@gmail.com

Francis P. Boscoe New York State Cancer Registry, Albany, NY 12237, USA, fpb01@health.state.ny.us

Jodi Bryon NYC Department of City Planning, New York, NY, USA, jbryon@planning.nyc.gov

Jayajit Chakraborty Department of Geography, University of South Florida, Tampa, FL 33620, USA, jchakrab@usf.edu

George Christakos Department of Geography, San Diego State University, San Diego, CA 92182–4493, USA, gchrista@mail.sdsu.edu

Ellen K. Cromley Department of Community Medicine and Health Care, University of Connecticut School of Medicine, Farmington, CT 06030-6325, USA, cromley@charter.net

Kyle Crowder Department of Sociology, Carolina Population Center, University of North Carolina at Chapel Hill, Chapel Hill, NC 27599-3210, USA, Kyle.Crowder@unc.edu

Stefan Dongus Vector Group, Liverpool School of Tropical Medicine, Liverpool, UK; Ifakara Health Institute, Dar es Salaam, United Republic of Tanzania, sdongus@ihi.or.tz

Liam Downey Department of Sociology and IBS Population Center, University of Colorado at Boulder, Boulder, CO 80309, USA, Liam.Downey@colorado.edu

Michael Emch Department of Geography, University of North Carolina, Chapel Hill, NC, USA, emch@email.unc.edu

Trevor Fuller Department of Geography, University of Illinois at Urbana-Champaign, Urbana, IL 61801, USA, tfuller3@illinois.edu

Debarchana Ghosh Department of Geography, Kent State University, 413 McGilvrey Hall, Kent, OH, USA, dghosh@kent.edu

Chris Goranson GIS Center of Excellence, New York City Department of Health and Mental Hygiene, New York, NY, USA, cgoranso@health.nyc.gov

Sue C. Grady Michigan State University, East Lansing, MI 48824, USA, gradys@msu.edu

Kristen Grady Earth and Environmental Sciences Doctoral Program, City University of New York Graduate Center, New York, NY, USA; Urban GISc Laboratory, Lehman College, Bronx, NY, USA, kgrady@gmail.com

Carol Hanchette Department of Geography and Geosciences, University of Louisville, Louisville, KY 40292, USA, carol.hanchette@louisville.edu

Zhiyong Hu Department of Environmental Studies, University of West Florida, Pensacola, FL 32514, USA, zhu@uwf.edu

Perry Hystad School of Population and Public Health, University of British Columbia, Vancouver, BC, Canada V6T 1Z3, phystad@gmail.com

Khadija Kannady Urban Malaria Control Programme, Dar es Salaam City Council, City Medical Office of Health, Dar es Salaam, United Republic of Tanzania, kadkhadija@yahoo.co.uk

C. Peter Keller Department of Geography, University of Victoria, Victoria, BC, Canada V8R 3W4, pkeller@uvic.ca

Bonnie Kerker Bureau of Epidemiology Services, New York City Department of Health and Mental Hygiene, New York, NY, USA, bkerker@health.nyc.gov

Gerry Killeen Vector Group, Liverpool School of Tropical Medicine, Liverpool, UK; Ifakara Health Institute, Coordination Office, Dar es Salaam, United Republic of Tanzania, gkilleen@ihi.or.tz

Jong-Hyung Lee Department of Physics, Chungbuk National University, Cheongju, Korea, popopo17@gmail.com

Johan Liebens Department of Environmental Studies, University of West Florida, Pensacola, FL 32514, USA, liebens@uwf.edu

Juliana A. Maantay Earth, Environmental, and Geospatial Sciences Department; MPH Program, Department of Health Sciences; Urban GISc Laboratory, Lehman College, City University of New York, Bronx, NY, USA; Earth and Environmental Sciences Doctoral Program, City University of New York Graduate Center, New York, NY, USA, Juliana.maantay@lehman.cuny.edu

Florence M. Margai Department of Geography, State University of New York, Binghamton, NY, USA, margai@binghamton.edu

Andrew Maroko MPH Program, Department of Health Sciences, Lehman College, City University of New York, Bronx, NY, USA; Urban GISc Laboratory, Lehman College, Bronx, NY, USA, andrew.maroko@lehman.cuny.edu

Sara McLafferty Department of Geography, University of Illinois at Urbana-Champaign, Urbana, IL, USA, smclaff@uiuc.edu

Victoria Mwakalinga Ifakara Health Institute, Dar es Salaam, United Republic of Tanzania, vmwakalinga@ihi.or.tz

Cathy Nonas Physical Activity and Nutrition Program, New York City Department of Health and Mental Hygiene, New York, NY, USA, cnonas@health.nyc.gov

Immaculata I.C. Nwokoro Department of Urban and Regional Planning, University of Lagos, Lagos, Nigeria, inwokoro@unilag.edu.ng; ifunanya66@yahoo.com

Joseph Oppong Department of Geography, University of North Texas, Denton, TX, USA, joseph.oppong@unt.edu

K. Ranga Rao Center for Environmental Diagnostics and Bioremediation, University of West Florida, Pensacola, FL 32514, USA, rrao@uwf.edu

Flavio Rojas University of North Carolina at Chapel Hill, Odum Institute for Research in Social Science Manning Hall, CB #3355, Chapel Hill, NC 27599, USA, frojas@email.unc.edu

Elisabeth D. Root Department of Geography, University of Colorado, Boulder, CO, USA, roote@colorado.edu

Eleanor M. Setton Department of Geography, University of Victoria, Victoria, BC, Canada V8R 3W4, elsetton@uvic.ca

Laura Smith NYC Department of City Planning, New York, NY, USA, lsmith@planning.nyc.gov

Marcel Tanner Swiss Tropical and Public Health Institute, Switzerland, marcel.tanner@unibas.ch

Chetan Tiwari Department of Geography, University of North Texas, Denton, TX, USA, chetan.tiwari@unt.edu

Jean-François Viel CNRS n° 6249 "Chrono-Environment", Faculty of Medicine, Besançon, France, jean-francois.viel@univ-fcomte.fr

Chih-Hsin Wang Department of Bioenvironmental Systems Engineering, National Taiwan University, Taipei 10617, Taiwan, r96622049@ntu.edu.tw

Junfeng Wang Department of Political Science, University of North Carolina at Pembroke, Pembroke, NC 28372, USA, jwang86@uis.edu

Phillip Williamson Department of Forensic and Investigative Genetics, University of North Texas Health Science Center, Fort Worth, TX, USA, phillip.williamson@unthsc.edu

Yu-Zhang Wu Department of Bioenvironmental Systems Engineering, National Taiwan University, Taipei 10617, Taiwan, wop@ms28.url.com.tw

Niko Yiannakoulias School of Geography and Earth Sciences, McMaster University, Hamilton, ON, Canada L8S4K1, yiannan@mcmaster.ca

Hwa-Lung Yu Department of Bioenvironmental Systems Engineering, National Taiwan University, Taipei 10617, Taiwan, hlyu@ntu.edu.tw

Part I
General Considerations in Geospatial Analysis of Environmental Health

Chapter 1
Environmental Health and Geospatial Analysis: An Overview

Juliana A. Maantay and Sara McLafferty

Abstract This chapter provides a brief overview of the use of geographic information science (GISc) in environmental health research and reviews the main themes and concepts highlighted in each chapter. We summarize applications of GISc in environmental hazard surveillance, exposure assessment and health outcomes surveillance. Challenges of using geospatial tools and methods are discussed. The final sections briefly review the contributions of each chapter and the connections among chapters.

Keywords GISc · Environmental health · Hazard surveillance · Exposure assessment · Outcomes surveillance · GISc limitations

1.1 Introduction

Environmental health research is at an exciting point in its use of geospatial technologies, and there are many researchers in a number of different disciplines working on innovative approaches. Heretofore, these studies have not been compiled into one volume, and this book is the first to focus specifically on the field of environmental health and geospatial sciences. We believe that this book is very timely and allows for an important scholarly contribution in updating and showcasing current perspectives on using geospatial methods for environmental health research. At the present time, research published on the topic of environmental health spatial sciences is dispersed throughout a number of public health, geography, epidemiology, sociology, and environmental science journals, and we believe there is a need to gather good illustrative examples of the state-of-the-science of this research in

J.A. Maantay (✉)
Earth, Environmental, and Geospatial Sciences Department; MPH Program, Department of Health Sciences; Urban GISc Laboratory, Lehman College, City University of New York, Bronx, NY, USA; Earth and Environmental Sciences Doctoral Program, City University of New York Graduate Center, New York, NY, USA
e-mail: Juliana.maantay@lehman.cuny.edu

J.A. Maantay, S. McLafferty (eds.), *Geospatial Analysis of Environmental Health*, Geotechnologies and the Environment 4, DOI 10.1007/978-94-007-0329-2_1,

a comprehensive volume which can be more readily and conveniently accessed by students and professionals.

This volume, *Geospatial Analysis of Environmental Health*, provides an overview of the major issues in the field of Environmental Health Spatial Sciences today. Each chapter contains original research which utilizes a geotechnical tool (e.g., Geographic Information Systems, remote sensing, Global Positioning Systems) to address an environmental health, environmental health justice, and/or environmental health disparities problem. The chapter authors are among the leading researchers and practitioners in the field of environmental health spatial sciences, and their work represents the wide range of topics currently being addressed, as well as newly emerging concerns in the field of environmental health geographics. The editors of this volume have themselves published widely on the subject, and some of their research is also included here.

We anticipate that the book will have broad appeal to professionals in all the affiliated disciplines connected to environmental health spatial sciences. Additionally, the book is structured so that it can be easily used as a course compendium text for a class on environmental health and Geographic Information Science (GISc). Our objective was to compile a thorough exposition of environmental health spatial sciences research, but we also wanted to address a need we believe exists, as borne out by our own experiences in teaching college undergraduate and graduate courses: the lack of a book specifically geared towards courses on environmental health and GISc. Such courses are rapidly becoming required or strongly recommended elective course offerings at many schools of public health, in MPH and doctoral public health programs, as well as in health geography, environmental sciences, and urban environmental planning programs. How can GISc and other geospatial technologies inform our analyses of environmental health?

1.2 The Role of GISc in Environmental Health Research

The World Health Organization estimates that environmental hazards account for one-quarter of the global burden of disease (WHO, 2008). Environmental hazards include biological factors such as bacteria in food or drinking water, and disease carrying insect vectors; chemical factors such as lead in soil and housing, or chemicals emitted from industrial sites; and physical factors related to, for example, the design of housing and transportation systems. Environmental health is concerned with the impacts on human health of these diverse environmental factors and the design of effective public health policies to mitigate these impacts (US Department of Health and Human Services, 1998). Key issues in environmental health include monitoring the uneven and changing distribution of hazards within the environment, understanding people's exposures to environmental hazards, and assessing the impacts on human health. Not just a field of scientific inquiry, environmental health is an essential body of knowledge for use in planning interventions to improve population health. Many of the most significant advances in human health have been achieved through environmental interventions – improvements in public water supply, sanitation, housing, and food supply. These continue to be critically important as we take steps to address the wide inequalities in health across places and populations, and as we attempt to understand the health implications of global and local economic, demographic, and environmental transformations.

The concept of environment is fundamental to environmental health. "Environment" refers to the lived space outside the person. It includes elements of the natural environment such as climate, soils, water, and insects, as well as the characteristics of the built environment such as housing, transportation, land use, open space, facilities and services. Although not typically encompassed in definitions of environmental health, socio-cultural factors including social interactions and institutions, have important roles in environmental health through their impacts on the built and natural environments and their effects on human vulnerability to environmental hazards.

Environmental health has been divided into three areas of inquiry that describe different components of environmental health surveillance (Thacker et al., 1996). In this context, surveillance refers to the process of understanding and monitoring the impacts on health of environmental hazards. *Hazard surveillance* describes the process by which environmental hazards are identified, monitored, and modeled; *exposure surveillance* examines how people are exposed to particular hazardous environmental agents and the biological processes through which the agent produces an unhealthy effect; *outcome surveillance* focuses on recording and monitoring the clinical manifestations of ill-health. These three components are clearly interconnected. Because the health effects of many environmental features are unknown, an important task for environmental health is to evaluate hypothesized connections between hazard, exposure and outcome to determine if empirical data support the existence of a health effect.

Space and location are important in each of the three types of environmental health surveillance (Cromley and McLafferty, 2011). Both environmental hazards and the human activities that put people at risk vary over space, and co-location of people and environmental hazards in space and time is a necessary, although not sufficient, condition for health impacts to occur. People's vulnerability to environmental hazards and their deleterious health effects also varies geographically. Moreover, people and hazards move through space creating continually evolving patterns of exposure and risk. Thus, geography and spatial relationships are central to understanding how environmental factors influence health.

With their emphasis on space and location, GISc and spatial analysis methods are uniquely suited to environmental health investigations, and they have played an important role in environmental health studies for several decades. Early GISc applications in environmental health include Openshaw, Charlton, and Craft's (1988) analysis of spatial clustering of childhood leukemia in relation to nuclear facilities in England, a study that laid the groundwork for the extensive body of research on GIS-based analysis of spatial disease clusters. Other early studies include McMaster's (1988) GIS assessment of community vulnerability to hazardous materials and Wartenberg, Greenberg, and Lathrop's (1993) use of GIS to characterize populations living near high-voltage transmission lines. By the early 1990s, GIS were also being used in vector-borne disease studies to determine the associations between environmental features and vector concentrations (e.g. Glass et al., 1994). Reviews of the literature on health applications of GISc first appeared in the mid-1990s (Clarke et al., 1996; Nyerges et al., 1997). Since that time, use of GISc has expanded rapidly both in environmental health research investigations and in public health planning.

How is GISc used in each of the three areas of environmental health surveillance? A detailed discussion is provided in Cromley and McLafferty (2011); here we offer a brief overview focusing on each type of surveillance. Additionally, the chapters in this book represent the rich diversity of potential GISc applications and the range of innovative data sources and methods of analysis.

1.2.1 Hazard Surveillance

GIS have long been used to map the uneven spatial distribution of environmental hazards and to manage the large geospatial data sets about hazard locations and intensities. Many governmental agencies routinely collect data on air, water, and soil contamination. Collected at discrete sampling points, these data can be entered in GIS via the process of geocoding and then mapped to display geographic variation. Sometimes environmental hazard data are made available to the public, as in the EPA's Toxics Release Inventory Explorer system, an on-line mapping system which can be used to create dynamic maps of chemical releases. For hazards ranging from soil lead, to particulates in the air, to disease carrying mosquitoes, researchers have used GIS to investigate where hazards exist in the environment and to model their spatial distributions (Glass et al., 1995; Guthe et al., 1992). Many of the chapters in this book, including chapters by Cromley (Chapter 12), Margai (Chapter 13), Grady (Chapter 15) and Yu et al. (Chapter 24), illustrate the use of GIS for hazard surveillance.

Although many environmental hazards are distributed continuously over space, for example air and soil contamination, hazard data are typically collected at a few discrete locations. To estimate concentrations of contaminants at locations where no measurements exist, researchers have used spatial interpolation methods such as inverse distance weighting and kriging (Aelion, Davis, Lawson, and McDermott, 2009). The end result is a map representing the intensity of environmental hazard as a continuous surface with peaks indicating areas of high intensity. Root and Emch's chapter in this book uses these methods not for hazard assessment, but to estimate a disease risk surface. These methods can also be extended to estimate the changing geographic concentrations of pollutants over time (see Chapter 24 by Yu et al.).

Using innovative new geospatial technologies to monitor environmental hazards has emerged as an important area of research in the past decade. Devices mounted on vehicles or carried in backpacks, and distributed "sensor grid" networks present exciting opportunities for real-time environmental monitoring. Satellite data also hold promise for near – real-time monitoring of environmental hazards. Orbiting the globe at frequent and regular intervals, satellites provide a continuous stream of remotely-sensed data recorded over a grid of "pixels" (small regularly-sized square areas) on the Earth's surface. Although satellite data have mostly been used to characterize vegetation, recent research suggests that they may be useful in monitoring particulate matter, an air pollutant linked to asthma and other respiratory concerns. The chapters by Hu et al. and Setton et al. in this volume shows the kinds of insights and challenges associated with using satellite data in environmental health investigations.

Because many environmental hazards flow through air, water, and human interactions from place to place, understanding how hazards move through time and space is important. Known as fate and transport, these processes can be modeled using GISc tools and methods. An early example is Chakraborty and Armstrong's (1995) use of GIS-based "plume" (dispersion) models to model atmospheric dispersion of chemicals around accident sites involving transport of hazardous materials. In this volume, Chapter 22 by Viel provides a fine example of fate and transport modeling. Models are also available for estimating dispersion of pollutants from non-point sources such as pesticide/herbicide applications. Hydrologic models of surface and groundwater systems and network models of municipal water supply can be used to trace the waterborne flows of pollutants (Root and Emch, 2010). GIS has also been applied to depict mobile hazards associated with, for example, traffic flows and transportation of hazardous wastes (Lovett et al., 2006). Advances in geospatial technologies and space-time methods are greatly enhancing our ability to model and monitor the spatial distributions and flows of environmental hazards.

Built environments include an array of features that can be hazardous or beneficial (health-promoting) for human health. Health promoting features include parks and recreational facilities, high-quality housing, and supermarkets that offer an array of food products at reasonable prices. These kinds of features can be mapped and visualized using the same suite of GIS methods as are applied to hazardous environmental features. Chapter 10 (Byron et al.) illustrates such an approach for studying access to supermarkets.

1.2.2 Exposure Surveillance

Exposure surveillance examines how people are exposed to environmental hazards and the processes through which exposure results in an adverse health effect. GIS applications mainly focus on the social and environmental processes that influence human contact with hazardous agents. Understanding the spatial distribution of hazards, discussed above, is a critical component of exposure assessment, but knowledge of the locations of people and their activities is also essential. GISc has proven to be a very valuable tool for linking the two types of data – population data and environmental hazard data – that are required in exposure assessment (Jerrett et al., 2005). When the two coincide in time and space, "exposure" occurs.

GIS techniques such as overlay and spatial buffering have been widely used for estimating populations exposed to environmental hazards. Overlay is the linking of mapped variables based on geographic location. In spatial buffering, GIS is used to identify the zone that falls within a particular distance of a point, line or area. Detailed discussions of these and other methods for exposure modeling are provided in Chapter 3 (Setton, Allen, Hystad and Keller) and Chapter 5

(Chakraborty and Maantay). Other chapters present case studies of the application of GISc methods for particular kinds of exposure assessment. These include Chapter 17 (Chakraborty) on air pollution exposure, Chapter 18 (Avina et al.) on exposure to Lyme disease in Texas, Chapter 19 (Downey and Crowder) on exposure to hazardous facilities.

Many GIS-based studies focus on people's residential location as a site of exposure; however, people also come into contact with environmental hazards (and health-promoting features) during the course of their everyday activities. Workplaces have long been recognized as sites of health-related exposures, but places like schools, shopping and recreational facilities, and transportation corridors are also important. Researchers are increasingly using models of time-space activity patterns to understand health-related exposures outside the home (Gulliver and Briggs, 2005). GPS-enabled cell phones are also being employed to record people's everyday activity patterns (Wiehe et al., 2008). An emerging approach is to use GPS-enabled personal monitoring devices to record, for example, exposure to air particulates in real-time (Adams et al., 2009). Although there are many challenges to analyzing real-time exposure data in GIS, the data offer a rich resource for environmental health assessment.

In analyzing exposure, it is also important to recognize social differences in population vulnerability to environmental hazards. Age, income, gender, and ethnicity affect people's activity patterns, thus affecting their encounters with and experiences of environmental hazards. Vulnerable populations may be more exposed to environmental hazards because they live in places where hazards are concentrated. Environmental (in) justice, the structuring of exposure to environmental hazards by class, race, and ethnicity, has been widely studied using GIScience tools and methods, and it is a central theme of many of the chapters in this volume. It has also been noted by researchers that less-affluent populations and communities of color are not only more likely to live in close proximity to environmentally burdensome facilities and thus be more exposed to pollution, but that the health effects of exposure to these burdens are further modified by socio-economic status, and "due to material deprivation and psychosocial stress [these populations] may be more susceptible to the health effects of air pollution," (O'Neill et al., 2003, 1861).

1.2.3 Outcomes Surveillance

Outcomes surveillance involves tracing the health impacts of environmental exposures. Researchers have used GIS to create maps of health outcomes and to analyze associations between outcome data and environmental hazards and exposures. For centuries, maps have been employed to display geographic variation in health outcomes and to reveal spatial concentrations, or "clusters", of health events. There are well-developed methods for mapping health data, and Maantay and Maroko's chapter (Chapter 2 by Maroko, Maantay, and Grady, this book) provides a detailed discussion focusing on environmental health applications. Other chapters provide examples of outcomes mapping for conditions such as asthma, West Nile virus,

Lyme disease, respiratory infections, and maternal and infant health. Another important component of outcomes surveillance is to analyze spatial clusters or "hotspots" of health outcomes. Hotspots are important because they may signal the presence of an environmental hazard that is responsible for the spatial clustering of ill-health. Many spatial statistical methods have been developed for detecting hotspots, and these methods are often employed in investigating environmental health outcomes (see Chapter 23 by Yiannakoulias).

How are health outcomes related to environmental hazards and exposures? Answering this important question requires linking data on hazards, exposures and outcomes, a task that can be accomplished effectively with GIS. Sometimes this linking is part of an exploratory analysis to determine if the incidence or prevalence of a health outcome is associated with an environmental hazard. Such exploratory analyses may provide important clues about disease etiology by indicating potential environmental triggers. Exploratory studies of outcomes, hazards and exposures are well represented in this book covering a diverse range of infectious and degenerative diseases and reproductive health concerns.

GIS is a very valuable tool in these investigations because it enables us to link data from different sources, measured at different levels of geography and for different geospatial units. For example, we might have data on asthma cases by zip code, measurements of air pollution at monitoring sites, and indicators of housing quality by census block. The data management and spatial analysis capabilities of GIS make it possible to transform these data to consistent geospatial units so that health outcomes and environmental hazards can be directly compared. Procedures for transforming data from one set of geographic units to another include spatial aggregation, areal interpolation, dasymetric mapping and many others.

Along with assessing hazards, exposures and health outcomes, geospatial technologies have an important role in planning public health interventions to mitigate environmental health concerns. Such interventions include environmental modifications to reduce disease transmission and exposure to hazards; medical strategies including vaccination and treatment; mobility strategies to alter human activity patterns and interactions; and behavioral strategies that focus on knowledge, education and experiences (McLafferty, 2010). A large body of research shows that public health interventions are most effective if they are geographically targeted to the places and population most in need. Although most of the chapters in this book do not address specific public health interventions, they provide maps and geospatial analyses that can inform the design of such interventions. Chapter 6 by Fuller and Chapter 16 by Dongus et al. also highlight "participatory" GIS as a means of enhancing community involvement in public health planning.

In summary, the links between environmental hazards, exposures and health outcomes are fundamentally spatial, depending on the interactions between people and hazards in space and time. The data management, geovisualization and spatial analysis capabilities of GIS make them a very valuable tool for environmental health assessment. From basic procedures such as mapping and overlay to more complex kinds of spatial analysis, GIS enable researchers and public health analysts to characterize health-related hazards and opportunities in people's everyday environments and to evaluate hypothesized impacts on health. New geospatial technologies including global positioning systems, environmental sensors and satellite imaging systems make it possible to monitor environmental hazards and exposures on an almost real-time basis, creating new frontiers for environmental health research.

1.3 Limitations of Geospatial Methods for Environmental Health Research

As we have seen above, there are many benefits and advantages derived from using geospatial technologies for environmental health research, and indeed, many research questions simply would not be easily or accurately answerable without the use of GISc and other spatial analysis methods. However, there are also a number of important limitations and drawbacks inherent in GISc analyses, particularly when used for epidemiological purposes. Knowledge of these limitations will hopefully inform and inspire the next wave of data and methodological breakthroughs in research, and so it is useful to review these limitations to understand what still has to be done to improve methods and data for environmental health research.

Additionally, in order to design robust and meaningful research projects, it is necessary to understand exactly what GISc and other geospatial methods are *not* able to do, so that we can avoid overreliance on possibly faulty models, misinterpretations of analyses, misleading results, and data and methods that foster false precision. The limitations are discussed below under the broad categories of data deficiencies; data aggregation issues; accuracy of locational data; technological limitations; lack of temporal data on residential history and daily locations; and constraints in using exposure proxies. These topics are re-visited in more detail in many of the chapters.

1.3.1 Data Deficiencies

Many of the limitations and shortcomings of geospatial analyses of environmental health are due to data deficiencies and constraints, particularly the lack of comprehensive health data, and the lack of access to patient-level health data. In attempting to connect environmental factors and health outcomes, the nuances cannot be gauged, and certainly causality cannot be ascribed, without fine-grained health outcome data. Due to issues of patient confidentiality, these data are not readily available, thus restricting the analyses to the use of health data aggregated by relatively large geographic units, making the spatial correspondence between environmental factors and health outcomes difficult to pinpoint, and therefore conclusions more tenuous.

In addition to lack of high resolution health data, many health outcomes are not systematically tracked at any level in the US., although this differs in other countries. Records on health outcomes such as diabetes, obesity, asthma, and cardiovascular disease, for example, are kept independently by private doctors, clinics, and out-patient facilities, and are not routinely compiled into one master database.

In lieu of disease incidence data, researchers generally have to resort to hospitalization data, which in the US is typically maintained by the state, and is available to researchers only on an aggregated basis, and more rarely at the individual patient-level. However, hospital admissions data is a poor proxy for actual prevalence of the disease, since hospitalizations for a given health outcome represent only a small segment of the population affected by the disease or condition, albeit the most serious cases.

Data about general population numbers and characteristics can also be problematic. Population data are necessary in order to calculate disease rates and the extent of the population affected by environmental exposures. The primary source for population data in most countries is the national census, and the US Census, for instance, serves as a useful source of data related to socio-demographic characteristics such as age, race, ethnicity, income, and educational level, although it is often recognized as being ambiguous, incomplete, and inconsistent (Maantay et al., 2010: 13). It is acknowledged as having a systemic undercounting of certain population groups, which makes it less useful for some types of health research, and it is only a static snapshot of population numbers and residential locations every 10 years, which renders the validity of research on time periods at the end of the census period somewhat dubious. Additionally, the census data is aggregated by enumeration units which may be too large to adequately account for environmental exposures, especially in hyper-heterogeneous micro-environments.

Useful information about current environmental conditions is also very often in short supply and of questionable accuracy. Data about air pollution emissions from industrial and municipal facilities, for instance, are often based on self-reported estimates rather than actual monitored amounts. Additionally, data on pollution emissions from facilities are generally annual averages. This fails to capture accidental releases which may result in contaminant concentrations much higher than the thresholds deemed to be safe to human health, but which may appear to be within acceptable limits when bundled into an annual average.

Information on ambient air quality, even when based on actual monitored data, can be less than optimally useful, since the monitors may be so sparsely located as to make interpolation of pollutant concentration values assigned to unmonitored locations inaccurate, and drawing inferences about ambient levels of pollution from federal or state environmental agency monitors can therefore be quite misleading. Many air pollution dispersion models require very detailed data as inputs, such as hourly meteorological conditions, smokestack height, exit velocity and temperature of the emissions, topographical factors and nearby building data to calculate downwash. Groundwater flow and transport models require equally detailed and complex data sets. These data are rarely readily available, making quantitative analyses about existing environmental conditions and predictions about future conditions under different scenarios very cumbersome and expensive to produce. Chapter 3 by Setton et al. reviews some of these issues and, as evidenced by a number of other case studies and methods chapters in this volume, data availability and data access drive research design in many cases.

1.3.2 Data Aggregation Issues and Other Concerns with Spatial Data

Much of the health, population, and even environmental data are gathered and reported by geographic units such as census tracts, postal codes, counties, and so forth. We generally have no choice but to use data aggregated at these levels, which are provided in whatever unit of aggregation the primary data collection is carried out, to suit the purposes of the primary data gathers, not necessarily for the convenience of the secondary data users. The only other option is to perform our own inventories and surveys, and gather the data on our own, which is not usually feasible or practicable, especially for population and health data. However, using data aggregated by administrative or political units, while expedient and convenient, poses a number of difficulties.

A well-known problematic factor in geographic analyses is the "Modifable Areal Unit Problem," or MAUP. "The issue here is that the aggregation units used are arbitrarily with respect to the phenomena under investigation, yet the aggregation units used will affect statistics determined on the basis of data reported in this way," (O'Sullivan and Unwin, 2003, 30). In numerous studies it has been shown that by altering the unit of aggregation, or even the configuration of the boundaries of the units, results can differ sharply (Anderton et al., 1994; Cutter et al., 1996; Glickman and Hersh, 1995; McMaster et al., 1997; Openshaw and Taylor, 1979). For example, after analyzing data aggregated by census tract, one is likely to get different results if the analysis is repeated using the same data aggregated by a smaller or larger unit of analysis. The differences could be dramatic: in one instance, the analysis may show an environmental impact upon a population, and in the next instance, show no impact whatsoever, with the only change being the level of aggregation used. The unit of analysis used is therefore of paramount importance, and depending upon the unit used, very different patterns and relationships may be exhibited.

Another problem of aggregating data by artificial administrative units is that environmental features such as air and water do not respect the boundaries of these units, and will not stop at the borders, but will continue on through in a continuous fashion. Data aggregated by administrative units do not take natural features into account, and for expediency's sake, many studies are by necessity based on the assumption that environmental phenomena are contained within discrete bounded units. Contrariwise, population and health data are not, as a rule, collected at the watershed level or by air quality zone, so there, again, is a conflict of non-coinciding data units that must be resolved when working with all these types of data sources. This can be particularly troublesome when dealing with areas of suspected air or water quality impact, since the boundaries of the impact zones are likely to be more amorphous and fluid than the administrative or political units that capture the population and health data. Methods have been developed to make the units mesh sufficiently so that data can be used from disparate shapes and sizes of units whose boundaries do not coincide. These methods of reapportioning the data into new areal units include areal interpolation, filtered areal weighting, transformation of discrete data into continuous surfaces, and various forms of dasymetric mapping (Eicher and Brewer,

2001; Flowerdew and Green, 1994; Goodchild et al., 1993; Gotway and Young, 2002; Holt et al., 2004; Maantay et al., 2007; Martin, 2006; Mennis and Hultgren, 2005). Some of these methods are discussed further in Chapters 2 and 5.

A number of other, somewhat inter-related aspects of spatial data also need to be considered when undertaking statistical analysis with a geographical focus. These include the problems of ecological fallacy; the non-uniformity of space; scale; edge effects; and spatial autocorrelation. We will touch briefly on them here, although a more thorough description of the concepts will be necessary in order to develop effective and accurate research designs.

"Ecological fallacy," a concept somewhat related to the MAUP issue, refers to the fact that things that are true or valid at one level of data aggregation may not hold true at a higher or lower level of aggregation. For instance, there may be a relationship between asthma hospitalization rates and low-income populations at the zip code level, but we cannot assume from this that low-income individuals are necessarily more prone to be hospitalized for asthma. We can only conclude from the relationship that zip codes comprised of high proportions of low-income people also tend to have high asthma hospitalization rates. We cannot infer anything about individuals, nor can we ascribe any health outcome to people at certain income levels, except at the unit of aggregation that was used for the analysis. In general, the larger the unit of aggregation, the more likely it is that bias will be introduced due to heterogeneity across and within these units (Maroko et al., 2009) and ecological fallacy may result.

"Non-uniformity of space" is the idea that phenomena vary over space and are not homogeneously distributed, especially with respect to the underlying geography and its attributes. For instance, we might be able to detect what appear to be clusters of a particular disease within a city – "hotspots" – where it looks like there may be concentrations of the disease. However, this clustered distribution may merely reflect the underlying population density of the city, with the understanding that where there are more people, there is a greater likelihood of there being a higher incidence of disease, as well. We cannot infer from this that there is something in that locality's environment "causing" the disease cluster, aside from a higher number of potential susceptible receptors (in other words, more people).

"Scale," or the geographic extent of the study area, is another factor that will influence the outcome of GISc analyses, and how the study area is bounded needs to be considered carefully when designing a study. "To reflect a potential environmental health-based concept of risk, the boundaries should relate to exposure or risk from the site; however, a single boundary reflecting all variations in toxicity and contaminant fate and transport for each chemical present plus variabilities in the duration of human exposure and vulnerability would be virtually impossible... The scale of analysis chosen is often dictated by expediency, determined by how existing data bases are aggregated," (Zimmerman, 1993, 650).

"Edge effects" is a concept related to scale that addresses the fact that study extents are bounded and have edges, and there will not be as many data observations in all directions at the edges of the study area as there will be in the center of the study area. Of course, in reality, the phenomena under investigation do not stop at the boundaries of the study area – this is an artifact of how the study area is defined, and the fact that usually our data values cease at the boundaries of our study area. Therefore, the skewed distribution of the values over space has to be taken into account, although techniques to do this are only beginning to be developed (Yamada and Rogerson, 2003).

"Spatial autocorrelation," often called the First Law of Geography (Tobler, 1979), simply means that near things tend to be more closely related to each other than they are to distant things. Values of points close to each other in geographic space will be similar, and the values of distant points will be less similar. Traditional statistics rely on an assumption of random samples, which makes it difficult to apply conventional statistical tests to spatial data, since spatial data are not randomly distributed, but correlated based on distance. This problem and its possible solutions are discussed in several of the chapters, notably in Chapter 17 by Chakraborty.

1.3.3 Accuracy of Locational Information

Geospatial analysis for environmental health studies depends on accurate locational information. One of the main functions of GISc is the ability to plot locations of geographic entities, such as polluting facilities or residences of patients, in order to investigate spatial correspondence amongst the features. However, there are many points in the process where inaccuracies of various types can creep in, both spatial inaccuracies, as well as inaccuracies with non-spatial attribute information. We often receive locational information about polluting facilities, for instance, in the form of an address table, which then has to be spatialized by a process called geocoding. This entails taking street addresses, transforming them to x, y coordinates, and plotting them on a map.

Geocoding is a fairly automated process, but there are several points along the way where the process can break down and result in inaccurate or incomplete mapping. First, not all the facilities may have "matchable" addresses. If the address has been entered into the table incorrectly, if there is a spelling error in the street or number, or if the geocoding program used does not recognize the street or building number in its address-matching database, the program will be unable to plot the location. This is then referred to as an "unmatched" location, and generally speaking a match rate of 85% or better is considered very good. But that means up to 15% or more of the addresses do not appear on the map or show up in incorrect locations. Secondly, the geocoding process usually maps addresses using a mathematical algorithm to figure out where on a street segment that particular building number is likely to be, which is not the same as mapping the location to an actual building number. This yields results within a certain threshold of accuracy, but can be off by 50–100 ft or more, having implications for fine-gained impact analysis. Aside from geocoding issues, often times the address data itself is faulty, with the address shown being the location of the owner's business office, company headquarters, mailing address, home address, or someplace other than the actual site of the polluting facility. This clearly would impact the validity of the analysis.

In addition to geocoding, there are other ways of transferring locational data to maps, such as digitizing hard copies of maps with a digitizing tablet; scanning paper maps; or manually locating points or polygons by heads-up (or on-screen) digitizing. None of these can claim infallible accuracy, either. The salient point here is that all locational data created through geocoding or any other means of transforming data into digital maps have the potential to be inaccurate, with the possibility of serious ramifications to the results of any health analysis based on the locations as plotted.

1.3.4 Technological Limitations

Researchers and analysts new to GIScience often do not appreciate the very real technical limitations of geospatial technologies. GISc enthusiasts and software manufacturers frequently extol the virtues of GIS, unwittingly giving novices an unrealistic and overly-rosy perception of what GIS can do. Although it can do many complex things, it cannot do everything, and the mechanical use of a technology without a thorough understanding of it is also no substitute for a well-thought out and well-crafted research design.

One of the most obvious technical limitations is that although many computational processes are automated in GIS, the user/analyst still has to make any number of important decisions and be able to justify assumptions, which require some knowledge of statistics and quantitative reasoning. One can, of course, just rely on the default settings, but this may lead to incorrect analyses and difficulties in interpreting the results accurately. It is also crucial that the analysts possess a deep familiarity with their datasets and geographic study locations, without which all the technology in the world will not compensate. Chapter 2 by Maroko et al. emphasizes the necessity of the researcher/analyst having or obtaining an intimate knowledge of their data and study extent.

There is a fairly high level of skill required in order to go beyond basic mapping functions. In the past decade, GIS software has been improved and functionality has been expanded significantly, and the interfaces have become much more user-friendly and simplified. Gone are the days of needing to know complex computer programming codes to operate most GIS software packages, and these have by and large been replaced by the familiar drop-down menus and plain language options now common in desk top applications. A benefit of this is that mapping technology is now easier to use and therefore available to many more people, including non-experts, making the technology more democratic, as described in Fuller's Chapter 6. However, many aspects of using geotechnologies are not intuitive, especially when applied to the increasingly complex analyses required by environmental health research. Obtaining the correct data in the proper formats and being able to manipulate and prepare the data for analysis requires a fairly high degree of expertise and proficiency with large datasets, as well as statistical and geographic computer software, which is another potential drawback for conducting such research using GISc. The learning curve for much of this work is steep, and the technical support, specialized expertise, and hardware and software necessary to carry it out may not be accessible or affordable, especially in less affluent countries, thereby limiting who can perform this type of research.

The development of mathematical or cartographic models for use in environmental health research is also fraught with difficulties, as there are many environmental and health processes that we simply do not understand well enough to model, or which do not lend themselves to quantification. Such processes, when shoe-horned into an inappropriate, incomplete, or inaccurate model format, will not be well-represented by the model, and therefore the model results must be viewed with skepticism and less than full confidence that they reflect reality. If these are sub-models that are then inserted into larger models, there is a cascade effect of the results being further and further removed from reality. Even when environmental and health processes are well understood, modeling these processes requires many assumptions to be made by the analyst, and often yields few solid and certain results.

Although we have mentioned above the difficulties with obtaining data necessary for environmental health research, remote-sensing technology has the potential to provide us with almost unlimited amounts of data about the earth. Satellites orbiting the earth and geostationary satellites are streaming millions of pixel's worth of information back to us each day – so much data, in fact, that only a small percentage of it can be utilized at the present time, given person-power limitations, and the time and skill required for image processing. Unfortunately, remotely-sensed data is typically of a resolution too coarse to be used for many environmental health studies.

"Remote sensing (RS) could be a tremendous asset to health geography as remotely sensed images are updated frequently and cover large, often not easily accessible, areas. It would hypothetically be possible to use RS data to estimate exposure by incorporating GIS population and health data. Unfortunately, RS data are often unreliable over urban areas due to high spatial and spectral variability, as well as the irregular size, shape and orientation of objects... Additionally, there is a scalar mismatch in the definition of "high spatial resolution" between remote sensing scientists and health geographers. An area like NYC may be viewed as very small and homogeneous by remote sensing standards, whereas it is often viewed extremely large and heterogeneous by the standards of health geographers," (Maroko, 2010, 104).

Nevertheless, there are excellent applications of using remotely-sensed images for health research, including using the data for estimating open space and green spaces; for modeling vegetation as a component in vector-borne disease studies; to calibrate and validate land use regression models; as ancillary data sets in dasymetric mapping to spatially disaggregate variables; for global studies of issues like vegetation change and disease spread; and for estimating air particulate concentrations (See Chapter 20 by Hu et al., Albert et al., 2000; Anyamba et al., 2002; Bavia et al., 2001; Beck et al., 2000; Brooker et al., 2002; Brooker and Michael, 2000; Estrada-Peña, 1998; Goetz et al., 2000; Green and Hay, 2002; Hay, 2000; Lo and Quattrochi, 2003; Lobitz et al., 2000; Rogers, 2000; Rogers et al., 1996; Seto et al., 2002; Tatem and Hay, 2004; Tran et al., 2002; Ward et al., 2000). But unfortunately, remotely-sensed data has not proved to be central to environmental health research except mainly in studies dealing with continental, regional, or other large geographic extents. In just a few years, resolution of some types of remotely-sensed data has improved from pixels representing kilometers to those whose resolution is measured in meters or smaller, and there is every reason to believe that the resolution will continue to be enhanced in the future, to the point where it is more useful for analysis of hyper-heterogeneous urban environments. The use of remote sensing has increased dramatically in the past decade for answering many health-related research questions, and will likely assume a greater importance in future years.

1.3.5 Drawbacks Pertaining to Temporal Data on Residential History and Daily Locations

A major stumbling block with trying to find spatial correspondence between health outcomes and the presences of environmental burdens is the fact that many, if not most, diseases are not acute. In other words, people most often do not develop symptoms upon first exposure to an environmental hazard. Some health outcomes, like many forms of cancer, have very long latency periods, and in order to draw the links between health outcome and hazard, we must have more information than the current address of the person. The necessity of residential history is the subject of Chapter 4 by Boscoe, and while it is becoming increasingly important to take this into account for accurately portraying the risks of environmental exposures, residential history is rarely available, and usually not at the spatial extent required.

Another temporal issue with data is the lack of information on day-time locations of people, which is a major impediment in measuring disparities in proximity or exposure to environmental health hazards accurately and comprehensively. Except for perhaps young children and the elderly, most people do not spend the majority of their time in the home, but go further afield to schools, workplaces, etc., all of which can result in exposures that right now are not being accounted for in most environmental health studies. However, researchers are increasingly using real-time data sets derived in some cases from personal monitoring devices in order to establish exposures, as mentioned above. This type of information will become crucial in more accurately estimating exposures.

Because we rely so heavily on census data to tell us the locations of populations, and census data only reports on where people live, we are restricted to the so-called "night-time" locations of the majority of the population. New methods must be developed, and new data sources tapped, to enable us to calculate environmental health burdens for the workplace populations, and in general, to account for the mobility of people as they move through space engaging in the various activities of their daily lives.

1.3.6 Exposure Proxies and Misclassification of Exposures

In environmental health research using GISc, we frequently use exposure proxies as crude indicators of individual exposure to ambient conditions, or of body or target organ doses. As discussed above, very often there is no way to obtain measurements of actual pollutant concentrations, and even if there were, for many pollutants there is no definite threshold value established for human health safety. So we must rely on proximity to the environmental hazard, in many cases, to draw connections between environmental burdens and health effects. However, from the point of view of drawing causal links between health outcomes and environmental burdens, relying on exposure proxies such as proximity buffers to indicate where health risk is greatest is rife with uncertainties, because it is unknown how much of the pollution hazard (e.g., air pollution emissions) is reaching and impacting any specific individual, how much of it is being absorbed into the individual's body through breathing, ingestion, or dermal contact, and how much is reaching a sensitive organ and potentially causing an adverse health outcome. Chapter 5, by Chakraborty and Maantay, discusses the implications of proximity analysis in more detail.

A fundamental concern with mapping environmental health is that it does not yield definitive findings about actual exposure levels or health outcomes for the population in proximity to the noxious facilities or land uses. The difficulties of linking proximity and exposure make these studies less useful in conclusively demonstrating (and measuring) the correspondence between the location of potential environmental burdens, exposures, and health effects. Additionally, exposure assignment is frequently based on one address, such as address at birth, diagnosis, or death, which potentially introduced exposure misclassification by not accounting for residential mobility, as mentioned in the discussion above about the drawbacks pertaining to temporal data.

To summarize and conclude the discussion on limitations, environmental health research using GISc and other geospatial techniques is still in its infancy, but by studying the limitations of the methods and the currently available data, we can learn a great deal about what we must work on in order to move forward. We are hopeful that this book will help in that regard to advance research.

1.4 The Structure of the Book

Geospatial Analysis of Environmental Health is divided into three major sections, starting first with a section on general considerations in using geospatial analysis for environmental health, including some of the main uses and issues. The second section contains case study examples using geospatial analysis in examining environmental health issues, and we coordinated these chapters so that they loosely follow the table of contents of some of the leading environmental health textbooks currently in use, featuring those topics that are most frequently covered in such books. The third section reviews some of the principal methodological techniques of geospatial analysis for environmental health research, including some of the current innovative methods, like geographically weighted regression, spatio-temporal analysis, distance decay techniques, and Bayesian analysis. By necessity, there is a certain amount of overlap amongst the topics in all three sections, and we have cross-referenced many of the chapters with other chapters in this book.

1.4.1 Section I: General Considerations in Geospatial Analysis of Environmental Health

Section I consists of a suite of six chapters that reviews and illustrates some of the important concerns and issues in using geospatial analysis for environmental health research, starting with this chapter, Chapter 1, which outlines the development and the principal benefits and limitations of geotechnologies and environmental health, and introduces some of the key concepts. Many of these are discussed in more detail and amplified further in the chapters that follow.

In Chapter 2, *Using GeoVisualization and Geospatial Analysis to Explore Respiratory Disease and Environmental Health Justice in New York City*, Maroko et al. illustrate how complex issues in environmental health justice analysis can benefit from geovisualization and exploration within a Geographic Information Science (GISc) framework. Geovisualization is the process of using spatial data to conduct exploratory data analysis to "see the unseen" in these large data sets. With GISc it is possible to look at the data in many new and novel ways, and from multiple perspectives. By mapping the data, reclassifying it, manipulating the data, examining its statistics, and utilizing various other approaches in exploratory data analysis, we can often find relationships and linkages that would not be apparent in any other way. Chapter 2 uses both a hypothetical data set and a real-world example of respiratory disease to illustrate these concepts, and serves as a basic primer for conducting exploratory spatial data analysis through geovisualization and geospatial analysis.

Outdoor Air Pollution and Health – A Review of the Contributions of Geotechnologies to Exposure Assessment, Chapter 3 by Setton et al., is a comprehensive overview of exposure assessment with geotechnologies, focusing on air pollution, and reviews two major types of studies: epidemiological studies that look for associations between health outcomes and exposures to air pollution; and exposure studies that attempt to predict or explain who is exposed, and by how much. In the first case, epidemiological studies provide important information on which air pollutants are associated with harmful health effects. In the second case, population exposure studies provide important information for reducing exposures to pollutants known or suspected to be harmful to particular population groups and communities, and for prioritizing regulatory policy in terms of potential number of people affected.

In Chapter 4, *The Use of Residential History in Environmental Health Studies*, Boscoe surveys the current uses and application of residential history data in the field of environmental health, considers the implications of incorporating residential history data into disease surveillance, and discusses the reasons why it poses more challenges than are commonly assumed. Residential histories are used in environmental health investigations to establish the likelihood of past exposures and to ascertain past social and economic conditions. The long latency of many diseases makes it crucial to know the residential history of cases. Many studies only use information on current residential address, but this practice can be very misleading and will provide less than accurate results for many health outcomes. Because past environmental exposures are believed to impact current health, the incorporation of residential history information makes it more likely that an environment-disease link will be identified correctly.

Proximity Analysis Methods for Exposure Assessment in Environmental Health Justice Research, Chapter 5, by Chakraborty and Maantay, provides an historical overview of methods, models, and data used to measure proximity to environmental hazards and potential exposure to their adverse health effects in the environmental justice (EJ) research literature. It explores how the assessment of disproportionate proximity and exposure has evolved from comparing the prevalence of minority or low-income residents in geographic entities hosting pollution sources, to the use of discrete fixed-distance impact buffer zones as proxies for exposure, to more refined techniques that utilize continuous distances, pollutant fate-and-transport models, and estimates of health risk from toxic exposure. It also reviews analytical techniques used to determine the characteristics of populations residing in areas potentially exposed to environmental hazards, as well as emerging geospatial techniques, such as Geographically Weighted Regression, that are often more appropriate for spatial analyses than conventional statistical methods. A number of these themes are addressed in detail in later chapters illustrating the use of these methods.

Section I concludes with Chapter 6, by Fuller, on an increasingly important aspect of using geospatial technologies in environmental health studies – the participation by non-specialists in map creation, spatial analysis, and data collection. For instance, the recent emergence of the Volunteered Geographic Information (VGI) phenomenon has, to a large extent, democratized the way maps are used, interpreted, and created (Elwood, 2008). Michael Goodchild (2007) states "there has been an explosion of interest in using the Web to create, assemble, and disseminate geographic information provided voluntarily by individuals. Sites such as Wikimapia and OpenStreetMap are empowering citizens to create a global patchwork of geographic information, while Google Earth and other virtual globes are encouraging volunteers to develop interesting applications using their own data." The new web sites and other digital sources that allow the input of geo-referenced data and/or interactive mapping accessible to almost anybody has the potential to provide more timely and up-to-date data, as well as permit laypeople to become key players in producing and using geographic knowledge about their communities.

There has also been an interest by community organizations and advocacy groups in "counter-mapping," an activity tangentially related to VGI that entails mapping as a means for progressive change from and against dominant power structures (Maantay, 1996; Peluso, 1995). Although counter-mapping has been used most often and very effectively by indigenous peoples to reclaim traditional rights to land and resources, urban communities have also used it to good effect. According to Nancy Peluso, the goal of counter-mapping "is to appropriate the state's techniques and manner of representation to bolster the legitimacy of "customary" claims to resources," (Peluso, 1995:384). In the case of urban and non-indigenous rural populations, these claims to resources consist of the right to breathe clean air, or the right to refuse to have sewage sludge applied to their open lands. Mapping has proved to be an effective means of communicating to decision-makers and elected officials about disproportionate environmental burdens and health impacts. "Street

Scientists" have also pointed out the value of citizen mapping and spatial analysis as part of the effort to legitimize the work of community scientists and reach parity with the technical-scientific experts, by using community data, community know-how, and local knowledge bases in order to effect constructive change that the technical-scientific experts alone would not be able to do (Corburn, 2005; Maantay, 1996; Maantay and Ziegler, 2006).

Chapter 6, *Their Data, Our Cause: An Exploration of the Form, Function, and Deployment of Maps among Environmental Justice Groups*, by Fuller, surveys the use of GIS by a selection of community-based environmental justice organizations. His inquiry into how GISc is being used by these groups points out some of the drawbacks and difficulties in community-based GISc as well as the benefits that can accrue in the process. This topic of participatory GISc is picked up again in Section III's Chapter 16 by Dongus et al. on malaria vector control in Tanzania using locally produced maps and spatial inventories. In many parts of the developing world, the contribution of local people in environmental health spatial analysis is proving to be of vital importance. Since a lack of governmental resources may prevent such analyses from being undertaken through official channels or by technical "experts," geospatial health surveys and mapping by the local community are often the only means of accomplishing the task.

1.4.2 Section II: Impacts on Environmental Health (Topical Case Studies)

Section II contains 9 chapters, representing case study examples covering a wide array of environmental hazards and risks, and using various geospatial methods to examine and analyze them. These topical issues are at the heart of environmental health concerns, and by and large constitute the initial impetus for developing geospatial methods for health research.

1.4.2.1 Zoonotic and Vector-Borne Diseases

Vector-borne and zoonotic diseases, such as malaria, Lyme disease, viral meningitis, hantavirus, Dengue Fever, Yellow Fever, and rabies, among others, continue to be serious environmental health concerns in most parts of the world, including, increasingly, the more developed countries. Many of the vector-borne and zoonotic diseases long thought to be confined primarily to the developing world have made a dramatic resurgence in the developed world, in part due to the increase in population movement and displacement, migration, and international travel, and in part due to the effects of global climate change, which are redefining and expanding the habitable ranges of some of the vectors. Ghosh's *Geospatial Analysis of West Nile Virus (WNV) Incidences in an Urban Environment* (Chapter 7), analyzes the association of urban environmental features that facilitate the viral activities of the mosquito-borne WNV infection in the Minneapolis-Saint Paul metro area, and addresses the question of how urban morphology affects human health. Using a combination of factorial ecology, geospatial techniques, and hierarchical cluster analysis, urban landscape classes are derived from the environmental and built environment risk-factors hypothesized to be associated with WNV transmission. The infection rate among birds, mosquitoes, and human cases are then compared to these urban classes to better understand the characteristics of those landscape classes most conducive to the proliferation of the virus. The analysis of vector-borne disease is covered again in Chapter 18 by Avina et al., who investigate Lyme disease in Texas, and in Chapter 16 by Dongus et al., who use participatory mapping as a strategy in controlling the malaria vector in Tanzania.

1.4.2.2 Toxic Metals and Elements

When we think of environmental hazards, very often the first thing that springs to mind is toxic elements and metals such as lead, cadmium, chromium, arsenic, mercury, and nickel, which are all considered by the US. Agency for Toxic Substances and Disease Registry to be "major toxic metals with multiple health effects." Most toxic metals are carcinogenic, and the effects of heavy metal poisoning can be fatal. They can also cause respiratory problems; adverse pregnancy outcomes; severe developmental disorders in children; degenerative diseases of the nervous system, including Alzheimer's disease, Parkinson's disease, and multiple sclerosis (MS); problems with skeletal development and maintenance (e.g. osteoporosis); kidney disorders; blood disorders; and many other serious acute and chronic health impacts. The metals are ubiquitous in our environment, in developed as well as developing countries, and are introduced into the environment and our bodies through both natural pathways (such as volcanoes or rainwater dissolving metals that are present in rocks and ores), as well as through the combustion of coal and crude oil, battery manufacturing, or the ingestion of fish and other food that have been contaminated by toxic elements and metals in the water.

In many landscapes, toxic metals and other toxic substances have contaminated the environment through a number of pathways, including unsafe manufacturing processes and fossil fuel combustion by power plants and vehicles, which can result in deposition of toxic substances to the soil. These hyper-contaminated lands, primarily located in urban and suburban areas, have been termed "brownfields," and in the past two decades many municipalities have become interested in reclaiming and cleaning up these oftentimes abandoned industrial lands for much needed housing or other development projects in order to remove the hazard, make the properties productive again, and restore them to the property tax rolls. In Chapter 8, *The Health Impacts of Brownfields in Charlotte NC: A Spatial Approach*, Wang describes such a situation in Charlotte, North Carolina, where brownfields have been identified for redevelopment, but no health impact study on the effects of living in close proximity to a brownfield has been undertaken by the city. This chapter analyzes the possible association between proximity to brownfield sites and the reproductive health outcome of low birth weight in infants, by means of a geospatial analysis of the relationship amongst the health outcome, the socio-demographic characteristics of the populations, and location, density, and size of the brownfield sites.

1.4.2.3 Water Quality

Water quality is another leading environmental health concern, from the multiple standpoints of access to and availability of clean drinking water; the dangers of water-borne contaminants affecting the food supply; and also the threat of general environmental degradation and habitat destruction for other species caused by water pollution. Water can be polluted by toxic metals or chemicals from industrial and agricultural uses, from flooding and stormwater runoff, as well as by disease-causing bacteria, such as the cholera bacterium and e. coli, often from untreated sewage discharged to water supplies. The adverse health impacts of water pollution are well-known, and can cause both chronic and acute effects, sometimes life-threatening. Dehydration from diarrheal diseases is the second leading cause of death for children under the age of five globally. Nearly 1.5 million children, mainly in developing countries, die each year due to diarrhea, often caused by water contamination. This accounts for about 20% of all child deaths (UNICEF/WHO, 2009). Root and Emch (Chapter 9, *Regional Environmental Patterns of Diarrheal Disease in Bangladesh: A Spatial Analytical and Multilevel Approach*) describe the spatial distribution of childhood diarrhea to examine its correspondence to land use types characterized by flood inundation levels, household and community characteristics, and information about water supply and sanitation infrastructure. The topic of water-borne diseases is visited again in Chapter 22, where Yiannakoulias uses cluster detection methods to investigate schistomsomiasis infection in Kenya, which is caused by a water-borne parasite.

1.4.2.4 Food Safety/Food Security

Food safety, food security, and access to healthy foods are issues that have far-reaching effects on health, including hunger and malnutrition; disease outbreaks from bacteria and parasite-infected foods; various vitamin deficiency diseases; diabetes and the obesity epidemic now occurring in many developed countries (and in those developing countries that are unfortunately copying the poorer aspects of Western eating habits). Access to healthy foods has become a hot button issue, with the realization by public health officials and the general public that many people, even those in densely settled urban areas, lack ready access to fresh fruits and vegetables and other healthy food choices, while at the same time suffering no lack of fast food establishments and corner shops, generally selling less healthy alternatives.

These areas of healthy food deprivation have been termed "food deserts," and have been deemed at least partially responsible for the growing obesity problem. One mitigating factor in some inner-city areas has been the development of food-producing community gardens and other applications of urban agriculture. Regularly-held farmers' markets and cooperative agreements with local farmers called Community Supported Agriculture (or CSAs) have also helped to combat-the problem of food deserts, and have encouraged people in all walks of life and living in a wide spectrum of communities to become "locavores," eating produce and other food products "in season" that are grown locally. This consumption of locally produced food has a significant health benefit due to the freshness of the foods and the higher vitamin content, by virtue of having been ripened in place and shipped to nearby places, avoiding a long transit where much of the nutritional value can be lost. Eating locally grown foods also has additional environmental benefits of cutting down on long-distance truck traffic to haul food stuffs from one end of the country to the other, as well as cutting back on wasteful practices like airplane loads of fresh raspberries bound for US markets from southern continents in January. Reduction in plastic packaging also results from the locavore lifestyle, making some inroads into the solid waste conundrum we face.

However, many of us still need to do at least some if not a majority of our food shopping in a food store, and by all accounts, a full-service supermarket provides the best option for both selection and pricing. Additionally, many supermarkets now have jumped on the organic and locally-grown bandwagon, too, making it possible to shop in a convenient way while still adhering to smart environmental and health practices. Yet supermarkets are not evenly distributed across the landscape, making it difficult for many people to shop effectively for healthy foods, especially those in economically-depressed areas that are shunned by major supermarket chains. *Developing a Supermarket Need Index*, Chapter 10, by Byron et al., details the difficulties in measuring access to supermarkets, and the development of a supermarket need index which takes into account not only access to full-service supermarkets, but also the prevalence of diet-related diseases by neighborhood. Their analysis reveals that three million New York City residents are living in "food deserts," with all that implies for their health and well-being.

1.4.2.5 Air Quality

In Chapter 11, *Asthma, Air Quality, and Environmental Justice in Louisville, Kentucky*, Hanchette et al. discuss air quality, specifically criteria pollutants and volatile organic compounds, and their relationship with childhood asthma hospitalization rates. Spatial clustering of areas having high asthma rates and poor air quality point to a problem of health inequities, since these areas tend to be populated by the less affluent and by people of color. The case study described in Chapter 11 is an example of how adverse health outcomes, disproportionate pollution burdens, and environmental justice intersect to create health disparities, which in this case especially affect children. Asthma is a disease with uncertain and probably multiple causes, exacerbated by many different triggers, and it is debilitating as well as sometimes fatal. It is the third leading cause of childhood hospitalization in the US, and tends to particularly affect those living in industrialized areas and places with high traffic volumes. In many urban areas, children suffering from asthma are ubiquitous, and regardless of whether the high asthma hospitalization rates are due to environmental causes or are primarily the result of issues related to poverty and other socio-demographic factors, this asthma epidemic points to a health and environmental justice crisis.

A meta-analysis of EJ literature (Maantay et al., 2010) reveals that a large proportion of EJ studies to date focus on the linkages amongst exposure to air pollution, the socio-demographic characteristics of the proximate population, and respiratory illness, as opposed to any other of the many possible combinations of environmental hazards and health outcomes. Because of these very visible and wide-ranging impacts, it is no wonder that so many environmental health justice studies focus on air pollution and respiratory disease, and asthma in particular, despite the fact that air pollution is an especially complicated environmental hazard to quantify, measure, and accurately assess exposure. Air quality is one of the most stringently regulated environmental media in much of the developed world, and although great strides have been made in past decades to reduce air pollution, it remains a serious threat, and is one of the root causes of health disparities and disproportionate impacts. In addition to respiratory diseases, air pollution has also been connected to strokes and other cardiovascular diseases, Non-Hodgkins Lymphoma, various types of adult onset cancers, childhood brain cancers, childhood leukemia, and adverse pregnancy outcomes, such as stillbirths, pre-term births, low birth weight babies, neural tube defects, congenital malformations, chromosomal abnormalities, and spina bifida (Maantay et al., 2010). Air quality analysis features prominently in this volume, especially in later Chapters 20, 21, 22, and 24, each of which bring a different method to the examination of the health impacts of air pollution.

1.4.2.6 Solid and Liquid Waste

Solid waste disposal is a growing environmental concern, especially in these NIMBY "Not In My Backyard" times we live in. An inordinate amount of solid waste is generated by residents and businesses every day, which is part and parcel of the overconsumption-waste cycle prevalent in most developed countries. But few people will voluntarily live near the final resting place of the garbage. For years, many municipalities have been landfilling their solid waste in parts of their own cities, which tended to be in the less affluent areas or in communities of color, where the community's lack of political capital negated or muted their attempts at resistance. When the municipalities finally started running out of room within their cities, they then began exporting their waste to other (usually poorer) regions in their own states, or to other poorer parts of the country, where this was often viewed by local officials (and sometimes the residents) as a sound economic development strategy, bringing welcome jobs to depressed regions. When all else fails, the garbage is exported to other countries, generally in the less developed world. Who, after seeing them, can forget the pictures of the infamous "garbage barge," carrying over 3,000 tons of solid waste, which in the summer of 1987 circled desperately from New York City down the entire eastern seaboard of the US, thorough much of the Gulf Coast of Mexico, Central America, and various Caribbean Islands, looking for a place to dump its load? No one would accept it, and so it ended up in Brooklyn, where it was finally incinerated many months after embarking on its forlorn voyage. Incineration of solid waste, of course, produces its own deleterious environmental and health impacts in the way of air pollution from combustion, and the resulting production of toxic ash, which in turn needs disposal. Parenthetically, the garbage barge incident occurred while New York City was still (legally) dumping much of its sewage sludge and medical and industrial waste into an underwater site 12 miles off the coast in the Atlantic Ocean. This was subsequently banned by EPA, and NYC ceased ocean dumping in 1992, making its solid waste disposal problem even more acute since one of its major "sinks" was now off-limits.

Not only is landfilling the ultimate waste of land, since the land is generally never able to be returned to productive purposes (occasional recreational reuses such as conversion to parkland notwithstanding), but there are serious environmental risks to solid waste disposal, including polluted leachate from buried garbage impacting water quality; methane gas emissions from the decomposing garbage; increased air pollution due to air-borne particulate matter from the garbage itself when it is dumped; and the air quality, noise, and traffic congestion impacts of thousands of garbage-hauling trucks accessing the landfill each day. Additionally, solid waste is an environmental problem long before it reaches its ultimate landfill destination. One manifestation of this problem is that it creates significant impacts on those communities hosting solid waste transfer stations, such as municipal solid waste transfer stations (including those handling putrescible garbage), construction and demolition debris transfer stations, recycling facilities, and hazardous and medical waste transfer and disposal facilities, where solid waste is hauled in and then generally hauled out again after sorting and separation for shipping to different places. For instance, in the South Bronx NYC, which is part of the poorest Congressional District in the United States and has an almost 100% minority population, it is not uncommon for over 1,000 trucks per day to access one solid waste transfer station, and there are over 60 solid waste transfer stations of one type or another in this area (Maantay, 2007). Not coincidentally, this area has the highest asthma hospitalization and asthma death rates in the city.

Chapters 12 and 13 each deal with different aspects of the waste disposal issue. In Cromley's chapter, *The Impact of Changes in Municipal Solid Waste Disposal Laws on Proximity to Environmental Hazards*, we see how changes in policies intended to improve municipal solid waste disposal at the state level can actually have a deleterious impact on communities of color and the less affluent by concentrating the waste sites closer to them. Rather than keeping the waste within each locale's municipal borders, as in the past, the regulatory change redirected flows of waste to a more centralized network of transfer stations and trash-to-energy plants, usually located in closer proximity to the predominantly minority and less affluent areas. This brings up the interesting fact that environmental hazards are not necessarily static in space, and sometimes, as in this case, there is a changing geography of environmental hazards. Policy-makers need to take into account environmental justice impacts such as these when re-crafting policies and regulations.

1.4.2.7 Environmental Justice

Environmental Justice (EJ) is defined broadly as the problematic of the disproportionate distribution of environmental "goods" and "bads," with the burden of the "bads" and the dearth of the "goods" falling mainly on racial, ethnic, and religious minorities, immigrants, the less affluent, and other vulnerable groups, with concomitant deleterious impacts on health and quality-of-life for these populations. EJ concerns have been linked to the vital issues of health disparities; inequity in housing; residential segregation based on racial, ethnic, and religious characterization; reduction in quality-of-life factors and lack of urban amenities; increased vulnerability and risk of certain sub-populations to natural, technological, and human-made disasters; and environmental degradation.

In Chapter 13, *Global Geographies of Environmental Injustice and Health: The Case of Illegal Hazardous Waste Dumping in Côte d'Ivoire*, Margai shows us how multinational corporations operate in disposing of toxic waste in a less developed country, where the residents are even more vulnerable and less likely to have access to health care than the populations in those countries where the waste is generated. Unfortunately, this scenario is repeated in many places all over the world, and is one of the reasons that it is impossible to think of achieving environmental justice except at a global level. "Until all are free, then none are free," applies to environmental justice as well as to civil rights, because EJ achieved in one location may just mean that the problem (the environmental hazard) is pushed to an even poorer or more vulnerable place somewhere else in the city, state, country, or world.

The dumping of toxic waste has serious health consequences, especially because in many cases the nearby populations may be unaware of the dangers, and have no idea what they are dealing with. Even if they do know, oftentimes they have no choice but to use the water, breathe the air, and till the soil, all of which may be contaminated and slowly killing them. We don't have to look back very far in our own history in the US to understand that people often don't know the dangers they live with on an intimate and daily basis until it is too late and the health damages have begun to be felt. Most times, it is precisely the onset of health impacts that first alerts residents that something is terribly wrong in their living environment. In the infamous case of Love Canal in upstate New York, toxic waste was buried by the owners of the land, a chemical manufacturer, and then the land subsequently sold for a new housing development which was built on top of the dump site. This case, as well as a similarly tragic case in Times Beach, Missouri, became notorious in the media, a cause célèbre in environmental circles, and created a groundswell of grassroots agitation for better environmental protection. These cases and others like them led to the creation of the federal Comprehensive Environmental Response Compensation and Liability Act (CERCLA), commonly known as Superfund, after the fund set up within the Act to help pay for clean-up of the most egregious of these sites. Unfortunately, there are hundreds of housing developments in the US today which in the past 30–50 years were knowingly or unknowingly built on top of toxic waste dumps. In the developing countries, the dumping of toxic waste where people live is still occurring on a regular basis.

1.4.2.8 Health Disparities in Women

Environmental injustice and health inequities tend to affect women's health especially severely. Gender is an important determinant in health outcomes, particularly in less developed countries, as women there are usually more economically vulnerable and have less access to health care than men do, often because of cultural and religious imperatives. Due to the additional riskiness and frequency of childbearing and the lower status of women in many less developed countries, health outcomes can be particularly dire. As usual all over the world, the adverse impacts of gender and poor environmental quality are mediated by income and class, with more affluent and well-educated women having better health, thanks to access to care, clean water, safer sanitary provisions, and better housing conditions. In Chapter 14, *Environmental and Health Inequalities of Women in Different Neighbourhoods of Metropolitan Lagos, Nigeria*, Nwokoro and Agbola employ GIS to show the spatial variation of health status of women across neighborhoods, and describe the factors involved in the apparent health inequities of women in the different neighborhoods.

1.4.2.9 Urbanization and Impacts of the Built Environment

In recent years the public health field has renewed its historic focus on the impacts of urbanization and the built environment on health. Public health as a discipline began in the nineteenth century by making the linkages between the living environments of people and their health conditions. The nascent urban planning profession and the City Beautiful movements went hand-in-hand with public health goals. In the US., we obtained most of our housing regulations and land use zoning due to the public health professionals' outcry against over-crowded and unsanitary living conditions in dense urban settlements (Maantay, 2001). Tenement housing with inadequate or non-existent water supply, toilet facilities, ventilation, heat, and light were seen as the primary reason for the high rates of mortality and morbidity amongst the urban poor and working classes. Despite a century and a half of housing and zoning legislation expressly designed to protect and improve the health, lives, safety, properties, and welfare of the population, sub-standard conditions in the built environment continue to have adverse effects on public health. Research has also linked residential segregation based on race, ethnicity, or religion with adverse health outcomes (Morello-Frosch and Jesdale, 2006).

Grady's *Housing and Racial Disparities in Low Birth Weight: A GIS Risk Assessment*, Chapter 15, looks precisely at this issue of how housing conditions and deficiencies affect health, using low birth weight as an indicator. She argues that there are racial disparities involved in both of these factors, which is an important finding for policy- and decision-makers to understand and take into account in their work.

1.4.3 Section III – Geospatial Methods in Investigating Environmental Health

The book's final section focuses on advanced GISc methods and their use in analyzing environmental health issues. Although all chapters in the book utilize methods of some sort, what distinguishes the chapters in this section is their detailed coverage of methods and their emphasis on innovative new methods of spatial analysis.

Methods encompass both quantitative approaches that involve mathematical-statistical analysis of numerical data and qualitative methods including textual analysis, interpretive methods and participatory approaches to understanding. For most of its history, GISc has been associated with quantitative methods, and such methods continue to dominate the field. The 2- (or 3-)dimensional structure of GIS data makes it uniquely suited to various types of quantitative data analysis. Yet, there is also heightened interest in incorporating qualitative data such as video, photos and interviews in GIS and in using GIS to facilitate qualitative and participatory kinds of analyses. This section also includes some examples of national-scale studies that rely on large, dynamic spatial and spatiotemporal datasets.

Chapter 16, *Participatory Mapping as a Component of Operational Malaria Vector Control in Tanzania*, by Dongus et al., is the only chapter in this section emphasizing qualitative methods. The authors describe the use of participatory mapping to guide mosquito control efforts within small areas of the city of Dar es Salaam. As described earlier, participatory mapping involves incorporating local knowledge, maps and preferences in decision-making processes to give community members a voice in decisions that affect them. In Dongus et al.'s chapter, sketch maps of small neighborhood areas identifying features, households and land ownership are hand-drawn and linked with aerial photos. Not only are the maps used in tailoring malaria control strategies to detailed local conditions, but also they can be easily updated and shared with other health and social agencies.

The remaining chapters focus on quantitative spatial analytic methods. In Chapter 17, *Revisiting Tobler's First Law of Geography: Spatial Regression Models for Assessing Environmental Justice and Health Risk Disparities*, Chakraborty provides a detailed and comprehensive summary of spatial regression analysis, a method increasingly applied in environmental health research studies. Situating his chapter in relation to Tobler's Law, which serves as a foundation for geospatial analysis, Chakraborty discusses the theoretical underpinnings of spatial regression analysis and issues related to application and interpretation. An environmental justice example, investigating the uneven effects of vehicular pollution on racially- and economically-marginalizaed groups, is provided to guide the reader through a practical application. Note that several other chapters in this book use spatial regression analysis for modeling spatial relationships between environmental hazards and health outcomes.

There is growing interest in integrating models of hazard, exposure and outcome in environmental health studies. Using Lyme disease in Texas as a case study, Avina et al.'s chapter, *A Spatially Explicit Environmental Health Surveillance Framework for Tick-Borne Diseases*, (Chapter 18) describes how GISc can be used to facilitate such integration. Using GIS-based map algebra tools, data on ticks, tick habitats and population distribution are combined in estimating Lyme disease risk. The authors apply maximum entropy modeling and geographically-weighted regression to map and predict geographic variation in the prevalence of Lyme-infected ticks across the state.

Environmental justice analysis is the topic of Downey and Crowder's chapter, *Using Distance Decay Techniques and Household-Level Data to Explore*

Regional Variation in Environmental Inequality. Although many environmental justice studies rely on a simple "container" approach in modeling exposure to hazardous facilities, Downey and Crowder adopt a much more rigorous distance-based approach in analyzing racial and economic disparities in environmental burdens. Their grid-based method estimates a population's exposure to hazardous emissions based on distance to facilities and volume of emissions. They apply these computationally-intensive methods in a national-scale analysis of environmental justice, and findings show substantial regional differences in the unequal burdens of hazardous emissions among demographic and racial groups.

Studies of health and environment increasingly rely on information gathered from satellites, but in most applications the data are used in analyzing vegetation or "green" spaces. The chapter by Hu et al., *Merging Satellite Measurement with Ground-Based Air Quality Monitoring Data to Assess Health Effects of Fine Particulate Matter Pollution*, discusses an emerging area of application – use of satellite data on aerosol optical depth (AOD) to estimate fine particulate matter concentrations. Fine particulates are thought to trigger asthma and other respiratory conditions in susceptible populations, so analyzing the associations between particulates and health is an important research endeavor. Using geographically-weighted regression and Bayesian hierarchical modeling, Hu et al. analyze the associations between local AOD measurements, fine particulate concentrations and health outcomes. Looking ahead, their research raises the exciting possibility that other important environmental indicators might be tracked and modeled with satellite data.

Bayesian statistical methods are the subject of two papers in this section, Chapter 21 by Rojas (*Poverty Determinants of Acute Respiratory Infections in the Mapuche Population of Ninth Region of Araucania, Chile (2000–2005): A Bayesian Approach with Time-Space Modelling*) and Chapter 24 by Yu et al. (*Spatiotemporal Analysis of $PM_{2.5}$ Exposure in Taipei (Taiwan) by Integrating PM_{10} and TSP Observations*). Some statisticians believe that Bayesian methods offer a superior approach to spatial and temporal modeling than more traditional frequentist statistical methods. Rojas uses a Bayesian time-space nested model to assess the relationships between poverty and health in the Araucania region of Bolivia. Poverty and spatial segregation emerge as key determinants of ill-health among the highly vulnerable Mapuche population. In the Yu et al. chapter, Bayesian space-time methods serve as a tool for estimating local concentrations of particulates in Taiwan over space and time. Although mathematically challenging, these chapters show that the environmental analyst's geospatial toolkit is rapidly expanding to include more accurate and rigorous spatial statistical methods.

Modeling dispersion of air pollutants is the topic of Chapter 22 (*GIS and Atmospheric Diffusion Modeling for Assessment of Individual Exposure to Dioxins Emitted from a Municipal Solid Waste Incinerator*) by Viel. As in Downey and Crowder's chapter, the goal is to estimate concentrations of pollutants in areas near hazardous facilities; however, Viel's chapter examines this topic at a local scale. Viel uses GIS-based plume models to estimate dioxin concentrations around municipal solid waste incinerators in Besancon, France. Dioxin is a known carcinogen, and the health impacts of dioxin released from municipal facilities are of great concern. To make accurate predictions, Viel uses a plume model of atmospheric dispersion which incorporates local weather patterns, including prevailing winds and the impacts of topography. Reflecting atmospheric processes, plume models are superior to distance-based methods for air pollution modeling. Viel's chapter also takes the next step, analyzing associations between modeled dioxin concentrations and cancer outcomes via multilevel modeling.

Identifying spatial clusters of health events is a critically important task for public health agencies concerned with improving environmental health. A spatial cluster is a geographical area in which the incidence or prevalence of disease is unusually high (Waller and Gotway, 2004). Clusters are important because, for example, they may signal an unexpected outbreak of infectious disease, or the presence of hazardous material(s) in the environment resulting in adverse health effects. Chapter 23, *Synthesizing Waterborne Infection Prevalence for Comparative Analysis of Cluster Detection Methods*, by Yiannakoulias, develops and implements a new method for spatial cluster detection (the "greedy growth scan") that can identify irregularly-shaped clusters. Most existing methods seek out clusters that have compact geometric shapes (e.g. circular, rectangular); Yiannakoulias' method focuses on elongated, "snake-shaped" clusters that might emerge along waterways, roads, or powerlines. Using GIS-based synthesized data on Schistosomiasis infection in two regions of Kenya, he compares the new method's ability to detect clusters with that of the commonly used spatial scan statistic, finding significant advantages of the new method in detecting outbreaks of water-borne disease.

1.5 Recurrent Themes

In conclusion, these chapters highlight a number of important themes in research with GISc. Many of the chapters emphasize the multi-factorial nature of environmental health issues and the fact that they involve a complex mix of social, ecological, and physical environmental causes. Other recurrent themes are the impacts of the built environment on health, and the challenge of linking environmental exposure and health outcomes. A number of the chapters develop innovative methods for addressing these challenges.

Environmental justice and the associated issue of health disparities are recurrent themes running through many of the chapters in this book. The editors strongly believe that innovative geospatial tools and methods can be effectively employed in understanding and tackling environmental justice and health disparities, two of the most significant public health challenges of our time. The chapters in this volume provide an important step in that direction.

Author Biographies

Juliana A. Maantay is Professor of Urban and Environmental Geography at Lehman College, City University of New York (CUNY), Director of Lehman's Geographical Information Science (GISc) Program, and Director of the Urban GISc Lab. Her research interests include using GISc for spatial analyses of environmental health justice issues, land use and health impacts, urban hazards and risk assessment, and community-based participatory research.

Juliana A. Maantay, a native of New York City, has had a life-long interest in the urban environment, especially that of her home-town. She considers herself very fortunate to have a job that allows her to combine this fascination of urban culture, history, and the built environment with her passion for social justice issues. Dr. Maantay has over 20 years experience as an urban and environmental planner and policy analyst with governmental agencies, non-profit "think tanks," community environmental advocacy organizations, and private sector consulting firms, and has been active in environmental justice research and advocacy for more than 15 years.

In addition to her primary duties in the Department of Earth, Environmental, and Geospatial Sciences at Lehman College, she is a faculty member in the Earth and Environmental Sciences Doctoral Program at the CUNY Graduate Center, the Public Health Doctoral Program, the MPH Program at Lehman, and is a research scientist with the National Oceanic and Atmospheric Administration's Center of Remote Sensing Science and Technology (NOAA-CREST) at City College.

Her recent research on environmental health justice has been published in the *American Journal of Public Health, Environmental Health Perspectives, Health and Place, International Journal of Health Geographics, Applied Geography*, and *The Journal of Law, Medicine, and Ethics*, and her work also appears in a number of edited volumes on urban public health issues and health inequities. Her book, *GIS for the Urban Environment* (2006, Environmental Systems Research Institute, ESRI Press), promotes the ethical use of GIScience for environmental awareness and community empowerment, and is currently used as a textbook for GISc, public health, environmental planning, and urban design courses throughout the world, as well as winning an award from the Cartography and GIScience Society/American Congress on Surveying and Mapping. Dr. Maantay has been invited to present her environmental health justice research at the New York Academy of Sciences, the United States Environmental Protection Agency, the National Research Council of the National Academy of Sciences, the National Institute of Environmental Health Sciences, and the United Nations, among others.

She is currently a Co-PI in grant-funded projects with NOAA, NIEHS, USDA, and the National Center for Minority Health and Health Disparities, conducting research on the relationship between air pollution and respiratory and cardiovascular disease, diabetes/obesity and location of active recreational spaces and healthy food choices, urban agriculture and health, segregation, disease, and health inequities, new dasymetric methods of population mapping, interdisciplinary urban health research methods, community-based participatory research using GISc, and geovisualization of health data.

Juliana A. Maantay earned a B.Sc. from Cornell University, a Master of Urban Planning (M.U.P.) from New York University, an M.A. in Geography/Geographic Information Systems from Hunter College/CUNY, and a Ph.D. in Urban Environmental Geography from Rutgers University.

Sara McLafferty is Professor of Geography at the University of Illinois at Urbana-Champaign. Her research investigates place-based disparities in health and access to health services and employment opportunities for women, immigrants and racial/ethnic minorities in the United States. She has also written widely about the use of GIS and spatial analysis methods in exploring health and health care inequalities and in assisting community efforts to improve health.

With Ellen Cromley, she is the author of *GIS and Public Health*, Guilford Press. She also coedited *A Companion to Health and Medical Geography* (with Tim Brown and Graham Moon) and *Geographies of Women's Health* (with Isabel Dyck and Nancy Lewis). She has published in a wide range of geography, epidemiology, and urban studies journals and currently serves on the editorial boards of the *Annals of the Association of American Geographers, Geographical Analysis, Transactions in GIS* and is an Associate Editor of *Health and Place*.

Sara McLafferty holds a B.A. from Barnard College and M.A. and Ph.D. degrees from the University of Iowa. Before coming to Illinois, she taught for almost 2 decades in New York City at Hunter College-CUNY and Columbia University.

References and Further Reading

Aberley D (1993) Boundaries of home: mapping for local empowerment. New Society Publishers, Gabriola Island, BC

Adams C, Riggs F, Volckens J (2009) Development of a method for personal, spatiotemporal exposure assessment. J Environ Monit 11(7):1331–1339

Aelion C, Davis H, Liu Y, Lawson A, McDermott S (2009) Validation of Bayesian kriging of arsenic, chromium, lead, and mercury surface soil concentrations based on internode sampling. Environ Sci Technol 43(12):4432–4438

Albert DP, Gesler WM, Levergood B (2000) Spatial analysis, GIS, and remote sensing applications in the health sciences. In Albert DP, Gesler WM, Levergood B (eds) GIS and remote sensing: applications in the health sciences. Ann Arbor Press, Chelsea, MI

Anderton D, Anderson J, Oakes J, Fraser M (1994) Environmental equity: the demographics of dumping. Demography 31:229–248

Anyamba A, Lithicum KJ, Mahoney R, Tucker CJ, Kelley PW (2002) Mapping potential risk of Rift Valley fever outbreaks in African savannas using vegetation index time series data. Photogram Eng Remote Sensing 68(2):137–145

Bavia ME et al (2001) Use of thermal and vegetation index data from earth observing satellites to evaluate the risk of schistosomiasis in Bahia, Brazil. Acta Tropica 79(1):79–85

Beck LR, Lobtiz BM, Wood BL (2000) Remote sensing and human health: new sensors and new opportunities. Emerg Infect Dis 6(3):217–227

Bentham G (1988) Migration and morbidity: implications for geographical studies of disease. Soc Sci Med 26:46–54

Briggs DJ, Elliott P (1995) The use of geographical information systems in studies on environment and health. World Health Stat Q 48:85–94

Brooker S, Hay SI, Tchuente LT, Ratard R (2002) Using NOAA-AVHRR data to model human helminth distributions on planning disease control in Cameroon, West Africa. Photogram Eng Remote Sensing 68(2):175–179

Brooker S, Michael E (2000) The potential of geographical information systems and remote sensing in the epidemiology and control of human helminth infections. In: Hay S, Rogers D, Randolph S (eds) Remote sensing and GIS in public health. Advances in Parasitology, vol 47. Academic Press, London, UK, pp 245–288

Chakraborty J, Armstrong M (1995) Using geographic plume analysis to assess community vulnerability to hazardous accidents. Comput Environ Urban Syst 19(5):1–17

Clarke K, McLafferty S, Tempalski B (1996) On epidemiology and geographic information systems: a review and discussion of future directions. Emerg Infect Dis 2(2):85–92

Cliff AD, Haggett P, Smallman-Raynor M (1998) Deciphering global epidemics: analytical approaches to the disease records of world cities. Cambridge University Press, Cambridge, 1888–1912

Corburn J (2005) Street science: community knowledge and environmental health justice. The Massachusetts Institute of Technology (MIT) Press, Cambridge, MA

Cromley E, McLafferty S (2011) GIS and public health, 2nd edn. Guilford, New York, NY

Croner CM, Sperling J, Broome FR (2000) Geographic information systems (GIS): new perspectives in understanding human health and environmental relationships. Stat Med 15:1961–1977

Cutter SL, Holm D, Clark L (1996) The role of geographic scale in monitoring environmental justice. Risk Anal 16:517–526

Davis SI (2001) Brownfields and other distressed communities: assessing neighborhood characteristics. J Environ Health 64(1):47–48

Diggle PJ (2000) Overview of statistical methods for disease mapping and its relationship to cluster detection. In: Elliot P et al (eds) Spatial epidemiology: methods and applications. Oxford University Press, Oxford, UK

Eicher C, Brewer C (2001) Dasymetric mapping and areal interpolation: implementation and evaluation. Cartogr Geogra Inf Sci 28:125–138

Elliot, P et al (eds) (2000) Spatial epidemiology: methods and applications. Oxford University Press, Oxford, UK

Elwood S (2008) Volunteered geographic information: future research directions motivated by critical, participatory, and feminist GIS. GEO J 72(3–4):173–183

Estrada-Peña A (1998) Geostatistics and remote sensing as predictive tools of tick distribution: a cokriging system to estimate *Ixodes scapularis* (Acari: Ixodidae) habitat suitability in the United States and Canada from advanced very high resolution radiometer satellite imagery. J Med Entomol 35(6):989–995

Flowerdew R, Green M (1994) Areal interpolation and types of data. In: Fotheringham AS Rogerson PA (eds) Spatial analysis and GIS. Taylor and Francis, London, UK, pp 121–145

Fotheringham AS, Brunsdon C, Charlton M (2000) Quantitative geography: perspectives on spatial data analysis. Sage Publications, London, UK

Gatrell AC (2001) Geographies of health: an introduction. Blackwell Publishing Professional, Oxford, UK

Gatrell, AC, and Löytönen M (eds) (1998) GIS and health. Taylor and Francis, London

Glass G, Schwartz B, Morgan J, Johnson D, Noy P, Israel E (1995) Environmental risk factors for Lyme disease identified with GIS. Am J Public Health 85(7):944–948

Glass G, Amerasinghe F, Morgan J, Scott T (1994) Predicting *Ioxdes scapularis* abundance on white-tailed deer using geographic information systems. Am J Trop Med Hyg 51(5): 538–544

Glickman TS, Hersh R (1995). Evaluating environmental equity: the impacts of industrial hazards on selected social groups in Allegheny County, Pennsylvania. Discussion paper 95–13, Washington, DC: Resources for the Future

Goetz SJ, Prince SD, Small J (2000) Advances in satellite remote sensing of environmental variables for epidemiological applications. Adv Parasitol 47:289–307

Goodchild MF (2007) Citizens as sensors: the world of volunteered geography. GEO J 69(4): 211–221

Goodchild M, Anselin L, Deichmann U (1993) A framework for the areal interpolation of socioeconomic data. Environ Plan A 25:383–397

Gotway C, Young L (2002) Combining incompatible spatial data. J Am Stat Assoc 97:632–648

Green RM, Hay SI (2002) The potential of pathfinder AVHRR data for providing surrogate climatic variables across Africa and Europe for epidemiological applications. Remote Sensing Environ 79(2–3):166–175

Gulliver J, Briggs D (2005) Time-space modeling of journey-time exposures to traffic-related air pollution using GIS. Environ Res 97(1):10–25

Guthe W, Tucker R, Murphy E, England R, Stevenson E, Luckhardt J (1992) Reassessment of lead exposure in New Jersey using GIS. Environ Res 59(2):318–325

Hay SI (2000) An overviewof remote sensing and deodesy for epidemiology and public health applications. Adv Parasitol 47:1–35

Holt JB, Lo CP, Hodler TW (2004) Dasymetric estimation of population density and areal interpolation of census data. Cartogr Geogra Inf Sci 31:103–121

Jerrett M, Arain A, Kanaroglou P, Beckerman B, Potoglou D, Sahsuvaroglou T et al (2005) A review and evaluation of intraurban air pollution models. J Expo Anal Environ Epidemiol 15:185–204

Johnson S (2006) The ghost map: the story of London's most terrifying epidemic – and how it changed science, cities, and the modern world. Riverhead Books, Penguin Group (USA) Inc, New York, NY

Koch T (2005) Cartographies of disease: Maps, mapping, and medicine. ESRI Press, Redlands, CA

Koch T, Denike K (2001) GIS Approaches to the pboblem of disease clusters: a brief commentary. Soc Sci Med 52:1751–1754

Lawson A, Kulldorff M (eds) (1999) Disease mapping and risk assessment for public health. Wiley and Sons, Chichester, UK

Leyland A, Goldstein H (2001) Multilevel modeling of health statistics. Wiley, New York, NY

Lo CP, Quattrochi DA (2003) Land-use and land-cover change, urban heat Island phenomenon, and health implications: a remote sensing approach. Photogram Eng Remote Sensing 69(9): 1053–1063

Lobitz B, Beck L, Huq A, Wood B, Fuchs G, Faruque A, Colwell R (2000) Climate and infectious disease: use of remote sensing for detection of *Vibrio cholera*e by indirect measurement. Proc Natl Acad Sci USA 97:1438–1443

Lovett A, Parfitt J, Brainard J (2006) Using GIS in risk analysis: a case study of hazardous waste transport. Risk Anal 17(5):625–633

Maantay JA (1996) Bringing mapping (and power!) to the people: using "counter-mapping" to assess, analyze, challenge, and change environmental, social, and health conditions in the South Bronx, New York. Discussion paper. New York: The Center for a Sustainable Urban Environment

Maantay JA (2001) Zoning, equity, and public health. Am J Public Health 91(7):1033–1041

Maantay JA (2002) Mapping environmental injustices: pitfalls and potential of geographic information systems (GIS) in assessing environmental health and equity. Environ Health Perspect 110(Supp 2):161–171 Special issue: Advancing Environmental Justice Through Community-Based Participatory Planning

Maantay JA (2007) Asthma and air pollution in the bronx: methodological and data considerations in using GIS for environmental justice and health research. Health Place 13:32–56 Special issue: Linking Population Health, Critical Theory, and Geographical Information Science

Maantay JA, Charaborty J, Brender JD (2010) Proximity to environmental hazards: environmental justice and adverse health outcomes. US Environmental Protection Agency (US EPA), Washington, DC

Maantay JA, Maroko A, Herrmann C (2007) Mapping population distribution in the urban environment: The cadastral-based expert dasymetric system (CEDS). Cartogr Geogr Inf Sci 34(2):77–102

Maantay JA, Ziegler J (2006) GIS for the urban environment. Environmental Systems Research Institute, Redlands, CA

Maheswaran R, Craglia M (2004) GIS in public health practice. CRC Press, Boca Raton, FL

Malone JB et al (2001) Satellite climatology and the environmental risk of *Schistosoma mansoni* in Ethiopia and East Africa. Acta Tropica 79 (1):59–72

Maroko AR (2010) Chronic exposure to fine particulate matter and heart failure in New York City: a methodological exploration of environmental justice and health. Doctoral Dissertation, City University of New York, The Graduate Center, Earth and Environmental Sciences Ph.D. Program

Maroko AR, Maantay JA, Sohler N, Grady K, Arno P (2009) The complexities of measuring access to parks and physical activity sites in New York City: a quantitative and qualitative approach. Int J Health Geogr 8(34):1–23

Martin D (2006) An assessment of surface and zonal models of population. Int J Geogr Inf Syst 10(8):973–989

McLafferty S (2010) Placing pandemics: geographical dimensions of vulnerability and spread. Eurasian Geogr Econ 51(2):143–161

McMaster R (1988) Modeling community vulnerability to hazardous materials using geographic information systems. In: Proceedings, third international symposium on spatial data handling. Sydney, pp 143–156

McMaster RB, Leitner H, Sheppard E (1997) GIS-based environmental equity and risk assessment: methodological problems and prospects. Cartogr Geogr Inf Syst 24(3):172–189

Mennis J, Hultgren T (2005) Dasymetric mapping for disaggregating coarse resolution population data. In: Proceedings of the 22nd Annual International Cartographic Conference, July 9–16, 2005, A Coruna, Spain

Morello-Frosch R, Jesdale W (2006) Separate and unequal: residential segregation and estimated cancer risks associated with ambient air toxics in US. metropolitan areas. Environ Health Perspect 114:1–8

Nyerges T, Robkin M, Moore T (1997) Geographic information systems for risk evaluation: perspectives on applications to environmental health. Cartogr Geogr Inf Syst 24(3):123–144

Omumbo JA, Hay SI, Goetz SJ, Snow RW, Rogers DJ (2002) Updating historical maps of malaria transmission duration in East Africa using remote sensing. Photogram Eng Remote Sensing 68(2):161–166

O'Neill MS, Jerrett M, Kawachi I, Levy JI, Cohen AJ, Gouveia N, Wilkinson P, Fletcher T, Cifuentes L, Schwarz J (2003) Health, wealth, and air pollution: advancing theory and methods. Environ Health Perspect 111(16):1861–1870

Openshaw S (1984) The modifiable areal unit problem. CATMOG 38. GeoBooks, Norwich, UK

Openshaw S, Charlton M, Craft A (1988) Searching for leukemia clusters using a geographical analysis machine. Pap Reg Sci Assoc 64:95–106

Openshaw S, Taylor P (1979) A million or so correlation coefficients: three experiments on the modifiable areal unit problem. In: Wrigley N (ed) Statistical applications in the spatial sciences. Pion, London, pp 127–144

O'Sullivan D, Unwin DJ (2003) Geographic information analysis. (Chapter 2, the pitfalls and potential of spatial data). John Wiley and Sons, New York, NY

Peluso N (1995) Whose woods are these? counter-mapping forest territories in Kalimantan, Indonesia. Antipode 27(4):383–406

Rogers DJ (2000) Satellites, space, time and the African trypanosomiases. In: Hay S, Randolph S, Rogers D (eds) Remote sensing and GIS in public health. Advances in Parasitology, vol 47. Academic Press, London, UK, pp 130–171

Rogers DJ, Hay SI, Packer MJ (1996) Predicting the distribution of tsetse flies in West Africa using temporal Fourier processed meteorological satellite data. Ann Trop Med Parasitol 90: 225–241

Root E, Emch M (2010) Tracing drinking water to its source: an ecological study of the relationship between textile mills and gastroschisis in North Carolina. Health Place 16(5):794–802

Rushton G, Elmes G, McMaster R (2000) Considerations for improving geographic information research in public health. URISA J 12(2):31–49

Rushton G, Lolonis P (1996) Exploratory spatial analysis of birth defect rates in an urban population. Stat Med 7:717–726

Seto E, Xu B, Liang S, Gong P, Wu W, Davis G, Qiu D, Gu X, Spear R (2002) The use of remote sensing for predictive modeling of schistosomiasis in China. Photogram Eng Remote Sensing 68(2):167–174

Stillwell J, Clarke G (eds) (2004) Applied GIS and spatial analysis. John Wiley and Sons, New York, NY

Tatem AJ, Hay SI (2004) Measuring urbanization pattern and extent for malaria research: a review of remote sensing approaches. J Urban Health 81(3):363–376

Thacker S, Stroup D, Parrish R, Anderson H (1996) Surveillance in environmental public health: issues, systems, and sources. Am J Public Health 86(5):633–638

Tim U (1995) The application of GIS in environmental health sciences: opportunities and limitations. Environ Res 71(2):75–82

Tobler W (1979) Smooth pycnophylactic interpolation for geographical regions. J Am Stat Assoc 74:519–536

Tran A, Gardon J, Weber S, Polidori L (2002) Mapping disease incidence in suburban areas using remotely sensed data. Am J Epidemiol 156(7):662–668

UNICEF/WHO (2009) Diarrhoea: why children are still dying and what can be done. UNICEF, New York, NY

Vine MF, Degnan D, Hanchette C (1997) Geographic information systems: their use in environmental epidemiologic research. Environ Health Perspect 105:598–605

Waller LA, Gotway CA (2004) Applied spatial statistics for public health data. Wiley, Hoboken, NJ

Ward M, Nuckols J et al (2000) Identifying populations potentially exposed to agricultural pesticides using remote sensing and a geographic information system. Environ Health Perspect 108:5–12

Wartenberg D (2001) Investigating disease clusters: why, when, and how? J R Stat Soc Ser A Stat Soc 164:13–22

Wartenberg D, Greenberg M, Lathrop R (1993) Identification and characterization of populations living near high-voltage transmission lines: a pilot study. Environ Health Perspect 101(7): 626–631

Wiehe S, Carroll A, Haberkorn K, Hoch S, Wilson J, Fortenbery J (2008) Using GPS-enabled cell phones to track the travel patterns of adolescents. Int J Health Geogr 7(May 21):22

World Health Organization (2008) Ten facts on preventing disease through healthy environments. http://www.who.int/features/factfiles/environmental_health/en/index.html. 4 June 2008

Yamada I, Rogerson P (2003) A comparison of edge effect correction methods applied to k-function analysis. Geogr Anal 35:97–110

Yasnoff W, Sondik E (1999) Geographic information systems (GIS) in public health practice in the new millennium. J Public Health Manag Pract 5(4):ix-xii

Zimmerman R (1993) Social equity and environmental risk. Risk Anal 13(6):649–666

Chapter 2
Using Geovisualization and Geospatial Analysis to Explore Respiratory Disease and Environmental Health Justice in New York City

Andrew Maroko, Juliana A. Maantay, and Kristen Grady

Abstract The goal of this chapter is to illustrate how complex issues in environmental health justice analysis can benefit from geovisualization and exploration within a Geographic Information Science (GISc) framework. Individual health outcome variables, such as hospitalizations due to respiratory disease, can be very difficult to interpret without a geographic context; and interactions amongst variables such as disease, socio-demographic characteristics, or environmental exposures, further complicate an accurate interpretation of the data. Data exploration and visualization through mapping and spatial analysis often provides a more robust understanding of the data, as well as improved clarity in viewing the phenomena under study, which will lead to better design of further analyses and additional hypothesis generation, in an iterative fashion. In the first part of this chapter, we use a hypothetical data set to illustrate some of the data exploration, geovisualization, statistical methods, and geospatial analyses. In the second part of the chapter, we use a worked example of respiratory disease and socio-demographic variables in New York City to assess potential environmental justice impacts, in order to further demonstrate the importance of geovisualization and geospatial analysis in achieving a better understanding of environmental health issues.

Keywords Geovisualization · Geostatistics · New York City · Asthma · Environmental Justice · Environmental Health

2.1 Introduction

The goal of this chapter is to illustrate how complex issues in environmental health justice analysis can benefit from geovisualization and exploration within a Geographic Information Science (GISc) framework. Individual health outcome variables, such as hospitalizations due to respiratory disease, can be very difficult

A. Maroko (✉)
MPH Program, Department of Health Sciences, Lehman College, City University of New York, Bronx, NY, USA; Urban GISc Laboratory, Lehman College, Bronx, NY, USA
e-mail: andrew.maroko@lehman.cuny.edu

J.A. Maantay, S. McLafferty (eds.), *Geospatial Analysis of Environmental Health*, Geotechnologies and the Environment 4, DOI 10.1007/978-94-007-0329-2_2,

to interpret without a geographic context; and interactions amongst variables such as disease, socio-demographic characteristics, or environmental exposures, further complicate an accurate interpretation of the data. Data exploration and visualization through mapping and spatial analysis often provides a more robust understanding of the data, as well as improved clarity in viewing the phenomena under study, which will lead to better design of further analyses and additional hypothesis generation, in an iterative fashion.

Geovisualization is defined as "the use of computer technology for exploring data in visual form.... and the use of computer graphics for acquiring a deeper understanding of data" (Visvalingam, 1994). Visualization can also be thought of as "the interplay between technology and the human mind" (Davies and Medyckyj – Scott, 1994). The impetus behind such a geovisualization process is to "see the unseen" in these increasingly large and complex datasets, where, without computational exploratory mapping, it is unlikely that we would be able to ferret out many of these "unseen" relationships (Orford, 2005).

Maps are, and always have been, rich sources of data. GISc increases the richness of the data, and the functions of GISc make it possible to look at the data in many different ways and from various viewpoints. We can manipulate the data, examine its statistics, plot graphs of it, classify and reclassify it with different schemes of class breaks and classification methods, and look at multiple views of the data at the same time. This kind of geovisualization has become easier and more productive since the advent of accessible forms of GISc software and their computerized cartographic capabilities (Kraak and Orneling, 1996; MacEachern and Kraak, 1997).

In the first part of this chapter, we use a hypothetical data set to illustrate some of the data exploration, geovisualization, statistical methods, and geospatial analyses. In the second part of the chapter, we use a worked example of respiratory disease and socio-demographic variables in New York City to assess potential environmental justice impacts, in order to further demonstrate the importance of geovisualization and geospatial analysis in achieving a better understanding of environmental health issues.

2.2 Environmental Health Justice

Environmental Justice (EJ), as a research framework, is the attempt to document and address the disproportionate environmental and health burdens borne by the poor, people of color, and other vulnerable populations. In a broader context, EJ theory encompasses everything that is unsustainable about the world we have created, including rampant population growth, industrialization, pollution, consumption patterns, energy use, food production, and resource depletion. "The EJ movement has sought to redefine environmentalism as much more integrated with the social needs of human populations, and, in contrast with the more eco-centric environmental movement, its fundamental goals include challenging the capitalist growth economy, as well" (Pellow and Brulle, 2005, 3).

Environmental Justice, both as an advocacy movement and as a field of research, came into being over 20 years ago, and ever since that time, Geographic Information

Science has been used to examine the spatial realities of environmental injustice (Boer et al., 1997; Bowen et al., 1995; Burke, 1993; Chakraborty and Armstrong, 1997; Chakraborty et al., 1999; Maantay et al., 1997; Maantay, 2002; Morello-Frosch et al., 2001; Neumann et al., 1998; Perlin et al., 1995; Pollock and Vittas, 1995; Sheppard et al., 1999).

GISc methods have been used in environmental justice research primarily to analyze the spatial relationships between sources of pollution burdens and the socio-demographic characteristics of potentially affected populations, and for the most part, researchers have found strong associations between race, class, and environmental burdens. More recently, health outcomes and exposure measures have also been included in order to draw more definite connections between pollution, poor people, communities of color, other vulnerable populations, and adverse health outcomes. GISc technology is particularly well-suited for EJ research because it allows for the integration of multiple data sources (e.g., location of polluting facilities, population characteristics, and disease rates), representation of geographic data in map form, and the application of various spatial analytic techniques (e.g., buffering) for proximity analysis (Zandbergen and Chakraborty, 2006).

2.3 Data Exploration Example Using Hypothetical Data Set

GISc can be invaluable in data exploration. Looking at data spatially allows a much more holistic and complete view of the phenomena or processes under study. Spatial analysis requires data with a geographic identifier: any data that has a locational component (e.g., a street address, latitude/longitude, zipcode, census tract) can be mapped and analyzed with GISc.

When starting the process of data analysis, it is traditional to run some basic descriptive statistics (e.g. mean, median, mode, standard deviation, etc.) on the pertinent datasets. Although these numbers can provide some extremely useful summaries of the data, they do not provide a spatial understanding – that is, they do not show how values vary from place to place. For instance, if you are presented with a dataset that contains thousands of samples (e.g., census tracts), each of which has information regarding the population and a disease of interest, an a-spatial analysis may not provide you with all the information that is required for a complete interpretation, vis-à-vis, how disease varies across geographic space.

To illustrate this point, a hypothetical study area was created containing a 30×30 grid of cells ($n = 900$), each of which could be considered the geographic unit of analysis (resolution) in a GISc study. Each cell is 1,000 by 1,000 ft, and contains a value for population data as well as disease data. A random population between 500 and 600 was given to each unit. Disease data, however, was not randomly distributed. Instead, "disease centers" were defined around two locations. Geographic units that were greater than 5,000 ft from either of the source points were assigned a random disease rate between 1 and 2 cases per 100 persons. As the distance from a cell to either of the centers decreases, the rates increase. In other words, the rates in the geographic units proximal to the disease centers were calculated as higher than distant geographic units.

If this data were to be looked at tabularly (Table 2.1) or with a graph (Fig. 2.1), relatively few inferences could be made regarding the nature of the phenomenon, how it is distributed in space, and any relationship between the areas and the rates. However, by mapping the data, spatial patterns can emerge, and provide more explanatory power in terms of the geographic context. The relative clarity of the spatial patterns depends on many things, such as the choice of data classification method (e.g. natural breaks, quantile, etc.) and the type of thematic map (e.g. dot density for number of cases, choropleth for rates, etc.), as discussed below.

Data Classification Process: When mapping quantitative information, it is usual to classify the numerical data into ranges of numbers. This is done for convenience and for ease of reading and interpreting the information on the map, since it is often impossible or impractical to represent every unique value in the data with a different unique symbol on the map.

Table 2.1 Tabular view of the first eleven records in the hypothetical dataset

Cell id	Population	Number of cases	Disease rate (×100)
1	571	10	1.7513
2	553	8	1.4467
3	558	9	1.6129
4	529	7	1.3233
5	530	7	1.3208
6	577	10	1.7331
7	501	5	0.9980
8	578	33	5.7093
9	576	10	1.7361
10	581	11	1.8933
11	571	10	1.7513
...	...	...	...

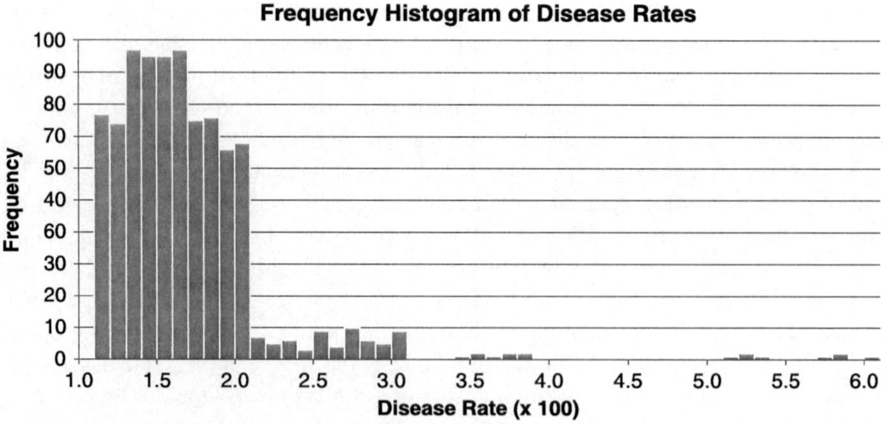

Fig. 2.1 Graph (histogram) of the hypothetical dataset

There are a number of data classification methods that are commonly used, and which one is selected will depend to a large degree on the dataset being mapped – its content, configuration, and its "shape." This entails, for instance, whether or not the data has a normal distribution, and other particularities of the data set, which can be ascertained in general terms by running some basic descriptive statistics. The data classification method ultimately chosen will affect the emphasis of the mapped data, and some cartographers maintain that the "best" way to present quantitative data is by classifying it in several different ways, allowing the analyst to piece them together and form a composite picture of different aspects of the data.

The data classification methods briefly described here are the ones most generally available with standard GIS software packages – equal interval, quantile, natural breaks, and standard deviation – and these tend to cover most circumstances. There are other useful methods, such as arithmetic, geometric, harmonic, and nested mean, which can be used in a GIS software program by manually calculating and setting the class breaks.

Equal Interval: in some ways this is the easiest classification method to understand and use, and maps created with equal interval class breaks have an inherent logic and intuitive feel about them. With this method, the data range is broken into classes, each containing the same interval. For instance, if we were interested in mapping percent minority population by census tract for a city, and our data range was 1–100 (1 being the minimum value and 100 being the maximum value) and we wanted five classes, our classes would be 1–20; 21–40; 41–60; 61–80; 81–100. This method works well for a dataset that has a normal distribution of values. The drawback to Equal Interval comes if your dataset contains values that primarily fall in just one or two of the classes. The resulting map would show only one or two classes leaving other classes unrepresented. As such, little useful information about the spatial distribution of the variable would be visible.

Quantile: this method produces a map in which every class has an equal number of areal units or observations. Let us imagine that we have the same dataset as above with percent minority population (data range: 1–100) which are aggregated into 200 census tracts (i.e. $n = 200$). If we want five classes again, then each class will have 40 census tracts in it, and that will determine the class breaks. If the data values are arrayed in order of magnitude, the class breaks would be drawn so that each class includes a set of 40 census tracts, consecutively based on the data values. Although this method produces a map that can be more visually interesting than that produced by equal interval (because by definition, each class will have some units in it) it can also be misleading to the map viewer, since the map will show each data class with equal weight (the same number of areal units per class) even if the values are not that different between classes. This method tends to work best when data is aggregated by areal units that are roughly the same size. Additionally, outliers may be de-emphasized in the quantile method, due to the grouping of values by ordinal ranking.

Natural Breaks: very often, this is the default classification method in GIS software. The assignment of class breaks in this method is very dependent on the dataset you are working with. It uses an algorithm to create classes which are as homogeneous as possible internally, while maximizing differences amongst the classes. It does this by organizing an array of the data, and then finding the "natural" break points or discontinuities in the array, thus combining the values that are similar into classes. Very often, this is the most realistic or "true" view of the data, since each class is internally consistent. The drawback is that map viewers may have a more difficult time interpreting maps made with natural breaks, since the class ranges in the map legend may appear random and arbitrary, and some classes may be overly inclusive, and other classes may contain very few data values.

Standard Deviation: The standard deviation method of classification groups the data values into classes based on the mean and the standard deviation of the data set. Each class represents one (or one half, one third, etc.) standard deviation above or below the mean (the arithmetic average) of the dataset. The standard deviation classification can result in classes containing class break values outside the actual range of the data, due to the way standard deviations are constructed. The standard deviation classification method is especially useful when performing longitudinal studies (comparing different time periods) or for comparisons amongst datasets that vary widely in their mean, median, or other measurement of central tendency. In standard deviation classification, the mean and the standard deviation are the basis of the class boundary formation, and so the mean and standard deviation are more comparable across datasets with varying ranges.

The process of data classification is itself a type of data exploration. It allows the analysts to familiarize themselves with the data, and see if and how the map changes based on the classification method used. This can reveal information about the data and affect the ease of data interpretation. Maps made with the exact same data but using different classification methods will, more often than not, look markedly different. In addition to the classification method selected, factors such as how many classes are created and how many areal units are included in the data set, will also have an impact on the map's appearance and its interpretation. Below is a graphic representation of how different classification methods would group the data from our 900 sample hypothetical dataset of disease rates differently in terms of class breaks (Fig. 2.2) as well as number of samples per class (frequency, Fig. 2.3).

Types of Thematic Maps: The types of maps usually used for data exploration and geovisualization are, broadly speaking, thematic maps, and they can be either quantitative or qualitative in nature. Thematic maps express a "theme" of information, as opposed to a reference map, which is used for way-finding and identifying actual locations. Although it is common for a thematic map to only represent one "theme" or variable, effective maps can also be made to show multiple variables, often by employing several types of thematic maps in one. The following types of thematic maps are used most frequently for data exploration and visualization:

- Dot density
- Choropleth
- Proportional symbol
- Isoline
- Continuous surface (interpolation of point data)
- Cartogram

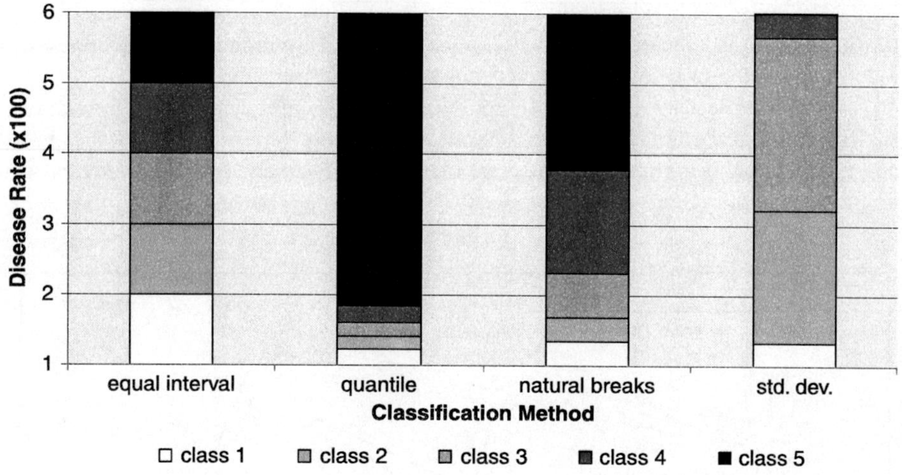

Fig. 2.2 Class breaks of disease rates using four different classification methods (hypothetical data)

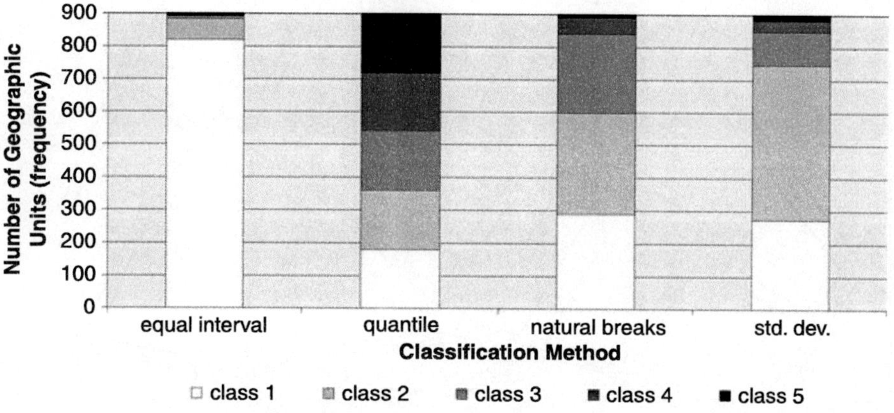

Fig. 2.3 Frequency of geographic units per class using four classification methods (hypothetical data)

Dot Density Maps: A dot density map is used to plot the absolute numbers of things – people, cases, or events – as aggregated by a geographic area. Usually, each dot represents a certain number of the things being mapped, e.g., one dot equals 10 cases of tuberculosis. The dots are not intended to correspond to the actual locations of these things, but rather are a random distribution of the points within each area of aggregation. Dot density maps are very useful for obtaining a "snapshot" of the distribution of the variable within the larger geography, and are often used for data exploration with the dots "on top" of a choropleth map showing rates or percentages of another variable. This is an easy way to investigate the potential spatial correspondence of two or more variables, or different aspects of the same variable. Dot maps have been in use for hundreds of years, for instance, John Snow's dot map of cholera deaths in Soho in nineteenth century London (Fig. 2.4), which is a modified use of a dot map, where one dot equals one event at its actual location.

In their simplest form, one dot equals one case. Figure 2.5 depicts how altering the dot value changes our perception and possible interpretation of the data, and reveals why it is important to select the dot value wisely. As with many cartographic decisions, there is no necessarily "right" choice, but through a process of

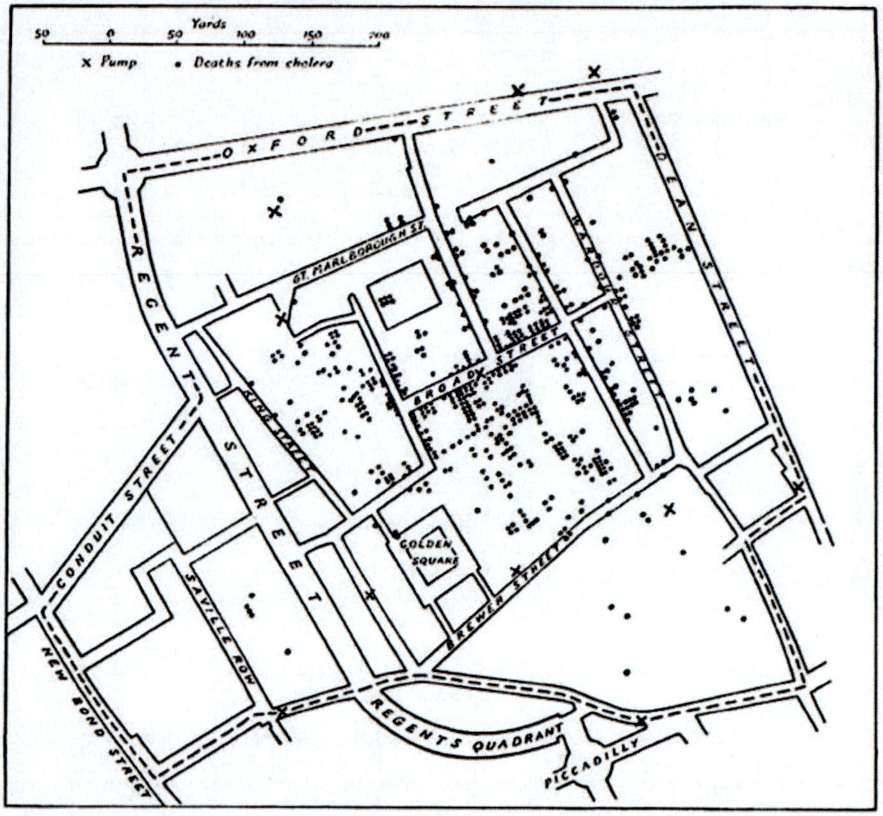

Fig. 2.4 By plotting deaths from cholera on to a map and revealing the geographical relationship between deaths and the location of the Broad Street pump, Dr. Snow had showed how maps could provide a unique insight into the patterns, processes, and relationships of spatial phenomena. The relationship between polluted drinking water and cholera was not self-evident and had to be graphically displayed before the connection could be made. (Orford, 2005:190–191). This seminal thematic dot map by Dr. John Snow of deaths from Cholera in 19th century London, as reproduced in E.W. Gilbert's 1958 article, is often thought to be the inspiration for the discovery of the waterborne nature of the disease's transmission (although this commonly-held idea of the map's role in this discovery is disputed as apocryphal by several authors, notably Tom Koch in *Cartographies of Disease*, and Orford in Visualization and Cartography). Map Source: "Pioneer Maps of Health and Disease in England," *Geographical Journal*, 124 (1958), 172–183

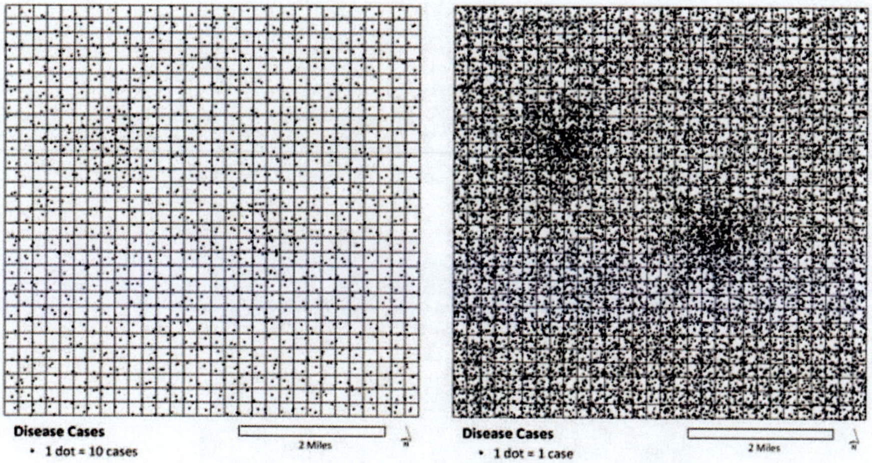

Fig. 2.5 Dot density maps of hypothetical data showing how dot value selection can affect ease of visual interpretation (*left*) shows 1 *dot* = 50 cases (*right*) shows 1 *dot* = 5 cases

"intelligent" trial and error, one can find the optimal way to present the data graphically in order to show the greatest detail and meaning in the data, and to distinguish any spatial pattern.

Choropleth maps: Choropleth maps, also called graduated color maps, are used to show rates, percentages, or ratios of a phenomenon, as aggregated by some geographical unit, such as a census tract, zip code, county, or state. In a choropleth map, each unit receives a color, pattern, or tone that designates the value range of the variable. These colors or tones are graduated in hue or intensity to denote the relative magnitude of the variable. Intuitively, the progression of shades makes sense: as the rate or percentage increases, the color deepens. It is easy to compare two choropleth maps showing different variables in order to ascertain geographic distributions and any spatial correspondence amongst the mapped variables, provided the maps have the same extent and unit of analysis. Using the natural breaks classification method on the hypothetical dataset, a choropleth map was created that renders the areas with unusually high disease rates easily identifiable (Fig. 2.6).

Proportional symbol: A proportional symbol map, also called a graduated symbol map, shows relative or absolute amounts of the variable by using a symbol placed at the corresponding point on the map, or at the centroid of each unit of aggregation. Data can be utilized in the form of absolute numbers or counts, or in the form of percentages, ratios, or proportions. Proportional symbol maps can also depict ordinal, or ranked, data, such as small, medium and large, or high, medium, and low density. These maps are useful to obtain an overview of the variable, and are particularly informative when used in combination with other types of thematic maps. Using proportional symbols on the hypothetical data, areas of high disease rates are again easily visible (Fig. 2.7).

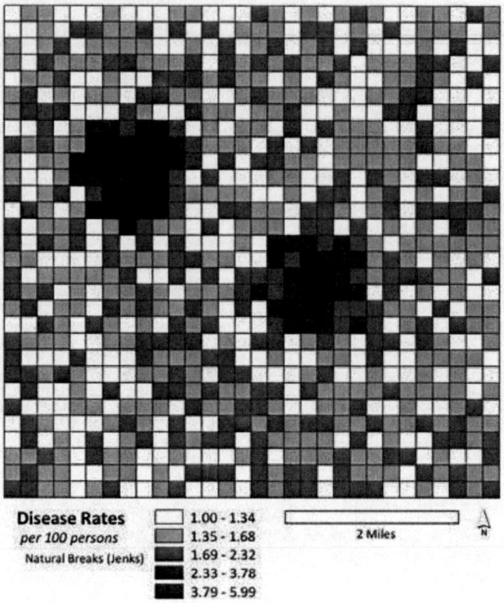

Fig. 2.6 Choropleth map of hypothetical data using "natural breaks" classification

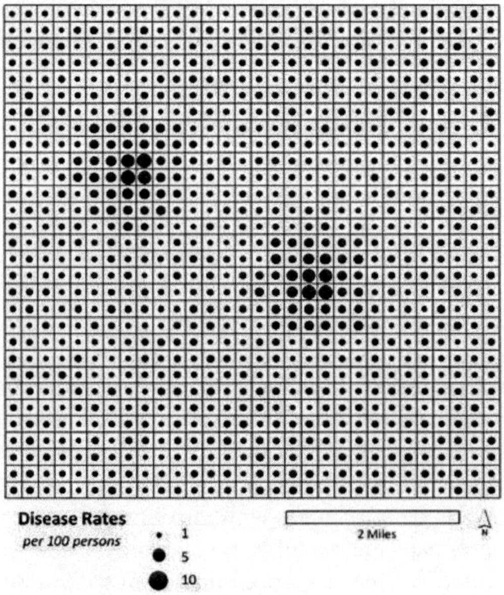

Fig. 2.7 Proportional symbol map of the hypothetical data

Continuous Surface: When data is presented as a continuous surface, spatial trends can often be intuitively and easily seen. A surface can be created by interpolating values between sampled points, essentially estimating values between the known samples, thereby "smoothing" the raw data. Although continuous surfaces are most often used to represent environmental data such as temperature, rainfall, air quality, and elevation, they can be used for nearly any type of dataset. However, caution must be used when interpolating sparse data (i.e. few sample points) or data that is inherently delineated by discrete boundaries (e.g. legal jurisdictions). There are many ways to achieve this interpolation, including inverse distance weighting, spline, and Kriging. Although an in depth discussion of each technique is beyond the scope of this chapter, the nature of the data being analyzed and the desired product will often dictate which methodology is most appropriate. In our hypothetical dataset, ordinary Kriging was used to estimate the values of points between the centroids of the original polygons. The output is in raster format (gridded) and can be viewed as either a true continuous surface where there are no classified divisions in the data (i.e. "stretched") or as a classified surface utilizing the same classification schemes discussed earlier (Fig. 2.8). Just as with other techniques, classification choices, as well as additional choices which must be made regarding interpolation, can profoundly affect the appearance and ease of interpretation of the product.

Isoline: Isoline maps, or contour maps, are made from a continuous variable, by connecting places of equal value of the variable. All points along any given line are presumed to have the same value for the particular variable being mapped. As isolines are essentially an alternate way to represent a continuous surface, discrete occurrences of a phenomenon (e.g. land use) should not be mapped using this method. Although topography is the most common variable used with isolines,

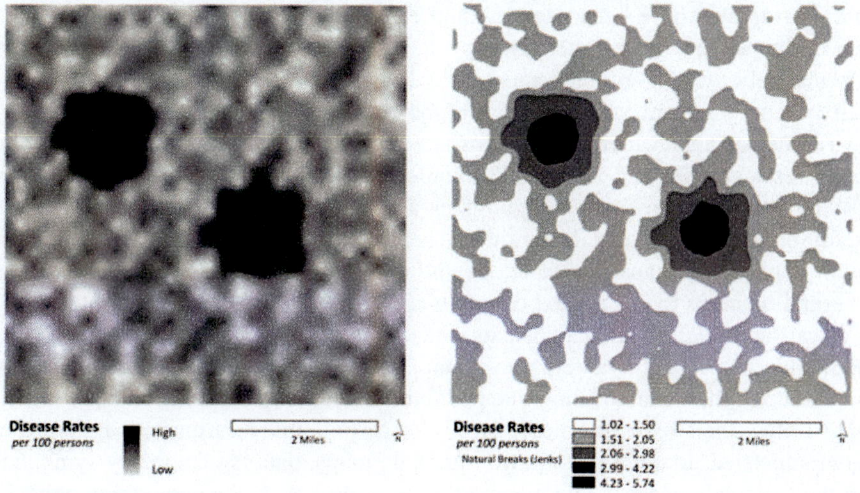

Fig. 2.8 Continuous surface of hypothetical dataset (*left*) shows a "stretched" surface and (*right*) shows a classified continuous surface

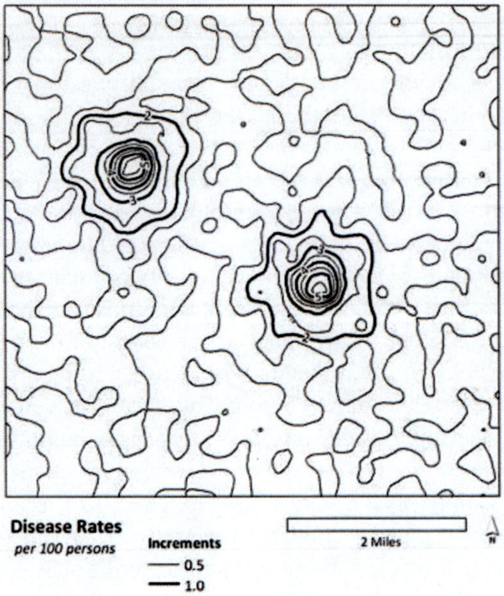

Fig. 2.9 Isolines of the hypothetical dataset derived from the continuous surface

any continuous data, such as most environmental variables as well as some human variables (e.g. median housing costs, population density, and disease rates) can be mapped in this fashion. For the hypothetical dataset, isolines were created based on the interpolated surface described above (Fig. 2.9).

Cartograms: Cartograms are not very commonly used in public health mapping, but deserve a wider exposure, as they can be quite effective in helping to visualize data. Cartograms have been used to good effect, for instance, in Danny Dorling's World Mapper cartogram series (Dorling, accessed 2010), which includes maps that informatively present complex and varied data in a "snapshot" way, and are especially useful for displaying data at a global extent. They have also been used in environmental health literature to illustrate the spatiality of environmental public health research (Sui and Holt, 2008; Houle et al., 2009). Cartograms can help in visually comparing regions and variables across regions at a glance, but are less useful for determining actual quantities. Therefore, they are primarily used as working or exploratory maps to indicate potential areas for more detailed study, and as presentation maps to inform and communicate to the public about different issues.

A cartogram, also called a "density-equalizing map," "equal-area map," "isodemographic map," and "value-by-area map," shows land areas sized to reflect the magnitude of the variable being mapped. Normally, the geographic units shown on maps reflect their real geographic size. Not cartograms – cartograms ignore true geographic size. In other quantitative thematic maps, data is mapped by symbolizing the variable's quantity and placing the symbol in or on the geographic units. In the cartogram, the size of the geographic unit itself is intended to communicate the variable's quantity. For instance, in a cartogram of world population by country, the geographic size of the countries would be drawn in proportion to their population size, not to their geographic size. China, India, the United States, and Indonesia, the world's four most populous countries, would therefore be drawn with the largest geographic extents. The size of the country itself on the map thereby represents the variable of population size. The cartogram does not lose data by classification or generalization, as do choropleth maps, for instance. However, the map user may find it difficult to understand the cartogram, depending on the map user's existing knowledge of and familiarity with the geography being portrayed. Many mapmakers choose to include a small inset map with the cartogram. The inset map can be used to remind the map user of the real relationship and physical sizes of the geographic units being mapped, so the information embedded in the altered sizes of the geographic units in the cartogram can be more easily interpreted. The geographic shapes on many cartograms can be highly generalized, often with box-like forms that make no attempt to conform to true shape, resulting in a map that looks more like an organizational chart (Maantay and Ziegler, 2006). Cartograms are often used in combination with other thematic mapping methods, such as choropleth or proportional symbol maps, which add richness and nuance to the data presentation.

The cartograms contained in this chapter are density-equalizing cartograms, in which areal units are drawn proportionally to the salient population characteristics. The benefit of this type of cartogram is to overcome the inevitability in non-cartogramatic maps for a variable to "show high incidence in cities and low incidence in rural areas, solely because more people live in cities," (Gastner and Newman, 2004, 7499). The maps below (Fig. 2.10) show the juxtaposition of the choropleth and cartogram of the hypothetical data disease rates. Notice how the two "disease centers" are not only brought out by the shading, but are also enlarged cartogramatically for a dramatic visual effect.

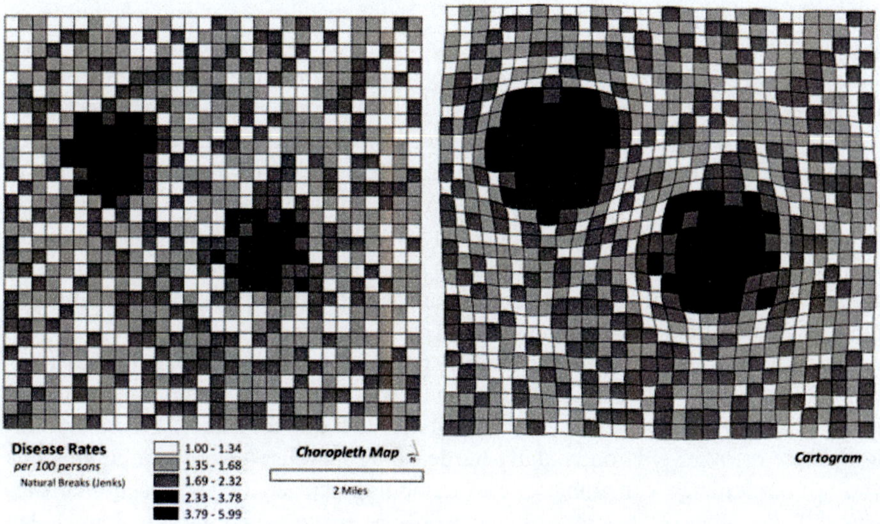

Fig. 2.10 Choropleth (*left*) and cartogram (*right*) of the hypothetical dataset

Geostatistics is a branch of statistics which focuses on the analysis of spatial data. As was demonstrated with the simple hypothetical example above, the locational dimension of a dataset is tremendously important in understanding environmental and health concerns. What follows is not meant to be an exhaustive discussion of geostatistics or geospatial analysis by any means, but rather a simple demonstration of how this type of quantitative exploration can help the researcher to better understand the nature of what is under investigation. This is done by working through only a few techniques: Moran's I and Local Indicators of Spatial Autocorrelation (LISA) are introduced below; and Geographically Weighted Regression is introduced in the next section. Many other measures and analyses could have been used (e.g. Getis-Ord G or Gi*, spatial autoregressive models, etc.), but that is beyond the scope of this introductory chapter. For more information see Chapters 17 and 20 (Chakraborty; Hu et al).

Spatial autocorrelation is an important concept in spatial analysis. If the value in one geographic unit is correlated with the values in neighboring units, a variable can be considered spatially autocorrelated (Cliff and Ord, 1973). Moran's I is a global measure for autocorrelation which ranges from –1 to 1. When values approach 0, there is no spatial autocorrelation; as the Moran's I approaches 1 or –1, there is positive (clustering) or negative (dispersion) autocorrelation, respectively. A standardized Z score can be used to assess significance (with the null hypothesis representing a random spatial distribution). In order to calculate a Moran's I, a spatial weights matrix must first be defined. This matrix defines the spatial relationship among the samples (e.g. census tracts). There are many options regarding the definition of the spatial weights matrix, the most common of which include polygon contiguity, simple distance threshold, distance decay, and k-nearest neighbors. For more information regarding spatial autocorrelation, see Chapter 17 by Chakraborty, in this book.

The Moran's I for our dataset suggests a statistically significant clustering of disease rates ($I = 0.62$ when first order contiguity is used, Z $Score = 26.4$). This clustering can be further explored using local indicators of spatial autocorrelation (LISA). A LISA can be used to quantify spatial autocorrelation locally by calculating a Moran's I and an associated significance level for each spatial unit. The sum of all of the LISAs will be proportional to the global measure of spatial autocorrelation (Anselin, 1995). When our dataset is looked at in this fashion, once again using first order contiguity, the areas with local clusters of high disease rates can be clearly seen ($p<0.01$) (Fig. 2.11). It is important to note that what is being seen in this map are the local clusters of similar or dissimilar rates (neighboring geographic units in this case), rather than the rates themselves.

These types of visualization and simple spatial analysis methods can lead to interesting findings and raise interesting new questions. For instance, in our hypothetical data, what may be causing these clusters of elevated disease rates? Is there a pollution source or other environmentally burdensome land use? Are there populations who are particularly vulnerable due to demographic, genetic, or social characteristics? These types of questions often can only be answered with a good knowledge of the study area's physical and social environments.

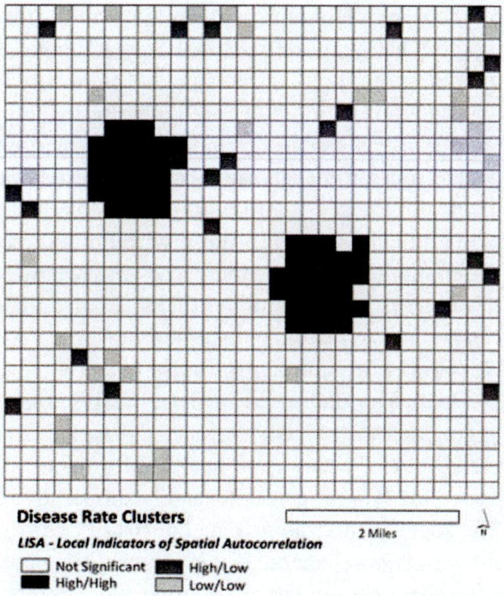

Fig. 2.11 Local Indicators of Spatial Autocorrelation (LISA) clusters of disease rates in the hypothetical dataset. High/high suggests local clusters of high disease rates (high values surrounded by neighbors of similarly high values). Low/low indicates local clusters of low disease rates. High/low and low/high suggest local statistical outliers (high values surrounded by low values or low values surrounded by high values, respectively)

When the number of variables or the complexity of the relationships increase, it can be very useful to explore the phenomena through more complex statistics rather than simply cartographically visualizing the individual variables. There are many ways to look at this data statistically, and each dataset may lend itself to one technique or another. A common approach when trying to quantify relationships among variables is regression analysis. Similar to the previously discussed exploratory techniques, regressions can be approached a-spatially or spatially. Ordinary least squares regression (OLS) results in summary statistics for the entire study area (i.e. global), however if the relationship(s) being examined are geographic in nature and have a truly spatial component, it can be beneficial to perform the regressions geographically (e.g. spatial regressions) or even locally (e.g. geographically weighted regression). The remainder of this chapter will go through a worked real-world example demonstrating the utility of data visualization and geostatistics.

2.4 Respiratory Disease and Environmental Health Justice in New York City

Respiratory disease rates can be a major concern in urban environments. One of the causes of high rates of respiratory disease may be poor air quality due to close proximity of residential areas to areas with high vehicular traffic, industrial land uses, high population densities, and other environmentally burdensome land uses (Aylin et al., 2001; Edwards et al., 1994; Maantay et al 2008; Smargiassi et al., 2009). Respiratory disease is an environmental justice concern that combines issues of socio and economic vulnerability with unequal environmental exposure. Some research has suggested that not only are lower-income populations and communities of color more likely to live in close proximity to environmentally burdensome facilities and thus be more exposed to pollution, but that the health effects of exposure to these burdens are further modified by socio-economic status, and "due to material deprivation and psychosocial stress [these populations] may be more susceptible to the health effects of air pollution," (O'Neill et al., 2003, 1861). Therefore, vulnerable populations, such as those with limited income or educational attainment, may suffer more from the same exposures when compared to other groups. What follows serves as an example of how geovisualization and exploratory spatial analysis can be used to examine respiratory disease vis-à-vis socio-demographic variables in New York City in order to assess potential environmental justice impacts.

Data: The data used in these analyses are all publically available information aggregated to the census tract, and includes socio-demographic information from the 2000 US. census and health information (hospital admissions) from New York Statewide Planning and Research Cooperative System (SPARCS) via Infoshare.org. The socio-demographic data consist of adults (> 25) who do not have a high school diploma, non-Hispanic black persons, Hispanic persons, and persons below the federal poverty level. Other variables were explored as well; however, due to concerns such as excessive colinearity, they were not included in these analyses. All of these variables were examined as rates (e.g. percentage of population that is below poverty thresholds, per census tract). As can be seen by examining the maps (choropleth and cartograms), the socio-demographic variables are not evenly distributed across NYC (Figs. 2.12 and 2.13).

Health outcomes were represented by the number of individuals who were hospitalized for respiratory illness, geocoded to the census tract of their residence. The SPARCS data was queried using ICD-9 codes for acute respiratory infections (ICD-9: 460–466), chronic obstructive pulmonary diseases and allied conditions (ICD-9: 490–496), and pneumonoconioses and other lung diseases from external agents (ICD-9: 500–508). Five years of data were combined (1998–2002, inclusive) and averaged in order to stabilize the rates. To further stabilize the rates, census tracts with fewer 250 persons or fewer than 20 people hospitalized for respiratory disease over the 5 years were excluded from the analysis. It is important to note that "persons hospitalized" is not necessarily equivalent to incidence or prevalence. Even though our previous research studies have used hospitalization records for health outcomes, the data has limitations and may be biased due the way in which NYC residents utilize hospitals and how existing disease is managed due to differences in health insurance coverage, physical access to primary care treatment and prevention, education regarding maintenance of health conditions, and other social and economic issues (Maantay 2007; Maantay et al., 2007, 2009a, b). The NYC respiratory hospitalization data was then age-adjusted using New York State as the standard population. As can be seen in Fig. 2.14 (left), there are high rates of respiratory hospitalizations in certain areas of NYC, which are indicated by the dramatically ballooned sizes of the affected census tracts. Some of these areas correspond to areas of high poverty and high minority population as is shown

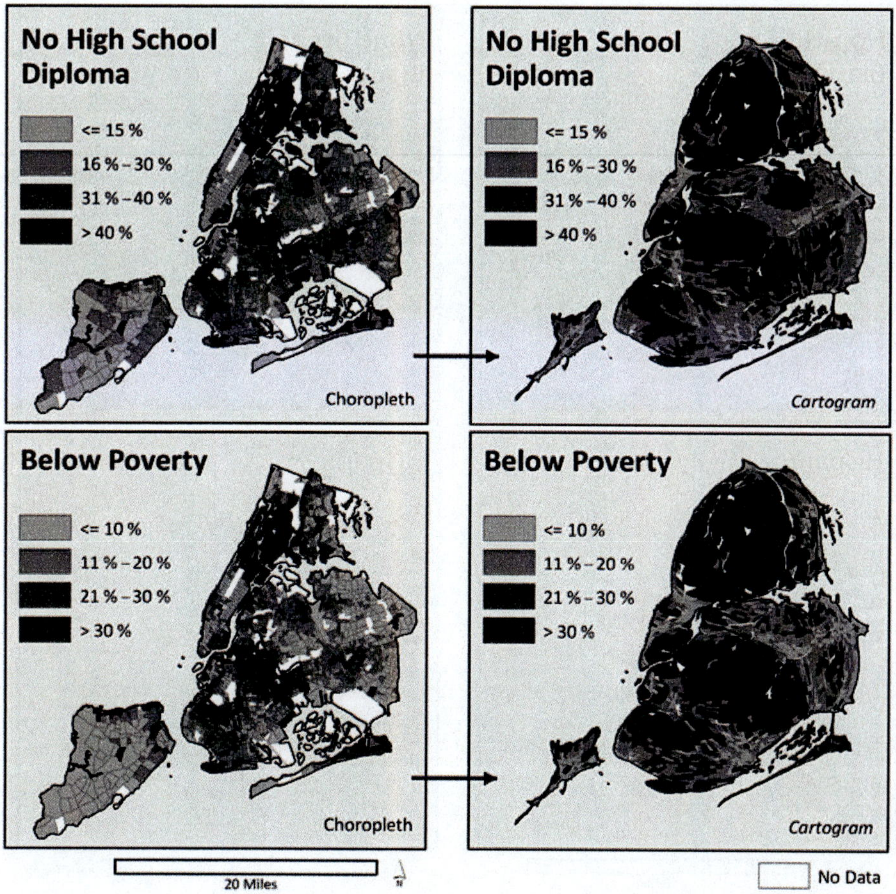

Fig. 2.12 Percentage of adults without a high school diploma and percent below poverty by census tract in NYC using choropleth mapping and cartograms

in Figs. 2.11 and 2.12. Once again, the utilization of cartograms can allow for an effective visualization of the data (Fig. 2.14, right).

Analysis: When the number of variables or the complexity of the relationships increase, it can be very useful to explore data statistically to augment the visualization of the variables individually. With a study such as this one, multiple ordinary least squares regression (OLS) is a natural choice. Age-adjusted rates for persons hospitalized for respiratory disease were used as the dependent variable, and percent Hispanic, percent non-Hispanic black, percent of adults without a high school education, and percent of people living below poverty were used as the independent variables. Note that there is no exposure estimate in the model, simply the health outcome versus the socio-demographic variables. Once again, rates were stabilized by excluding census tracts with fewer than 250 people or fewer than 20 people hospitalized between 1998 and 2002, inclusive (n = 1,880 tracts). The hospitalization

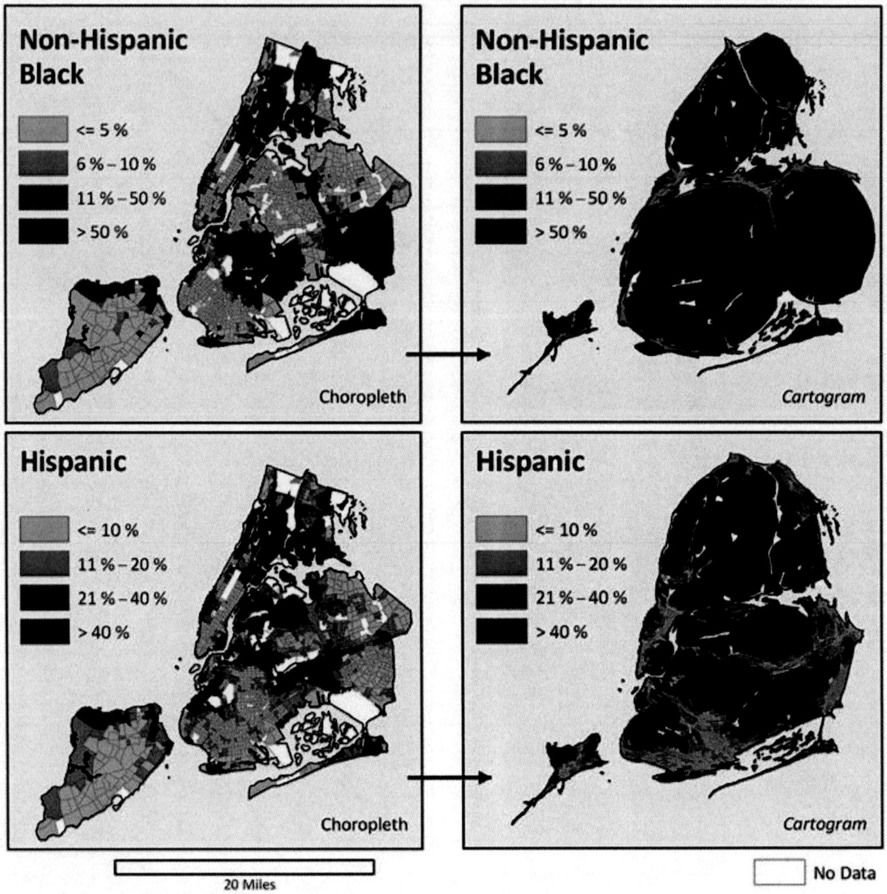

Fig. 2.13 Percent non-hispanic black and percent Hispanic by census tract in NYC using choropleth mapping and cartograms

data was log10 transformed in order to approximate a normal distribution in the residuals.

Ultimately, the model had an R^2 value of 0.50, suggesting that approximately 50% of the variance in the hospitalization rates can be explained by the socio-demographic measures. All of the SES variables showed a significant ($p<0.01$) positive association with respiratory hospitalization rates except for the education variable which was not significant (Table 2.2).

It can be informative to map out some of the OLS results, particularly the residuals, in order to get a more complete geographic understanding of the results. As can be seen in the map, there appears to be clustering of tracts with similar values. For example, the Whitestone/College Point neighborhoods in Queens show a group of tracts where the OLS underestimated the hospitalization rate (area of interest in Fig. 2.15). The Moran's I statistic ($I = 0.23$ with first order contiguity, $Z\ Score = 16.93$) statistically confirms spatial autocorrelation. This suggests that

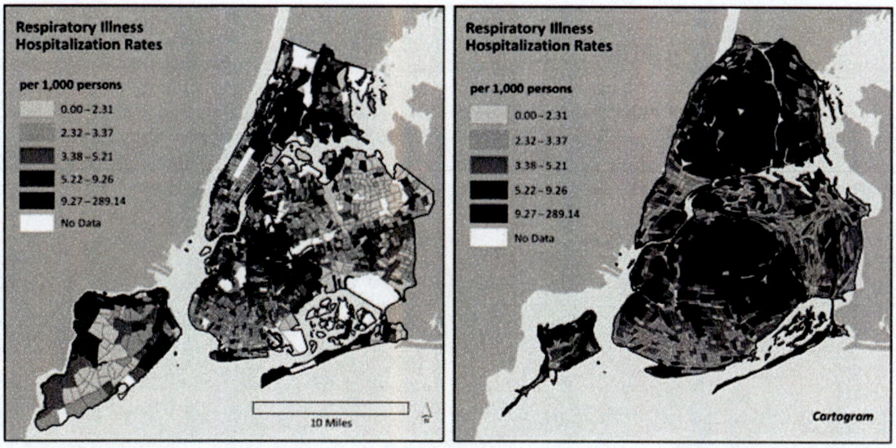

Fig. 2.14 Age-adjusted respiratory hospitalization rates in NYC visualized using a choropleth map (*left*) and cartogram (*right*)

Table 2.2 OLS parameter estimates

Parameter	Estimate	Std err	*t* value
Intercept[a]	0.29848	0.0128328	23.26
Pct without HS diploma	−0.00020	0.0007151	−0.29
Pct below poverty[a]	0.00919	0.0006683	13.75
Pct non-Hispanic black[a]	0.00276	0.0001960	14.07
Pct Hispanic[a]	0.00533	0.0003606	14.78

[a] $p<0.01$.

the residuals of the OLS were not randomly distributed which is a violation of OLS assumptions.

It may be illuminating, then, to examine the local nature of relationships between respiratory health and the socioeconomic variables. Geographically weighted regression (GWR) is a technique developed by Fotheringham et al. (1998) which quantifies locally varying relationships among data, rather than computing a global relationship as OLS does (similar to the distinction between Moran's I and a LISA). Instead of calculating global parameter estimates based on one regression, GWR performs many local regressions, each of which is influenced by the surrounding data resulting in a set of summary statistics for each regression point. In this way, GWR shows local variations in the regression relationships and is able to account for potential spatial non-stationarity, where the relationships among the independent and dependent variables vary over space. Locally varying relationships may suggest a number of things, including possible model misspecification, sampling variation, or simply a relationship that intrinsically varies over space (Fotheringham et al., 2002). In this study, GWR was used as an exploratory tool to discover

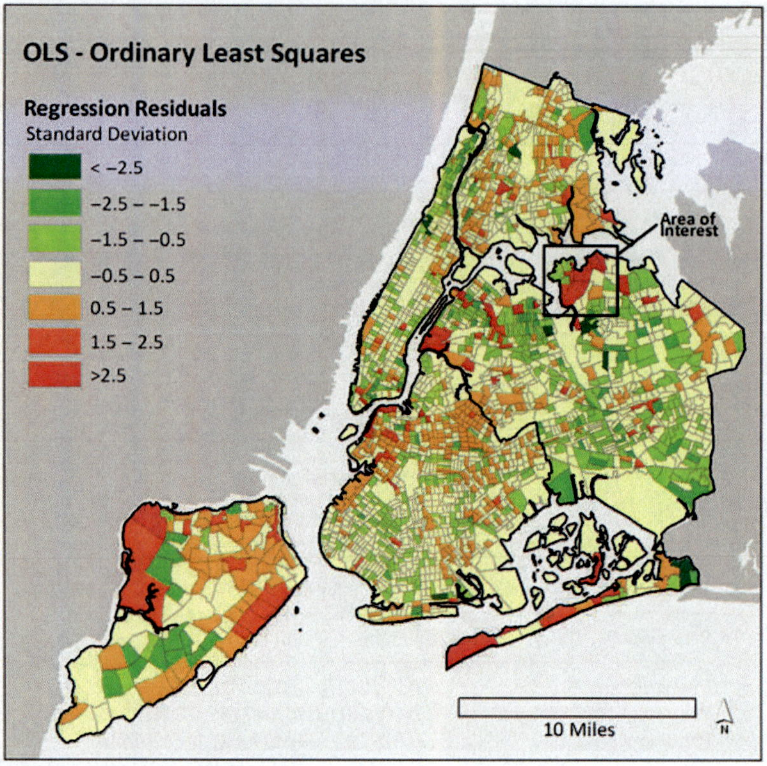

Fig. 2.15 OLS regression model residuals

how the relationships may vary and hypothesize as to why they may be spatially heterogeneous.

The GWR model was specified in the same way as the OLS, with respiratory rate as the dependent variable, and race/ethnicity, education, and poverty as the independent variables. GWR requires that the analyst specify a kernel bandwidth (distance radius) for the model identifying the size/radius of the area in which the local regression model is estimated. The two most common choices are a fixed (Gaussian) or adaptive (bi-square) kernel. The fixed kernel uses a constant radius, whereas the adaptive kernel involves varying the radius across the study area. The adaptive kernel works in a similar way to k-nearest neighbors, selecting a certain number of samples per local regression. As such, the adaptive kernel is able to "grow" when the samples are sparse, and "shrink" when there is a high density of sample points. Both fixed and adaptive bandwidths weight distant samples less heavy than proximal ones.

In our case study, a fixed bandwidth (rather than adaptive kernel) was assigned using an iterative process within the GWR3 software designed to minimize the Akaike Information Criterion (AIC) – a diagnostic statistic which describes the performance of the model. The result was a bandwidth of just under 1.5 miles.

Table 2.3 Comparison of OLS and GWR model diagnostics

Diagnostic	OLS	GWR
Residual sum of squares	104.3	72.2
Standard error	0.236	0.202
Akaike information criterion	−88.8	−570.8
R^2	0.501	0.655
R_a^2	0.500	0.635

Table 2.4 Five number summaries for GWR parameters. A Monte Carlo test of the local parameter estimates reveals significant spatial variability for all of the variables ($p<0.01$)

Parameter	Minimum	1st quartile	Median	3rd quartile	Maximum
Intercept	−0.2300	0.2366	0.2802	0.3661	1.3016
Pct without HS diploma	−0.0671	−0.0014	0.0007	0.0042	0.0767
Pct below poverty	−0.0423	0.0050	0.0081	0.0105	0.0620
Pct non-Hispanic Black	−0.0861	0.0023	0.0030	0.0037	0.0655
Pct Hispanic	−0.0424	0.0041	0.0057	0.0075	0.0304

Diagnostics of the GWR results suggest that it performed better than the global estimate as can be seen by the decrease in AIC, increase in R^2, and changes in other diagnostics such as residual sum of squares and standard error (Table 2.3).

Examining parameter estimates can be a bit more difficult in a GWR than an OLS. Global OLS regression results in one set of summary statistics for each parameter; however GWR has a set for each regression point, which in this case is each census tract centroid. Although it is not uncommon to report the results as a five number summary (Table 2.4), it is much more revealing to examine the geographically-varying parameters cartographically. This can include mapping values for the local R^2, error, parameter estimates, Cook's distance, or *t*-values.

When the *t*-values of the GWR results are mapped by interpolating the values between regression points (creating a continuous surface), the spatial variability in the relationships becomes visible. For instance, the relationship between percent non-Hispanic black and respiratory hospitalization rate are relatively stable while adjusting for the other variables, with only one area showing a negative association (area of interest in Fig. 2.16). The relationships with percent Hispanic and percent below poverty behave similarly to one another. They are somewhat stable, although both show significant negative relationships near the Whitestone/College Point neighborhoods of Queens (areas of interest in Figs. 2.17 and 2.18). Mapping the *t*-values for the percent of adults without a high school diploma, a variable that had no significance in the OLS, reveals that although the majority of NYC does not show a significant relationship, distinct areas of negative and positive associations can be identified (Fig. 2.19). The area with the positive relationship is, once again, around the Whitestone/College Point neighborhoods of Queens. The educational

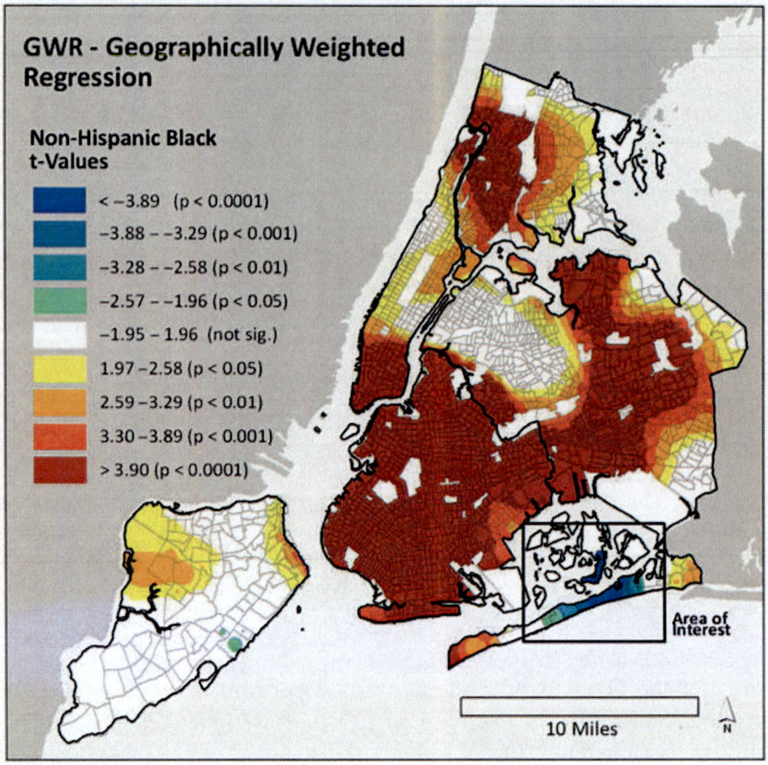

Fig. 2.16 GWR output – percent non-Hispanic black *t*-values

attainment variable must be interpreted with caution however, since the relationships are relatively weak and inconsistent. Also, excessive colinearity with other variables (e.g. poverty) may be skewing the results.

The potential spatial non-stationarity revealed by the GWR can identify areas of interest for further investigation. For instance, it may be useful to conduct a more qualitative assessment of the Whitestone/College Point neighborhoods, which suffer from high hospitalization rates and have high OLS residuals, in order to hypothesize about possible causes of the varying relationships of respiratory hospitalizations with percent Hispanic, poverty, and educational attainment. It is possible that these discrepancies are due to particular environmental conditions (built, physical or social) occurring in the area, which is information not available in the data sets used for the geovisualization. Such information could form the basis for additional questions and hypotheses which could be examined using more in-depth qualitative and quantitative methods.

Results Summary: As can be seen in Tables 2.2, 2.3, and 2.4, both the OLS and GWR suggest that there are measurable associations between hospitalizations for respiratory disease and selected socio-demographic variables. More specifically, as the percent of the population that is below the poverty level, non-Hispanic

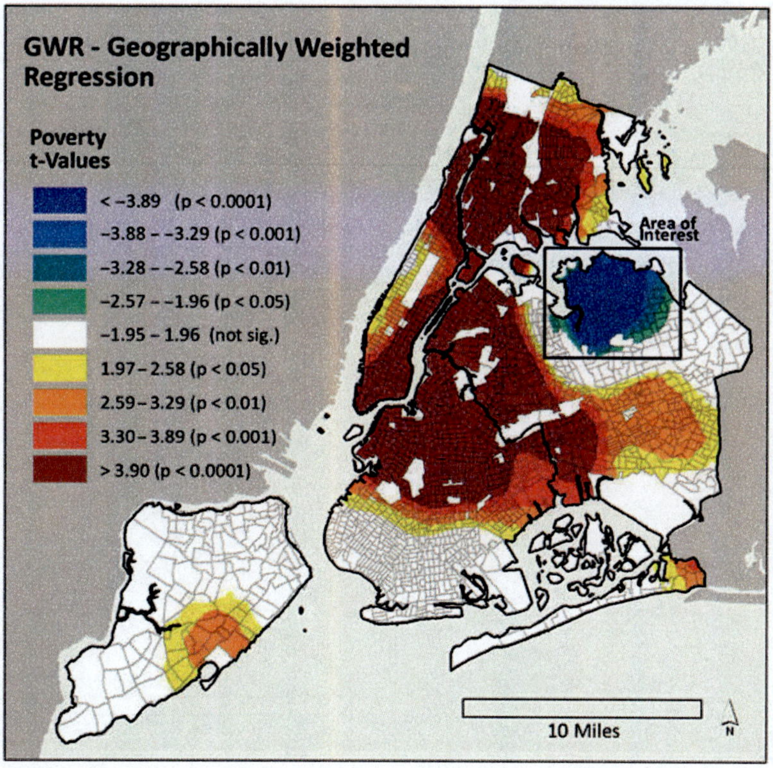

Fig. 2.17 GWR output – poverty rate *t*-values

black, or Hispanic increases, so do the hospitalization rates for respiratory disease. Approximately 50% of the variance of the hospitalization data is predicted by race/ethnicity and poverty status in the OLS model (AIC: –88.9) – approximately 64% is explained in the GWR (AIC: –570.8). Even though the OLS results in a high R^2, according to the AIC the local analysis (GWR) outperforms the global analysis (OLS). This appears to be mainly due to the non-stationary nature of the relationships around the Whitestone/College Point neighborhoods of NYC.

2.5 Conclusions

Geovisualization is an "intelligent" trial and error iterative process, involving various types of thematic mapping, data classification schemes, and geospatial analysis and geostatistics. It can be extremely useful in developing further hypotheses for testing, as well as guiding future analyses. It can also, in its own right, be helpful in answering questions about complex geospatial problems.

However, interpretation of the geovisualized results will have a higher degree of reliability if the analyst has a holistic and intimate knowledge of not only the data

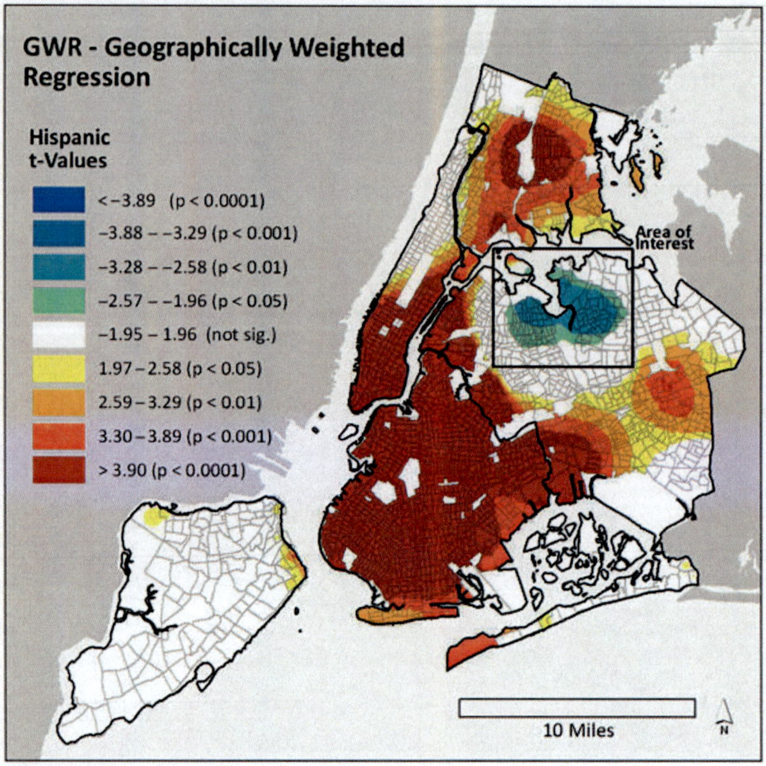

Fig. 2.18 GWR output – percent Hispanic/Latino *t*-values

itself, but also the geography of the study area and familiarity with the built, natural, and social environments. Otherwise, the explanatory power of geovisualization will be limited, and explications of any anomalous situations will be highly speculative, if they are possible at all.

As shown in the example of respiratory disease and environmental health justice in New York City, geovisualization techniques identified areas of concentrations of individual variables and potential spatial co-incidences amongst them, as well as geostatistical trends and anomalies. For instance, although most of NYC exhibits a positive relationship between census tracts with high proportion of non-Hispanic Black residents, Hispanic residents, and those below poverty with respiratory disease hospitalization rates, there are areas which suggest a spatially varying relationship. This is clearly seen when the results of the GWR are mapped, but may have otherwise not been detected.

The geospatial analysis gave us some answers, but also presented new questions, which led us to plans for a more detailed qualitative and quantitative analyses to ferret out some of the spatially inconsistent associations which only became observable with the geovisualization of the data and statistical results.

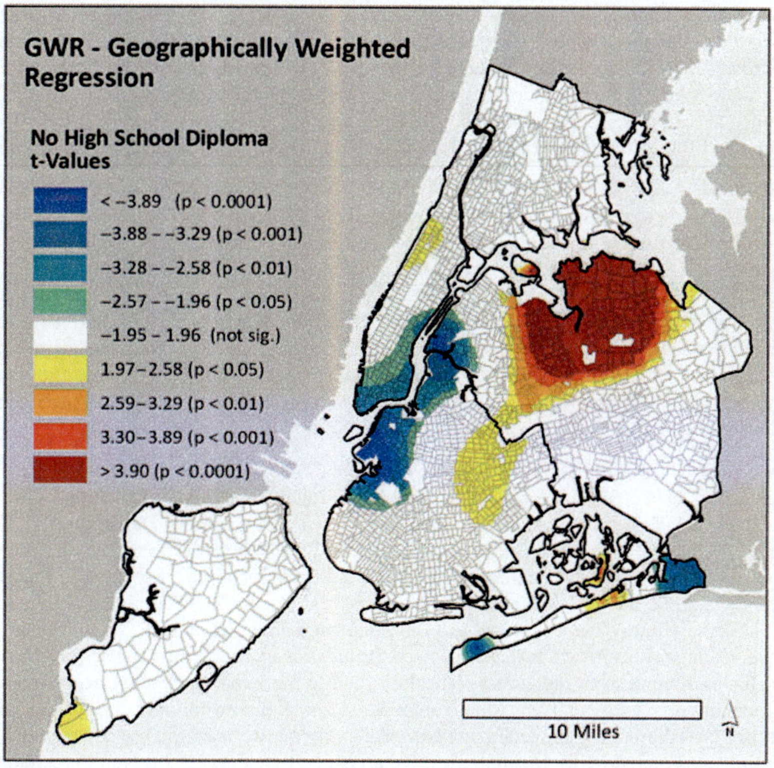

Fig. 2.19 GWR output – rate of adults with no high school diploma *t*-values

Acknowledgments This chapter resulted from on-going research, partially supported by the National Oceanic and Atmospheric Administration's Cooperative Remote Sensing Science and Technology Center (NOAA-CREST) under NOAA grant number NA17AE162. The National Institute of Environmental Health Sciences of the National Institutes of Health also provided critical support for this research under grant number 2 R25 ES01185-05. The National Center for Minority Health and Health Disparities, National Institutes of Health also helped fund this research as part of the Bronx CREED (the Bronx Center to Reduce and Eliminate Racial and Ethnic Health Disparities) under grant P60-MD0005-03. The statements contained within this paper are not the opinions of the funding agencies or the US. government, but reflect the authors' opinions. Thanks are also due to the member organizations of the South Bronx Environmental Justice Partnership, who understood the relevance of this research to environmental health justice and gave their unstinting encouragement and assistance in the effort.

Author Biographies

Andrew Maroko is an Assistant Professor in the Department of Health Sciences at Lehman College, City University of New York (CUNY). He received his Ph.D. in geography from The Graduate Center of CUNY, a GISc Graduate Certificate from Lehman College, CUNY, and his B.A. in biology from Rutgers University. His research interests include the spatial analysis of environmentally-linked diseases and environmental health justice, focusing on the associations between public health and socio-demographics in an explicitly geographical context. Much of his work examines methodological issues related to the quantification of the associations described above (e.g. geo-statistics, exposure modeling, and dasymetric disaggregation). He is a member of the Urban GISc Lab at Lehman College and teaches in the MPH program and in both the Health Sciences and Geography departments, and is an active collaborator in the South Bronx Environmental Justice Partnership, as well as both the Bronx CREED (the Bronx Center to Reduce and Eliminate Ethnic and Racial Health Disparities), and the CUNY Institute for Health Equity at Lehman College.

Juliana A. Maantay is Professor of Urban Environmental Geography, the Director of the GISc Program, and Director of the Urban GISc Lab at Lehman College, City University of New York. She also is a faculty member in Lehman's MPH program, the Earth and Environmental Sciences Ph.D. program at the CUNY Graduate Center, and the doctoral program (DrPH) in Public Health, as well as a research scientist with NOAA-CREST, the National Oceanic and Atmospheric Administration Center of Remote Sensing Science and Technology. Prior to her academic career, Dr. Maantay worked as an urban and environmental planner and policy analyst with governmental agencies, non-profit organizations, and private sector consulting firms, and has been active in environmental health justice research and advocacy for more than 15 years.

Kristen Grady (Ph.D. Student, Earth and Environmental Sciences, City University of New York (CUNY) Graduate Center; B.A. 2008, Geography, Hunter College) is an Adjunct GISc Lab Technician in the Environmental, Geographic and Geological Sciences Department at CUNY Lehman College, and a Research Assistant at several CUNY institutes and centers, including the Urban GISc Lab at CUNY Lehman College, the Center for Urban Research at the CUNY Graduate Center, and the CUNY Institute for Sustainable Cities at CUNY Hunter College. Kristen is also a USDA Fellow in the Department of Environmental, Geographic and Geological Sciences at Lehman College. Recent GIS research projects include measuring access to parks in New York City, analyzing food deserts and access to healthy food in the Bronx in a study of urban agriculture and community gardening, mapping future flood zones due to climate change-induced sea level rise for New York City, and analyzing New York City's historical land use changes. Kristen's research interests include GIS Technologies, Urban Mental Health, Geographic Perception, and Cartography.

References

Anselin L (1995) Local indicators of spatial association – LISA. Geogr Anal 27:93–115

Aylin P, Bottle A, Wakefield J, Jarup L, Elliott P (2001) Proximity to coke works and hospital admissions for respiratory and cardiovascular disease in England and Wales. Thorax 56(3): 228–233

Boer J, Pastor M Jr et al (1997) Is there environmental racism? the demographics of hazardous waste in Los Angeles County. Soc Sci Q 78(4):793–810

Bowen W, Salling M et al (1995) Toward environmental justice: Spatial equity in Ohio and Cleveland. Ann Assoc Am Geogr 85:641–663

Burke L (1993) Race and environmental equity: a geographic analysis in Los Angeles. GEO Inf Syst October: 44–50

Chakraborty J, Armstrong M (1997) Exploring the use of buffer analysis for the identification of impacted areas in environmental equity assessment. Cartogr Geogr Inf Syst 24(3):145–157

Chakraborty J, Schweitzer L et al (1999) Using GIS to assess the environmental justice consequences of transportation system changes. Transact GIS 3(3):239–258

Cliff AD, Ord J (1973) Spatial autocorrelation. Pioneer, London

Davies C, Medyckyj-Scott D (1994) Introduction: the importance of human factors. In: Hearnshaw HM, Unwin DJ (eds) Visualization in geographical information systems. Wiley, New York, NY, pp 189–192

Dorling D, Newman M, Allsopp G, Barford A. http://www.worldmapper.org/. Worldmapper Project Retrieved 1/6/2010

Edwards J, Walters S, Griffiths R (1994) Hospital admissions for asthma in preschool children: relationship to major roads in Birmingham, United Kingdom. Arch Environ Health 49(4): 223–227

Fotheringham SA, Brunsdon C et al (2002) Geographically weighted Regression: the analysis of spatially varying relationships. Wiley, Chichester, UK

Fotheringham SA, Brunsdon C, Charlton ME (1998) Geographically weighted regression: a natural evolution of the expansion method for spatial data analysis. Environ Plan A 30: 1905–1927

Gastner M, Newman M (2004) Diffusion-based method for producing density equalizing maps. Proc Natl Acad Sci 101(20):7499–7504

Houle B, Holt J et al (2009) Use of density-equalizing cartograms to visualize trends and disparities in state-specific prevalence of obesity: 1996–2006. Am J Public Health 99(2):308–312

Koch T (2005) Cartographies of disease: maps, mapping, and medicine. ESRI Press, Redlands, CA

Kraak MJ, Orneling FJ (1996) Cartography: visualization of spatial data. Pearson Education Ltd, Harlow

Maantay J (2002) Mapping environmental injustices: pitfalls and potential of geographic information systems in assessing environmental health and equity. Environ Health Perspect 110(2):161–171

Maantay J (2007) Asthma and air pollution in the Bronx: methodological and data considerations in using GIS for environmental justice and health research. Health and place. Linking Environ Justice Popul Health Geogr Inform Sci 13:32–56, special issue

Maantay J, Maroko A, Alicea C, Strelnick A (2009a) Geographic information systems, environmental justice, and health disparities: the need for an interdisciplinary approach to study asthma and air pollution in the Bronx, New York. In: Freudenberg N, Saegert S Klitzman S (eds) Urban health and society: interdisciplinary approaches to research and practice. Jossey Bass, pp 93–126

Maantay JA, Maroko AR, Porter-Morgan H (2008) A new method for population mapping and understanding the spatial dynamics of disease in urban areas. Urban Geogr 29(7):724–738

Maantay J, Maroko A, Herrmann C (2007) Mapping population distribution in the urban environment: the cadastral-based expert dasymetric system (CEDS). Cartogr Geogr Inf Sci 34(2):77–102 special issue: Cartography 2007: Reflections, Status, and Prediction

Maantay J, Timander L, Graziosi G, Meyer L (1997) The Bronx Toxic Release Inventory Report: US. Environmental Protection Agency Region 2 and Hostos Community College/Center for a Sustainable Urban Environment [CSUE]

Maantay J, Tu J, Maroko A (2009b) Loose-coupling an air dispersion model and a geographic information system (GIS) for studying air pollution and asthma in the Bronx, New York City. Int J Environ Health Res 19(1):59–79

Maantay JA, Ziegler J (2006) GIS for the urban environment. Environmental Systems Research Institute (ESRI), Redlands, CA

MacEachren AM, Kraak MJ (1997) Exploratory cartographic visualization: advancing the agenda. Comput GEO sci 23:64–81

Morello-Frosch R, Pastor M et al (2001) Environmental justice and southern California's "riskscape" – the distribution of air toxics exposures and health risks among diverse communities. Urban Aff Rev 36(4):551–578

Neumann CM, Forman DL et al (1998) Hazard screening of chemical releases and environmental equity analysis of populations proximate to toxic release inventory facilities in Oregon. Environ Health Perspect 106(4):217–226

O'Neill M, Jerrett M, Kawachi I, Levy J, Cohen A, Gouveia N et al (2003) Health, wealth, and air pollution: advancing theory and methods. Environ Health Perspect 111(16):1861–1870

Orford S (2005) Cartography and Visualization. In: Castree N, Rogers A, Sherman D (eds) Questioning geography: fundamental debates. Blackwell Publishing, Oxford

Pellow D and Brulle R (eds) (2005) Power, justice, and the environment: a critical appraisal of the environmental justice movement. MIT Press, Cambridge

Perlin SA, Setzer RW et al (1995) of industrial air emissions by income and race in the United States: an approach using the toxic release inventory. Environ Sci Technol 29(1):69–80

Pollock PH, Vittas ME (1995) Who bears the burden of environmental pollution? race, ethnicity, and environmental equity in Florida. Soc Sci Q 76(2):294–309

Sheppard E, Leitner H et al (1999) GIS-based measures of environmental equity: exploring their sensitivity and significance. J Expo Anal Environ Epidemiol 9:18–28

Smargiassi A, Kosatsky T, Hicks J, Plante C, Armstrong B, Villeneuve P (2009) Risk of asthmatic episodes in children exposed to sulfur dioxide stack emissions from a refinery point source in Montreal, Canada. Environ Health Perspect 117(4):653–659

Sui D, Holt J (2008) Visualizing and analysing public-health data using value-by-area cartograms: toward a new synthetic framework. Cartographica 43(1):3–20

Visvalingam M (1994) Visualization in GIS, Cartography and ViSC. In: Hearnshaw HM Unwin DJ (eds) Visualization in geographical information systems. Wiley, New York, NY, pp 18–25

Zandbergen PA, Chakraborty J (2006) Improving environmental exposure analysis using cumulative distribution functions and individual geocoding. Int J Health Geogr 5:23

Chapter 3
Outdoor Air Pollution and Health – A Review of the Contributions of Geotechnologies to Exposure Assessment

Eleanor M. Setton, Ryan Allen, Perry Hystad, and C. Peter Keller

Abstract An individual's exposure to air pollution is affected by the variability of pollution concentrations at different locations and times. The most accurate measures of exposure incorporate personal monitoring, but as the number of people included in a study increases, the feasibility of conducting individual-level monitoring quickly decreases, thus requiring the development of more practical approaches, sometimes at the cost of capturing these key sources of variability. In this chapter, we focus on how geotechnologies contribute to characterizing variable air pollution levels and individual geographic mobility with respect to exposure assessment for epidemiological and exposure determinants studies. Rather than an exhaustive literature review, we provide a general discussion of geotechnology uses, include representative examples from epidemiological or exposure determinants studies, and identify limitations and future potential applications. Approaches discussed include simple proximity-to-source or monitor methods, dispersion models, spatial interpolation techniques, and the potential for using satellite-derived air pollution estimates and global positioning systems. The use of time-activity patterns and travel surveys is also included. Finally, an annotated list of recommended review articles is provided for readers interested in more in-depth treatment of many of the topics presented in this chapter.

Keywords Interpolation · Land-use regression · Global positioning systems · Remote sensing · Dispersion models · Exposure simulation

List of Acronyms and Abbreviations

AERMOD	American Meteorological Society/US Environmental Protection Agency (EPA) Regulatory Model with Plume Rise Model Enhancements
ASPEN	Assessment System for Population Exposure Nationwide
CALGRID	California Photochemical Grid Model

E.M. Setton (✉)
Department of Geography, University of Victoria, Victoria, BC, Canada V8R 3W4
e-mail: elsetton@uvic.ca

J.A. Maantay, S. McLafferty (eds.), *Geospatial Analysis of Environmental Health*, Geotechnologies and the Environment 4, DOI 10.1007/978-94-007-0329-2_3,

CALIOP	Cloud Aerosol LIdar with Orthogonal Polarization
CAMx	Comprehensive Air quality Model with Extensions
CMAQ	Community Multi-scale Air Quality modelling system
GASP	GOES Aerosol Smoke Products
GIS	Geographic Information System
GOES	Geostationary Operational Environmental Satellite
GPS	Global Positioning System
HAPEM	Hazardous Air Pollution Exposure Model
MISR	Multi-angle Imaging SpectroRadiometer
MODIS	MODerate Resolution Imaging SpectroRadiometer
USEPA	United States Environmental Protection Agency

How does air pollution affect health? Which health effects are associated with which pollutants and at what concentrations do health effects occur? What determines why some people are more or less exposed than others? These questions have been, and continue to be, the subject of much interest around the world. Over time, methods for studying these questions have become increasingly sophisticated, but the approach remains relatively simple. To examine health effects: conduct an epidemiological study looking for variations in health outcomes given variations in exposure levels. To examine determinants of exposure: conduct a study looking for variations in exposure levels given variations in individual or population characteristics. No matter the question, the common need is some measure or estimate of exposure – it is in this realm that geotechnologies contribute.[1]

Exposure assessment includes widely varying methods such as using biomarkers, in which samples of blood, hair, and other biological samples are analyzed for indicators of exposure, to simply classifying an individual as exposed or not exposed depending on the city in which they live or their proximity to pollution sources. Exposure measurement methods can be generally described as either direct or indirect (Duan, 1982; Ott, 1985). For exposure to air pollutants, direct measurements of exposure are most often gathered by monitoring each subject individually over time with portable personal monitors. Sometimes the monitors are capable of measuring and logging pollutant concentrations every few minutes, in other cases, the monitors absorb or collect pollutants on a filter, thus providing a single measure for the entire duration over which the monitor is exposed to the air. The resulting measure of exposure reflects the varying pollution levels encountered in different locations, and subjects often keep detailed time-activity diaries to support the identification of potential determinants of exposure. In contrast (and at least in theory), exposure to air pollution also can be measured indirectly by placing monitors at each

[1]Exposure assessment is also a key component of risk assessment, which is conducted to identify or predict exposed populations and the level of health risk attributable to that exposure. We do not discuss risk assessment as a specific topic here, given that the exposure assessment methods for outdoor air pollution are generally similar to those used for epidemiological and exposure determinants studies, with similar associated issues, application of geotechnologies and emerging trends.

of the locations (called microenvironments) in which a subject spends time, then calculating the total exposure of an individual by weighting the microenvironmental measurements by the amount of time spent in each microenvironment.

Assessing exposure using either the direct or indirect approach captures two important aspects of exposure, namely (1) pollution concentrations vary over space and time, and (2) individuals move about through space and time. As the number of people included in a study increases, the feasibility of conducting individual-level monitoring, either directly or indirectly, quickly decreases, thus requiring the development of more practical approaches, but sometimes at the cost of capturing these key sources of variability in exposure.

In this chapter, we focus on how geotechnologies contribute to characterizing variable air pollution levels and individual geographic mobility with respect to exposure assessment for epidemiological and exposure determinants studies. Rather than an exhaustive literature review, we provide a general discussion of geotechnology uses, include several (but by no means all) representative examples from epidemiological or exposure determinants studies, and identify limitations and future potential applications. Also included is an annotated list of recommended articles that provide reviews on many of the topics included in this chapter but in more detail than is possible here. For our purposes, geotechnologies are defined as any instrument that produces geo-referenced data, such as Global Positioning Systems or remote sensing platforms, as well as air quality modelling applications that have spatial inputs and outputs, and, of course, geographic information systems (GIS), including the spatial analytical tools included therein.

3.1 Using Geotechnologies to Estimate Variations in Outdoor Pollution Levels

An air quality monitoring instrument for every subject or in every microenvironment visited by a subject may be ideal, but often not feasible. Instead, air quality models and geographic information systems are increasingly being used to produce spatial estimates of exposures or pollution concentrations. In this section we describe approaches ranging from the basic identification of subjects' proximity to sources or monitors to more elaborate spatial models that predict pollution levels at all locations within a study area. The emerging use of remotely sensed data, and their potential to augment (or even replace) ground-level monitoring is also discussed. Given the range of possible methods, this section concludes with examples of integrating several methods into a single exposure assessment, and briefly considers the issue of air pollution infiltration into residences.

3.1.1 Proximity to Sources or Monitors

Studies employing the proximity to source approach are common, given digital maps of sources and the residential location of each subject. In the simplest case, study subjects can be categorized as either exposed or not exposed, based on their proximity to a particular source of pollution. This implicitly assumes that all subjects inside a zone defined by a specific distance have the same level of exposure all the time (hence the exact pollution level is not required), and that it is higher than the exposure experienced by subjects outside the zone. More recently, GIS has been used to incorporate modifying factors such as source density or size, and simple wind direction indicators.

The proximity approach is often used in studies of traffic-related air pollution. In a study of biomarkers of inflammation and cellular immune function for 115 women in Seattle, exposure metrics included a simple presence/absence classification for roads of a given type (freeway, arterial, or truck route) within 150 m of a residence, as well as a classification of subjects into groups based on distance from roads of a given type, i.e., 0–50 m, 50–100 m, and so on, with those living more than 500 m considered as unexposed (Williams et al., 2009). Other studies have incorporated traffic volume information from local government transportation planning departments to provide additional indicators of traffic density, sometimes by vehicle types (i.e., heavy duty vehicles), and sometime employing distance weighting to improve the basic proximity metric, for example, de Medeiros et al. (2009) (perinatal mortality in Sao Paulo, Brazil); Margolis et al. (2009) (pulmonary function of 214 children in Fresno, California); and Kim et al. (2008) (respiratory symptoms in 1,080 children in San Francisco). These last two studies also incorporated local weather and wind direction information in order to account for short term variations in regional pollution levels (Margolis et al., 2009) and to identify locations downwind from sources (Kim et al., 2009). Notably, Kim et al. (2009) also conducted a field monitoring campaign to evaluate how well the indicators characterized spatial gradients in traffic-related air pollution, which is highly recommended but seldom performed when using general indicators of exposure.

The proximity approach is also commonly used for point sources of air pollution. For example, the effects of exposure to fumes from aviation fuels and heavy traffic associated with airports on hospitalizations for respiratory symptoms have been studied in New York state (Lin et al., 2008). All residents within five miles of three large airports (Rochester, LaGuardia, and MacArthur International) were considered to be exposed, while all residents living between 5 and 12 miles from the airports were classified as unexposed. A basic downwind indicator was also employed. In Spain, a recent study of lung, laryngeal, and bladder cancer in residents of communities near industrial combustion facilities assigned exposure based on distance from the community centre to the nearest facility, and incorporated a classification based on the type of fuel used (coal, fuel oil, natural gas) (Garcia-Perez et al., 2009).

Rather than using proximity to sources, a large number of studies have used measurements made at nearby government-operated air quality monitoring stations to estimate exposure.[2] At its simplest, it is assumed that all subjects residing in the

[2]Non-GIS-based applications of this approach compare health outcomes of long-term air pollution exposure among populations residing in different cities using a single central monitor in each area

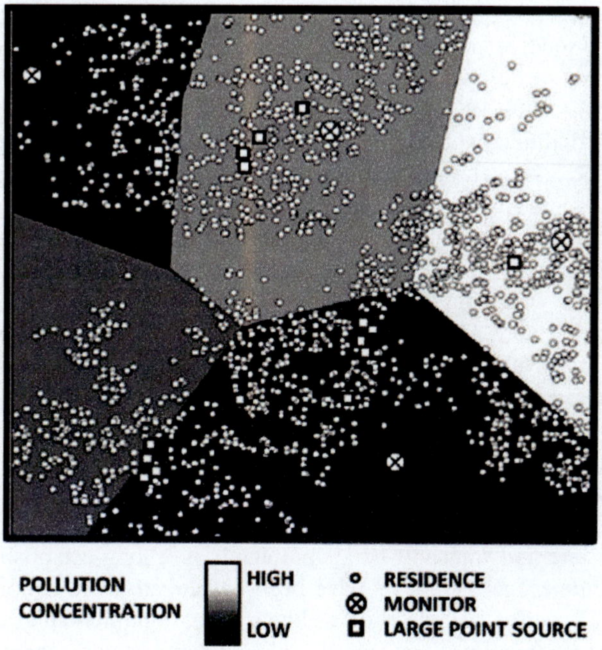

Fig. 3.1 Assigning pollutant concentration at nearest monitor to indicate exposure

monitor area experience the same pollution levels all the time, and a GIS is usually used to identify the nearest monitor, given digital maps of monitor and subject locations (Fig. 3.1). A recent study of air pollution effects on birth weight in Los Angeles County assigned exposure to each subject using the pollution concentration at the air quality monitor nearest to home, and also included a distance from monitor category (within one mile, more than one mile) that allowed an investigation of how distance from the monitor affected the results (Wilhelm and Ritz, 2005). In the same region, Ritz et al. (2006) studied the effects of air pollution on infant mortality. Pollution concentrations from the nearest monitor were averaged over 2 weeks, 1 month, 2 months and 6 months prior to death, and assigned based on subjects' residential zipcode location. The choice of nearest monitor was modified using information on wind flow patterns and geographic features.

The use of GIS to assign exposure based on proximity or nearest monitors is no longer considered to be innovative, due to increasingly available digital datasets

---

as the exposure measure, for example Frye et al. (2003), Peters et al. (1999), Raizenne et al. (1996), and Laden et al. (2000). Similarly, another large area of air pollution and health effects research examines acute health effects using time-series models that relate day to day changes in health events (i.e., hospitalizations or mortalities) in large populations to day to day changes in pollution concentrations measured at a single, central air quality monitor, for example, Breitner et al. (2009), Moura et al. (2009), and Vedal et al. (2009).

(or datasets that can be quickly geocoded using postal code or street address information) and the relative simplicity of the approaches. While these types of studies continue to grow in number, it is clearly recognized that there may be a significant oversimplification of the true spatial variability in pollutant concentrations and therefore individual exposure may be misrepresented. Moreover, the proximity approach may be less useful in determining associations between a single pollutant and a health outcome when the source produces multiple pollutants with similar effects. For example, traffic produces both air and noise pollution, both of which may be linked to cardiovascular morbidity (Allen et al., 2009), and so it may be difficult to assess which pollutant has the greater influence. While some of the examples provided above have employed methods to capture some aspect of the variability, others have attempted to address this major limitation through the use of more sophisticated spatial models.

3.1.2 Dispersion Models

Modelling the fate and transport of air pollutants is a long-established field, and over time, a wide range of models have been developed, from simple plume dispersion models used for a single point or line source, to numerical grid models that incorporate interactions among numerous pollutants and produce three-dimensional spatial estimates for relatively large regions. While it is impossible to adequately cover this complex topic here, a few examples of health studies that have used air quality models are provided here, along with a discussion of typical limitations.

Several recent studies have used dispersion models to study air quality and respiratory health in children. Smargiassi et al. (2009) used AERMOD to assess exposure to refinery-related sulphur dioxide in Montreal, Canada, and its effects on hospitalizations for asthmatic episodes in young children. Asthma hospitalization rates were assessed for areas around a number of point sources in the Bronx, US., using dispersion model results to define exposure areas and associated emission rates (Maantay et al., 2009). In Sweden, a birth cohort of 4,089 children was evaluated for respiratory symptoms, function and allergies based on exposure to air pollution (Nordling et al., 2008). The study used a dispersion model to develop concentration estimates at different scales: 500 × 500 m for regional/countryside areas; 100 × 100 m for urban areas, and 25 × 25 m for inner-city areas. Detailed data were required for inner-city areas, including traffic flow, speed, heavy traffic share, and number of stops per km for each street segment.

Health outcomes for relatively large regions have also been analysed with respect to air pollution through the use of air quality models. Scoggins et al. (2004) used CALGRID, a photochemical air quality model designed to estimate regional ozone and its precursors, to produce a 3 × 3 km grid of estimates of annual average nitrogen dioxide for the Auckland region of New Zealand. Concentrations in census area units were calculated using an area-weighted average of the portions of all 3 × 3 km grid cells falling with the census area unit, and mortality rates for each unit were compared with respect to variations in pollutant concentration.

Input data required for these models include, at a minimum, source characteristics (i.e., geographic coordinates, and for point sources, stack height and emissions temperature among other parameters), the emission rate, meteorological information (wind speed and direction, atmospheric stability, mixing height, and temperature) and locations of receptors at which pollutant concentrations will be estimated. More advanced models include parameters on chemical interactions to characterize the formation of secondary pollutants, require a high level of training to run properly, and can be time-consuming to execute on current desktop computers.

3.1.3 Geographic Information System-Based Spatial Models

The use of GIS for producing spatial estimates of pollution concentrations has developed somewhat in parallel to, although more recently than, that of dispersion modelling, due to the increasing availability of off-the-shelf, easy-to-use GIS applications, and lower input data requirements. All of the GIS-based approaches are based on monitoring data, and employ the basic spatial analytical tools inherent to GIS; however, in some cases additional statistical software applications are used to develop models, using GIS-derived inputs.

3.1.3.1 Spatial Interpolation

Interpolation of values between monitor locations is an approach that can be completed entirely within a GIS. One common technique is inverse distance weighted interpolation. For any given location in the study area (i.e. each study subject's residential address or postal code), a weighted average concentration is developed, with the highest weight given to the nearest monitors, thereby producing a continuous pollution surface (Fig. 3.2). There are a number of pre-defined weighting options available, one of which must be selected by the user, as must a minimum number of monitors to include or a threshold distance to exclude monitors that are too distant. Kriging is a more sophisticated interpolation approach that allows the selection of included monitors to vary from point to point, depending on the spatial autocorrelation structure of the monitoring data, and has the added benefit of producing information about the uncertainty of the estimate at any given location. A further refinement, called co-kriging, incorporates additional information from spatially correlated predictor variables (e.g., elevation).

A study of adverse birth outcomes conducted in Vancouver, Canada, employed (among other measures) an estimate based on inverse distance weighting of the monitored levels for a range of pollutants (Brauer et al., 2008). The number of monitors available varied from seven (fine particulates) to 24 (ozone). Monitored levels were aggregated into monthly averages; these formed the basis for assigning exposure for each month of pregnancy at subjects' residential postal code locations, using inverse distance weighting based on the three closest monitors within 50 km. At a much larger geographic scale, Liao et al. (2009) made use of the United States Environmental Protection Agency (US EPA) air quality monitoring data for the continental US. and kriging to estimate daily levels of particulates for 57,422 female

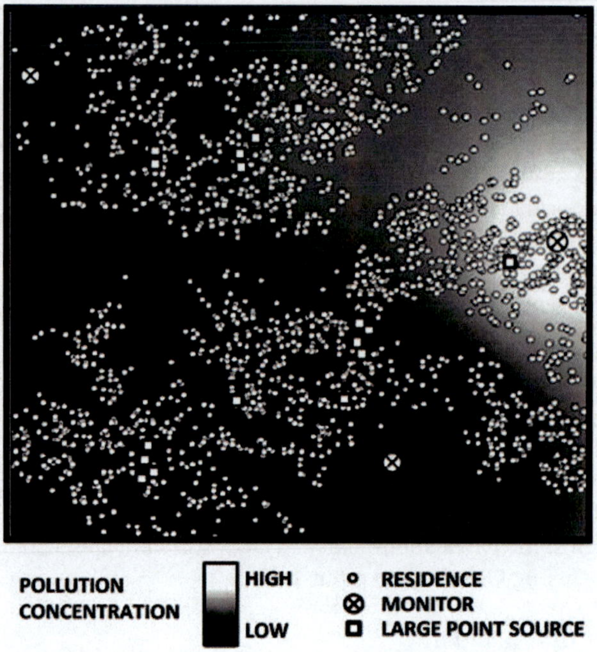

Fig. 3.2 An inverse-distance-weighted interpolation of monitored levels to estimate a pollution concentration at each residence

subjects in 24 states. Exposure metrics for each subject included average particulates on the day of an electrocardiogram, as well as an average over previous 2 days, 30 days, and 365 days.

A critical consideration in applying these interpolation methods is the number of monitors present in the study area. For kriging, the general rule is that at least 50 data points (monitors) are required to reliably measure the spatial autocorrelation functions (Hengl, 2007). Otherwise, inverse distance weighted interpolation may be used for a smaller number of monitors, but fewer monitors will capture less local variability in pollutant concentrations, and so may not provide much improvement over a nearest monitor approach. Another consideration is the “edge effect”, which refers to poor quality interpolation results at locations near the perimeter of the monitored area. Ideally, the interpolation would include a monitored area somewhat larger than the area containing the study population, thus increasing the number of monitors required. For some pollutants, particularly those associated with traffic, variability over distances of several hundred meters can be high depending on proximity to roads, and a vast number of appropriately located monitors would be required to capture this variability adequately for any sizeable study area.

3.1.3.2 Land-Use Regression

Land use regression is a method of spatial interpolation that has become very popular in recent years and therefore deserves particular attention here. Typically, temporary monitors (usually between 40 and 100) are deployed throughout the study area, with locations chosen to represent the full range of pollutant concentrations expected. Techniques for selecting sampling locations have ranged in sophistication from informal site selection based on local knowledge of the study area (e.g., (Lebret et al., 2000)) to location allocation methods that identify areas of expected pollution variability and relatively high population density (Kanaroglou et al., 2005). For each measurement location, a series of predictor variables are produced for a range of buffer distances using GIS, and usually include length of roads by road class, area of residential or commercial development, population density, and so on. Theoretically, these predictor variables represent emission sources for the pollutant of interest. Given the measured pollution concentration and the predictor variables in GIS format for each monitor, standard statistical software applications are use to develop a regression-based model. This model, in the form of an equation, is then applied via GIS to calculate the pollution concentration for each point of a fine grid covering the study area. The result is an estimate of concentrations at a high spatial resolution, often in the range of 5–10 m, so that it is possible to assign a unique exposure to each study subject at a given location (Fig. 3.3). The number of studies using the land use regression approach is rapidly increasing. Hoek et al. (2008) identify 25 recent studies and provide an excellent review of emerging applications and improvements. In general, studies comparing the land use regression approach with other methods suggest that results are as good as, or better in some cases, as those derived with dispersion or numerical models, but with much less complicated input data.

The land use regression approach has been applied most often to estimating annual average concentrations of pollutants, given that the predictor variables do

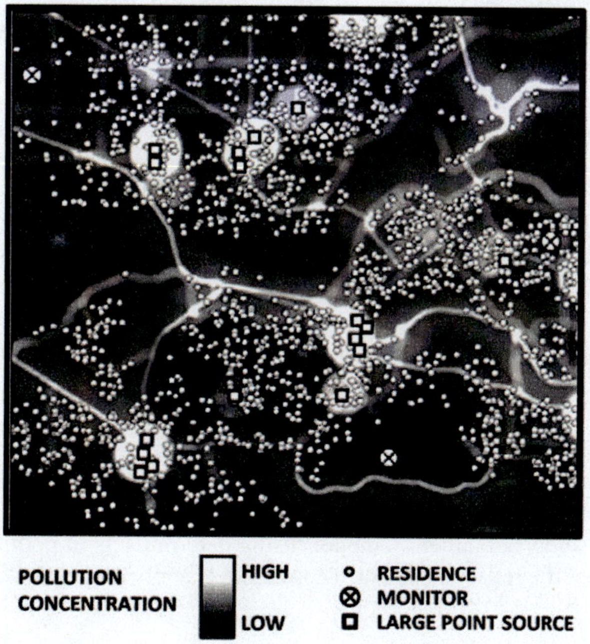

Fig. 3.3 A land use regression-based estimate of pollution levels providing a concentration for each residence

not change much over the course of a year (or even several years), but some have attempted to adjust for shorter term variations in pollution levels by adjusting the annual average up or down using local monitoring data from existing long-term stations, for example, Nethery et al. (2008). Increasingly sophisticated methods are being used to consider space-time interactions (i.e. spatial differences in temporal patterns and/or temporal changes in spatial patterns) in exposure models. Yanosky et al. (2008) describe a method that uses measured concentrations, meteorological data, and land-use regression-type variables to estimate monthly exposure to particulates over a number of years within a generalized additive model framework.

3.1.4 Satellite Data

Remotely sensed data from instruments mounted on satellites above the Earth's atmosphere have been used in a variety of air quality applications over the past three decades. Most simply, images of large weather systems provide synoptic information about current and near-future weather patterns and are used for short-term weather and air quality forecasting. Here, we focus on the application of satellite-derived estimates for more general exposure assessment to particulates, ozone, nitrogen dioxide, sulphur dioxide and methane. A key advantage of these satellite-based estimates is the continuous spatial coverage for vast areas of the globe, thereby reducing reliance on extensive ground-level monitoring. Disadvantages can include the relatively coarse spatial and temporal resolution of the data, the interference of cloud, snow and ice, and the difficulty in determining how much of the pollutant is actually at ground-level (in the cases of ozone and particulates).

With typical spatial resolutions of 4–10 km or more, satellite-derived atmospheric data do not currently meet the increasing demand for high resolution (10–100 s of metres) estimates of pollution needed to support exposure assessment at the postal code or street address level (Fig. 3.4). In terms of temporal resolution, many satellites currently producing data useful for air pollution estimates have sun-synchronous orbits, i.e., they are overhead at about the same local time on every pass, providing a consistent illumination angle, and therefore a single instantaneous snapshot of conditions at the time of overpass. How pollution levels derived from these data relate to the hourly, daily, seasonal or annual averages at that same location is a key consideration in determining how useful satellite-derived estimates are for exposure assessment. An exception is the GOES/GASP product, which is collected from a geo-synchronous satellite, and therefore collects data over the same location every 30 min; however, the data are not as precise as those gathered by other satellites with less temporal coverage (MODIS or MISR) (Prados et al., 2007).

With one exception (CALIOP), the on-board instruments currently used are passive sensors, meaning that they record the amount of solar radiation reflected from the earth's surface and atmosphere or the amount of infrared (thermal) emission (Martin, 2008). Estimates of air pollution are therefore not based on direct measures, but rather on models using determinants of pollution concentrations that best fit the reflectance data. These models apply only when there is a clear view from the

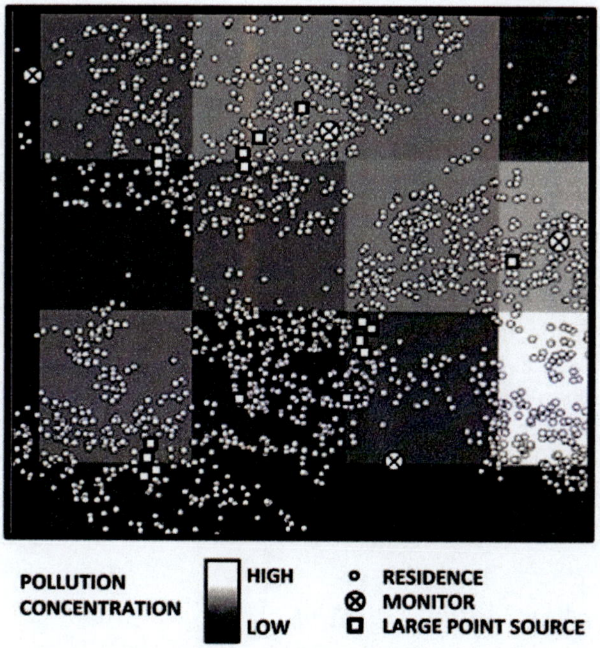

Fig. 3.4 Typical resolution of satellite-based pollution concentration estimates

top of the atmosphere to the Earth's surface, and so when clouds are present, no pollution estimate is possible. Snow and ice can also interfere, and pollution estimates may not be available in winter for regions with cold climates.

Since the satellite-based data relate to the entire column of atmosphere, it is not always possible to determine the pollution concentration at ground-level, which is of most interest for exposure assessment. Some pollutants, particularly nitrogen dioxide and sulphur dioxide are relatively confined to near-surface boundary layer of the atmosphere, while this is not the case for ozone, carbon monoxide, and particulates (Martin, 2008). In the case of fine particulates, good agreement between satellite-derived estimates and ground-level monitoring data, at least in eastern North America, has been achieved recently by incorporating information on atmospheric stability and/or mixing height (Liu et al., 2004, 2007; Paciorek et al., 2008; van Donkelaar et al., 2006).

Satellite-derived air pollution concentration estimates are not yet commonly used in conjunction with individual-level health-related data; however, this is a rapidly emerging area of research. A number of studies looking specifically at smoke from wildfires have employed satellite-based data to aid in the development of estimates useful for exposure assessment. Delfino et al. (2009) make use of MODIS images of forest fire plumes as an additional indicator of elevated fine particulate levels in their study of the impact of wildfires in California on respiratory and cardiovascular hospital admissions. The satellite data were used to predict particulate values at monitors with missing data, then complete datasets (measured and estimated) were used as inputs for inverse distance weighted interpolation, kriging, and co-kriging in support of assigning exposures at the zip code level (Wu et al., 2006). Henderson et al. (2008) also employed MODIS data to estimate exposure to wildfire smoke in south-eastern British Columbia, Canada. Data included active fire detects, aerosol optical thickness estimates, and true colour images. The fire detection data, which include fire location and radiative power, were used to develop an emissions inventory for input to a dispersion model. MODIS aerosol optical depth values and true colour images were used to evaluate the dispersion model results. More recently, MODIS data have been used in support of an ecological study of fine particulates and chronic ischemic heart disease – mean aerosol optical depth (the MODIS data component that relates to particulates) for counties in the eastern United States was seen to be related to race and age standardized county-level mortality rates (Hu and Rao, 2009). While the study should be considered as exploratory due to the ecological design, it does illustrate the current value in developing simple health-related air quality indicators using remotely-sensed data, and the immense future potential for using satellite-based air pollution estimates as spatial resolution becomes finer.

3.1.5 Integrating Multiple Methods

Clearly, there are many potential methods for estimating air pollution concentrations, in support of exposure assessment. The emerging question is: which method is “best”? The answer, of course, depends on many factors, including the size and geographic location of the study population, the pollutant of interest and its associated scale of spatial variation, and data quality and availability, among others. Even so, there may be numerous possible approaches. Recent studies comparing estimates using different methods for the same study area suggest the following conclusions: (1) epidemiological analyses using different exposure methods for the same study population may yield different estimates of effects (Bell, 2006; Brauer et al., 2008; Ryan et al., 2007); and, (2) different methods capture different aspects of a pollutant’s spatial variation (Marshall et al., 2008). In the first case, there are no standard guidelines, although the trend appears to be including various metrics and addressing differences explicitly in the interpretation of the results. In the second case, one way forward is to integrate a combination of methods to develop a single individual-level exposure measure, as the following studies illustrate.

The individual-level air pollution exposure assessment for the Netherlands Cohort Study (approximately 120,800 subjects) consists of three separate, additive components: (1) regional background, based on inverse distance weighted interpolation of black smoke and nitrogen dioxide levels measured at national air quality monitoring stations; (2) urban levels, based on a land use regression approach including address density as a predictor variable; and (3) local influences, indicated by living with 50 m of a major urban road and/or within 100 m of a freeway (Hoek et al., 2001; 2002). More recent studies of this cohort have expanded the land use regression-based urban component with additional variables, and added traffic intensity to the local component (Beelen et al., 2007, 2008). Estimates of long-term exposure to nitrogen dioxide on heart rate variability in a Swiss cohort study of 1,408 adults incorporate several approaches. For each sub-region in the study area, regression models with inputs from a dispersion model are used to characterize urban background pollution levels. GIS-based residence-specific variables (e.g., distance to major street, traffic volume within 200 m, etc), temperature and space-time interactions are also included, but vary among sub-regions in order to capture more local influences. The sub-region-specific model outputs (bi-weekly nitrogen dioxide concentration estimates) were then aggregated to annual averages and assigned to residential locations (Dietrich et al., 2008).

3.1.6 Estimating Residential Infiltration of Outdoor Air Pollutants

Finally, an area in which GIS has the potential to contribute to the study of outdoor air pollution and health deserves some attention. Few large epidemiological studies have considered the movement of outdoor pollution to indoor environments as part of their exposure assessment. Exposure assessments have focused on estimating outdoor pollutant concentrations at or near residences; however, when at home, most people are indoors, where, depending on local weather, windows and doors may be closed and outdoor-generated air pollution levels may be relatively low. Recent work has examined how housing characteristics affect the amount of outdoor pollution that infiltrates indoors. Hystad et al. (2006) acquired residence-specific housing characteristics from spatial property tax assessment data and included them in a model of particulate infiltration, an approach that could be applied in large epidemiological studies, given the residential address of each subject.

3.2 Using Geotechnologies to Incorporate Individual Mobility

Although not discussed in detail in the previous section, the majority of epidemiological studies of large populations use estimates of air pollution concentration at the residential address of each subject, since individual addresses are generally available from large health databases, can be readily geocoded, and most people spend a majority of their time at home (Klepeis et al., 2001). Still, it is widely recognized that exposure occurring away from home may be important, and the use of a residence-only based exposure may introduce errors that undermine subsequent health effects estimates (Huang and Batterman, 2000), particularly for working adults, school-aged children, and others who spend significant time away from home. Some health studies with relatively small study populations have used home and school locations (Brunekreef et al., 1997; Hirsch et al., 1999; Kramer et al., 2000), and some have incorporated work locations (Beeson et al., 1998), but this is not currently widely practiced. In this section, the emerging potential for using global positioning systems to track individual mobility is discussed, followed by an overview of how mobility is incorporated in population-level exposure simulation models.

3.2.1 GPS and Time-Activity Patterns

The last decade has seen rapid growth in the use of GPS for personal and business uses; however, large scale use of GPS technology for air pollution exposure assessment in epidemiological studies has not yet occurred, likely due to the relatively high administrative cost of deploying personal monitors of any kind, including GPS, for populations larger than a few hundred people. There have been recent uses of GPS, however, among relatively small study populations.

An early evaluation of GPS technology for personal location tracking was conducted by Phillips et al. (2001) in Oklahoma, primarily as an aid to validating time-activity diaries completed by study subjects. Technical specifications for the GPS included position logging every 5–15 min, storage for 16–24 h of data, and battery life requiring at most 2 battery changes per 24 h period. The selected GPS unit weighed 2 kg and had an external antenna roughly the size of a deck of playing cards. Out of 25 trials, the GPS units operated properly only about 30% of the time, but when operating, the GPS data did appear to add useful detail to the time-activity diaries filled out by the participants. Since then, several research groups have developed customized GPS units to overcome some of the issues with battery life, portability, and speed of signal acquisition (Elgethun et al., 2003; Rainham et al., 2008), but commercially available GPS have improved significantly as well. In a study of 62 pregnant women in Vancouver, Canada, Nethery et al. (2008) used commercial GPS units (about half the size of a deck of playing cards, including an internal antenna) fitted with a long-life battery pack to record subjects' geographic mobility. Subjects also completed time-activity diaries and wore monitors measuring fine particulates, nitric oxide and nitrogen dioxide. Overall, many of the issues identified by Phillips et al. (2001) remained: loss of signal and/or battery failure limited the number of complete records. Interestingly, Nethery et al. (2008) overlayed the GPS tracks on a land use regression-derived estimate of nitrogen dioxide to develop an exposure estimate, and these were seen to be more accurate than exposures assigned solely on subjects' residential locations.

GPS-enable cell phones are quickly becoming more common and have also been evaluated for use in tracking individual locations (Wiehe et al., 2008), but perhaps of most interest is the emerging integration of GPS with air pollution sensors and cell phones or other mobile wireless devices. Widely covered by the popular media was the 2006 release of homing pigeons outfitted with GPS units, real time air pollution sensors, and cell phone transmitters by a researcher in California (da Costa et al., 2009). Similarly (although using human subjects), Kanjo et al. (2008) describes the use of a cell phone as a noise monitor, and as a data logger via a bluetooth connection to a small portable carbon monoxide monitor, both with associated GPS location data.

3.2.2 Modelling Population Exposures and Determinants

Up to this point, the discussion and examples have focused mainly on how geotechnologies are being used to produce or improve air pollution estimates for epidemiological studies. Modelling population-level exposures and determinants via simulation is a related branch of exposure assessment that is more concerned with understanding the probable range of exposures experienced in a particular population (i.e., of an entire urban region, or a demographic subgroup), in support of setting regulatory thresholds and prioritizing exposure reduction efforts.

The exposure simulation approach has been used most extensively by the US EPA, beginning as early as the mid-1980s (Johnson, 1995; McCurdy, 1995) and by researchers in Europe in this decade as part of the EXPOLIS study (Hanninen et al., 2003; Kruize et al., 2003) and is based on the indirect exposure approach. Briefly, instead of using measurements of pollution levels in each location a person might visit, a range of possible values for typical locations (i.e., indoor at home, outdoors, indoor at work, and so on) can be substituted. The range of pollution values might come from limited monitoring in representative locations, or may be based on a variety of spatial models of pollution levels, including many of the aproaches discussed earlier. Similarly, instead of collecting a unique time-activity diary for each subject, a set of population representative time-activity diaries (including age, sex and other demographic information) can be substituted. By randomly choosing a time-activity diary, and randomly selecting from the ranges of possible pollution values at typical locations, a *probable* exposure can be calculated for a "simulate" person. With enough repetitions of this procedure, a distribution of probable exposures can be simulated, and used to estimate the mean and variance of probable exposure, and other meaningful statistics for comparison purposes (Fig. 3.5). The geographic mobility of the population is captured by running the model in many sub-regions of the study area (often census tracts) and including a work flow matrix (usually based on census or transportation planning data). Additional refinements can be made by matching the time-activity patterns available for selection to the census tract population based on demographic characteristics, and by including age- and activity-specific breathing rates. The output is a single probability distribution of all possible exposures.

The US EPA has developed a number of simulation models. The Hazardous Air Pollutant Exposure Model (HAPEM) is a relatively simple example, and was first developed in 1985 to estimate exposure to non-reactive pollutants from mobile sources. The current model, HAPEM4, has been used to predict annual average exposure levels for each census tract in the US, as part of the National-Scale Air Toxics Assessment program. For each census tract in the US., HAPEM4 randomly selects daily time-activity patterns for a summer weekday, a non-summer weekday, and a weekend from a database, then combines them into a single time activity pattern, using weights based on the number of summer weekdays, non-summer weekdays, and weekends per year. This is done 100 times for each of 10 pre-defined demographic groups, and then 30 of these aggregated time-activity patterns are randomly selected to represent the population of the demographic group in that census

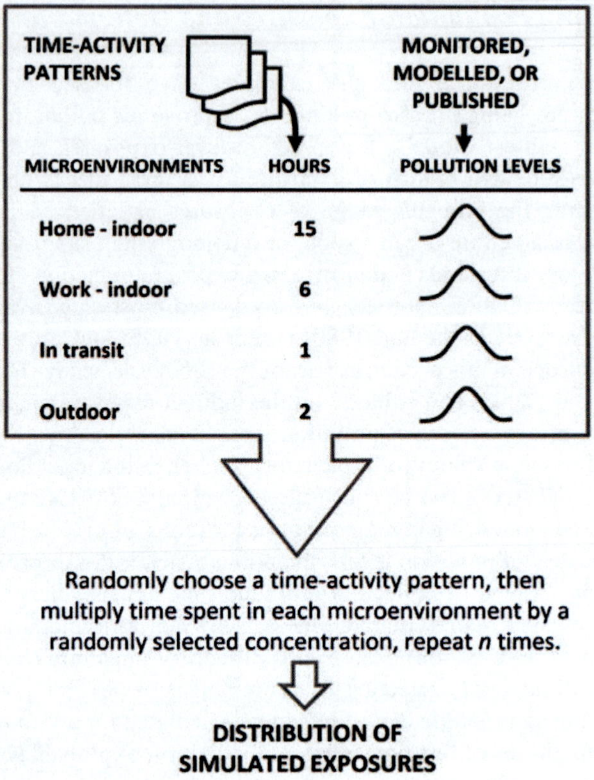

Fig. 3.5 Procedure for simulating exposure using time activity patterns

tract (Cook et al., 2007; United States Environmental Protection Agency 2008). The US EPA "Assessment System for Population Exposure Nationwide" (ASPEN) dispersion-based model is used to estimate daily pollution levels for each census tract in the assessment, which are then aggregated into annual averages. US Census Bureau data on work flows among census tracts is used to characterize the geographic mobility of people who work in census tracts in which they do not reside. Results of the HAPEM4 model have been used map median exposure concentration estimates at the county-level for the entire US., and support the assessment of cancer and non-cancer health risks due to the inhalation of a broad range of air pollutants. Similarly, the more detailed SHEDS model (described more fully in Burke et al. (2001)) has been used to simulate population exposure to air pollutants, using spatial estimates of concentrations derived by interpolation of existing monitoring data (Kibria et al., 2002), by the Models 3/CMAQ numerical air quality model (Georgopoulos et al., 2005), and by a combination of CMAQ and AERMOD air quality models (Isakov et al., 2009).

Some improvements have been made to the simulation approach as described above, mainly in the use of the shortest path GIS function. Gulliver and Briggs

(2005) describe the Space-Time Exposure Modelling System, which is used to simulate the exposures of 50 school children in Northampton, UK. Each subject provided a time-activity diary with home and school locations; GIS was used to identify the shortest walking route and extract pollution concentrations developed by combining dispersion and traffic models. More recently, Setton et al. (2008) incorporated additional spatial data on residential and commercial building locations, and GIS-based shortest path analysis to refine a simulation of exposures to traffic-related air pollution. An estimate of nitrogen dioxide concentrations developed using the land use regression approach was used as a basis for estimating exposure distributions in each of 382 census tract in the Greater Vancouver Regional District of British Columbia, Canada. Their study identified spatial variations in exposure that were associated specifically with geographic location instead of individual demographic characteristics.

An alternative approach, based on the use of origin-destination surveys in conjunction with spatial models of pollutant concentrations, has recently been employed by researchers in the United States to investigate exposure determinants. Marshall et al. (2006) estimate intake rates (a measure of exposure) for

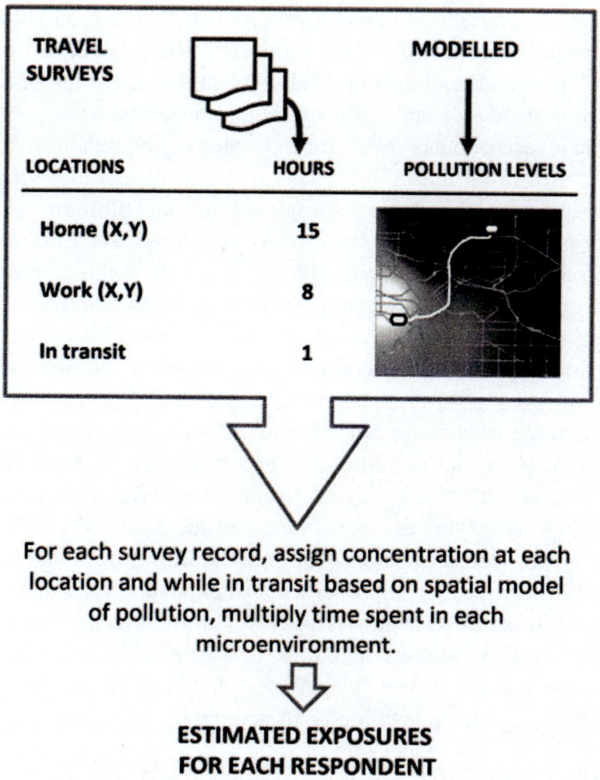

Fig. 3.6 Procedure for estimating exposure using travel surveys

each of 28,746 person-days included in the Southern California Association of Government transportation survey (conducted in 2000). The survey was conducted to support transportation planning, and so included not only typical time-activity and demographic information, but also the geographic locations for trip origins and destinations, as well as the start time of each trip and its duration. These geocoded surveys were then overlaid with pollution estimates (at a 2×2 km resolution) produced using the CAMx air quality model (Fig. 3.6). Like the US EPA models, representative breathing rates according to activity were employed, and an intake rate calculated for each survey record (Marshall et al., 2006).

3.3 Conclusion

Geotechnologies have been, and will continue to be, used in a wide variety of ways to support epidemiological and exposure studies of outdoor air pollution, and while some of those uses have been recognized for several decades, others are just now beginning to emerge as technology and spatial data resources advance.

Cohort, case-control and cross-sectional epidemiological studies require individual-level exposure estimates, and depending on the health outcome of interest, relatively large populations may be required to support statistical inference. The larger the population, however, the more difficult it becomes to conduct the "gold standard" personal monitoring, and it is this reality that has lead to the increasingly common use of surrogate exposures developed via geotechnologies. In the past decade, research has focused on developing increasingly higher resolution spatial estimates of pollution, ostensibly to better reflect the spatial variability in concentrations. Given this drive for high-resolution pollution estimates, some approaches that have merit are not being fully embraced, for example, the use of satellite-derived data, which are currently not available at high resolution scales. The integration approach should be more commonly used than it currently is, as it allows for using different models or datasets to capture different scales of pollutant variability, and thus would allow for the integration of satellite-based data. The emerging trend of using several exposure metrics serves in some way as a sensitivity analysis, in that the robustness of the health effect results in the face of different exposure measured is explicitly addressed; however, there is some danger in producing voluminous results which may be difficult to interpret and compare, given a lack of clear theory underlying the exposure assessment and why differences might exist. Little focus has been brought to bear on including geographic locations away from home, and it may be that improvements in exposure assessment can be realized by linking administrative health databases with income tax records that would allow for the identification of work locations, or with education databases that identify school locations. Even in the face of these kinds of improvements regarding geographic mobility, given the difficulty in accurately characterizing spatial variations in pollution over short periods of time (even with complex air quality or spatio-temporal models) and developing individual-level mobility data required to match the same temporal scale, these approaches may be better suited to studies of long-term exposures or to populations that are less mobile in general (small children or seniors, for example).

Estimating population-level exposures via simulation is an interesting area of research, but one that fills a relatively narrow niche, given its suitability for regulatory purposes rather than health effects studies. The underlying model of exposure has remained the same since its inception, and improvements have consisted mainly of incorporating "better" (i.e., higher resolution) spatial estimates of pollution concentrations and refining the input distributions of population characteristics. The recent use of geocoded origin-destination surveys by Marshall et al. (2006) illustrates the novel use of a dataset that may exist in any number of large cities across the globe, and there is some potential for growth in this area.

Arguably, the highest potential for improving exposure assessments is in the realm of personal monitoring with integrated GPS. It is difficult to predict how technology will improve in the next decade, but given the advances seen in the last decade, it does not seem unreasonable to envision small, dependable air pollution sensors with GPS and wireless communication functionality at relatively low costs. Whether these kinds of equipment become inexpensive enough to deploy with large populations remains to be seen, but their use for validating exposure assessments developed via other approaches could prove invaluable.

3.4 Recommended Reviews

For the purposes of brevity, we have organized this chapter based on two key components of exposure to outdoor air pollution: variability in pollutant concentrations over time and space; and variability in individuals' geographic locations over time and space. Some aspects of exposure to outdoor air pollution have thus been omitted, but should not be considered as unimportant, including which approaches are better suited to which study designs, and also to which pollutants, given the different spatial variability among outdoor air pollutants. We leave it to the reader to further explore these issues, and recommend the following articles as a starting point.

Briggs D (2005) The role of GIS: coping with space (and time) in air pollution exposure assessment. J Toxicol Environ Health-Part A-Curr Issues 68:1243–1261.

This review summarizes potential applications of GIS for air pollution exposure assessment. Topics include simple proximity methods, interpolation methods such as inverse distance weighting and kriging, and 'dynamic modeling' including dispersion models and the incorporation of human time-location patterns into exposure assessments.

Chen H, Goldberg M, Villeneuve P (2008) A systematic review of the relation between long-term exposure to ambient air pollution and chronic diseases. Rev Environ Health 23(4):244–297

This comprehensive review of epidemiological studies provides detail on the exposure assessment methods used in each reviewed study.

Hoek G, Beelen R, de Hoogh K, Vienneau D, Gulliver J, Fischer P et al (2008) A review of land-use regression models to assess spatial variation of outdoor air pollution. Atmos Environ 42:7561–7578

A summary of 25 land use regression (LUR) studies in North America and Europe, including descriptions of air pollution data sources, monitoring site locations and numbers, predictor variables, and model performance. The authors conclude that in urban areas LUR performs as well or better than geostatistical methods and dispersion models. Suggestions for areas of further research include LUR model 'transferability,' inclusion of additional data sources, consideration of temporal variability, and personal exposure validation studies.

Huang YL, Batterman S (2000) Residence location as a measure of environmental exposure: a review of air pollution epidemiology studies. J Expo Anal Environ Epidemiol 10:66–85

This review summarizes 45 epidemiologic studies in which residential location was used to estimate environmental exposures, including air pollutants, metals, and pesticides. Studies are categorized based on the availability of measurements and/or dispersion models and whether statistically significant associations with health endpoints are reported. The authors conclude that residence location is most appropriately used in combination with other sources of data, such as pollution measurements and/or models of pollutant emissions and transport.

Jerrett M, Arain A, Kanaroglou P, Beckerman B, Potoglou D, Sahsuvaroglu T et al (2005) A review and evaluation of intraurban air pollution exposure models. J Expo Anal Environ Epidemiol 15:185–204

A thorough review of interpolation methods including simple proximity approaches, interpolation models (such as inverse distance weighting and kriging), LUR models, dispersion models, integrated meteorological-emissions models, and hybrid approaches combining multiple interpolation techniques. Methods are described and evaluated, and the applicability of each method to health effects research is discussed.

Martin RV (2008) Satellite remote sensing of surface air quality. Atmos Environ 42:7823–7843

This review discusses satellite remote sensing of surface air pollutants including aerosols, ozone (O_3), nitrogen dioxide (NO_2), carbon monoxide (CO), formaldehyde (HCHO), and sulphur dioxide (SO_2). Topics include a general overview of remote sensing and satellites, as well as applications of satellite remote sensing to forecasting air quality events, surface air quality evaluation, and emissions assessments.

Nieuwenhuijsen M, Paustenbach D, Duarte-Davidson R (2006) New developments in exposure assessment: the impact on the practice of health risk assessment and epidemiological studies. Environ Int 32:996–1009

Advances in environmental exposure assessment since the mid-1990s are described. Exposure biomarkers, deterministic and GIS-based models (including LUR, satellite remote sensing), and statistical methods (including source apportionment, Monte Carlos simulation, and Bayesian statistics) are discussed.

Reid N, Misra PK, Amman M, Hales J (2007) Air quality modeling for policy development. J Toxicol Environ Health-Part A-Curr Issues 70:295–310

This review discusses atmospheric dispersion and transport models and distinguishes between models used for research purposes and those applicable to policy development. The focus is on the latter, and these models are distinguished into those applied on the local (e.g. Gaussian plume models), regional (e.g. Lagrangian models), and global spatial scales (e.g. Eulerian models). Examples of models used in policy development are given.

Touma JS, Isakov V, Ching J, Seigneur C (2006) Air quality modeling of hazardous pollutants: Current status and future directions. J Air Waste Manag Assoc 56: 547–558

This paper describes models for hazardous air pollutants (generally categorized as source-based dispersion models and grid-based chemical transport models), a discussion of modeling challenges, and recommendations for addressing those challenges. Model applications at the national, regional, and neighbourhood scale are discussed.

Author Biographies

Eleanor M. Setton, *Ph.D., Adjunct Assistant Professor in the Department of Geography at the University of Victoria, in Victoria, British Columbia, Canada.* Her research interests lie mainly in investigating links between environmental quality and human health. This includes monitoring and modelling environmental quality, with a current focus on air pollution as well as carcinogenic pollutants in all media. She is also interested in how exposure measurement error affects statistical analyses of dose-response relationships. Her use of a spatial perspective and geomatics as a methodological basis for research serves to highlight innovative applications of accepted techniques.

Ryan Allen, *Ph.D., Assistant Professor, Faculty of Health Sciences, Simon Fraser University, British Columbia, Canada.* Dr. Allen's research is focused on air pollution exposure assessment and epidemiology. Areas of particular interest include improved exposure assessment in large epidemiologic studies, estimating the impact of pollution from outdoor sources on indoor environments, and studying the links between combustion-derived air pollution and adverse respiratory and cardiovascular health effects.

Perry Hystad, *Ph.D. Candidate, School of Population and Public Health, University of British Columbia, Canada.* Mr. Hystad is Ph.D. candidate supported by two national and one provincial research fellowship. His research focuses on environmental health risks, cancer primary prevention and spatial epidemiology. He is a researcher with CAREX Canada, and a member of the Expert Environmental Working Group for the Canadian Cancer Cohort.

Dr. C. Peter Keller, *Professor, Geography, Dean of Social Sciences, University of Victoria, British Columbia, Canada.* Dr. Keller is recognized as a leader in developing GIS-based decision support systems with a particular emphasis on policy development and change. Recent research in the environmental health field includes using high resolution GIS datasets for exposure and risk assessment, and the development of online health atlases.

References

Allen RW, Davies H, Cohen MA, Mallach G, Kaufman JD, Adar SD (2009) The spatial relationship between traffic-generated air pollution and noise in 2 US cities. Environ Res 109:334–342

Beelen R, Hoek G, Fischer P, van den Brandt PA, Brunekreef B (2007) Estimated long-term outdoor air pollution concentrations in a cohort study. Atmos Environ 41:5552 (vol 41, pg 1343, 2007)

Beelen R, Hoek G, van den Brandt PA, Goldbohm RA, Fischer P, Schouten LJ et al (2008) Long-term effects of traffic-related air pollution on mortality in a Dutch cohort (NLCS-AIR study). Environ Health Perspect 116:196–202

Beeson WL, Abbey DE, Knutsen SF (1998) Long-term concentrations of ambient air pollutants and incident lung cancer in California adults: results from the AHSMOG study. Environ Health Perspect 106:813–822

Bell ML (2006) The use of ambient air quality modeling to estimate individual and population exposure for human health research: a case study of ozone in the Northern Georgia Region of the United States. Environ Int 32:586–593

Brauer M, Lencar C, Tamburic L, Koehoorn M, Demers P, Karr C (2008) A cohort study of traffic-related air pollution impacts on birth outcomes. Environ Health Perspect 116:680–686

Breitner S, Stolzel M, Cyrys J, Pitz M, Wolke G, Kreyling W et al (2009) Short-term mortality rates during a decade of improved air quality in erfurt, Germany. Environ Health Perspect 117:448–454

Brunekreef B, Janssen NAH, de Hartog J, Harssema H, Knape M, van Vliet P (1997) Air pollution from truck traffic and lung function in children living near motorways. Epidemiology 8: 298–303

Burke JM, Zufall MJ, Ozkaynak H (2001) A population exposure model for particulate matter: case study results for PM2.5 in Philadelphia, PA. J Expo Anal Environ Epidemiol 11:470–489

Cook R, Strum M, Touma JS, Palma T, Thurman J, Ensley D et al (2007) Inhalation exposure and risk from mobile source air toxics in future years. J Expo Sci Environ Epidemiol 17:95–105

da Costa B, Hazegh C, Ponto K. 2009. PigeonBlog. In http://www.beatrizdacosta.net/pigeonblog.php

de Medeiros APP, Gouveia N, Machado RPP, de Souza MR, Alencar GP, Novaes HMD et al (2009) Traffic-related air pollution and perinatal mortality: a case-control study. Environ Health Perspect 117:127–132

Delfino RJ, Brummel S, Wu J, Stern H, Ostro B, Lipsett M et al (2009) The relationship of respiratory and cardiovascular hospital admissions to the southern California wildfires of 2003. Occup Environ Med 66:189–197

Dietrich DF, Gemperli A, Gaspoz JM, Schindler C, Liu LJS, Gold DR et al (2008) Differences in heart rate variability associated with long-term exposure to NO2. Environ Health Perspect 116:1357–1361

Duan N 1982. Models for human exposure to air pollution. American association for the advancement of science abstracts of papers of the national meeting 148:59

Elgethun K, Fenske RA, Yost MG, Palcisko GJ (2003) Time-location analysis for exposure assessment studies of children using a novel global positioning system instrument. Environ Health Perspect 111:115–122

Frye C, Hoelscher B, Cyrys J, Wjst M, Wichmann HE, Heinrich J (2003) Association of lung function with declining ambient air pollution. Environ Health Perspect 111:383–387

Garcia-Perez J, Pollan M, Boldo E, Perez-Gomez B, Aragones N, Lope V et al (2009) Mortality due to lung, laryngeal and bladder cancer in towns lying in the vicinity of combustion installations. Sci Total Environ 407:2593–2602

Georgopoulos PG, Wang SW, Vyas VM, Sun Q, Burke J, Vedantham R et al (2005) A source-to-dose assessment of population exposures to fine PM and ozone in Philadelphia, PA, during a summer 1999 episode. J Expo Anal Environ Epidemiol 15:439–457

Gulliver J, Briggs DJ (2005) Time-space modeling of journey-time exposure to traffic-related air pollution using GIS. Environ Res 97:10–25

Hanninen O, Kruize H, Lebret E, Jantunen M (2003) EXPOLIS simulation model: PM2.5 application and comparison with measurements in Helsinki. J Expo Anal Environ Epidemiol 13:74–85

Henderson SB, Burkholder B, Jackson PL, Brauer M, Ichoku C (2008) Use of MODIS products to simplify and evaluate a forest fire plume dispersion model for PM10 exposure assessment. Atmos Environ 42:8524–8532

Hengl T 2007. A practical guide to geostatistical mapping of environmental variables. Joint Research Centre, Institute for Environment and Sustainability, European Commission. European Communities, Luxembourg.

Hirsch T, Weiland SK, von Mutius E, Safeca AF, Grafe H, Csaplovics E et al (1999) Inner city air pollution and respiratory health and atopy in children. Eur Respir J 14:669–677

Hoek G, Beelen R, de Hoogh K, Vienneau D, Gulliver J, Fischer P et al (2008) A review of land-use regression models to assess spatial variation of outdoor air pollution. Atmos Environ 42:7561–7578

Hoek G, Brunekreef B, Goldbohm S, Fischer P, van den Brandt PA (2002) Association between mortality and indicators of traffic-related air pollution in the Netherlands: a cohort study. Lancet 360:1203–1209

Hoek G, Fischer P, van den Brandt P, Goldbohm S, Brunekreef B (2001) Estimation of long-term average exposure to outdoor air pollution for a cohort study on mortality. Epidemiology 12: 508

Hu ZY, Rao KR (2009) Particulate air pollution and chronic ischemic heart disease in the eastern United States: a county level ecological study using satellite aerosol data. Environ Health 12:26

Huang YL, Batterman S (2000) Residence location as a measure of environmental exposure: a review of air pollution epidemiology studies. J Expo Anal Environ Epidemiol 10:66–85

Hystad P, Setton E, Keller P, Allen R, Cloutier-Fisher D, Foster L et al (2006) Predicting ambient PM2.5 infiltration for individual residences at a regional scale. Epidemiology 17:S113

Isakov V, Touma JS, Burke J, Lobdell DT, Palma T, Rosenbaum A et al (2009) Combining regional- and local-scale air quality models with exposure models for use in environmental health studies. J Air Waste Manag Assoc 59:461–472

Johnson TR (1995) Recent advances in the estimation of population exposure to mobile source pollutants. J Expo Anal Environ Epidemiol 5:551–571

Kanaroglou PS, Jerrett M, Morrison J, Beckerman B, Arain MA, Gilbert NL et al (2005) Establishing an air pollution monitoring network for intra-urban population exposure assessment: a location-allocation approach. Atmos Environ 39:2399–2409

Kanjo E, Benford S, Paxton M, Chamberlain A, Fraser DS, Woodgate D et al (2008) MobGeoSen: facilitating personal geosensor data collection and visualization using mobile phones. Pers Ubiquit Comput 12:599–607

Kibria BMG, Sun L, Zidek JV, Le ND (2002) Bayesian spatial prediction of random space-time fields with application to mapping PM2.5 exposure. J Am Stat Assoc 97:112–124

Kim JJ, Huen K, Adams S, Smorodinsky S, Hoats A, Malig B et al (2008) Residential traffic and children's respiratory health. Environ Health Perspect 116:1274–1279

Kim SY, Sheppard L, Kim H (2009) Health effects of long-term air pollution influence of exposure prediction methods. Epidemiology 20:442–450

Klepeis NE, Nelson WC, Ott WR, Robinson JP, Tsang AM, Switzer P et al (2001) The national human activity pattern survey (NHAPS): a resource for assessing exposure to environmental pollutants. J Expo Anal Environ Epidemiol 11:231–252

Kramer U, Koch T, Ranft U, Ring J, Behrendt H (2000) Traffic-related air pollution is associated with atopy in children living in urban areas. Epidemiology 11:64–70

Kruize H, Hanninen O, Breugelmans O, Lebret E, Jantunen M (2003) Description and demonstration of the EXPOLIS simulation model: two examples of modeling population exposure to particulate matter. J Expo Anal Environ Epidemiol 13:87–99

Laden F, Neas LM, Dockery DW, Schwartz J (2000) Association of fine particulate matter from different sources with daily mortality in six US cities. Environ Health Perspect 108:941–947

Lebret E, Briggs D, van Reeuwijk H, Fischer P, Smallbone K, Harssema H et al (2000) Small area variations in ambient NO2 concentrations in four European areas. Atmos Environ 34: 177–185

Liao DP, Whitsel E, Duan YK, Lin HM, Quibrera PM, Smith R et al (2009) Ambient particulate air pollution and ectopythe environmental epidemiology of arrhythmogenesis in women's health initiative study, 1999–2004. J Toxicol Environ Health A Curr Issues 72:30–38

Lin S, Munsie JP, Herdt-Losavio M, Hwang SA, Civerolo K, McGarry K et al (2008) Residential proximity to large airports and potential health impacts in New York State. Int Arch Occup Environ Health 81:797–804

Liu Y, Franklin M, Kahn R, Koutrakis P (2007) Using aerosol optical thickness to predict ground-level PM2.5 concentrations in the St. Louis area: a comparison between MISR and MODIS. Remote Sensing Environ 107:33–44

Liu Y, Park RJ, Jacob DJ, Li QB, Kilaru V, Sarnat JA (2004) Mapping annual mean ground-level PM2.5 concentrations using Multiangle Imaging Spectroradiometer aerosol optical thickness over the contiguous United States. J Geophys Res Atmos 109:D22206

Maantay JA, Tu J, Maroko AR (2009) Loose-coupling an air dispersion model and a geographic information system (GIS) for studying air pollution and asthma in the Bronx, New York City. Int J Environ Health Res 19:59–79

Margolis HG, Mann JK, Lurmann FW, Mortimer KM, Balmes JR, Hammond SK et al (2009) Altered pulmonary function in children with asthma associated with highway traffic near residence. Int J Environ Health Res 19:139–155

Marshall JD, Granvold PW, Hoats AS, Mckone TE, Deakin E, Nazaroff WW (2006) Inhalation intake of ambient air pollution in California's South Coast Air Basin. Atmos Environ 40: 4381–4392

Marshall JD, Nethery E, Brauer M (2008) Within-urban variability in ambient air pollution: comparison of estimation methods. Atmos Environ 42:1359–1369

Martin RV (2008) Satellite remote sensing of surface air quality. Atmos Environ 42:7823–7843

McCurdy T (1995) Estimating human exposure to selected motor vehicle pollutants using the NEM series of models: lessons to be learned. J Expo Anal Environ Epidemiol 5: 533–550

Moura M, Junger WL, Mendonca GAES, de Leon AP (2009) Air quality and emergency pediatric care for symptoms of bronchial obstruction categorized by age bracket in Rio de Janeiro, Brazil. Cadernos de Saude Publica 25:635–644

Nethery E, Leckie SE, Teschke K, Brauer M (2008) From measures to models: an evaluation of air pollution exposure assessment for epidemiological studies of pregnant women. Occup Environ Med 65:579–586

Nordling E, Berglind N, Melen E, Emenius G, Hallberg J, Nyberg F et al (2008) Traffic-related air pollution and childhood respiratory symptoms, function and allergies. Epidemiology 19: 401–408

Ott WR (1985) Total human exposure. Environ Sci Technol 19:880–886

Paciorek CJ, Liu Y, Moreno-Macias H, Kondragunta S (2008) Spatiotemporal associations between GOES aerosol optical depth retrievals and ground-level PM2.5. Environ Sci Technol 42: 5800–5806

Peters JM, Avol E, Navidi W, London SJ, Gauderman WJ, Lurmann F et al (1999) A study of twelve southern California communities with differing levels and types of air pollution – I. Prevalence of respiratory morbidity. Am J Respir Crit Care Med 159:760–767

Phillips ML, Hall TA, Esmen NA, Lynch R, Johnson DL (2001) Use of global positioning system technology to track subject's location during environmental exposure sampling. J Expo Anal Environ Epidemiol 11:207–215

Prados AI, Kondragunta S, Ciren P, Knapp KR (2007) GOES aerosol/smoke product (GASP) over North America: comparisons to AERONET and MODIS observations. J Geophys Res Atmos 112:D15201

Rainham D, Krewski D, McDowell I, Sawada M, Liekens B (2008) Development of a wearable global positioning system for place and health research. Int J Health Geogr 7:59

Raizenne M, Neas LM, Damokosh AI, Dockery DW, Spengler JD, Koutrakis P et al (1996) Health effects of acid aerosols on North American children: Pulmonary function. Environ Health Perspect 104:506–514

Ritz B, Wilhelm M, Zhao YX (2006) Air pollution and infant death in southern California, 1989–2000. Pediatrics 118:493–502

Ryan PH, LeMasters GK, Biswas P, Levin L, Hu SH, Lindsey M et al (2007) A comparison of proximity and land use regression traffic exposure models and wheezing in infants. Environ Health Perspect 115:278–284

Scoggins A, Kjellstrom T, Fisher G, Connor J, Gimson N (2004) Spatial analysis of annual air pollution exposure and mortality. Sci Total Environ 321:71–85

Setton EM, Keller CP, Cloutier-Fisher D, Hystad PW (2008) Spatial variations in estimated chronic exposure to traffic-related air pollution in working populations: a simulation. Int J Health Geogr 7:39

Smargiassi A, Kosatsky T, Hicks J, Plante C, Armstrong B, Villeneuve PJ et al (2009) Risk of asthmatic episodes in children exposed to sulfur dioxide stack emissions from a refinery point source in montreal, Canada. Environ Health Perspect 117:653–659

United States Environmental Protection Agency (2008) Further Technical Details about HAPEM4. In

van Donkelaar A, Martin RV, Park RJ (2006) Estimating ground-level PM2.5 using aerosol optical depth determined from satellite remote sensing. J Geophys Res Atmos 111:D21201

Vedal S, Hannigan MP, Dutton SJ, Miller SL, Milford JB, Rabinovitch N et al (2009) The denver aerosol sources and health (DASH) study: overview and early findings. Atmos Environ 43:1666–1673

Wiehe SE, Carroll AE, Liu GC, Haberkorn KL, Hoch SC, Wilson JS et al (2008) Using GPS-enabled cell phones to track the travel patterns of adolescents. Int J Health Geogr 7:22

Wilhelm M, Ritz B (2005) Local variations in CO and particulate air pollution and adverse birth outcomes in Los Angeles County, California, USA. Environ Health Perspect 113:1212–1221

Williams LA, Ulrich CM, Larson T, Wener MH, Wood B, Campbell PT et al (2009) Proximity to traffic: inflammation, and immune function among women in the seattle, Washington, Area. Environ Health Perspect 117:373–378

Wu J, Winer AM, Delfino RJ (2006) Exposure assessment of particulate matter air pollution before, during, and after the 2003 Southern California wildfires. Atmos Environ 40:3333–3348

Yanosky JD, Paciorek CJ, Schwartz J, Laden F, Puett R, Suh HH (2008) Spatio-temporal modeling of chronic PM10 exposure for the nurses' health study. Atmos Environ 42:4047–4062

Chapter 4
The Use of Residential History in Environmental Health Studies

Francis P. Boscoe

Abstract Residential histories – listings of the places and dates where people have lived over their lives – are useful for assessing lifetime proximity to environmental hazards. When past residences are ignored, as is the norm, results are biased against finding an association between exposure and disease. I conducted a comprehensive review of 26 published environmental epidemiological studies using residential histories to assess current practice. Most often, studies collect all of a person's exact lifetime addresses resided in for at least 1 year, and exclude missing data – reasonable, though not necessarily optimal, choices. Residential histories are complex and time-consuming to collect, and must be researcher-initiated, as they are not an element of any population-based disease surveillance systems in the United States. Indeed, surveillance systems often have difficulty collecting even basic demographic items. As such, residential histories are best suited for focused research studies involving direct contact with subjects through interviews or questionnaires.

Keywords Residential history · Disease surveillance · Life course epidemiology · Geocoding · Mobility · Migration

4.1 Background

4.1.1 Introduction

When conducting an environmental health investigation, it is useful to have residential histories, a listing of the places and dates where people have lived throughout their lives. This information can be used to estimate past exposures to hazards, which can then be linked to health outcomes. Such exposure estimates are necessarily crude ones, especially since information on occupation, daily activity patterns, and other behavioral and lifestyles variables is often lacking. Still, this approach is an improvement over the usual alternative, which is to use only the

F.P. Boscoe (✉)
New York State Cancer Registry, Albany, NY 12237, USA
e-mail: fpb01@health.state.ny.us

J.A. Maantay, S. McLafferty (eds.), *Geospatial Analysis of Environmental Health*, Geotechnologies and the Environment 4, DOI 10.1007/978-94-007-0329-2_4,

current address or the address at the time of disease diagnosis. A single address may say very little about disease risk or etiology, given that cancer and certain chronic diseases have long latency periods. Also, as I will explain, the use of a single address is generally biased against finding associations between environment and disease. As a consequence, some scientists and advocates have called for residential history to be more routinely incorporated into case-control studies (Han et al., 2004; Jacquez et al., 2007; Urayama et al., 2009) and to be collected by population-based disease surveillance systems (Chittleborough et al., 2006; Office on Women's Health and Department of Health and Human Services, 2003).

This chapter surveys the current usage and application of residential history data in the field of environmental health. I begin with a brief summary of how past environmental exposures are believed to impact current health. Next, I explain how the incorporation of residential history information makes it more likely that an environment-disease link will be identified if one truly exists. I then review 26 recently published environmental health studies making use of residential history information in order to assess the kinds of research questions being asked and whether a standardized approach is being applied. Finally, I discuss the difficulties of collecting residential history in population-based surveillance systems, in part by giving the example of my own lengthy residential history. I conclude that residential histories are best suited for focused research studies that allow direct contact with subjects. My hope is that this chapter will provide a useful point of reference for researchers, advocates, and public health professionals interested in better measuring how past exposures may influence present health outcomes.

4.1.2 How Past Environmental Conditions Impact Current Health

Broadly speaking, one's lifetime residences – from in utero through infancy, childhood, adolescence and adulthood – influence current health in two primary ways. First, one's socioeconomic environment, as measured by levels of affluence, employment, education, safety, and quality of physical surroundings, directly bear on the prevalence of chronic and psychosocial conditions ranging from heart disease, stroke, diabetes and cancer to obesity, depression and teenage pregancy (Ben-Shlomo and Kun, 2004; Barker, 1998; Jelleyman and Spencer, 2008; Power et al., 2005). Second, specific chemical exposures from the air, soil, and water can directly contribute to ill health, particularly in the form of cancer and birth defects (United Nations Environment Programme, 2000). The quality of the socioeconomic environment and the physical environment are often correlated, though this is not always so.

These two forms of environmental exposure have given rise to two distinct research streams that approach residential history very differently (Mackenbach and Howden-Chapman, 2003). Researchers focused on the socioeconomic environment typically seek general indicators rather than exact residential addresses. When interviewing subjects, they typically collect historical information about items such as income, occupation, housing tenure, home ownership versus rental, neighborhood characteristics, and physical condition of housing. For some data items, this approach is less than optimal as better accuracy could be obtained by collecting addresses and linking them to historic census data. For example, people may better remember their address at the time they began school than detailed neighborhood characteristics from that time. Still, a review of over forty studies of this type identified only one that made specific use of past addresses, while all of the rest relied on self-reported descriptive information (Chittleborough et al., 2006).

Studies focused on chemical exposures, in contrast, rely extensively on residential addresses. In these studies, addresses are converted into point locations and predicted chemical exposures assigned based on these locations. Predicted exposures are usually derived from data collected by government agencies independently of the study, though researchers may collect their own primary environmental data by taking air, soil or water samples. Obtaining exposure information through interview or questionnaire is rarely possible because contaminants are so often invisible, as with radiation and trace contaminants in drinking water. For reasons of simplicity and economy, studies typically only consider a single address per person. This is most often the address at the time of the study, the address at time of disease diagnosis, or the address at time of birth. This chapter is focused on the relatively few studies that allow for multiple addresses per person.

Though it seems obvious that one's current health would be related to lifetime historic exposures, this is a relatively recent consensus (Ben-Shlomo and Kuh, 2002; Blane et al., 2007). Even within this "life course epidemiology" paradigm, the exact mechanisms by which past exposures influence current health remain a matter of debate. There are two major explanatory models. The critical period or biological programming model holds that specific periods of adverse exposure (whether socioeconomic or chemical) influence long-term health independently of eventual adult circumstances. The cumulative model holds that the intensity and duration of adverse exposure throughout life affects health in a dose-response fashion (Ljung and Hallqvist, 2006; Pensola and Martikainen, 2003). For example, in a study of asthma development, the critical period model might emphasize exposures to secondhand smoke below age one, while the cumulative model might equally weight all childhood exposure. These two models are not incompatible, and numerous hybrids have been offered (Ben-Shlomo and Kuh, 2004; Blane et al., 2007).

4.1.3 Why Residential History Matters

To illustrate how residential history adds clarity to an analysis, suppose that naturally occurring arsenic found in well water in certain regions of the United States promotes bladder cancer in a dose-response fashion such that people with the highest exposures have a doubled risk.[1] Without residential history information, it is

[1]The highest arsenic levels in the US are not especially high by world standards, particularly compared to those in developing countries where industrial pollution is the major source. The positive

unlikely that a retrospective epidemiological study would be able to accurately measure this risk Because population mobility in the United States is high and regions with high arsenic levels are few, it is likely that many of the people presently exposed to arsenic have not been exposed for most of their lives. For these people, their true exposure is lower than their address would imply. If participants in a study only lived in a high-arsenic area for 25% of their lives on average, a measured relative risk of 1.25 would be expected, all other things being equal. Warner and others illustrate this concept in detail using the example of radon and lung cancer (Warner et al., 1996). Like arsenic, radon is not a widespread phenomenon nationally and exposure depends not just on local geology but on the structural properties of the residence. If a current residence shows high radon readings, it is likely that other lifetime residences would have had little or no radon exposure, substantially reducing the apparent risk for most people.

Another potential source of bias results from the fact that even correctly assigned exposures are heavily weighted toward exposures occurring at older ages. Because the median age of diagnosis of bladder cancer is 68, a study using the address at time of diagnosis will mostly identify exposures occurring in the years just preceding age 68. If in utero or childhood or young adult exposure is more relevant than recent exposure, then this would yield a further downward bias in observed risk.

These biases may be partially offset in the presence of selective geographic migration to and from polluted and/or economically deprived areas (Rogerson and Han, 2002; Boyle 2004; Connolly et al., 2007). In general, those with greater means are more likely to migrate from such areas and are also likely be in better health than those remaining behind. Conversely, those migrating into such areas are likely to be in poorer health than those they replaced. This would tend to falsely inflate apparent associations between health and local pollution sources.

Without consideration of residential histories, it is impossible to quantify these complex sources of bias. When residential histories are available, this ceases to be an issue.

4.2 Geotechnology and Residential History

Geotechnical tools naturally lend themselves for use in studies incorporating residential history. Common to all such studies is the need to convert residential addresses into point locations, a process known as geocoding. Geocoding involves comparing the text elements of an address against a database containing a complete set of addresses and their corresponding latitudes and longitudes. Traditionally this has been accomplished with either specialized software or by using a module within geographic information system (GIS) software. More recently, free online geocoding solutions have emerged that directly or indirectly make use of the databases behind powerful mapping web sites such as Google Maps or Yahoo! Maps

---

association between bladder cancer and arsenic is well-established in developing countries, but less so in the United States (Navarro Silvera and Rohan, 2007; Chu and Crawford-Brown, 2007).

(University of Southern California, 2009). GIS software also provides an ideal platform for managing and manipulating data with space and time elements and for assigning exposures based on location. Most of the studies to be discussed later in this chapter make explicit reference to the use of a GIS. GIS software is not essential, though. With some knowledge of computational geometry – the study of algorithms and data structures involving points, lines, and areas (de Berg et al., 1998) – it is possible to perform the necessary analyses using any robust statistical software package.

Jacquez and colleagues have been actively advancing the geotechnical toolkit for analyzing residential history data (Jacquez et al., 2005; Jacquez et al., 2006; Jacquez et al., 2007; Meliker and Jacquez, 2007). For example, their Q statistics measure global, local and focused clustering that accounts for residential mobility. Global clustering describes the tendency of a data set as a whole to be clustered, local clustering identifies specific locations where clustering occurs, and focused clustering measures clustering around specific points of interest, such as industrial sites. The Q statistic measures the tendency of a case to be closer to another case versus a control, or more specifically, to be closer to m cases out of its n nearest neighbors. Whether the ratio m/n is statistically meaningful is determined through simulating large numbers of plausible data configurations. Using a preliminary data set from an ongoing case-control study, the authors have identified the existence of a cluster dating to 1929. Further enhancements to their methods address the estimation of induction and latency periods, age-specific susceptibility, and critical exposure time points. Whether these methodological advances will have wide applicability remains to be seen, but they illustrate that the full range of uses for residential history data is yet to be realized.

4.3 Review of Environmental Health Studies Using Residential History

I conducted a comprehensive review of published environmental epidemiological studies where residential history was explicitly mentioned in the abstract. Searches were conducted on PubMed and Web of Science using the terms "residential history", "residential mobility", and "population mobility", with forward and backward citation searches conducted on relevant articles. Papers were limited to those published since 2002 in order to ensure that recent advances in geocoding and geographic information systems were available to the researchers. The primary focus was on North American studies, but a Swedish, Taiwanese, and two multinational studies were included to provide additional diversity and context. Ultimately, twenty-six distinct studies were identified. This is but a tiny fraction of the total number of studies relating disease and exposure that are limited to a single residential location.

I noted the disease or condition being studied, the hypothesis being tested, and whether an association was found (Table 4.1). I also recorded the type of study and number of subjects, how the residential history was obtained (interview,

Table 4.1 Studies reviewed for this chapter

	Study location	Hypothesis	Findings
1. Gammon et al. (2002)	Long Island, New York	Long-term residents (15 + years) had different breast cancer risk factors than short-term residents	Long-term residents differed in many characteristics, including being more likely to be older, white, less educated, older at menopause, never or past smoker, lower income, more children, never breastfed, higher BMI, ever drank alcohol, never used oral contraceptives, ever used hormone replacement, and ever had a mammogram
2. O'Leary et al. (2004)	Long Island, New York	Proximity to hazardous waste sites containing pesticides, residence on former farmland, and public water supply are associated with breast cancer	A weak but significant association was found for living within 1 mile of a hazardous waste site. No other associations found
3. Han et al. (2004), Han et al. (2005), Bonner et al. (2005)	Erie and Niagara Counties, New York	Early environmental exposures may be related to breast cancer risk, especially for pre-menopausal women	Cases were more clustered than controls at time of birth and menarche for pre-menopausal cancer. Premenopausal cases aged 35–44 showed evidence of clustering; clustering also found in specific geographic locations. TSP concentrations at birth were associated with breast cancer for postmenopausal women
4. McKelvey et al. (2004), Brody et al. (2004)	Cape Cod, Massachusetts	Duration of residence is related to increased breast cancer risk; historic pesticide exposure is related to breast cancer risk	Duration of residence: suggestive but not strong evidence of a higher risk, after controlling for numerous risk factors. Pesticide exposure: No association found
5. Brody et al. (2006)	Cape Cod, Massachusetts	Drinking public water contaminated by wastewater is associated with breast cancer risk	No association found
6. Schoenfeld et al. (2003)	Long Island, New York	EMF radiation is associated with breast cancer	No relationship found

Table 4.1 (continued)

	Study location	Hypothesis	Findings
7. Vieira et al. (2005, 2008, 2009)	Cape Cod, Massachusetts	Incidence of breast, colorectal, lung, bladder, kidney and pancreatic cancers is associated with residential location	Evidence of spatially and temporally varying risk in breast, kidney, and bladder cancer incidence. No clear findings for other sites
8. Hurley et al. (2005)	California	Residential mobility can impact a cohort study such as the relationship between pesticides and breast cancer in the California Teacher's Study	Persons with California birthplace and high SES were more residentially stable. Current residence "may reasonably reflect some aspect of longer term chronic exposures". Average number of residences was 8.9 with half between 6 and 10
9. Fears et al. (2002)	Philadelphia and San Francisco	Lifetime UVB exposure and time spent outdoors are associated with melanoma risk among non-Hispanic whites	For lifetime UVB exposure, there was a strong and consistent association for both men and women. For time spent outdoors, association was strong for men and for women who develop a suntan. Findings were stronger than for comparable studies that did not incorporate residential history
10. Solomon et al. (2004)	Seattle-Puget Sound area	Melanoma is associated with sun exposure	Association found for women but not for men
11. Tatalovich et al. (2006)	Los Angeles, California	Melanoma incidence is associated with lifetime UV exposure	Both cumulative exposure and average annual exposure were positively correlated with melanoma incidence, with the cumulative exposure association very strong
12. Lea et al. (2007)	Connecticut	Lifetime UVB exposure is associated with melanoma risk among whites	Strong dose-response relationship was found
13. Kricker et al. (2007)	International (US, Canada, Australia)	UVB exposure is associated with increased risk for multiple primary melanomas	Weak evidence of lifetime effect, stronger evidence of effect based on residence at birth and age 10
14. Knobeloch et al. (2006)	Rural Wisconsin	Arsenic exposure from well water is associated with skin cancer	Significant association found, magnified among current and former smokers

Table 4.1 (continued)

	Study location	Hypothesis	Findings
15. Von Behren et al. (2008)	Northern California	Childhood leukemia is associated with residential traffic density	No association found
16. Kasim et al. (2005, 2006)	Canada	Two studies using the same data set: (1) Lifetime exposure to environmental tobacco smoke is associated with adult leukemia (2) Chlorination disinfection by-products are associated with adult leukemia	Environmental tobacco smoke was associated with chronic lymphocytic leukemia (CLL) but not other leukemia subtypes. Disinfection by-products were positively associated with chronic myeloid leukemia, and inversely associated with CLL
17. Yu et al. (2006)	Taiwan	Residential exposure to petrochemicals is associated with leukemia	Among those age 20–29, a significant association was found but no association was found among those 10–19
18. Urayama et al. (2009)	Northern California	Residential mobility can impact analysis of childhood leukemia	Nearly 2/3 of children were movers, often moving outside the county, state or country. 25% of mothers moved in the year before birth. Movers have older age at diagnosis, lower maternal age, lower education and SES. They were more urban and more likely to be smokers. Moves within the same SES and urbanicity are typical, though there is a trend toward upward mobility
19. Choi et al. (2006)	Florida, New Jersey, New York excluding New York City, and Pennsylvania	There is an association between residential proximity to Toxics Release Inventory (TRI) facilities during pregnancy and brain cancer	For brain cancers under age 5, significant associations were found for those living within 1 mile of a TRI facility or a facility releasing carcinogens. No associations found when an exposure index was applied
20. Bassin et al. (2006)	United States	There is an association between fluoride in drinking water and osteosarcoma	Association was found among males but not females

Table 4.1 (continued)

	Study location	Hypothesis	Findings
21. Johnson et al. (2003)	Canada	Residential proximity to heavy industry is associated with non-Hodgkin lymphoma	Overall, no association found. Possible associations for proximity to copper smelters and sulfite pulp mills, but numbers very small
22. Meliker et al. (2007)	Michigan	Residential mobility can influence environmental exposure to arsenic. Part of a bladder cancer study, but link between disease and exposure was not evaluated in this study	Considerable temporal variability in exposure within individuals was demonstrated
23. Steinmaus et al. (2003)	Nevada and California	Arsenic exposure from drinking water is associated with bladder cancer	No associations found
24. Diez-Roux et al. (2008)	United States	Exposure to airborne particulate matter contributes to development of atherosclerosis	Weak association found. Questionnaire was repeated 1 year later with very high test-retest reliability of the exposure measure
25. Selander et al. (2009)	Stockholm, Sweden	Road traffic noise is associated with myocardial infarction	Modest association found
26. Brauer et al. (2008)	Vancouver, British Columbia	Traffic-related air pollution is associated with low birth weight, preterm birth	Positive associations were found

questionnaire, review of historical records), the time points of interest (entire life, last 10 years, birth and age five, etc.), the level of geographic detail (coordinates taken from an aerial photograph, street address, zip code, etc.), how incomplete or partial information was handled, whether there was a minimum threshold for residence duration, and whether there were any notable methodological innovations. I identified several instances where multiple published papers made use of the same data set, and these were counted only once. The papers by Jacquez et al. discussed previously were excluded given their methodological emphasis.

In terms of geographic location, Long Island (New York) and Cape Cod (Massachusetts) appeared three times each, reflecting the influence of the Long Island Breast Cancer Study Project (Winn, 2005) and ongoing research at the Silent Spring Institute (2009). The other studies were in widely scattered locations, including a number of multi-city studies. Cancer was by far the most common condition studied, accounting for 23 of the 26 studies. Of these, there were seven breast cancer studies, six skin cancer, four leukemia, two bladder, one each of non-Hodgkin lymphoma, brain, and osteosarcoma, and one that examined multiple cancer sites. The non-cancer studies included one each of atherosclerosis, myocardial infarction, and adverse birth outcomes. Twenty-three of the studies followed a case-control design, and there were three cohort studies. Sample sizes, combining cases and controls, ranged from 300 to 6,000.

Environmental hazards spanned the gamut from ultraviolet radiation (in most of the skin cancer studies) to hazardous waste to traffic to various forms of contamination of drinking water. Several studies did not identify any specific hazard, but compared long-term and short-term residents, whether cases had different demographic characteristics than controls, or evaluated overall mobility patterns.

All but one of the studies obtained residential histories through interviews or questionnaires. In general, interviews refer to information collected orally, and questionnaires to information collected in writing or electronically, though some authors used the terms interchangeably. An interview was the more popular method, with telephone interviews favored over in-person interviews. One Canadian study compiled residential histories entirely through administrative records, and the Swedish study used administrative records to augment its primary data collection. There were no United States studies where administrative or surveillance data were used for this purpose. Many authors noted the labor-intensive nature of interviews and questionnaires.

Most studies (16) collected address over the entire life course, with a few extending this to the year before birth. Five studies considered only the duration of current residence, two were limited to addresses in the past 20 years, and two collected only addresses at birth and in the year before birth. One study obtained addresses only at specific age points.

Seventeen of the studies collected exact addresses. One requested only county of residence, one city of residence, and five were unspecified. The two Canadian studies both used postcodes, which in most of the country describe a precise geographic location.

The most common minimum residence time requested was 1 year, used in eight studies. Eight studies used something shorter than this: 6 months (4), 1 month (2), or no minimum (2). Three studies used ten or more years at a residence, and for the remaining seven studies this information was not specified.

For time periods where a person could not recall a previous address, ten studies excluded these person-years from the analysis. Three of these ten provided detail on the extent of missing data. Four studies imputed or interpolated missing information, and two claimed to have no missing data. The remaining ten studies did not specify how missing data were handled; presumably it was excluded, because it seems unlikely that an interpolation or imputation process would be used and not mentioned.

To the degree there is a consensus across these studies, it is to collect all of a person's exact lifetime addresses resided in for at least 1 year beginning at birth, and to exclude data that could not be recalled. With continued advances and awareness of methods for handling missing data, the latter issue should be expected to diminish (Buhi et al., 2008). An open question is whether the minimum residential duration of 1 year and a start time of birth is adequate or should be tightened or relaxed. As I will show, collecting residence data from before birth (and even early childhood) and on short-term residences poses difficult challenges.

4.4 Difficulties in Collecting Residential History Information

There is a popular perception, at least among those who have never attempted it, that a residential history is an easy matter to compile, something that could be done in a few minutes in a waiting room, or perhaps over the phone with a receptionist. An attempt to compile my own residential history reveals this as a fallacy. If I, as someone with a focused research interest in this area, have difficulty with this task, then it is safe to assume that plenty of Americans would have even more difficulty.

The year before birth presents my first challenge. In early 1968, I was conceived while my father was stationed at a military base in Germany. I believe this was in Würzburg, but I am not certain. My mother returned to the states when she was about 5 or 6 months pregnant, and I do not know where she lived for the next few months. Of course, I could ask her, but that is not something I could assume I could do on demand at a doctor's office or hospital, particularly if I was at a more typical age to be diagnosed with cancer, when it is unlikely she would still be alive. My mother, now in her 60s, no longer has the ability to ask her late parents exactly where they lived when they were young.

I believe my father was discharged just before I was born and the family reunited in the first of several apartments in the outer suburbs of Philadelphia, Pennsylvania. I think there were three apartments in total from August, 1968 to circa 1972 as the family grew to three children; I know of two town names. I believe I could locate one of the apartments using Google Earth, assuming it is still standing, since it was on a major road and for years afterward my parents would pass it from time to time and point it out. With this one possible exception, I am analytically nearly useless in terms of having exposures assigned to me before about age four.

From that point through age seventeen I had three more addresses in rural eastern Pennsylvania, all of which I remember exactly. The first two were rural route addresses, and the names of the roads would probably allow geocoding to within one-mile accuracy. The third was in a village of 900 people – my first truly geocodeable address, at the age of 15.

In 1986, I went off to college in Pittsburgh, Pennsylvania. Throughout my college years and those immediately following I moved every single year, motivated by a constant quest for low rent, dependable roommates, and the need for band rehearsal space. I count seven places in 6 years, including a summer sublet. I can definitely recall all of the streets, though not all of the house numbers. Also during this period I spent one summer on the Greenland ice sheet doing ice-core research, where there is no address to be recorded, but I could provide an approximate latitude and longitude.

In 1992, after 2 years with a regular paycheck as an entry-level environmental engineer and tiring of the nomadic lifestyle, I bought a house. It was virtually the cheapest one I could find, in a struggling neighborhood close to Three Rivers Stadium in Pittsburgh. Fixing it up proved to be somewhat beyond my talents, and before I knew it I met my future wife and decided to go to graduate school. I never even spent that much time in the Pittsburgh house, as the engineering job took me to hazardous waste sites near Harrisburg, Pennsylvania and Oklahoma City for many months at a time. The corresponding Days Inns and La Quinta Inns would seem to qualify for my residential history, though with exact addresses unknown. Certainly my air quality, traffic, ultraviolet radiation, and drinking water exposures all changed considerably during these work assignments, even ignoring any exposure from the hazardous waste sites themselves.

In the fall of 1994 I began my master's degree in geography in Kent, Ohio. For the next 2 years I shuttled between Kent and my future wife's apartment in Pittsburgh. At first I rented an apartment in Kent, but then discovered that they allowed students to stay in one of the dorms, hotel-style, for under $10 a night (alas, the name of the dorm escapes me). Also during this time, I spent a summer on an archaeology crew in Hurricane, West Virginia.

After Kent, I moved on to get my Ph.D. in geography at Penn State. Here there were two addresses, an initial apartment and then a larger one that was needed once my son was born. My dissertation research then took me to Philadelphia, where I spent a year before getting hired by the New York State Department of Health in Albany, New York in 1999, where I remain today.

Even with long-term, stable employment, residential stability has continued to be elusive. We have had six addresses in 10 years, moving from an initial rental to a starter home to something a bit larger to accommodate my retired father-in-law's extended visits. These were sandwiched around a 6 month posting in Washington, DC at the National Cancer Institute.

All told, that is 20–30-plus addresses in 40 years, depending on how the shorter-term addresses are counted, with just over half of them geocodeable to precise latitude and longitude coordinates. While I have been more mobile than the average American – a widely cited figure is ten or eleven addresses per lifetime (Long, 1988) – I do not believe my situation is unusual compared with other academics or professional scientists. Certainly I am not near the fringe of mobility represented by military personnel, migrant farm workers, and the indigent.

While my evidence for the difficulty of developing a residential history is anecdotal, there have been researchers who have considered the issue more formally. These studies have consistently found that a substantial majority of subjects can recall basic details about their living environment after 50 years with useful accuracy, as measured by comparing recalled information with historic records (Blane, 1996). Recall can be further improved by providing a temporal reference system such as a timeline or calendar, with major personal or historic events included, and by interviewing spouses jointly (Blane, 1996; Zahm et al., 2001). Few of these studies evaluated the recall of addresses, however. In one of the few studies that did, subjects had good recall for street names but not specific house numbers (Berney and Blane, 1997). By extension, some of the geographical precision afforded by Canadian postal codes would be offset by peoples' difficulty in remembering them.

The exact implications of incompletely recalled residential histories in environmental studies are unclear, and I am unaware of any examples of sensitivity analyses having been performed. Put simply, though, research results from a partial residential history should fall between those based on a complete residential history and those based only on a single address. Such results would be expected to be biased toward the null hypothesis, but to a lesser degree than when only a single address is used. So it would seem that even my messy and incomplete residential history merits inclusion in a case-control study.

This does not mean it also merits inclusion in a population-based disease surveillance system. The main problem with expanding such a system to include residential histories is the difficulty and complexity of data collection. Demographic information currently collected typically includes name, birth date, gender, race/ethnicity, birth place (at the level of state or country), marital status, and current address. Some data systems also specify the collection of occupation, religion, tobacco use, and alcohol use. The fields that are used by hospitals for patient identification and billing (name, address, birth date, gender, and various identification numbers) tend to be highly complete and accurate. Beyond this, quality is less certain. Race and ethnicity are generally dependable when self-reported, but this information may be assigned by a clerk based on name or appearance, or omitted altogether. Standards are inconsistent – in some data systems, race and ethnicity are treated as separate constructs, but in others they are conflated. Cancer registries in the United States use a detailed scheme which incorporates specific Asian, Pacific Islander, and Hispanic subgroups, raising additional complications (Pinheiro et al., 2009; Schymura et al., 2007). Though straightforward and unambiguous, place of birth is typically missing in over half of case reports submitted to central cancer registries. In terms of marital status, useable distinctions can be made between never-married, married, and formerly married, but finer distinctions of divorced, separated and widowed are problematic. Occupation, religion, and behavioral variables are seldom collected and reported at all.

Residential histories are much more complex than these basic demographic data elements that have proven difficult to collect. Country or state of birth is a one-word response translatable into a 3-digit code; my residential history comprises over 30 addresses, cities, states, zip codes, start dates and end dates. It is hard to imagine a hospital clerk working with anxious patients to jog the names of their freshman dormitories or past military posts. Accordingly, there are no examples of surveillance systems in the United States that attempt to collect residential history.

"Years at current residence" has been offered as a compromise: a simple, easy-to-collect variable that may be well-suited for surveillance systems. This variable could allow epidemiologic studies to be limited to long-term residents of the areas under study and for recent in-migrants to be excluded (O'Leary et al., 2004). Unfortunately, even with this simpler approach, there remains the problem of how to obtain corresponding information on individuals without disease. It is not helpful to count those with breast cancer who have not moved in 10 years unless you also know the total population of those who have not moved in 10 years. Census data relating to mobility and migration are too crude and limited to be invoked for this purpose. What is needed is the identification, contacting, and interviewing of a suitable population control group, which negates much of the advantage of using surveillance systems toward this purpose.

To summarize, disease surveillance systems are designed to collect basic information about the entire population for the purposes of disease control, identifying trends, detecting emerging problems, and guiding policy and interventions. Such systems are designed to be quick, inexpensive (at least on a per capita basis), and coarse. In a surveillance context, knowing where people currently live is more relevant than knowing where they have previously lived. Research studies, in contrast, are deliberative, costly, and detailed. This is the best forum for investigating questions of disease etiology, and here is where the collection of residential histories is a more natural fit.

4.5 Conclusions

Residential histories add value to environmental health studies by reducing bias that tends to understate the relationship between exposure and disease. Despite the clear advantages of incorporating residential history data into a study, relatively few have done so. This chapter will hopefully serve as a catalyst towards promoting wider incorporation of these data. Residential histories do have limitations – collecting the information is labor-intensive and subject to recall gaps, especially for the prenatal and early childhood periods and for those with highly mobile lifestyles. For researchers lacking GIS experience, the task may seem daunting. But continuing improvements in the availability of reference data, computing speed, and statistical methodology has made the incorporation of residential history data into analyses easier than ever.

This chapter has made a conceptual argument for the usefulness of residential history data. Further work is needed to provide more specific guidance on how these data should best be collected and analyzed: whether it is safe to ignore short-term residences, how to approach missing or imprecise data, and under what conditions it makes sense to limit analysis to the residentially stable. It is important to keep in mind that residential location can never be more than a crude proxy for exposure. Going forward, biomonitoring, the measurement of chemicals in human tissues and fluids, offers the promise of much more accurate exposure assessment for a wide range of chemical insults (Angener et al., 2007). But there will long remain a demand for residence-based analyses, particularly for questions involving otherwise-unmeasured past exposures and for diseases with long latency periods.

Acknowledgments I would like to thank Maria Schymura of the New York State Cancer Registry for originally getting me to think about these issues. Daikwon Han of Texas A&M University provided some insight on the subject of recall bias.

Author Biography

Dr. Francis P. Boscoe joined the New York State Cancer Registry as a Research Scientist in 1999, 1 year before receiving his Ph.D. in geography from Pennsylvania State University. He is also an Assistant Professor in the Department of Epidemiology and Biostatistics at the University at Albany, where he teaches a course in Geographic Information Systems (GIS) and Public Health and chairs the departmental M.S. and Ph.D. admissions committee. His research focuses on geographic patterns in cancer incidence, spatial and demographic analytical methods, and quality and measurement issues in public health data. He has written 30 peer-reviewed articles and book chapters and actively serves on multiple committees for the North American Association of Central Cancer Registries (NAACCR).

References

Angener J, Ewers U, Wilhelm M (2007) Human biomonitoring: state of the art. Int J Hyg Environ Health 210(3–4):201–228

Barker DJP (1998) Mothers, babies and health in later life, 2nd edn. Churchill Livingstone, Edinburgh

Bassin EB, Wypij D, Davis RB, Mittleman MA (2006) Age-specific fluoride exposure in drinking water and osteosarcoma (United States). Cancer Causes Control 17:421–428

Ben-Shlomo Y, Kuh D (2002) A life course approach to chronic disease epidemiology: conceptual models, empirical challenges and interdisciplinary perspectives. Int J Epidemiol 31: 285–293

Ben-Shlomo, Y and Kuh, D (eds) (2004) A life course approach to chronic disease epidemiology, 2nd edn. Oxford University Press, Oxford

Berney LR, Blane DB (1997) Collecting retrospective data: accuracy of recall after 50 years judged against historical records. Soc Sci Med 45:1519–1525

Blane D, Netuveli G, Stone J (2007) The development of life course epidemiology. Rev Epidemiol Sante Publique 55:31–38

Blane DB (1996) Collecting retrospective data: Development of a reliable method and a pilot study of its use. Soc Sci Med 42:751–757

Bonner MR, Han D, Nie J, Rogerson P, Vena JE, Muti P, Trevisan M, Edge SB, Freudenheim JL (2005) Breast cancer risk and exposure in early life to polycyclic aromatic hydrocarbons using total suspended particulates as a proxy measure. Cancer Epidemiol Biomarkers Prev 14: 53–60

Brauer M, Lencar C, Tamburic L, Koehoorn M, Demers P, Karr C (2008) A cohort study of traffic-related air pollution impacts on birth outcomes. Environ Health Perspect 116:680–686

Brody JG, Aschengrau A, McKelvey W, Swartz CH, Kennedy T, Rudel RA (2006) Breast cancer risk and drinking water contaminated by wastewater: a case control study. Environ Health 5(28):28

Buhi ER, Goodson P, Neilands TB (2008) Out of sight, not out of mind: strategies for handling missing data. Am J Health Behav 32:83–92

Boyle P (2004) Population geography: migration and inequalities in mortality and morbidity. Prog Hum Geogr 28:767–776

Brody JG, Aschengrau A, McKelvey W, Rudel RA, Swartz CH, Kennedy T (2004) Breast cancer risk and historical exposure to pesticides from wide-area applications assessed with GIS. Environ Health Perspect 112:889–897

Chittleborough CR, Baum FE, Taylor AW, Hiller JE (2006) A life-course approach to measuring socioeconomic position in population health surveillance systems. J Epidemiol Commun Health 60:981–992

Choi HS, Shim YK, Kaye WE, Ryan PB (2006) Potential residential exposure to toxics release inventory chemicals during pregnancy and childhood brain cancer. Environ Health Perspect 114:1113–1118

Chu HA, Crawford-Brown D (2007) Inorganic arsenic in drinking water and bladder cancer: a meta-analysis for dose-response assessment. Int J Environ Res Public Health 4:340–341

Connolly S, O'Reilly D, Rosato M (2007) Increasing inequalities in health: is it an artefact caused by the selective movement of people? Soc Sci Med 64:2008–2015

de Berg, M, van Kreveld, M, Overmars, M and Schwarzkopf, O (eds) (1998) Computational geometry: algorithms and trends, 2nd ed. Springer-Verlag, Berlin

Diez Roux AV, Auchincloss AH, Franklin TG, Raghunathan T, Barr RG, Kaufman J, Astor B, Keeler J (2008) Long-term exposure to ambient particulate matter and prevalence of subclinical atherosclerosis in the Multi-Ethnic Study of Atherosclerosis. Am J Epidemiol 167: 667–675

Fears TR, Bird CC, Guerry D, Sagebiel RW, Gail MH, Elder DE, Halpern A, Holly EA, Hartge P, Tucker MA (2002) Average midrange ultraviolet radiation flux and time outdoors predict melanoma risk. Cancer Res 62:3992–3996

Gammon MD, Neugut AI, Santella RM, Teitelbaum SL, Britton JA, Terry MB, Eng SM, Wolff MS, Stellman SD, Kabat GC, Levin B, Bradlow HL, Hatch M, Beyea J, Camann D, Trent M, Senie RT, Garbowski GC, Maffeo C, Montalvan P, Berkowitz GS, Kemeny M, Citron M, Schnabe F, Schuss A, Hajdu S, Vincguerra V, Collman GW, Obrams GI (2002) The Long Island Breast Cancer Study Project: description of a multi-institutional collaboration to identify environmental risk factors for breast cancer. Breast Cancer Res Treat 74: 235–254

Han D, Rogerson PA, Bonner MR, Nie J, Vena JE, Muti P, Trevisan M, Freudenheim JL (2005) Assessing spatio-temporal variability of risk surfaces using residential history data in a case control study of breast cancer. Int J Health Geogr 4:9

Han D, Rogerson PA, Nie J, Bonner MR, Vena JE, Vito D, Muti P, Trevisan M, Edge SB, Freudenheim JL (2004) Geographic clustering of residence in early life and subsequent risk of breast cancer (United States). Cancer Causes Control 15:921–929

Hurley SE, Reynolds P, Goldberg DE, Hertz A, Anton-Culver H, Bernstein L, Deapen D, Peel D, Pinder R, Ross RK, West D, Wright WE, Ziogas A, Horn-Ross PL (2005) Residential mobility in the California teachers study: implications for geographic differences in disease rates. Soc Sci Med 60:1547–1555

Jacquez GM, Kaufmann A, Meliker J, Goovaerts P, AvRuskin G, Nriagu J (2005) Global, local and focused geographic clustering for case-control data with residential histories. Environ Health 4:4

Jacquez GM, Meliker J, Kaufmann A (2007) In search of induction and latency periods: space-time interaction accounting for residential mobility, risk factors and covariates. Int J Health Geogr 6(35):35

Jacquez GM, Meliker JR, Avruskin GA, Goovaerts P, Kaufmann A, Wilson ML, Nriagu J (2006) Case-control geographic clustering for residential histories accounting for risk factors and covariates. Int J Health Geogr 5(32):32

Jelleyman T, Spencer N (2008) Residential mobility in childhood and health outcomes: a systematic review. J Epidemiol Commun Health 62:584–592

Johnson KC, Pan S, Fry R, Mao Y (2003) Residential proximity to industrial plants and non-Hodgkin lymphoma. Epidemiology 14:687–693

Kasim K, Levallois P, Abdous B, Auger P, Johnson KC (2005) Environmental tobacco smoke and risk of adult leukemia. Epidemiology 16:672–680

Kasim K, Levallois P, Johnson KC, Abdous B, Auger P (2006) Chlorination disinfection by-products in drinking water and the risk of adult leukemia in Canada. Am J Epidemiol 163:116–126

Knobeloch LM, Zierold KM, Anderson HA (2006) Association of arsenic-contaminated drinking-water with prevalence of skin cancer in Wisconsin's Fox River Valley. J Health Popul Nutr 24:206–213

Kricker A, Armstrong BK, Goumas C, Litchfield M, Begg CB, Hummer AJ, Marrett LD, Theis B, Millikan RC, Thomas N, Culver HA, Gallagher RP, Dwyer T, Rebbeck TR, Kanetsky PA, Busam K, From L, Mujumdar U, Zanetti R, Berwick M (2007) Ambient UV, personal sun exposure and risk of multiple primary melanomas. Cancer Causes Control 18: 295–304

Lea CS, Scotto JA, Buffler PA, Fine J, Barnhill RL, Berwick M (2007) Ambient UVB and melanoma risk in the United States: a case-control analysis. Ann Epidemiol 17:447–453

Ljung R, Hallqvist J (2006) Accumulation of adverse socioeconomic position over the entire life course and the risk of myocardial infarction among men and women: results from the Stockholm heart epidemiology program (SHEEP). J Epidemiol Commun Health 60: 1080–1084

Long L (1988) Migration and residential mobility in the US. Russell Sage Foundation, New York, NY

Mackenbach JP, Howden-Chapman P (2003) New perspectives on socioeconomic inequalities in health. Perspect Biol Med 46:428–444

McKelvey W, Brody JG, Aschengrau A, Swartz CH (2004) Association between residence on Cape Cod, Massachusetts, and breast cancer. Ann Epidemiol 14:89–94

Meliker JR, Jacquez GM (2007) Space-time clustering of case-control data with residential histories: insights into empirical induction periods, age-specific susceptibility, and calendar year-specific effects. Stoch Environ Res Risk Assess 21:625–634

Meliker JR, Slotnick MJ, Avruskin GA, Kaufmann A, Fedewa SA, Goovaerts P, Jacquez GJ, Nriagu JO (2007) Individual lifetime exposure to inorganic arsenic using a space-time information system. Int Arch Occup Environ Health 80:184–197

Navarro Silvera SA, Rohan TE (2007) Trace elements and cancer risk: a review of the epidemiologic evidence. Cancer Causes Control 18:7–27

O'Leary ES, Vena JE, Freudenheim JL, Brasure J (2004) Pesticide exposure and risk of breast cancer: a nested case-control study of residentially stable women living on Long Island. Environ Res 94:134–144

Office on Women's Health, Department of Health and Human Services 2003. Workshop on Breast Cancer and the Environment, Final Report. Washington, DC.

Pensola TH, Martikainen P (2003) Cumulative social class and mortality from various causes of adult men. J Epidemiol Commun Health 57:745–751

Pinheiro PS, Sherman RL, Fleming LE, Gomez-Marin O, Wohler B, MacKinnon JA, Levin GM (2009) HOIA: an alternative hispanic origin identification algorithm for cancer registries. J Registry Manag 35:149–155

Power C, Graham H, Due P, Hallqvist J, Joung I, Kuh D, Lynch J (2005) The contribution of childhood and adult socioeconomic position to adult obesity and smoking behaviour: an international comparison. Int J Epidemiol 34:335–344

Rogerson PA, Han D (2002) The effects of migration on the detection of geographic differences in disease risk. Soc Sci Med 55:1817–1828

Schoenfeld ER, O'Leary ES, Henderson K, Grimson R, Kabat GC, Ahnn S, Kaune WT, Gammon MD, Leske MC (2003) Electromagnetic fields and breast cancer on Long Island: a case-control study. Am J Epidemiol 158:47–58

Schymura MJ, Boscoe FP, Hsieh M-C, Williams MA, Henry K 2007. "Working With Asian and Pacific Islander Data: II. Coding Issues with Pacific Islanders", in North American Association of Central Cancer Registries Annual Conference.

Selander J, Nilsson ME, Bluhm G, Rosenlund M, Lindqvist M, Nise G, Pershagen G (2009) Long-term exposure to road traffic noise and myocardial infarction. Epidemiology 20:272–279

Silent Spring Institute. 2009. Silent Spring Institute. Available at: http://www.silentspring.org/. Accessed 1 May 2009

Solomon CC, White E, Kristal AR, Vaughan T (2004) Melanoma and lifetime UV radiation. Cancer Causes Control 15:893–902

Steinmaus C, Yuan Y, Bates MN, Smith AH (2003) Case-control study of bladder cancer and drinking water arsenic in the western United States. Am J Epidemiol 158:1193–1201

Tatalovich Z, Wilson JP, Mack T, Yan Y, Cockburn M (2006) The objective assessment of life-time cumulative ultraviolet exposure for determining melanoma risk. J Photochem Photobiol B 85:198–204

United Nations Environment Programme. 2000. "Toxic Chemicals and Hazardous Waste" [accessed on May 5,2009] Available at: http://www.unep.org/GEO2000/english/0037.htm

University of Southern California GIS Research Laboratory. 2009. "Geocoding" [accessed on December 1, 2009] Available at: https://webgis.usc.edu/Services/Geocode/Default.aspx

Urayama KY, Von BJ, Reynolds P, Hertz A, Does M, Buffler PA (2009) Factors associated with residential mobility in children with leukemia: implications for assigning exposures. Ann Epidemiol 19(11):834–840

Vieira V, Webster T, Weinberg J, Aschengrau A (2009) Spatial analysis of bladder, kidney, and pancreatic cancer on upper Cape Cod: an application of generalized additive models to case-control data. Environ Health 8(3):3

Vieira V, Webster T, Weinberg J, Aschengrau A, Ozonoff D (2005) Spatial analysis of lung, colorectal, and breast cancer on Cape Cod: an application of generalized additive models to case-control data. Environ Health 4(11):11

Vieira VM, Webster TF, Weinberg JM, Aschengrau A (2008) Spatial-temporal analysis of breast cancer in upper Cape Cod, Massachusetts. Int J Health Geogr 7(46):46

Von BJ, Reynolds P, Gunier RB, Rull RP, Hertz A, Urayama KY, Kronish D, Buffler PA (2008) Residential traffic density and childhood leukemia risk. Cancer Epidemiol Biomarkers Prev 17:2298–2301

Warner KE, Mendez D, Courant PN (1996) Toward a more realistic appraisal of the lung cancer risk from radon: the effects of residential mobility. Am J Public Health 86:1222–1227

Winn DM (2005) Science and society: the Long Island Breast Cancer study project. Nat Rev Cancer 5:986–994

Yu CL, Wang SF, Pan PC et al (2006) Residential exposure to petrochemicals and the risk of leukemia: using geographic information system tools to estimate individual-level residential exposure. Am J Epidemiol 164:200–207

Zahm SH, Colt JS, Engel LS, Keifer MC, Alvarado AJ, Burau K, Butterfield P, Caldera S, Cooper SP, Garcia D, Hanis C, Hendrikson E, Heyer N, Hunt LM, Krauska M, MacNaughton N, McDonnell CJ, Mills PK, Mull LD, Nordstrom DL, Outterson B, Slesinger DP, Smith MA, Stallones L, Stephens C, Sweeney A, Sweitzer K, Veron SW, Blair A (2001) Development of a life events/icon calendar questionnaire to ascertain occupational histories and other characteristics of migrant farmworkers. Am J Ind Med 40:490–501.

Chapter 5
Proximity Analysis for Exposure Assessment in Environmental Health Justice Research

Jayajit Chakraborty and Juliana A. Maantay

Abstract This chapter provides a historical overview and constructive critique of analytical approaches and methods that have been used to measure proximity to environmental health hazards and potential exposure to their adverse effects in the environmental justice (EJ) research literature. After providing an introduction to environmental health justice research and key findings, we examine how quantitative EJ analysis has emerged from comparing the prevalence of minority or low-income populations in spatial units hosting environmental hazards and circular buffer zones to more advanced techniques that utilize GIS, pollution plume models, and estimates of health risk from ambient exposure to multiple pollutants and emission sources. We also review spatial analytical approaches used in previous studies to determine the demographic and socioeconomic characteristics of people residing in areas potentially exposed to environmental hazards, as well as newly emerging geostatistical techniques that are more appropriate for spatial analysis of EJ than conventional statistical methods used in prior research. The concluding section focuses on highlighting the key limitations and identifying future research needs associated with assessment of environmental health justice.

Keywords Environmental health justice · Proximity analysis · Buffer analysis · Areal interpolation · Dispersion model · Spatial regression · Geographically weighted regression

5.1 Environmental Health Justice

Environmental Justice, both as a term in our vocabularies and as a movement, came into being more than 20 years ago. Narrowly interpreted, Environmental Justice (EJ) is the attempt to document and address the disproportionate environmental and health burdens borne by the poor and people of color. In a broader context, EJ theory encompasses everything that is unsustainable about the world we have

J. Chakraborty (✉)
Department of Geography, University of South Florida, Tampa, FL 33620, USA
e-mail: jchakrab@usf.edu

J.A. Maantay, S. McLafferty (eds.), *Geospatial Analysis of Environmental Health*, Geotechnologies and the Environment 4, DOI 10.1007/978-94-007-0329-2_5,

created, including rampant population growth, industrialization, pollution, consumption patterns, energy use, food production, and resource depletion. "The EJ movement has sought to redefine environmentalism as much more integrated with the social needs of human populations, and, in contrast with the more eco-centric environmental movement, its fundamental goals include challenging the capitalist growth economy, as well," (Pellow and Brulle, 2005, 3).

A definition of environmental justice is "the provision of adequate protection from environmental toxicants for all people, regardless of age, ethnicity, gender, health status, social class, or race" (Nordenstam, 1995:52), and the proximity of noxious land uses to populated areas is believed to jeopardize environmental health and justice. Although many researchers have focused on the disproportionate environmental burdens borne by the poor and communities of color, others have expanded the definition of environmental justice to include additional vulnerable populations, such as the very young, the elderly, the infirm and immune-compromised, pregnant women, immigrants, and future generations (Greenberg, 1993). The term "Environmental Health Justice" was born of the need to clarify and emphasize why Environmental Justice is important: proximity to environmental burdens is detrimental to human health, and results in health disparities and inequities for some populations.

5.1.1 The Role of GIS in Environmental Health Justice Research

Since the late 1980s and beginning in earnest in the early 1990s, Geographic Information Systems have been used to examine the spatial realities of environmental injustice (Boer et al., 1997; Bowen et al., 1995; Burke, 1993; Chakraborty and Armstrong, 1997; Chakraborty et al., 1999; Maantay et al., 1997; Maantay, 2002; Morello-Frosch et al., 2001; Neumann et al., 1998; Perlin et al., 1995; Pollock and Vittas, 1995; Sheppard et al., 1999).

GIS methods have been used in environmental justice research primarily to analyze the spatial relationships between sources of pollution burdens and the socio-demographic characteristics of potentially affected populations. GIS technology is particularly well-suited for EJ research because it allows for the integration of multiple data sources (e.g., location of polluting facilities and population characteristics), representation of geographic data in map form, and the application of various spatial analytic techniques (e.g., buffering) for proximity analysis (Zandbergen and Chakraborty, 2006).

With GIS, it has become increasingly prevalent to try to map instances of environmental injustice, usually by geographically plotting facilities or land uses suspected of posing an environmental and human health hazard or risk, and then determining the racial, ethnic, and socioeconomic characteristics of the potentially affected populations compared with a reference population. This often results in dramatic maps showing toxic facilities concentrated in areas with high proportions of African-Americans, Latinos, or Native Americans (Burke, 1993; Clarke and Gerlak, 1998; Glickman and Hersh, 1995; Maantay et al., 1997; United Church of Christ,

1987, 2007). Mapping became a favored method among researchers attempting to determine the existence of environmental injustice. Additionally, the wealth of environmental and socio-demographic data now available on the Internet, as well as the proliferation of websites with interactive mapping applications available, have brought environmental justice mapping within reach of virtually anyone.

Although such maps can be unusually effective in visually demonstrating the disproportionate spatial distribution of noxious or hazardous facilities, these maps have also been criticized for being misleading and inaccurate, and their findings have often been contradicted by other spatial analyses. Mapping a phenomenon such as environmental injustice is not a straightforward exercise, and the difficulties encountered in producing such spatial analyses leave the maps open to a variety of interpretations and second-guessing. Just as no map can be viewed as an objective embodiment of the real world, maps depicting environmental injustice are also social constructions, and therefore subjective and based on assumptions (Dorling and Fairbairn, 1997; Wood, 1992).

The efficacy of GIS to pinpoint environmental injustices and, especially, the health impacts of pollution, has been questioned, and many researchers who use GIS have commented upon the challenges and limitations inherent in this method of spatial analysis (Clarke et al., 1996; Dunn et al., 2001; FitzGerald et al., 2004; Jacquez, 2000; Jarup, 2004; Kulldorff, 1999; Maantay, 2002; McMaster et al., 1997; Moore and Carpenter, 1999; Richards et al., 1999; Rushton et al., 2000; Sheppard et al., 1999; Vine et al., 1997; Wall and Devine, 2000; Yasnoff and Sondik, 1999). These publications document many of the spatial and attribute data deficiencies and methodological problems of using GIS to make the connections amongst proximity to pollution, exposure, health outcomes, and the location of vulnerable populations. For instance, geographic considerations in research design that can significantly influence the analysis and results but can be difficult to optimize include the delineation of the study area extent, determining the level of resolution and the unit of spatial data aggregation, and estimating the areal extent of exposure, as well as the various problems encountered in trying to statistically analyze and summarize spatial data. Due to the principle of spatial autocorrelation, which states that data from locations near one another in space are more likely to be similar than data from locations remote from one another, spatial data is by its very nature not randomly distributed, as traditional statistical approaches require (Tobler, 1979). Spatial autocorrelation, which is an inherent characteristic of geographically referenced data, thus becomes an impediment to the application of conventional statistical tests. These limitations are discussed in more detail in Chapter 17.

However, it is feasible to develop methods and tools for producing more meaningful spatial analyses, and recently health geographers and other researchers have been using GIS techniques effectively to show the correspondence amongst factors such as proximity to hazardous facilities and land uses, adverse health outcomes, disproportionate exposure and risk, and health disparities. This chapter reviews and assesses the methods most commonly used by environmental health justice researchers, as well as some of the more recent cutting-edge methods.

5.1.2 Environmental Justice Research Findings

The independent variables for most EJ studies are the population characteristics of the proximate and the reference populations, which typically include both socioeconomic status (SES) and race/ethnicity, or, occasionally, one or the other. Variables serving as proxies for SES are median household income, per capita income, percent owner-occupied homes, percent with public assistance income, percent below poverty level, and percent without a high school diploma, or some combination of several. The US. Census is usually the source for this socio-demographic information, although it is recognized as being incomplete and inconsistent from census to census, and even within any given census. Racial categories especially are fraught with ambiguity, and have changed dramatically over the decades, making longitudinal studies difficult and inaccurate. Residential segregation is a less-commonly explored factor, but one which poses environmental justice implications that could likely shed light on the connections between race and SES, as well as exposure to hazardous sites (Morello-Frosch and Jesdale, 2006).

Air pollution is the most common environmental burden examined in EJ studies, which typically measure residential proximity to sites that are pollution sources, including high-volume roads, power plants, and industrial facilities emitting air pollution. The dependent variable has been measured in various ways (as outlined more fully in a following section): presence of hazards, number or density of hazards, distance to hazards, or a measure of its magnitude, in terms of quantity of pollutants, toxicity, or health risk.

Aside from air pollution, other EJ studies focus on environmental burdens such as proximity to industrial zones and the characteristics of populations living near intensification of major industrial zones (Maantay, 2001); access (and lack thereof) to parks and active recreational spaces (Maroko et al., 2009; Talen, 1997; Talen and Anselin, 1998; Wolch et al., 2005); proximity to flood-prone areas (Maantay and Maroko, 2009); Superfund sites (Baden et al., 2007; Cutter et al., 1996); hazardous waste transfer, storage, and disposal facilities (TSDFs) (Anderton et al., 1994; Boer et al., 1997; Bolin et al., 2002; Cutter et al., 1996; Fricker and Hengartner, 2001; Goldman and Fitton, 1994); solid waste landfills (Been and Gupta, 1996; Higgs and Langford, 2009; Mohai and Saha, 2007; and the US. GAO, 1995); and noise pollution from airports (Most et al., 2004). The preponderance of these studies also found a positive spatial correspondence between minority/socio-economic status and proximity to hazards.

In many of the EJ studies, race has a consistent spatial correspondence with the location of hazardous facilities and land uses, and a concomitant potential for disproportionate environmental exposures and disease burdens (Apelberg, 2005; Chakraborty, 2009; Grineski, 2007; Linder et al., 2008; McMaster et al., 1997; Mirabelli et al., 2006; Morello-Frosch et al., 2001; Morello-Frosch and Jesdale, 2006; Pastor et al., 2005; Perlin et al., 1995; Pollock and Vittas, 1995). A previous literature review by Maantay (2002) on studies conducted throughout the 1990s overwhelmingly found disproportionate burdens based on race, and most of the studies also found disproportionate burdens based on income. However, race tended to be predictive of disproportion even when controlling for SES.

The rest of this chapter explores how proximity to environmental hazards and potential exposure to their adverse health effects have been analyzed in previous EJ studies. More specifically, we examine how the assessment of differential proximity to environmental health hazards in quantitative EJ research has evolved from comparing the prevalence of minority or low-income populations in pre-defined units hosting pollution sources and discrete buffer zones to more refined methods that utilize precise distances between hazards and people, air dispersion models, and modeled estimates of health risk from ambient toxic exposure. Methods used to estimate the number and socio-demographic characteristics of people residing in areas potentially exposed to hazards are also discussed, as well as emerging geostatistical techniques that address specific limitations of conventional statistical methods.

5.2 Estimating the Boundaries of Adverse Environmental Exposure

A variety of spatial analytical approaches have been used in EJ research to measure proximity to environmental hazard sources and estimate the boundaries of areas potentially exposed to their adverse effects. In spite of specific differences, the methodologies that have been employed can be classified into three broad categories, as described in this section: spatial coincidence analysis, distance-based methods, and pollutant fate and transport modeling.

5.2.1 Spatial Coincidence Analysis

Spatial coincidence, in context of EJ research, refers to a technique that assumes potential exposure to environmental health hazards is restricted to the boundaries of pre-defined geographic entities or administrative units (e.g., ZIP codes, census tracts, or block groups) that contain such hazards. Several different methods have been used to quantify potential exposure to hazards within census unit, as shown in Table 5.1. The most widely used and traditional approach, referred to as *unit-hazard coincidence* (Mohai and Saha, 2006), utilizes the location of a hazard within each spatial unit as a surrogate for environmental exposure. The socio-demographic characteristics of units containing a hazard (host units) are statistically compared to all others (non-host units) in the study area to evaluate disproportionate proximity or exposure. Several influential and widely-cited EJ studies conducted at the national level have used the presence or absence of hazardous facilities within ZIP codes (e.g., United Church of Christ, 1987; Goldman and Fitton, 1994) or census tracts (e.g., Anderton et al., 1994; Been, 1995) to determine disproportionate risk burdens. National and state level studies have even used the county as a spatial unit for coincidence analysis (e.g., Hird, 1993; Daniels and Friedman, 1999).

Table 5.1 Methodology for spatial definition of proximity and potential exposure to environmental hazards

Approach	Risk indicator	Examples: author and year of study
Spatial coincidence analysis	Presence of hazard (unit-hazard coincidence)	United Church of Christ 1987; Burke, 1993; Hird, 1993; Anderton et al., 1994; Goldman and Fitton, 1994; Been, 1995; Been and Gupta, 1996; Cutter et al., 1996; Boer et al., 1997; Daniels and Friedman, 1999; Fricker and Hengartner, 2001; Boone, 2002; Taquino et al., 2002; Walker et al., 2005; Baden et al., 2007
	Total number or density of hazards	Burke, 1993; Cutter and Solecki, 1996; Ringquist, 1997; Tiefenbacher and Hagelman, 1999; Fricker and Hengartner, 2001; Mennis and Jordan, 2005
	Total quantity of emitted pollutants	Bowen et al., 1995; Kriesel et al., 1996; Boer et al., 1997; Tiefenbacher and Hagelman, 1999; Daniels and Friedman, 1999; Bolin et al., 2000
	Toxicity-weighted quantity of pollutants	Bowen et al., 1995; Perlin et al., 1995; McMaster et al., 1997; Brooks and Sethi, 1997; Bolin et al., 2000
Distance-based analysis	Discrete distance from hazards (fixed buffer)	Glickman, 1994; Zimmerman, 1994; US GAO 1995; Glickman and Hersh, 1995; Chakraborty and Armstrong, 1997; Neumann et al., 1998; Perlin et al., 1999, 2001; Sheppard et al., 1999; Bolin et al., 2000, 2002; Altas, 2002; Baden and Coursey, 2002; Boone, 2002; Pastor et al., 2004; Mohai and Saha, 2006, 2007; Walker et al., 2005; United Church of Christ, 2007; Kearney and Kiros, 2009; Mohai et al., 2009
	Continuous distance from hazards	Pollock and Vittas, 1995; Gragg et al., 1996; Stretesky and Lynch, 1999; Cutter et al. 2001; Margai, 2001; Mennis, 2002; Waller et al., 1999; Zandbergen and Chakraborty, 2006; Downey, 2006; Chakraborty and Zandbergen, 2007; Fitos and Chakraborty, 2010
Pollutant fate and transport modeling	Geographic plume analysis	Glickman, 1994; Glickman and Hersh, 1995; Chakraborty and Armstrong, 1997, 2001; Chakraborty et al., 1999; Chakraborty, 2001; Margai, 2001; Dolinoy and Miranda, 2004; Most et al., 2004; Fisher et al., 2006; Bevc et al., 2007; Maantay, 2007
	Plume-based health risk estimate	Morello-Frosch et al., 2001; Bouwes et al., 2001; Ash and Fetter, 2004; Apelberg et al., 2005; Pastor et al., 2005; Morello-Frosch and Jesdale, 2006; Sicotte and Swanson, 2007; Gilbert and Chakraborty, 2008; Linder et al., 2008; Chakraborty, 2009; Williams, 2010

The choice of spatial unit to represent the host area has been subject of considerable debate in the EJ literature (McMaster et al., 1997; Mennis, 2002) and researchers have examined how EJ results from the unit-hazard coincidence method vary across multiple spatial scales (Glickman and Hersh, 1995; Cutter et al., 1996; Taquino et al., 2002; Baden et al., 2007). Although these studies are not comparable because of specific dissimilarities in study area and hazard examined, their findings suggest that different units of analysis potentially lead to different conclusions regarding explanatory factors such as race or income. Data aggregated at higher levels such as a county or metropolitan area (coarse spatial resolution), however, have been documented to be less reliable as indicators of disproportionate burdens than data aggregated to smaller units (finer spatial resolution) such as census block groups (McMaster et al., 1997; Sheppard et al., 1999; Maantay, 2002). It is generally acknowledged that using the smallest practicable unit of analysis yields the most accurate results (Maantay, 2007), while the use of larger areal units often increases the strength and significance of statistical relationships between environmental risk indicators and socio-demographic variables (Cutter et al., 1996; Taquino et al., 2002).

Regardless of the areal unit selected, there are several limitations associated with the unit-hazard coincidence method. First, most applications do not usually distinguish between spatial units that host one environmental hazard and those in which multiple hazards are located. Second, this approach does not account for boundary or edge effects. These effects deal with the possibility that a hazardous facility could be so close to the boundary of its host spatial unit that a neighboring non-host unit could be equally exposed to pollution. Unless the hazard is located near the center of the spatial unit, the representativeness of the socio-demographic data used to analyze EJ becomes questionable. Third, the unit-hazard coincidence method assumes that the adverse impacts of environmental hazards are confined only to the boundaries of their host units. However, pre-defined geographic entities such as census units or ZIP code areas are unlikely to represent the shape or size of the area potentially exposed to the entire range of health hazards associated with a polluting facility.

Figure 5.1 depicts the location of industrial toxic facilities across census tracts in a hypothetical county and illustrates the limitations of the unit-hazard coincidence

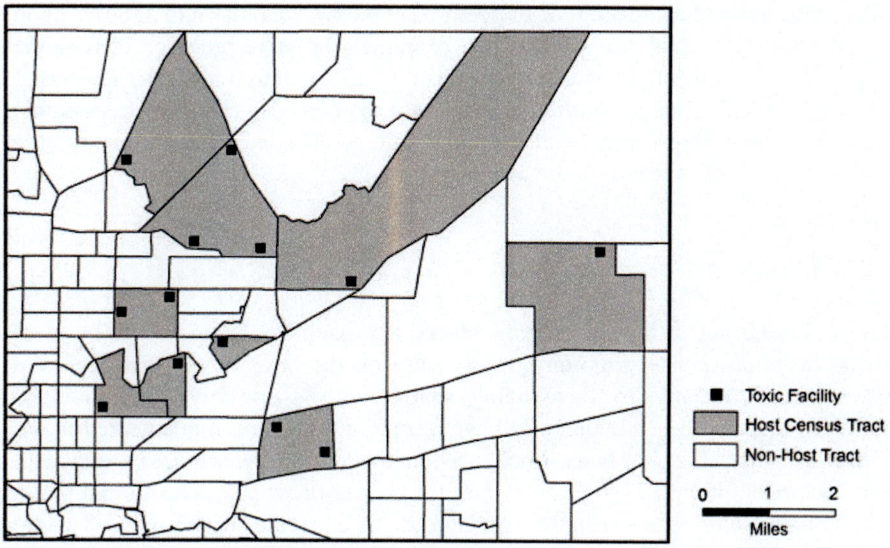

Fig. 5.1 Spatial coincidence analysis: selection of host census units

method. Most of these facilities are located near the boundaries of multiple tracts and closer to adjacent non-host tracts than to the far end of their own host tract. Given the spatial distribution of these toxic facilities, it appears unlikely that their adverse effects are confined solely to their host units. An additional limitation is that all host tracts are treated equally, although the number of facilities within each host tract ranges from one to three.

The inability to distinguish between host units based on the number or magnitude of hazards can be addressed by summing the number of facilities or the amount of pollutants released within each unit. Several EJ studies have extended the basic spatial coincidence approach by estimating the frequency of toxic facilities within census tracts (Burke, 1993; Fricker and Hengartner, 2001) and ZIP codes (Ringquist, 1997), as well as the number of airborne toxic releases within counties (Cutter and Solecki, 1996; Tiefenbacher and Hagelman, 1999). Since certain EPA databases such as the Toxic Release Inventory (TRI) provide detailed data on annual quantities of toxic chemicals released at each facility, a more refined assessment of the magnitude of pollution released within each host unit is possible. While several EJ studies have relied on the pounds of emitted pollutants from TRI facilities, others have used chemical toxicity indicators to weight annual release for each spatial unit, as indicated in Table 5.1. Since the TRI database does not include toxicity data for released chemicals, researchers have utilized surrogate measures such as threshold limit values (TLVs) to weight the pounds of emissions for each pollutant. Although TLVs are available for many TRI chemicals, it remains a problematic index for health risk and equity assessment because it was developed and intended to only assess occupational safety among a healthy worker population (Maantay, 2002).

The incorporation of data on the quality and quantity of pollution emitted from each hazard source have allowed researchers to distinguish between host spatial units on the basis of the magnitude of potential environmental risk and improve upon the unit-hazard coincidence method that examines the mere presence of hazards. Applications of spatial coincidence analysis that utilize emissions or toxicity data, however, are still limited by their inability to: (a) consider the exact geographic location of the hazard within the host spatial unit, and (b) determine the geographic extent of potential exposure to the hazard.

5.2.2 Distance-Based Analysis

Several limitations of the spatial coincidence approach can be addressed by measuring disproportionate proximity or exposure on the basis of the distance from environmental hazards to surrounding spatial units. A variety of simple and advanced distance-based techniques have been suggested and implemented in the EJ research literature. The most widely used method is *buffer analysis*, a spatial analytic technique provided by GIS software for creating new polygons around point, line, or area features on a map. A large number of EJ studies have used GIS-based circular buffers around point sources of hazards to identify areas and populations

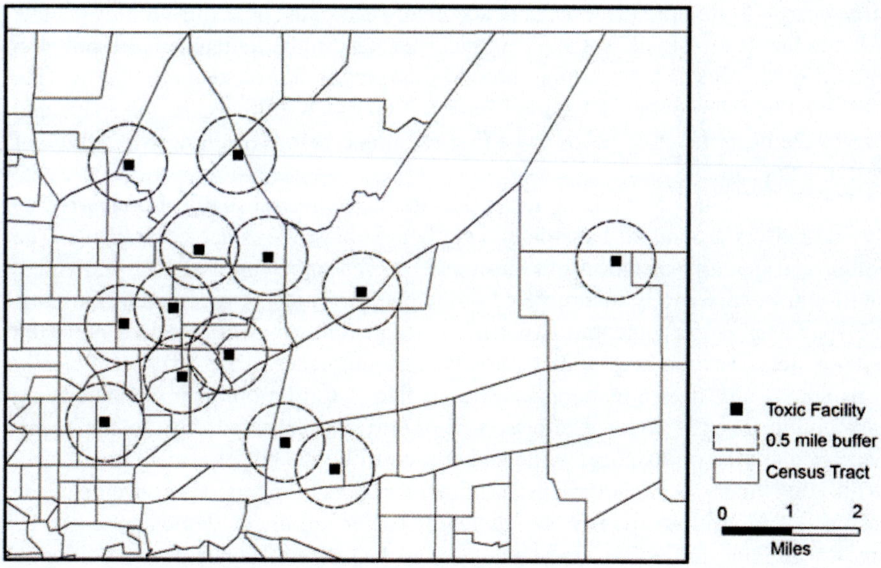

Fig. 5.2 Circular buffers of 0.5 mile radius around toxic facilities

exposed to their adverse effects (Table 5.1). The socio-demographic characteristics of areas lying inside buffer zones are statistically compared to the rest of the study area (outside the buffers) to determine disproportionate proximity or exposure to the hazards of concern. Figure 5.2 provides a typical example of buffer analysis, based on circles of radii one-half mile centered at each toxic facility. The radius of circular buffers in EJ studies have ranged from 100 yards (Sheppard et al., 1999) to 3 miles (Mohai and Saha, 2006). Distances of 0.5 and 1.0 mile from facilities of concern have been used most frequently (Glickman, 1994; Zimmerman, 1994; US GAO, 1995; Chakraborty and Armstrong, 1997; Neumann et al., 1998; Bolin et al., 2000; Baden and Coursey, 2002; Boone, 2002; Maantay, 2007; Mohai and Saha, 2007; Mohai et al., 2009; Kearney and Kiros, 2009). Instead of using a single radius or buffer, several studies have constructed multiple circular rings at increasing distances from hazard sources (e.g., Neumann et al., 1998; Perlin et al., 1999, 2001; Sheppard et al., 1999; Atlas, 2002; Pastor et al., 2004; Walker et al., 2005). To differentiate between buffers, some EJ studies have even estimated release volumes or toxicity-weighted emissions within a fixed radius of each TRI facility in the study area (e.g., Neumann et al., 1998; Bolin et al., 2002).

Buffer analysis provides a more accurate geographic representation of environmental exposure than the spatial coincidence because it does not assume that the adverse effects are confined to the boundaries of host spatial units. There are, however, several limitations associated with its application in EJ analysis. First, the facility or hazard representing the center of the circle is assumed to be small enough to be treated as a point. For undesirable land uses such as Superfund sites that are large in size, a circular buffer may not accurately depict the area surrounding the site if the radius is too small. Some hazardous sites need to be delineated as a polygon instead of a point and the buffer should be constructed around the polygon (Liu, 2001). Although previous EJ studies have not considered this issue, the shape and size of the hazard source needs to be first examined before deciding which type of buffer is appropriate.

A second limitation is that radius of the circular buffer is usually chosen arbitrarily and buffers around all hazards in a study area have the exact same radius. The nature and quantity of hazardous substances stored or released at each individual facility have been rarely incorporated in the determination of buffer radii to reflect the spatial extent of potential exposure. The operational parameters of emission releases (e.g., release height, exit velocity, exit temperature) are also not typically considered in the determination of the buffer size. A third problem is the underlying assumption that the adverse effects of a hazard are restricted only to the specified circular area or distance, while and areas outside the buffer remain unaffected. While this binary or dichotomous assumption makes comparisons convenient, the results are highly sensitive to the choice of buffer radius, as demonstrated in EJ studies utilizing more than one circle around facilities of concern. In addition, a discrete measurement (e.g., within 1 mile of a facility) is unlikely to reflect a more continuous or gradual reduction in environmental exposure with distance from the hazard. Using multiple circular buffers can overcome this limitation to a certain degree, but the determination of the number of buffers to use and choice of radii remain ambiguous and do not necessarily result in a more accurate representation of potential exposure (Zandbergen and Chakraborty, 2006).

Continuous distances, based on the calculation of the exact distance between locations of the potentially exposed population and environmental hazards, represent an alternative to the use of discrete buffer analysis. Several EJ studies have utilized the distance from the centroid of each census tract or block group to their nearest hazard source as an indicator of proximity (Pollock and Vittas, 1995; Gragg et al., 1996; Stretesky and Lynch, 1999; Margai, 2001; Mennis, 2002). The analysis of continuous distances can be enhanced by utilizing a cumulative distribution function (CDF). A CDF is typically depicted as a graph that provides the number or percentage of observations falling below every threshold value. Applied to any set of hazard sources, a CDF can be plotted to indicate how the size of a potentially exposed population (as a percentage of the total in the study area) increases with proximity.

Figure 5.3 depicts a pair of CDF curves that compare the location patterns of two racial subgroups with respect to a set of hazardous facilities in a hypothetical county. The CDFs representing the cumulative proportions of the non-White and White population in the county are depicted as a function of distance to the nearest facility, estimated using block group data. The collective percentage of people in both subgroups increases (0–100) as distance from the facilities increase. If the CDF curve for non-Whites is higher than the curve for Whites at any specific distance, the percentage of the county's non-White population residing within that distance of their nearest facility exceeds the percentage of the White population. The

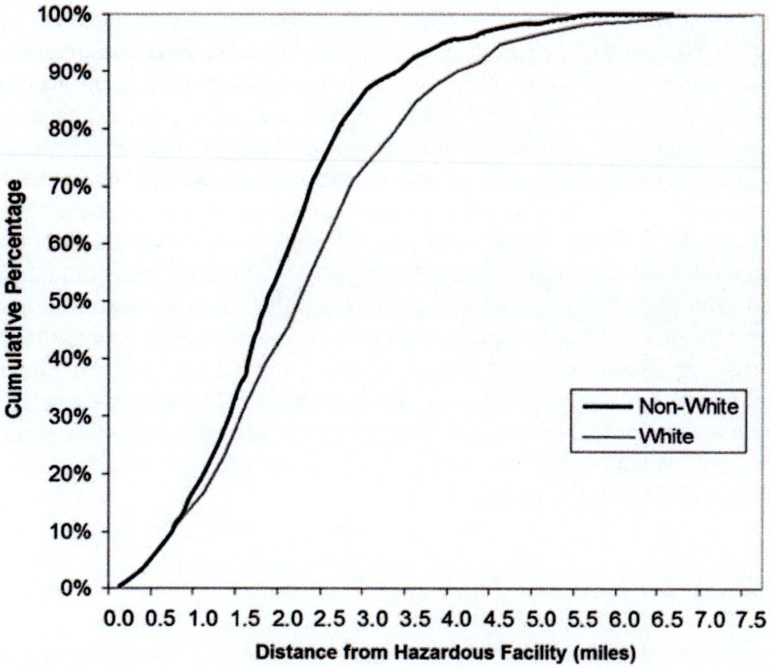

Fig. 5.3 Cumulative distribution functions for hazard proximity: comparing racial characteristics of the population

limitations of conventional buffer analysis based on arbitrary and discrete distance values can also be assessed from Fig. 5.3. A circular buffer of radius smaller than 0.75 mile would indicate almost similar percentages of non-White and White residents, or no evidence for disproportionate proximity. A buffer analysis based on a radius of 1 or 2 miles, in contrast, would yield a significantly higher non-White proportion and a different result. Since discrete buffer distances are typically chosen without knowledge of the actual empirical CDF, our example indicates that this approach could lead to an inaccurate characterization of environmental exposure and biased results. Several EJ studies have demonstrated that the CDFs are particularly well-suited for assessing disproportionate proximity because they overcome the limitations of choosing arbitrary and discrete buffer distances (Waller et al., 1999; Zandbergen and Chakraborty, 2006; Chakraborty and Zandbergen, 2007; Fitos and Chakraborty, 2010).

Instead of assuming that the adverse effects of a hazard decline with distance in a linear fashion, a few studies have utilized curvilinear distance decay functions to model residential proximity. Pollock and Vittas (1995) hypothesized three functional forms of exposure (linear, square root, and natural logarithm) with respect to distance from TRI facilities in Florida, and selected the natural logarithm of the distance to the nearest facility as a proxy for exposure. A GIS-based distance decay modeling technique was developed by Downey (2006) and applied to examine proximity to TRI facilities in Detroit. While this technique was found to be flexible enough to incorporate any appropriate distance decay function, several different curvilinear and reverse curvilinear functions were used to estimate neighborhood proximity to TRI activity. An inherent limitation of this distance decay approach is that researchers are unaware of the actual and precise rate at which the adverse impacts of a hazard decline with increasing distance. The mathematical functions used to calculate distance decay are typically based on assumptions about the distance decay process rather than precise knowledge (Downey, 2006).

Although distance-based approaches for EJ analysis have evolved from the use of discrete circular buffers to continuous functions, there are still limited by the fact that proximity many not always provide a valid proxy for exposure to pollution. Additionally, distance-based methods fails to consider directional biases in the distribution of environmental hazards by assuming that their adverse effects are identical and uniform in all directions. Although physical processes do not always operate in a symmetrical or isotropic manner, distance-based analyses assume that toxic exposure is not dependent on wind direction and the factors that influence the movement and dispersal of pollutants.

5.2.3 Pollutant Fate and Transport Modeling

To provide a more accurate spatial representation of the area potentially exposed to the adverse effects of a hazard, several EJ studies have used detailed information on toxic chemical emissions and local weather conditions to model the environmental fate and dispersal of pollutants released from the hazard source. *Geographic plume analysis* is a methodology that integrates air dispersion modeling with GIS to estimate areas and populations exposed to airborne releases of toxic substances (Chakraborty and Armstrong, 1996, 1997). Dispersion models typically combine data on the volume and physical properties of a released chemical with site-specific information and atmospheric conditions to estimate pollutant concentrations downwind from the emission source. This information is used to identify the spatial extent or boundary of the area potentially exposed to the chemical's spreading plume, or the plume footprint. The footprint represents the area where ground-level concentrations of the pollutant are predicted to exceed a user-specified limit (Fig. 5.4).

EJ studies using geographic plume analysis have often relied on ALOHA (Areal Locations of Hazardous Atmospheres), an air dispersion model developed by the National Oceanic and Atmospheric Administration (NOAA) and the EPA to support emergency responses to hazardous chemical accidents. The ALOHA model has been applied to generate, at each facility of concern, a single plume footprint (Chakraborty and Armstrong, 1996), a composite footprint based on historical weather patterns (Chakraborty and Armstrong, 1997, 2004), or plume-based circular buffers whose radii are based on worst-case chemical release scenarios (Chakraborty, 2001; Margai, 2001; Chakraborty and Armstrong, 2001). EJ studies have also utilized the Industrial Source Complex Short Term (ISCST) air dispersion model (Dolinoy and Miranda, 2004; Fisher et al., 2006; Maantay et al., 2009),

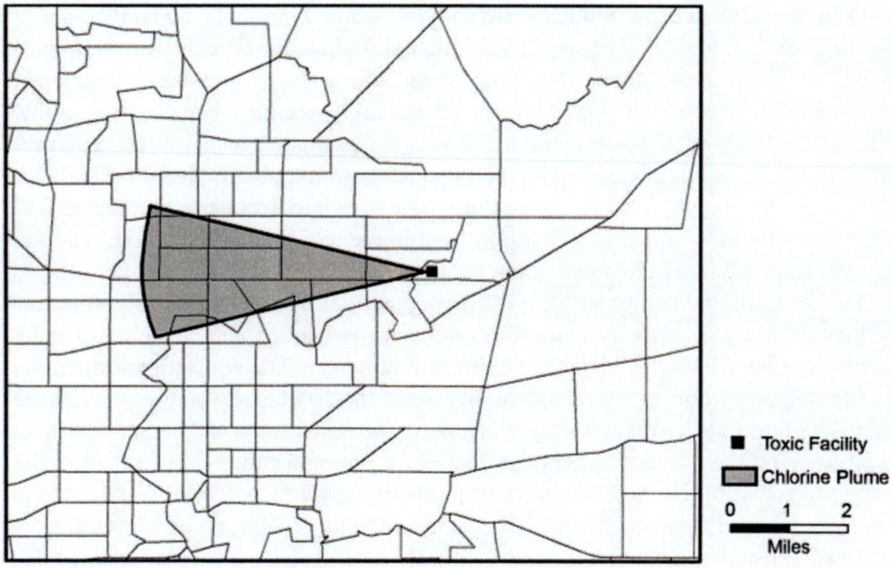

Fig. 5.4 A typical plume footprint for a hypothetical chlorine release scenario using the ALOHA model

ash deposition models (Bevc et al., 2007), and various noise pollution models (Chakraborty et al., 1999; Most et al., 2004).

The application of pollutant fate and transport modeling allows the concentration of toxic pollutants released from a hazard source and their estimated health risks to: (a) decline continuously with increasing distance from the release source; and (b) vary according to compass direction. Plume modeling thus addresses the problems of previous analytical approaches which assume that residing in either a census unit containing a hazard (spatial coincidence analysis) or within a specific distance from a hazard (distance-based analysis) leads to environmental exposure or adverse health risk. There are, however, certain limitations associated with this approach. First, dispersion models typically require large volumes of site-specific and facility-specific information, such as the facility's stack height and diameter, gas exit velocity and exit temperature, detailed emissions data on each chemical released (e.g., average hourly quantities and rates), and meteorological information (e.g., average monthly or hourly wind speed and direction, often for an entire year). The input data necessary for plume modeling is rarely available for all hazard sources in a study area. Second, some dispersion models such as ALOHA assume that topography is always flat and are unable to provide accurate concentration estimates when the atmosphere is stable or wind speeds are low. Third, the task of creating a plume modeling data set that includes all toxic facilities and chemical emissions in a large study area can be difficult, time-consuming, and expensive. As a consequence, a limited number of national or regional scale plume model data sets have been constructed and those that exist focus only on specific types of hazards.

Data sets derived from pollutant fate and transport modeling that cover the entire US includc the Risk-Screening Environmental Indicators (RSEI) and National-Scale Air Toxic Assessment (NATA). These national scale databases developed by the EPA are particularly suitable for EJ research, not only because they allow researchers to estimate the potential health risks associated with specific environmental hazards and analytical units, but also because the plume modeling and risk assessment techniques used to derive these data take into account a number of factors such as wind speed, wind direction, air turbulence, smokestack height and the rate of chemical decay and deposition.

The RSEI model can be used to estimate potential human health risks from air pollutants based on toxicity and atmospheric dispersion of chemicals emitted by facilities in the EPA's TRI database. For each individual TRI site and pollutant, the RSEI integrates information on the facility location, the quantity and toxicity of the chemical, fate and transport through the environment, the route and extent of human exposure, and the number of people affected for up to 44 miles (101 km) from the source of release. The ambient concentrations of each TRI pollutant is determined for each square kilometer of the 101-km by 101-km grid in which the facility is centered. EJ studies have merged risk scores from the RSEI grids with census socio-demographic data to analyze disproportionate exposure to TRI pollutants in the entire US (Bouwes et al., 2001; Ash and Fetter, 2004) and in Philadelphia (Sicotte and Swanson, 2007) and Tampa Bay (Williams, 2010). Since the toxic pollution plumes used to obtain the risk estimates can extend in any direction for up to 44 miles from a TRI facility, the RSEI modeling technique had the advantage of allowing hazards and emissions in one spatial unit to affect people living in other units.

The NATA was designed to guide air pollution reduction and related prioritization efforts, has also emerged as a valuable data set for estimating exposure concentrations and public health risks associated with inhalation of air toxics from different types of emission sources. While criteria air pollutants include common contaminants such as particulate matter or ozone, air toxics (also known as hazardous air pollutants) include 188 specific substances identified in the Clean Air Act Amendments of 1990 that are known to or suspected of causing cancer and other serious health problems, including respiratory, neurological, immune, or reproductive effects (US EPA, 2008). Tract level estimates of lifetime cancer risk from the 1996 NATA have been utilized for EJ analysis in Maryland (Apelberg et al., 2005), California (Pastor et al., 2005) and 309 metropolitan areas of the US (Morello-Frosch and Jesdale, 2006). Recent studies have used the 1999 NATA to examine the disproportionate distribution of cancer and respiratory risks in Florida (Gilbert and Chakraborty, 2008, 2011), and the metropolitan areas of Houston (Linder et al., 2008) and Tampa Bay (Chakraborty, 2009). An important advantage of the NATA is spatial compatibility with census socio-demographic data–the modeled risk estimates are available at the level of spatial units (e.g., tracts) that provide residential population data. Additionally, it provides health risk estimates for ambient exposure to multiple categories of emission sources. The NATA thus allows EJ analysis to extend beyond major stationary sources such as TRI facilities and include smaller emitters, as well as various off-road and on-road mobile sources.

Although plume modeling techniques and data sets represent a significant improvement over the spatial coincidence and distance-based approaches, these are often based on necessary and problematic assumptions and may not be as accurate as many researchers think. More importantly, their use is limited to studies that focus on specific types of public health risks and hazards. It is important to consider that health risks associated with exposure to pollution may not be the only set of risks imposed by environmental hazards. The presence of a hazard can also adversely affect nearby property values, perceptions of local health risks, psychological stress, local employment opportunities, sense of community and local economic activity (Downey, 2006). These potential negative consequences cannot be analyzed on the basis of plume modeling methods and data.

5.3 Estimating Population Characteristics of Proximate Areas

After delineating the boundaries of areas where people living could potentially be exposed to the adverse effects of environmental hazards, EJ studies have employed a variety of techniques to estimate the number and socio-demographic characteristics of people residing in such areas. These techniques can be classified into two basic categories, depending on the level of spatial aggregation of the residential population data: *point interpolation* and *areal interpolation*. When the addresses of all individuals or households relevant to the study are available and can be located on a map, *point interpolation* is the appropriate method. Street address information is typically used in conjunction with a detailed street network, and the geocoding tools of GIS software are utilized to determine an accurate location of each individual or household in the study area. The number and the socio-demographic characteristics of individuals or households potentially exposed to a hazard can be estimated based on points located inside a distance-based or plume-based buffer (point-in-polygon overlay). The earliest application of this method can be found in a study conducted by Mohai and Bryant (1992) on hazardous waste facilities in Detroit. Data from a probability sample survey were utilized to determine if the racial and economic status of respondents living within circular buffers of radii 1 and 1.5 miles were different from those residing outside the buffers. Subsequent EJ studies have relied on point interpolation to determine the number of self-identified individuals with special needs in Cedar Rapids, Iowa (Chakraborty and Armstrong, 2001), characteristics of survey respondents in Fort Lauderdale, Florida (Bevc et al., 2007), and the racial/ethnic status of school children in Orange County, Florida (Chakraborty and Zandbergen, 2007). A recent study (Mohai et al., 2009) utilized addresses of survey respondents in the American Changing Lives Study to examine the socio-demographic characteristics of residents living within one mile of TRI facilities.

Although point interpolation can be easily implemented, it requires data on the street addresses of all individuals relevant to the analysis. Data on demographic and socioeconomic characteristics of individuals or households not publicly available and can only be obtained through an extensive social survey that encompasses the entire study area. Consequently, EJ studies have relied mainly on socio-demographic information collected by the US. Census and other agencies that are aggregated at the level of pre-defined administrative boundaries or census units. If the area potentially exposed to a hazard is represented by a distance-based or plume-based buffer, the shape and size of the buffer area is unlikely to match the underlying census units that contain aggregated population data (see Fig. 5.2 or Fig. 5.4). A method of *areal interpolation* (polygon-on-polygon overlay) is necessary to transfer data from census units to the boundaries of areas potentially exposed to the adverse effects of a hazard. Several different areal interpolation techniques have been utilized in previous EJ studies. These are illustrated in Fig. 5.5, using a circular buffer around a single facility of concern.

The simplest method is *polygon containment*, where all spatial units or census polygons that either intersected or entirely enclosed by a distance-based or plume-based buffer are selected (Chakraborty and Armstrong, 1997). Also referred to as *adjacency analysis* (Most et al., 2004) or the *boundary intersection method* (Mohai and Saha, 2006), population characteristics of any given buffer zone are derived through a simple aggregation of all census units that are within or in contact with

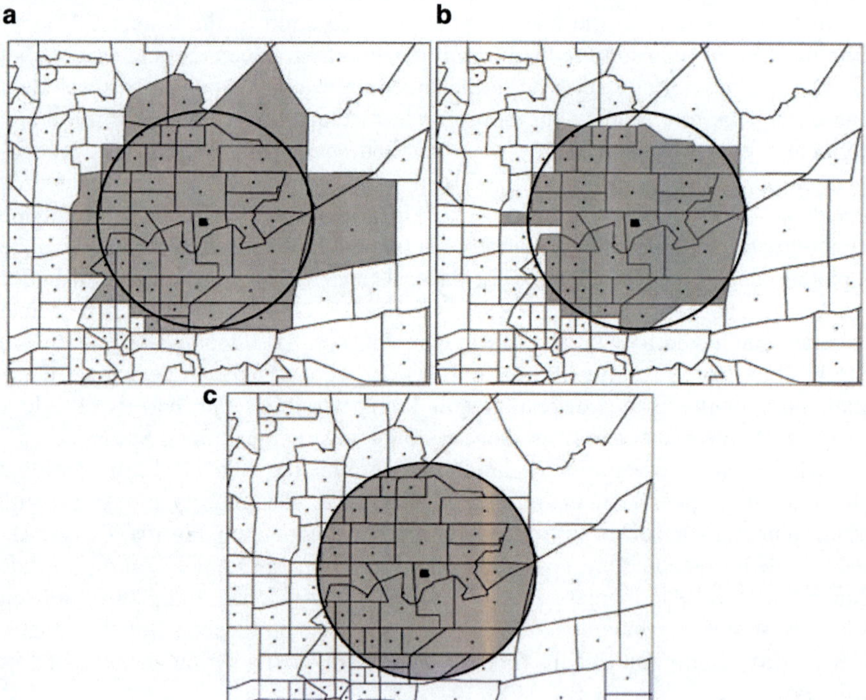

Fig. 5.5 Using areal interpolation to select census polygons within circular buffer. **a** Polygon containment. **b** Centroid containment. **c** Buffer containment

the buffer (Fig. 5.5). This method also does not make any distinction between census polygons that are completely enclosed and those that are partially enclosed by the original buffer. It could lead to an overestimation of the potentially exposed population if people residing in a partially enclosed census unit are concentrated in a location that falls outside the buffer boundary. A variation of this method is to use a cutoff criteria to exclude certain census units that are partially contained within the buffer. The most common application is to only include census units that have more than half of their area within the buffer zone, referred to as *50% area containment* method (Mohai and Saha, 2006, 2007).

The second method for estimating population characteristics within a buffer zone is known as *centroid containment* (Chakraborty and Armstrong, 1997; Maantay and Maroko, 2009). This technique selects only those census polygons whose geographic centers or centroids are located within the buffer, thus limiting the number of census units that can be included (Fig. 5.6). The effective buffer zone obtained through both polygon and centroid containment, however, will not resemble the original (distance-based or plume-based) buffer because it is based on the boundaries of the selected census polygons. Additionally, this method is likely to provide inaccurate estimates of the potentially exposed population if the actual residences of people in census units intersected by the buffer are not concentrated near their centroid.

The third and most widely-used method is *buffer containment* (Chakraborty and Armstrong, 1997). This method includes all census units lying within the buffer and a fraction of the population from units that are intersected by the buffer. Recent studies have referred to this approach as the *areal apportionment* method (Mohai and Saha, 2006; Kearney and Kiros, 2009). This particular method has the advantage

Fig. 5.6 Cadastral dasymetric mapping: using land parcels to estimate households within a buffer

of retaining the area and shape of the original circle or plume used to delineate the buffer zone. An areal weighting technique is typically utilized to determine the population characteristics of the buffer zone (Maantay and Maroko, 2009). Specifically, the population of each census unit is weighted by the proportion of its area that falls inside the circular or plume-based buffer. An important limitation is the assumption that the population of a census unit and all its characteristics are distributed uniformly within its boundary. This could thus to inaccurate estimates of if the actual residences of people within a census unit are concentrated in specific areas instead of being dispersed throughout the unit.

As an alternative to areal interpolation, one EJ study has utilized a hybrid method known as *cross-area transformation* (Most et al., 2004). The population characteristics of census units that are intersected or partially contained within a buffer are estimated by borrowing data from the completely contained census units, based on the remaining area of the buffer zone that falls outside the boundaries of all fully contained units. In addition to completely excluding census units intersected by the buffer, this method assumes that any census unit completely enclosed by a given buffer will adequately reflect the characteristics of people living in the larger area.

All of these methods have been widely employed in EJ analysis and no single best technique has emerged. The application of *dasymetric mapping* in combination with areal interpolation has been suggested as a promising approach (Holt et al., 2004; Zandbergen and Chakraborty, 2006; Maantay et al., 2007). This technique that uses ancillary information such as land use or land cover to redistribute spatial data in a more accurate and logical manner (Mennis, 2002). Recent studies suggest that cadastral dasymetric mapping represents a meaningful improvement on the use of the aggregated data when geocoded locations of individuals are unavailable (Maantay and Maroko, 2009; Maantay et al., 2008). Figure 5.6 depicts how boundaries of land parcels can be used to estimate households within a circular buffer. Additional details such as housing tenure, ownership, and values can also be utilized to assess socioeconomic characteristics of proximate households.

5.4 Geostatistical Methods

A variety of statistical methodologies have been used to determine if race/ethnicity or SES is related to indicators of proximity or exposure to environmental hazards. To compare the characteristics of areas potentially exposed to hazards with those that are not exposed, two-sample inferential tests of means and proportions have been utilized. Most EJ studies, however, have relied on linear correlation or multivariate regression analysis to measure the statistical significance of the association between environmental risk and relevant socio-demographic characteristics of the residential population. While least squares regression is an effective and widely used technique for measuring the strength and significance of relationships between the dependent and multiple explanatory factors, it is based on two assumptions that are rarely met by spatially distributed data and variables: *independence* and *homogeneity*.

The independence assumption of linear regression ignores the notion that locational proximity often results in value similarity when most socio-demographic variables are mapped. This fundamental concept was articulated by Tobler (1970: 236) as "everything is related to everything else, but near things are more related than distant things" and is known as Tobler's first law (TFL) of geography. The practical implication of TFL is that observations from nearby locations are often more similar than what can be expected on a random basis. This phenomenon is known as spatial dependence, and more formally as (positive) spatial autocorrelation. The presence of such autocorrelation can be problematic for standard statistical tests such as correlation and regression that assume independently distributed observations and errors. Although EJ analysis is based on spatial data and spatially distributed variables, most previous studies have assumed observations and error terms to be independent, thus violating one of the classical regression assumptions and ignoring spatial effects that could lead to incorrect inferences about the significance of explanatory variables representing race/ethnicity or SES.

A large body of literature in geographic analysis has focused on the development of methods that can be used to detect violations of the independence assumption and models that account for spatial autocorrelation (Cliff and Ord, 1981; Anselin and Bera, 1998; Anselin, 2005). Spatial regression models, such as simultaneous autoregressive (SAR) models, are statistical models that consider spatial dependence as an additional variable in the regression equation and estimate its effect simultaneously with effects of other explanatory variables. The use of spatial regression has increased in recent years, in part, due to the availability of GIS and user-friendly spatial analysis software programs such as *GeoDa* (Anselin, 2005) that are that capable of implementing the underlying spatial econometric techniques. Recent EJ studies have utilized spatial regression to explicitly consider the effects of spatial autocorrelation (Pastor et al., 2005; Grineski and Collins, 2008; Chakraborty, 2009). Chapter 17 of this volume demonstrates how SAR models can be used to address the limitations of conventional regression analysis and account for spatial dependence in EJ analysis.

In addition to independence, the classical linear regression model assumes a generating process that is considered to be stationary or homogeneous. When applied to a regression model, this means that a single set of global parameters is adequate to describe the process and there are no spatial variations in the relationships between the dependent and independent variables. The use of a single or "global" regression model for an entire study area, however, assumes model parameters do not vary spatially and ignores local differences in statistical associations between dependent and independent variables. Since conventional regression models used in EJ research do not account for this spatial variability and only provide global results for the whole study region, they have the potential to mask important geographic disparities in statistical relationships relevant to EJ. Geographically weighted regression (GWR) is a local spatial statistical technique used to analyze spatial nonstationarity, defined as when the measurement of relationships among variables differs from location to location (Fotheringham et al., 2002). Instead of generating a single global regression equation or one set of regression parameters for an entire study area, GWR produces a separate regression equation or a unique set of parameters for each observation or spatial unit in the study area. Maps generated from GWR analysis can then be used to investigate how regression diagnostics and the strength of statistical relationships differ from place to place within a study area, as demonstrated in recent EJ studies conducted in New Jersey (Mennis and Jordan, 2005), New York (Maroko et al., 2009), and Florida (Gilbert and Chakraborty, 2011).

Figure 5.7 illustrates, for example, how the nature and significance of the statistical association (values of t-statistic) between cumulative cancer risk from ambient exposure to hazardous air pollutants, derived from the 1999 NATA, and specific explanatory variables vary across census tracts in Florida. For each variable or GWR model coefficient, these maps indicate how the relation between cancer risk and the proportion of minority or impoverished residents could be significantly positive in some areas, significantly negative in other areas, and not significant at

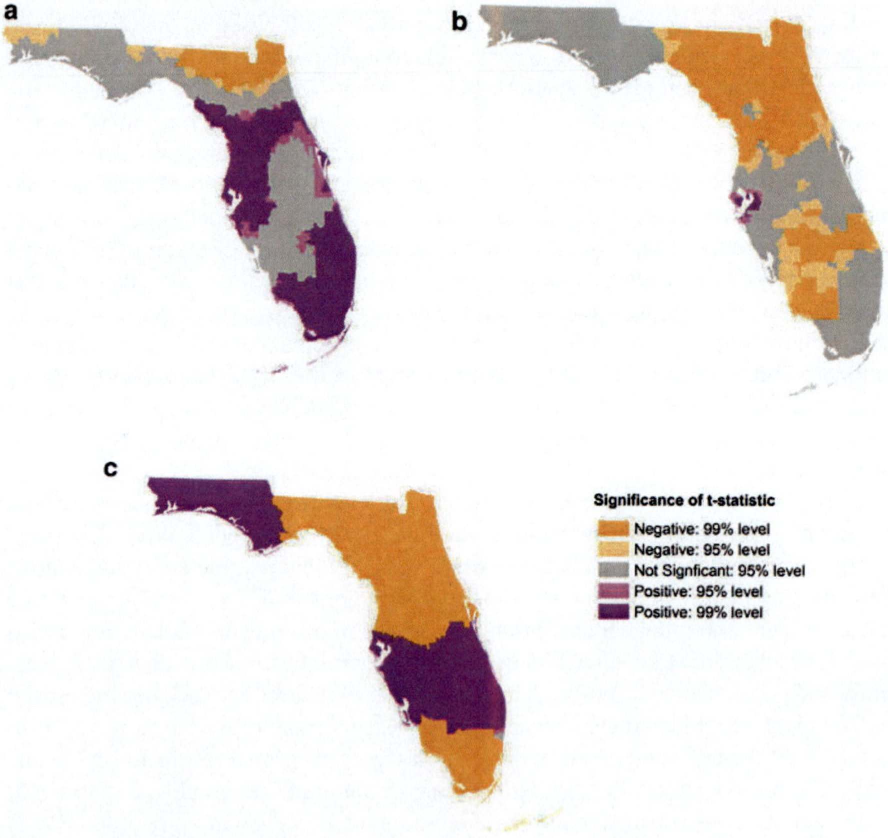

Fig. 5.7 Using geographically weighted regression to explore relationships between cumulative cancer risk from all sources of air toxics (1999) and explanatory variables (2000) in Florida: distribution of local t-statistic by census tract. **a** Proportion hispanic. **b** Proportion below poverty line. **c** Population per square mile

other locations, within the same study region (Florida). Traditional multivariate regression, however, is incapable of uncovering such spatial variation and could potentially lead to incorrect conclusions regarding disproportionate exposure to environmental hazards.

5.5 Concluding Discussion

This review has explored how the assessment of proximity and potential exposure to environmental hazards in quantitative EJ research has emerged from simple coincidence or distance-based methods to more sophisticated techniques that utilize pollutant fate and transport models or provide modeled estimates of health risks from cumulative exposure to multiple pollutants and emission sources. Methodological debates have also evolved from the choice of ZIP code or census tract for coincidence analysis to the selection of appropriate distance decay functions or geostatistical techniques that are more suitable for analyzing spatial data, variables, and relationships.

In spite of the methodological improvements in measuring disproportionate proximity and exposure to environmental hazards, EJ research still remains constrained by several limitations. First, a majority of studies have focused exclusively on night-time exposure by relying on socio-demographic data from the US. Census. Since census variables represent residential or night-time populations, they cannot be used to assess day-time risk. Most studies implicitly assume that people are non-mobile and are not exposed to pollution at non-residential locations. However, daily mobility typically results in residents moving to and from various locations, such as to work or to school. The journey-to-work commute between the suburbs and central cities could have important implications for EJ research and policy (Chakraborty, 2009). More affluent and White suburban residents, for example, could be spending a considerable amount of time during the day in census units near downtown locations that are exposed to adverse health risks. At the same time, minority residents who commute daily to suburban job locations may face lower levels of exposure for most of the day, thus reversing the inequity patterns reported in empirical studies. Future EJ research, however, should explore the use of additional data sources to construct temporally sensitive models that examine the day-time distribution of urban populations. Available data that can be used for this purpose include those on employment and people in day-time institutions such as schools and daycare centers (McMaster et al., 1997). Such information can be utilized to develop an independent model of the population distribution for the hours of 7 am to 5 pm to complement census residential data.

Another limitation in assessing disproportionate proximity or adverse health effects is the difficulty in obtaining data at a spatial resolution that is sufficiently detailed to reliably demonstrate the connection between environmental conditions and the socio-demographic characteristics of populations at risk. As mentioned previously, the lack of address-specific and individual level data forces most researchers to use aggregated census data tied to pre-defined geographic entities and rely on areal interpolation techniques that are based on simplistic assumptions about the population distribution. Although interpolation inaccuracies can be reduced by using spatial units that are smaller in size (i.e., census blocks), socioeconomic variables are not published by the US. Census at the block level of aggregation. Survey data has been proposed as an expensive, but useful alternative to examine individual or household level racial/ethnic and socioeconomic disparities in proximity to environmental hazards (Mohai et al., 2009). Future EJ research needs to incorporate local household surveys and techniques such as cadastral dasymetric mapping to enhance areal interpolation and estimate the characteristics of at-risk individuals more accurately.

It is also important to consider that although conventional statistical methods such as linear correlation or multivariate regression are used extensively to analyze EJ and health disparities, these techniques may not be appropriate for analyzing geospatial data because they violate classical statistical assumptions of independence and homogeneity. Instead of relying only on traditional statistical methods, future research needs to incorporate geostatistical techniques that are suitable for analyzing spatial data, variables, and relationships. Increased education in techniques such as SAR or GWR modeling is necessary to encourage new research incorporating these methods and assist researchers in developing new techniques that address the limitations of conventional approaches to environmental health justice analysis.

As scientists continue to examine proximity to environmental hazards and health disparities by race and SES in order to better understand the persistence of inequity in exposure to environmental hazards and associated health risks, Morello-Frosch, Pastor, and Sadd succinctly state their recommended objectives for future research that scientists should aim to "elucidate how institutional discrimination, uneven regional development, and a spatialized political economy shape distributions of environmental hazards, which in turn determine variations in community exposures and susceptibility to environmental hazards" (Morello-Frosch et al., 2001:572). Increased access to geographically detailed data sets such as EPA's NATA, which offers ongoing evaluations of air toxics and estimates of related health risk, as well as the refinement of GIS techniques, geospatial analyses, and dasymetric mapping, provide multiple opportunities for an increasing rigor in environmental health justice research that can even more effectively serve to advance just and equitable policy.

Author Biographies

Jayajit Chakraborty is an Associate Professor of Geography at the University of South Florida, Tampa. He received his Ph.D. in Geography from the University of Iowa, Iowa City, in 1999. His research interests lie at the intersection of geographic information science, environmental geography, and urban geography, and include air pollution, environmental health, racial/ethnic disparities, and vulnerability to environmental hazards.

Dr. Chakraborty is particularly interested in applications of geographic information science and spatial statistical techniques. He has authored articles in prominent academic journals such as *The American Journal of Public Health, Annals of the Association of American Geographers,*

Environment and Planning A, Health and Place, Risk Analysis, and *The Professional Geographer*. His research activities have been supported by grants from the US Environmental Protection Agency (EPA), National Science Foundation (NSF), and the Florida Department of Transportation. Dr. Chakraborty is also the editor of a book titled *Spatial and Environmental Injustice in an American Metropolis: A Study of Tampa Bay, Florida* (Cambria Press, 2010) and an editorial board member for the *International Journal of Population Research*. Additionally, he has served as a Chair of the Association of American Geographers Hazards Specialty Group (2005–2007) and was commissioned by the EPA to co-author and present a paper for their *Strengthening Environmental Justice Research and Decision Making Symposium* (2010).

Juliana A. Maantay is Professor of Urban Environmental Geography in the Earth, Environmental, and Geospatial Sciences Department, the Director of the GISc Program, and Director of the Urban GISc Lab at Lehman College, City University of New York. Prior to her academic career, Dr. Maantay worked as an urban and environmental planner and policy analyst with governmental agencies, non-profit organizations, and private sector consulting firms, and has been active in environmental health justice research and advocacy for more than 15 years.

Dr. Maantay's current research, funded by NIEHS, NOAA, and USDA, includes explorations of spatial correspondence between asthma and air pollution in the Bronx, residential segregation and health inequalities, new dasymetric methods of population mapping, interdisciplinary urban health research methods; community-based participatory research using GISc, West Nile Virus vector habitat characterization and risk assessment, spatial relationship between obesity/type 2 diabetes and access to recreational open space and healthy foods, and the benefits of urban agriculture in combating inner city "food deserts." She has published and presented widely on the built environment, environmental health, the geographies of health, zoning and land use planning, environmental modeling, cadastral systems, methodological innovations in Geographic Information Science, and is editing a special issue on health geographics for the *Journal of Applied Geography*. She is interested in ferreting out the connections between historical events and circumstances with current problems, believing that a better understanding of the historical process leading up to the current conditions can help promote constructive change.

References

Anderton D, Anderson J, Oakes J, Fraser M (1994) Environmental equity: the demographics of dumping. Demography 31:229–248

Anselin L (2005) Exploring spatial data with geoda: a workbook. Spatial Analysis Laboratory, Department of Geography, University of Illinois, Urbana-Champaign, IL

Anselin L, Bera A (1998) Spatial dependence in linear regression models with an introduction to spatial econometrics. In: Ullah A, Giles D (eds) Handbook of applied economic statistics. Marcel Dekker, New York, NY, 237–289

Apelberg BJ, Buckley TJ, White RH (2005) Socioeconomic and racial disparities in cancer risk from air toxics in Maryland. Environ Health Perspect 113(6):693–699

Ash M, Fetter TR (2004) Who lives on the wrong side of the environmental tracks? evidence from the EPA's risk-screening environmental indicators model. Soc Sci Q 78:793–810

Atlas M (2002) Few and far between? An environmental equity analysis of the geographic distribution of hazardous waste generation. Soc Sci Q 83(1):365–378

Baden BM, Coursey D (2002) The locality of waste within the city of Chicago: a demographic, social, and economic analysis. Resour Energy Econ 24:53–93

Baden BM, Noonan DS, Turaga RM (2007) Scales of justice: is there a geographic bias in environmental equity analysis? J Environ Plan Manag 50:163–185

Been V, Gupta F (1996) Coming to the nuisance or going to the barrios? A longitudinal analysis of environmental justice claims. Ecol Law Q 24(1):1–35

Been V (1995) Analyzing evidence of environmental justice. J Land Use Environ Law 11(1):1–36

Bevc CE, Marshall BK, Picou JS (2007) Environmental justice and toxic exposure: toward a spatial model of physical health and psychological well-being. Soc Sci Res 36:48–67

Boer JT, Pastor M, Sadd JL, Snyder LD (1997) Is there environmental racism? the demographics of hazardous waste in Los Angeles. Soc Sci Q 78:793–810

Bolin B, Matranga E, Hackett EJ, Sadalla EK, Pijawka KD, Brewer D, Sicotte D (2000) Environmental equity in a sunbelt city: the spatial distribution of toxic hazards in Phoenix, Arizona. Environ Hazards 2:11–24

Bolin B, Nelson A, Hackett E, Pijawka D, Smith S, Sadalla E, Sicotte D, Matranga E, O'Donnell M (2002) The ecology of technological risk in a sunbelt city. Environ Plan A 34: 317–339

Boone CG (2002) An assessment and explanation of environmental inequity in Baltimore. Urban Geogr 23(6):581–595

Bouwes N, Hassur SM, Shapiro MD (2001) Empowerment through risk-related information: EPA's Risk-Screening Environmental Indicators project. Working paper DPE-01-06. Amherst, MA: Political Economy Research Institute

Bowen WM, Salling MJ, Haynes KE, Cyran EJ (1995) Towards environmental justice: spatial equity in Ohio and Cleveland. Ann Assoc Am Geogr 85(4):641–663

Brooks N, Sethi R (1997) The distribution of pollution: community characteristics and exposure to air toxics. J Environ Econ Manag 32(2):233–250

Burke LM (1993) Environmental Equity In Los Angeles. National Center for Geographic Information and Analysis (NCGIA): Technical Report 93-6. Santa Barbara, CA: NCGIA

Chakraborty J (2001) Acute exposure to extremely hazardous substances: an analysis of environmental equity. Risk Anal 21:883–894

Chakraborty J (2009) Automobiles, air toxics, and adverse health risks: environmental inequities in Tampa Bay, Florida. Ann Assoc Am Geogr 99(4):647–697

Chakraborty J, Armstrong MP (1996) Using geographic plume analysis to assess community vulnerability to hazardous accidents. Comput Environ Urban Syst 19(5/6):1–17 49, 223–227

Chakraborty J, Armstrong MP (1997) Exploring the use of buffer analysis for the identification of impacted areas in environmental equity assessment. Cartogr GEO Inf Syst 24:145–157

Chakraborty J, Armstrong MP (2001) Assessing the impact of airborne toxic releases on populations with special needs. Prof Geogr 53:119–131

Chakraborty J, Armstrong M (2004) Thinking outside the circle: using geographical knowledge to focus environmental risk assessment investigations. In: Janelle D, Warf B, Hansen K (eds) WorldMinds: geographical perspectives on 100 problems. Kluwer Academic Publications, Dordrecht, The Netherlands, 435–442

Chakraborty J, Schweitzer L, Forkenbrock D (1999) Using GIS to assess the environmental justice consequences of transportation system changes. Transact GIS 3(3):239–258

Chakraborty J, Zandbergen P (2007) Children at risk: measuring racial/ethnic disparities in potential exposure to air pollution at school and home. J Epidemiol Commun Health 61:1074–1079

Clarke KC, McLafferty S, Templaski B (1996) On epidemiology and geographic information systems: a review and discussion of future direction. Emerg Infect Dis 2 (2):85–92

Clarke J, Gerlak A (1998) Environmental racism in southern Arizona. In: Camacho DE (ed) Environmental injustices, political struggles: race, class, and the environment. Duke University Press, Durham, NC

Cliff AD, Ord JK (1981) Spatial processes: models and applications. Pion Limited, London

Cutter SL, Holm D, Clark L (1996) The role of geographic scale in monitoring environmental justice. Risk Anal 16:517–526

Cutter S, Solecki WD (1996) Setting environmental justice in space and place: acute and chronic airborne toxic releases in the southeastern United States. Urban Geogra 17:380–399

Cutter SL, Hodgson ME, Dow K (2001) Subsidized inequities: the spatial patterning of environmental risks and federally assisted housing. Urban Geogr 22:29–53

Daniels G, Friedman S (1999) Spatial inequality and the distribution of industrial toxic releases: evidence from the 1990 TRI. Soc Sci Q 80:244–262

Dorling D, Fairbairn D (1997) Mapping: ways of representing the world. Addison Wesley Longman Ltd., Harlow

Dolinoy DC, Miranda ML (2004) GIS modeling of air toxics releases from TRI-reporting and non-TRI reporting facilities: impacts for environmental justice. Environ Health Perspect 112(17):1717–1724

Downey L (2006) Environmental racial inequality in Detroit. Soc Forces 85(2):772–796

Dunn EC, Kingham PS, Rowlingson B, Bhopal SR, Cocking S, Foy JWC, Acquilla DC, Halpin J, Diggle P, Walker D (2001) Analysing spatially referenced public health data: a comparison of three methodological approaches. Health Place 7:1–12

EPA (2008). National-scale air toxics assessment – glossary of key terms. http://www.epa.gov/ttn/atw/nata1999/gloss1.html. Accessed 10 Jan 2008

Fisher JB, Kelly M, Romm J (2006) Scales of environmental justice: combining GIS and spatial analysis for air toxics in West Oakland, California. Health Place 12:701–714

Fitos E, Chakraborty J (2010) Race, class, and wastewater pollution. In: Chakraborty J, Bosman MM (eds) Spatial and environmental injustice in an American metropolis: a study of Tampa Bay, Florida. Cambria Press, Amherst, NY, 145–151

Fitzgerald M, Schuurman N, Dragicevic S (2004) The utility of exploratory spatial data analysis in the study of tuberculosis incidences in an urban Canadian population. Cartogr 39(2):29–39

Fotheringham SA, Brunsdon C, Charlton ME (2002) Geographically weighted regression: the analysis of spatially varying relationships. Wiley, Chichester, UK

Fricker RD, Hengartner NW (2001) Environmental equity and the distribution of toxic release inventory and other environmentally undesirable sites in metropolitan NYC. Environ Ecol Stat 8:33–52

Gilbert A, Chakraborty J (2008) Health risks from hazardous air pollutants: analyzing environmental injustice in Florida. Florida Geogr 39:71–94

Gilbert A, Chakraborty J (2011) Using geographically weighted regression for environmental justice analysis: cumulative cancer risks from air Toxics in Florida. Soc Sci Res 40:273–286

Glickman TS (1994) Measuring environmental equity with geographical information systems. Renewable Resour J 12:17–21

Glickman TS, Hersh R (1995) Evaluating environmental equity: the impacts of industrial hazards on selected social groups in Allegheny County, Pennsylvania. Discussion paper 95–13, Washington, DC: Resources for the Future

Goldman BA, Fitton L (1994) Toxic wastes and race revisited: an update of the 1987 report on the racial and socioeconomic characteristics of communities with hazardous waste sites. Washington, DC: Center for Policy Alternatives

Gragg RD, Christaldi RA, Leong S, Cooper M (1996) The location and community demographics of targeted environmental hazardous sites in Florida. J Land Use Environ Law 12:1–24

Greenberg M (1993) Proving environmental inequity in siting locally unwanted land uses. Risk Issues Health Saf 4(3):235–252

Grineski SE (2007) Incorporating health outcomes into environmental justice research: The case of children's asthma and air pollution in Phoenix, Arizona. Environ Hazards 7:360–371

Grineski S, Collins T (2008) Exploring environmental injustice in the global South: maquiladoras in Ciudad Juárez. Popul Environ 29:247–270

Hird JA (1993) Environmental policy and equity: the case of Superfund. J Policy Anal Manag 1(2):323–343

Holt JB, Lo CP, Hodler RW (2004) Dasymetric estimation of population density and areal interpolation of census data. Cartogr Geogr Inf Sci 31(2):103–121

Jacquez GM (2000) Spatial analysis in epidemiology: nascent science or a failure of GIS? J Geogr Syst 2:91–97

Jarup L (2004) Health and environment information systems for exposure and disease mapping, and risk assessment. Environ Health Perspect 112(9):995–997

Kearney G, Kiros GE (2009) A spatial evaluation of socio demographics surrounding National Priorities List sites in Florida using a distance-based approach. Int J Health Geogr 8:33

Kriesel W, Centner TJ, Keeler AG (1996) Neighborhood exposure to toxic releases: are there racial inequities? Growth Change 27:479–499

Kulldorff M (1999) Geographical information systems (GIS) and community health: some statistical issues. J Public Health Manag Pract 5(2):100–106

Linder SH, Marko D, Sexton K (2008) Cumulative cancer risk from air pollution in houston: disparities in risk burden and social disadvantage. Environ Sci Technol 42(12):4312–4322

Maantay JA (2001) Zoning, equity, and public health. Am J Public Health 91(7):1033–1041

Maantay JA (2002) Mapping environmental injustices: Pitfalls and potential of geographic information systems (GIS) in assessing environmental health and equity. Environ Health Perspect 110(Supp 2):161–171

Maantay JA (2007 03/28) Asthma and air pollution in the Bronx: methodological and data considerations in using GIS for environmental justice and health research. Health and Place 13(1):32–56

Maantay JA, Timander L, Graziosi G, Meyers L (1997) The Bronx toxic release inventory report. Center for a Sustainable Urban Environment/US Environmental Protection Agency, Region 2, New York, NY

Maantay JA, Maroko A (2009) Mapping urban risk: Flood hazards, race, & environmental justice in New York. Appl Geogr 29(1):111–124

Maantay JA, Maroko A, Herrmann C (2007) Mapping population distribution in the urban environment: the cadastral-based expert dasymetric system (CEDS). Cartogr Geogr Inf Sci 34(2):77–102

Maantay JA, Maroko A, Porter-Morgan HA (2008) New method for population mapping and understanding the spatial dynamics of disease in urban areas. Urban Geogr 29(7):724–738

Maantay JA, Tu J, Maroko AR (2009) Loose-coupling an air dispersion model and a geographic information system (GIS) for studying air pollution and asthma in the Bronx, New York City. Int J Environ Health Res 19(1):59–79

Margai FL (2001) Health risks and environmental inequity: a geographical analysis of accidental releases of hazardous materials. Prof Geogr 53(1):422–434

Maroko AR, Maantay JA, Sohler NL, Grady K, Arno P (2009) The complexities of measuring access to parks and physical activity sites in New York City: a quantitative and qualitative approach. Int J Health Geogr 8(34):1–23

McMaster RB, Leitner H, Sheppard E (1997) GIS-based environmental equity and risk assessment: methodological problems and prospects. Cartogr Geogr Inf Syst 24(3):172–189

Mennis J (2002) Using geographic information systems to create and analyze statistical surfaces of population and risk for environmental justice analysis. Soc Sci Q 83:281–297

Mennis J, Jordan L (2005) The distribution of environmental equity: exploring spatial nonstationarity in multivariate models of air toxic releases. Ann Assoc Am Geogr 95(2):249–268

Mirabelli MC, Wing S, Marshall SW, Wilcosky TC (2006) Race, poverty, and potential exposure of middle-school students to air emissions from confined swine feeding operations. Environ Health Perspect 114(4):591–596

Mohai P, Bryant B (1992) Environmental racism: reviewing the evidence. In: Bryant B, Mohai B (eds) Race and the incidence of environmental hazards: a time for discourse. Westview Press, Boulder, CO, 163–176

Mohai P, Lantz PM, Morenoff J, House JS, Mero RP (2009) Racial and socioeconomic disparities in residential proximity to polluting industrial facilities: evidence from the Americans' Changing Lives Study. Am J Public Health 99(S3):649–655

Mohai P, Saha R (2006) Reassessing racial and socio-economic disparities in environmental justice research. Demography 43(2):383–399

Mohai P, Saha R (2007) Racial inequality in the distribution of hazardous waste: a national-level reassessment. Soc Probl 54:343–370

Moore D, Carpenter TE (1999) Spatial analytical methods and geographic information systems: use in health research and epidemiology. Epidemiol Rev 21(2):143–161

Morello-Frosch R, Jesdale W (2006) Separate and Unequal: Residential Segregation and Estimated Cancer Risks Associated with Ambient Air Toxics in US. Metropolitan Areas. Environ Health Perspect 114:1–8

Morello-Frosch R, Pastor M, Sadd J (2001) Environmental justice and Southern California's "riskscape" – the distribution of air toxics exposures and health risks among diverse communities. Urban Aff Rev 36:551–578

Most MT, Sengupta R, Burgener MA (2004) Spatial scale and population assignment choices in environmental justice analyses. Prof Geogr 56(4):574–586

Neumann CM, Forman DL, Rothlein JE (1998) Hazard screening of chemical releases and environmental equity analysis of populations proximate to toxic release inventory facilities in Oregon. Environ Health Perspect 106(4):217–226

Pastor M, Morello-Frosch R, Sadd J (2005) The air is always cleaner on the other side: race, space, and ambient air toxics exposures in California. J Urban Aff 27(2):127–148

Pastor M, Sadd JL, Morello-Frosch R (2004) Waiting to inhale: the demographics of toxic air releases in 21st century California. Soc Sci Q 85(2):420–440

Pellow DN, Brulle RJ (eds) (2005) Power, justice, and the environment: a critical appraisal of the environmental justice movement. MIT Press, Cambridge, MA

Perlin SA, Setzer RW, Creason J, Sexton K (1995) Distribution of industrial air emissions by income and race in the United States: an approach using the toxic release inventory. Environ Sci Technol 29(1):69–80

Perlin SA, Sexton K, Wong D (1999) An examination of race and poverty for populations living near industrial sources of air pollution. J Exp Anal Env Epdemiol 9:29–48

Perlin SA, Wong D, Sexton K (2001) Residential proximity to industrial sources of air pollution: interrelationships among race, poverty, and age. J Air Waste Manag Assoc 51:406–421

Pollock PH, Vittas ME (1995) Who bears the burden of environmental pollution? race, ethnicity, and environmental equity in Florida. Soc Sci Q 76(2):294–310

Richards TB, Croner CM, Rushton G, Brown CK, Folwer L (1999) Geographic information systems and public health: mapping the future. Public Health Rep 114:359–373

Ringquist EJ (1997) Equity and the distribution of environmental risk: the case of TRI facilities. Soc Sci Q 78:811–829

Rushton G, Elmes G, McMaster R (2000) Considerations for improving geographic information system research in public health. URISA J 12(2):31–49

Sheppard E, Leitner H, McMaster RB, Hongguo T (1999) GIS based measures of environmental equity: exploring their sensitivity and significance. J Expos Anal Environ Epidemiol 9:18–28

Sicotte D, Swanson S (2007) Whose risk in Philadelphia? proximity to unequally hazardous industrial facilities. Soc Sci Q 2007(88):515–534

Stretesky P, Lynch MJ (1999) Environmental justice and the predictions of distance to accidental chemical releases in Hillsborough County, Florida. Soc Sci Q 80(4):830–846

Talen E (1997) The social equality of urban service distribution an exploration of park access in Pueblo Colorado and Macon Georgia. Urban Geogr 18(6):521–541

Talen E, Anselin L (1998) Assessing spatial equity: an evaluation of measures of accessibility to public playgrounds. Environ Plan A 30:595–613

Taquino M, Parisi D, Gill DA (2002) Units of analysis and the environmental justice hypothesis: the case of industrial hog farms. Soc Sci Q 83(1):298–316

Tiefenbacher JP, Hagelman RR (1999) Environmental equity in urban Texas: race, income, and patterns of acute and chronic toxic air releases in metropolitan counties. Urban Geogr 1999(20):516–533

Tobler WR (1970) A computer movie simulating urban growth in the Detroit region. Econ Geogr 46:234–240

Tobler W (1979) Cellular geography. In: Steven, G., Gunnar, O. (eds) Philosophy in geography. Reidel, Dordrecht, NL

US GAO (Government Accountability Office) (1995) Demographics of people living near waste facilities. Government Printing Office, Washington, DC

United Church of Christ Commission for Racial Justice (1987) Toxic wastes and race in the United States: a national report on the racial and socio-economic characteristics of communities with hazardous waste sites. United Church of Christ, New York, NY

United Church of Christ (2007) Toxic wastes and race at twenty: 1987–2007. Justice and Witness Ministries, United Church of Christ, Cleveland, OH

Vine MF, Degnan D, Hanchette C (1997) Geographic information systems: their use in environmental epidemiologic research. Environ Health Perspect 105(6):598–605

Walker G, Mitchell G, Fairburn J, Smith G (2005) Industrial pollution and social deprivation: evidence and complexity in evaluating and responding to environmental inequality. Local Environ 10(4):361–377

Wall P, Devine O (2000) Interactive analysis of the spatial distribution of disease using a geographic information system. J Geogr Sys 2:243–256

Waller LA, Louis TA, Carlin BP (1999) Environmental justice and statistical summaries of differences in exposure distributions. J Exp Anal Env Epdemiol 9:56–65

Williams MM (2010) Health risks from point sources of industrial air pollution: modeling toxicity and exposure. In: Chakraborty J, Bosman MM (eds) Spatial and environmental injustice in an American metropolis: a study of Tampa Bay, Florida. Cambria Press, Amherst, NY, 145–151

Wolch J, Wilson J, Fehrenback J (2005) Parks and park funding in Los Angeles: an equity mapping analysis. Urban Geogr 25:4–35

Wood D (1992) The power of maps. Guilford Press, New York, NY

Yasnoff WA, Sondik EJ (1999) Geographic information systems (GIS) in public health practice in the New Millennium. J Public Health Manag Pract 5(4):ix–xii

Zandbergen PA, Chakraborty J (2006) Improving environmental exposure analysis using cumulative distribution functions and individual geocoding. Int J Health Geogr 5:23

Zimmerman R (1994) Issues of classification in environmental equity: how we manage is how we measure. Fordham Urban Law J 29(3):633–669

Chapter 6
Their Data, Our Cause: An Exploration of the Form, Function, and Deployment of Mapping Technologies among Community Environmental Justice Organizations

Trevor Fuller

Abstract The field of environmental justice offers many examples of the utility of maps and GIS for illustrating the disproportionate levels of environmental risk being endured by disadvantaged or marginalized racial, ethnic or income groups. However, these have predominantly focused on the distribution of environmental risk rather than focusing on the map-making parties themselves. This research directs attention towards the use of maps and GIS by "local" environmental justice organizations in their calls for environmental justice. I focus on activist engagement with maps and mapping technologies like GIS. Through a survey of community-based environmental organizations, I examine whether these organizations use maps and if so, how maps are produced including the sources of mapping knowledge used in the map-making process. By examining how and why such organizations map environmental hazards and use GIS, this chapter provides insights into the notion of GIS mapping as an empowering practice. I assess the types of maps produced and the data sources used in order to see more clearly potential restrictions on the power and ability of organizations to counter dominant narratives of their communities. This research reveals that while these organizations recognize the importance of maps to their efforts, there are significant differences in the resources and abilities of environmental justice organizations.

Keywords Environmental justice · Maps · GIS · Community organizations · Environmental hazards

6.1 Introduction

The environmental justice movement has long embraced maps as powerful visualizations of the environmental burdens faced by minority and low-income citizens. The field of environmental justice offers many examples of researchers examining the uneven spatial distributions of environmental hazards (negatives) and positive

---

T. Fuller (✉)
Department of Geography, University of Illinois at Urbana-Champaign, Urbana, IL 61801, USA
e-mail: tfuller3@illinois.edu

J.A. Maantay, S. McLafferty (eds.), *Geospatial Analysis of Environmental Health*, Geotechnologies and the Environment 4, DOI 10.1007/978-94-007-0329-2_6,

features in relation to the socio-demographic characteristics of places. Research has revealed many examples of maps and GIS illustrating the disproportionate levels of environmental risk being endured by disadvantaged or marginalized racial, ethnic or income groups.

In this chapter, I step away from the academic-centered debates regarding methodologies for visualizing environmental injustice to instead focus on activist engagement with maps and mapping technologies through GIS. By conducting a survey of community-based environmental organizations, I examine whether these organizations use maps and if so, how maps are produced including the sources of mapping knowledge used in the map-making process. By examining how and why such organizations map environmental hazards and use GIS, this chapter will provide insights into the notion of GIS mapping as an empowering practice. I assess the types of maps produced and the data sources used in order to see more clearly potential restrictions on the power and ability of organizations. In particular, this research seeks to answer the following research questions:

1. Do environmental justice organizations use maps in their efforts? What types of maps are produced and what sources of data are used?
2. Are these organizations predominantly dependent upon secondary sources of spatial hazards data, or do they collect their own?
3. How does community-collected environmental hazards data, differ from data from government sources?
4. How do these organizations perceive the value and power of mapping environmental hazards?

6.2 Literature Review

Environmental justice is defined as "the fair treatment and meaningful involvement of all people regardless of race, color, national origin, or income with respect to the development, implementation, and enforcement of environmental laws, regulations, and policies" (US EPA Office of Environmental Justice 2006). Concern regarding the unequal distribution of environmental hazards first began to enter public discourse in response to two high profile cases involving allegations by two predominantly African-Americans communities that were disproportionately exposed to environmental hazards (Bullard, 1990; United Church of Christ 1987; US GAO, 1983). This spurred academic research into environmental justice with the vast majority of studies demonstrating a disproportionate burden of environmental hazards on minority/low-income populations (Anderton et al., 1994; Bullard, 1990; Pulido, 2000; Pastor et al., 2001; Buzzelli et al., 2003). More recently, researchers have begun to examine residents' activism in response to environmental hazards (Chambers, 2007; Checker, 2008; Elliott et al., 1999, Wakefield et al., 2001, 2006). One of the tools community-based environmental groups use in responding to environmental injustice is GIS mapping. However, their use (or non-use) of mapping reflects diverse group and community characteristics.

6.2.1 Uneven Ability to Map

Environmental justice organizations, like other community organizations, often need to establish networks of support that cross multiple scales and ideologies, as they seek funding for training, technologies, and staff (Elwood, 2006). The ability of groups to engage with "scalar politics and creative alliances", including the ability to garner support for GIS assisted activities is uneven (Ghose, 2007). Environmental justice groups possess different abilities and resources with respect to creating and maintaining support networks and technological capabilities. Because of these unequal abilities to access resources (Elwood, 2006; Ghose, 2007; Rouse et al., 2007), we see wide variation in the funding and skill-set of environmental justice organizations. Groups that establish strong networks with particular actors gain a distinct advantage (Ghose, 2007). Uneven abilities to create networks of support and "professional-looking" visualizations of hazards affect not only groups' funding, but also their ability to effectively counter the efforts of more powerful actors. Gaining access to GIS is what sets the study organizations here apart from one another.

6.2.2 Co-optation/Resistance

Community groups also adopt different strategies in promoting environmental justice. Elwood (2006) refutes the notion that community groups must either assume the role of a "co-opted" group or a resistance group. Rather, groups can shift their politics in order to "simultaneously cultivate multiple roles". These multiple roles, utilized to effect change, play out through three different types of politics according to Elwood (2006): spatial, institutional, and knowledge. GIS can be used to convey a group's spatial knowledge in a way that refutes that of other actors, and it can be used to claim a particular physical area. Some community environmental justice organizations recognize the influence GIS has on planning decisions and the inherent "professionalism" it carries. These organizations are able to shape and display their protests in a manner amenable to government agencies and policymakers.

6.2.3 Mapping the Omissions/Emissions

What is mapped is also important. Harley (1989) discusses how maps are social constructions which are dependent upon the particular agenda of the map-maker. As such, maps often omit as much information as they reveal (Harley, 1989). Further, state-created maps are often accepted as legitimate as a map's 'authoritative power' comes from "the scientific aura and official status that accompanies its construction (Bassett, 1993, p. 2)". In this way, government-created maps of environmental hazards may not only present an incomplete picture of the hazard landscape, but they are often accepted and utilized by the very people living within that hazard landscape. However, the power of maps and GIS can also be harnessed to counter government-created maps (Peluso, 1995) that omit hazards, such as small environmental spills and illegal dumping. To redirect the power of maps for the benefit of citizens, participatory GIS and community mapping have developed, which can act as effective counters to the uneven distribution of power and knowledge (Peluso, 1995). McLafferty (2002) illustrated how residents produced a "counter-narrative" with the aid of outside GIS expertise.

In approaching questions of environmental justice, it is important to account for "people"s subjective experiences of everyday life' and their perceptions of environmental hazards (Knigge and Cope, 2006; Kwan, 2008). GIS provides a canvas on which the spatio-temporal components of context can be visualized. Given that environmental justice is rooted in the "lived spaces" of daily life, GIS offers community organizations opportunity to record their encounters with environmental hazards in their neighborhoods. The widespread use of geo-technologies means that more "ordinary people" are using GIS to create and visualize spatial data and to represent "everyday life" (Elwood, 2006). This visualization can be powerful as a "counter-narrative" and is often complemented by other media (photographs, oral histories, videos).

These literatures suggest that environmental justice organizations may recognize a power and value in maps for their efforts, but that adoption of mapping will vary among organizations. I also expect that although most organizations that use GIS will rely on secondary data, some will generate their own data as a way of creating counter-narratives.

6.3 Methodology

To assess the use of GIS and mapping by environmental justice organizations, this research relied upon two data collection methods: (1) an internet-based inventory of environmental justice organizations' web pages (see Table 6.1) and (2) an eight-item questionnaire regarding data sources and perceptions and use of maps. These two methods were applied in search of the abilities, perspectives, and perceptions of these organizations regarding the utility and power of maps and GIS in their efforts.

The internet inventory (Fig. 6.1) gathered information regarding whether each of the organizations use maps and if so, how they use them, who created them and with what data. 22 environmental justice organizations were selected for this inventory via a Google-powered search for "community environmental justice organizations." This revealed many listings, so the selection of groups was further refined to the "local" scale (i.e. the city-neighborhood level). The 22 groups have web pages and as such, are privileged over other organizations that may not have access to computer-based resources, including mapping technologies. Thus, the sample is probably not representative of all local environmental justice organizations in the US. Among this sample, organizations that do use maps were chosen as participants in the questionnaire.

Table 6.1 Community environmental justice organizations and maps

Environmental justice organization	Use maps	Public access	Public modify	Map maker	Data source(s)	Data types
CCEJ	Yes	Yes	No	King County	EPA	Toxics, demographics
Hopewell citizens for clean water/air	Yes	Yes	No	EPA	EPA's consultant	TCE
DWEJ	Yes	Yes	Yes	EPA enviro mapper	EPA	Toxic sites
West harlem ACT (WeACT)	Yes	Yes	No	WEACT and others, GIS specialist	EPA, Columbia center, WEACT	Air quality, access QOL, asthma, particulate matter
Pilsen environmental rights and reform organization (PERRO)	Yes	Yes	No	PERRO	EPA	Toxics, demographics
Little village environmental justice organization (LVEJO-El cilantro youth)	Yes	Yes	Yes	LVEJO	EPA, LVEJO	"Assets and toxins"
South jersey environmental justice alliance	Yes	Yes	No	NJ EPA	NJ EPA	Toxics, hazardous waste, demographics
El puente CHE institute	Yes	Yes	No	CHE, NYPIRG	CHE, NYPIRG, EPA	Brownfields/use
NYCEJA	Yes	Yes	No	NYCEJA	NYCEJA, EPA	Contaminated sites, demographics
Communities for a better environment	Yes	Yes	No	CBE	EPA	TRI sites
Women's environmental institute	Yes	Yes	No	WEI	MPCA	Contaminated/superfund
Invisible 5 Project	Yes	Yes	No	I-5 (audio/video tour)	State, Fed, I-5, google	Contaminated sites, demographics, histories

Table 6.1 (continued)

Environmental justice organization	Use maps	Public access	Public modify	Map maker	Data source(s)	Data types
Silicon Valley Toxics coalition	Yes	Yes	No	SVTC	EPA	Superfund, demographics
Environmental community organization	Yes	Yes	No	ECO/EPA	EPA	Contaminated sites
Greenaction	Yes	Yes	No	Greenaction, academics	EPA, state, Greenaction	Contaminated sites, demographics
Ironbound community collective	Yes	Yes	No	EPA, State, City	EPA, state, city of Newark	Brownfields, amenities, contaminated sites
Environmental health coalition	Yes	Yes	No	EHC	San Diego, Mexican agency INEGI	Contaminated sites, pollutant modeling, health data, amenities
Louisiana environmental justice community organization collective	Yes	Yes	No	EPA	EPA	Industrial plants, emission sites

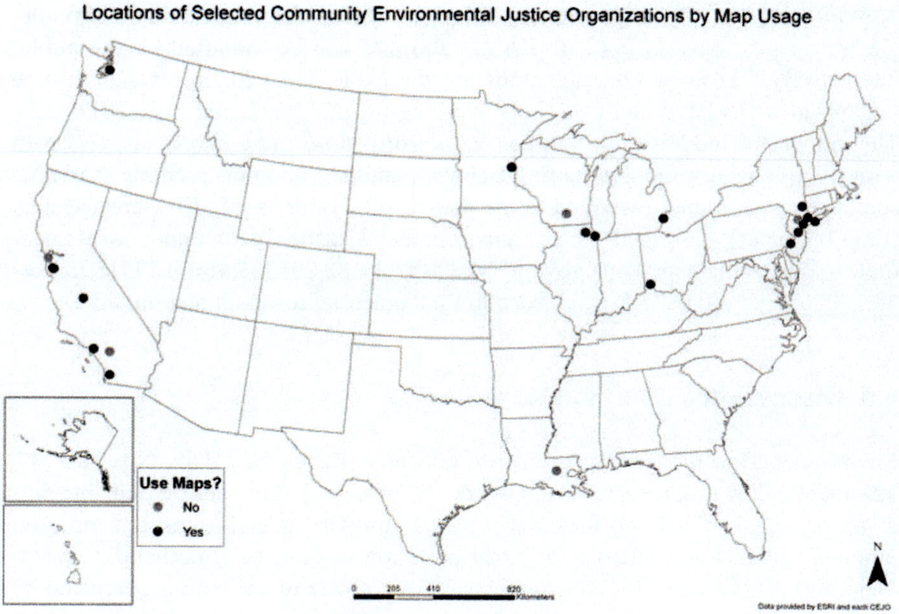

Fig. 6.1 Distribution of environmental justice organizations

The web-based inventory revealed that 82% (18 of 22) of the study organizations use maps (see Table 6.1). To probe in more depth the use of maps by these organizations, I distributed by email a questionnaire comprising eight open-ended questions. The questions were designed to elicit information about the organization's familiarity with mapping environmental hazards, the technology/data used, and the group's perceptions of the value of maps for their efforts. Eight of 18 organizations responded to the questionnaire for a 44% response rate. Results of the survey are discussed in the Discussion section.

6.4 Results

The types of maps produced by these organizations are similar in terms of environmental hazards displayed, such as Toxic Release Inventory sites, Superfund sites, and hazardous waste treatment, storage, and disposal facilities (TSDFs). What vary among these organizations are the mapmakers and data sources. Most of the organizations are the sole creators of the maps used although roughly 1/3 rely on outside groups or firms for map production. In terms of data sources, all of the study organizations use environmental hazards data provided by some level of government environmental agency. 55% of the study organizations only use data from the US. EPA or a government environmental agency. However, 44% of the organizations gather their own data to supplement government data.

While most of the maps used by these groups can be accessed by the public via web pages, only 11% (2 of 18) of the maps can be modified by the public interactively. Those two organizations are the Little Viejo Environmental Justice Organization (LVEJO) and Detroiters Working for Environmental Justice (DWEJ). The amount of modification that can be performed on these maps varies as well with LVEJO using a Google Earth framework and providing instructions as to how someone can add their own data to the map. DWEJ offers a US. EPA-created map (EnviroMapper) described as an "environmental justice geographic assessment tool" (http://epamap13.epa.gov/ej/EMej.asp?xl=-88.23333&yt=40.135189&xr=-88.190565&yb=40.083506) and has a limited potential for modification.

6.5 Discussion

Environmental justice organizations use GIS to visualize their daily "life spaces". The map(s) they produce are "doing work" by revealing the issues participants face as these organizations and their constituents struggle "to hold industry and government officials accountable for toxic pollution in their neighborhoods" (www.lvejo.org). Visualizing this data has power. This power of the map is perceived by some respondents as being a "mechanism for informing residents" (Greenaction) and a way to improve residents' personal knowledge of and response to environmental hazards. A map is seen as a tool for advocacy as well as for documenting environmental injustice. As one respondent stated, mapping offers: "an excellent opportunity to manipulate data and visually see cumulative impacts" (Ironbound).

Respondents perceive maps as having a power to influence. According to one respondent, "visual articulation of data can prompt a better understanding of the issues, and potentially assist change" (Invisible 5). Another impression of maps is that they "provide credibility for us when advocating for changes" (EHC). By displaying the disproportionate burden of environmental hazards within predominantly low-income/minority communities the injustice of this distribution might be better understood. While these organizations attribute a certain power to the map itself, the amount of power varies with the types of maps produced. Compare the mapping activities of LVEJO and the South Jersey Environmental Justice Alliance (SJEJA). The SJEJA uses maps and provides access to them on their web page. What it provides is a link to a page named "i-Map NJDEP", a mapping tool created and provided by the New Jersey Department of Environmental Protection (NJDEP). The tool is described by the NJDEP as a mapping tool that can be used by homeowners to "see what's in their backyard" (http://www.sjenvironmentaljustice.org/education/maps.htm). Once at the i-Map web page the NJDEP provides an additional link to a newer mapping tool titled "NJ-GeoWeb". The data available for mapping comes from NJDEP and is therefore restricted to the types of data that the agency wishes to make available. The mapping tool cannot be modified by the public other than through selection of data layers. No new data layers can be added by the public. In this top-down GIS, residents are told by an authoritative party what constitute environmental hazards. Local concerns not captured by NJDEP are omitted from the GIS maps.

In contrast to the SJEJA GIS, LVEJO provides two interesting examples that realize the “democratizing potential” (Harley, 1989) of mapping and GIS. The first is a “toxic tour”, which leads participants through the Chicago, Illinois neighborhood of Little Village, stopping at particular environmental hazards or parks/schools to not only discuss the impacts of hazards on the community, but also powerfully place the participant within an environmentally unjust landscape. This perspective may promote a deeper understanding of the environmental burdens residents endure. Residents also become more familiar with their neighborhood and the hazards they negotiate along their daily paths. These “toxic tours’ have been a tool of environmental justice organizations across the US. and this research included three organizations that offer such tours. LVEJO has taken an additional step by putting the map of the tour route on its web page so that the public can trace the route and sites visited. The second form of participatory mapping was created by LVEJO’s youth organization, El Cilantro. This map rests within a Google Earth frame with a satellite image of the neighborhood as a backdrop (http://www.elcilantro.org/?page_id=6). Using the open source Google Earth “MyMap” tool, map readers can collaborate with the organization to become co-producers of an environmental justice map. Citizens are able not only to add sites to the map, but also to write text describing each hazard and select symbology that adds to the overall message of the map. Organizations and citizens alike can contextualize local environmental injustices by contributing powerful qualitative data. In this way, this map of environmental injustice is continuously changing and “becoming” (Kitchin and Dodge, 2007).

While many of the study organizations rely partially, if not wholly, upon government sources of data, several of the organizations supplement such data with self-generated data, also known as “volunteered” geographic data. In this way, organizations recognize that the data and maps provided by government agencies are often replete with “omissions” (Harley, 1989). While government data are not necessarily seen as inaccurate, they are often seen as presenting a partial view of the environmental hazards landscape. Hazards that are seen as too small or innocuous in the eyes of environmental agencies might include illegal dumps, auto shops, metals/plating shops, or unregulated generators of small quantities of hazardous waste. Adding local knowledge of these facilities enlarges the maps of environmental hazards. Groups such as EHC have walked “neighborhoods such as San Diego’s Barrio Logan, and marked the sites of industries that used toxic materials on maps” (EHC). However, some organizations are restricted by lack of time and/or funding to engage in local mapping, as explained by one respondent, “Mapping can be very time-consuming and we haven’t got that kind of time” (Ironbound). Even if organizations are unable to use GIS to produce new maps, organizations still seize and attempt to politically recharge the government-created maps by placing them within an activist setting.

Most of the sampled organizations rely on the US. EPA’s Toxics Release Inventory database (http://www.epa.gov/TRI/), which is a reporting requirement for certain facilities as mandated by the Emergency Planning and Community Right-to-Know Act (EPCRA). Launched in 1988, the TRI database includes information regarding releases of approximately 650 chemicals. TRI is a robust repository of information that can be easily accessed and mapped by those with GIS resources. Although informative, TRI has important limitations that make its widespread use troubling (Maantay, 2007). Not only does TRI omit small hazardous facilities, but reporting requirements have changed over time. These changes "eliminated detailed reports from more than 5,000 facilities that release up to 2,000 pounds of chemicals very year" (Bullard et al., 2007, p. 13). Reporting of releases was changed from yearly to every 2 years and the threshold amount of contaminants needed to report a release was increased. These changes have undermined the value and accuracy of toxic release information across the US., leaving environmental justice organizations without crucial data. However, some organizations are able to counter this by gathering their own data and finding sites the government misses or ignores. Some organizations also account for omissions by contextualizing maps with qualitative data describing people's daily encounters with their unjust local landscapes. While many recognize the need to supplement and contextualize TRI data, some respondents lack the funds and expertise needed to do so. These environmental justice organizations not only recognize the value and power of maps but also the ways in which they can produce even more powerful maps through GIS visualizations of their "own" data.

Author Biography

Trevor Fuller is currently a Ph.D. candidate in the Department of Geography at the University of Illinois at Urbana-Champaign. He holds a B.A. in Natural Resources and Environmental Science from Purdue University as well as a M.A. in Geography from Indiana State University. Before pursuing his M.A. and Ph.D. in Geography, Trevor worked as an environmental scientist in both the private and public sectors, including 4 years at the Indiana Department of Environmental Management. Among other publications, he has co-authored a chapter in two editions of *Geospatial Technologies in Urban Environments: Policy, Practice and Pixels* (J. Gatrell and R. Jensen 2009). His current research interests revolve around environmental justice with particular interest in the drivers/inhibitors of resident activism and the role of mapping technologies in calls for environmental justice.

References

Anderton DL, Anderson A, Oakes J, Fraser M (1994) Environmental equity: the demographics of dumping. Demography 31:229–248

Bassett T (1993) Cartography, ideology and power: the world bank in northern cote d'Ivoire. Passages 5:1–9

Bullard R (1990) Dumping in dixie: race, class, and environmental quality. Westview Press, Boulder, CO

Bullard RD, Mohai P, Saha R, Wright B (2007) Toxic wastes and race at twenty 1987–2007: grassroots struggles to dismantle environmental racism in the United States. United Church of Christ Justice and Witness Ministry, Cleveland, OH

Buzzelli M, Jerrett M, Burnett R, Finklestein N (2003) Spatiotemporal perspectives on air pollution and environmental justice in Hamilton, Canada, 1985–1996. Ann AAG 93(3):557–573

Chambers S (2007) Minority empowerment and environmental justice. Urban Aff Rev 43(1): 28–54

Checker M (August 19, 2008) Bringing "Green Collar" jobs to the South Bronx. Gotham Gazette. Available at www.gothamgazette.com/article/communitydevelopment/20080819/20/2616

Elliott SJ, Cole D, Krueger P, Voorberg N, Wakefield S (1999) The power of perception: health risk attributed to pollution in an urban industrial neighborhood. Risk Anal 19(4):621–634

Elwood S (2006) Beyond cooptation or resistance: urban spatial politics, community organizations, and GIS-based spatial narratives. Ann Assoc Am Geogr 96(2):323–341

Ghose R (2007) Politics of scale and networks of association in public participation GIS. Environ Plan A 39(8):1961–1980

Harley JB (1989) Deconstructing the map. Cartographica 26(1):1–19

Kitchin R, Dodge M (2007) Rethinking maps. Prog Hum Geogr 31(3):331–344

Knigge L, Cope M (2006) Grounded visualization: Integrating the analysis of qualitative and quantitative data through grounded theory and visualization. Environ Plan A 38:2021–2037

Kwan M (2008) From oral histories to visual narratives: re-presenting the post-September 11 experiences of the Muslim women in the USA. Soc Cult Geogr 9(6):653–669

Maantay J (2007) Asthma and air pollution in the Bronx: methodological and data considerations in using GIS for environmental justice and health research. Health Place 13(1):32–56

McLafferty S (2002) Mapping women's worlds: knowledge, power, and the bounds of GIS. Gend Place Cult 9(3):263–269

Pastor M, Sadd J, Hipp J (2001) Which came first? toxic facilities, minority move-in, and environmental justice. J Urban Aff 23(1):1–21

Peluso N (1995) Whose woods are these? counter-mapping forest territories in Kalimantan, Indonesia. Antipode 27(4):383–406

Pulido L (2000) Rethinking environmental racism: white privilege and urban development in southern California. Ann Assoc Am Geogr 90:12–40

Rouse LJ, Bergeron S, Harris T (2007) Participating in the geospatial web: collaborative mapping, social networks and participatory GIS. In: Scharl A, Tochtermann K (eds) The geospatial web how geobrowsers, social software and the web 2.0 are shaping the network society. Springer, London, pp 153–158

United Church of Christ Commission for Racial Justice (1987) Toxic wastes and race in the United States: a national report on the racial and socioeconomic characteristics of communities with hazardous waste sites. Public Data Access, New York, NY

US EPA Office of Environmental Justice (2006) www.epa.gov/compliance/basics/ej.html

US Government Accounting Office (GAO) (1983) Siting of hazardous waste landfills and their correlation with racial and economic status of surrounding communities (GAO/RCED-83-168). US Government Printing Office, Washington, DC

Wakefield S, Elliott S, Cole D, Eyles J (2001) Environmental risk and (re)action: air quality, health, and civic involvement in an urban industrial neighbourhood. Health Place 7:163–177

Wakefield S, Elliott S, Cole D, Eyles J (2006) Taking environmental action: The role of composition, context, and collective. Environ Manage 37(1):40–53

Part II
Impacts on Environmental Health (Topical Case Studies)

Chapter 7
Geospatial Analysis of West Nile Virus (WNV) Incidences in a Heterogeneous Urban Environment: A Case Study in the Twin Cities Metropolitan Area of Minnesota

Debarchana Ghosh

Abstract West Nile virus (WNV) infected dead bird sites and human cases are frequently located in the densely populated, urban areas primarily because they are reported by people. However, the spatial pattern (i.e. morphology) of the urban landscape features could also contribute to the location of WNV incidences. This study has two objectives: (1) analyzing the association of urban environmental features that facilitated the viral activities of WNV infection in the TCMA from 2002 to 2007 and (2) comparing the spatial association between WNV infected mosquito pools and human cases with heterogeneous urban characteristics. It also addresses the question of how urban morphology affects human health. Using a combination of factorial ecology, geospatial techniques, and hierarchical cluster analysis, urban landscape classes are derived from the environmental and built environment risk-factors hypothesized to be associated with WNV transmission. The infection rate among, birds, mosquitoes, and human cases are then compared to these urban classes. Results indicate that the WNV infection rate is considerably higher in the urban class located just outside the cities of Minneapolis and Saint Paul. The dominant features of this class are close proximity to bogs and swamps, parks, sewerage system, waste water discharge sites, trails, high density of catch basins, moderate density of single family houses, and medium vegetation cover with stagnant waters. In general, the rate of infection decreases with increasing distance from the urban core. This is critical, in terms of vector control policies, because two out of four WNV carrying vectors, *Culex restuans* and *Culex pipiens* are predominantly urban mosquitoes.

Keywords West Nile virus · Urban environment · Factor analysis · Geographic Information Science · Cluster analysis

---

D. Ghosh (✉)
Department of Geography, Kent State University, 413 McGilvrey Hall, Kent, OH, USA
e-mail: dghosh@kent.edu

J.A. Maantay, S. McLafferty (eds.), *Geospatial Analysis of Environmental Health*, Geotechnologies and the Environment 4, DOI 10.1007/978-94-007-0329-2_7,

7.1 Introduction

West Nile virus (WNV) is a vector-borne infectious disease of global public health concern. It is transmitted to humans and other mammals by the bite of the infected mosquito, which acquires the virus by feeding on infected or dead birds. In the United States, the virus first appeared in New York in 1999 and since then it has spread rapidly, causing significant outbreaks in the densely populated urban areas. The metropolitan areas of Midwestern United States have exhibited strong spatial clustering of WNV infection among, birds, mosquitoes, and humans. In 2002, the states like Illinois and Michigan led the nation with 884 and 664 infected human cases respectively. Spatially, most of these disease incidents were clustered around the urban areas of Chicago and Detroit. In 2003, when the WNV infection reached a level of epidemic in Minnesota (MDH, 2003), significant clusters of infected dead birds and human cases were found in the urban areas of the Twin Cities Metropolitan Area (TCMA).

These outbreaks occur in a diverse mixture of buildings (old and new), sprawling development, transportation routes, vegetation (open green space, parks, trees, shrubs etc.), pockets of natural areas (lakes, reservoirs, golf courses, unpaved trails, etc), and people. However the role of these contextual factors is often neglected in research related to disease transmission. It is important to understand how these urban features affect disease pattern, especially for multi-host pathogens, such as WNV, where birds and mosquitoes and their interaction with the natural and built environment are fundamental to disease transmission to humans. Ruiz *et al.,* conducted an comparative analysis of association of WNV infected human cases and urban landscape features in Chicago and Detroit (Ruiz et al., 2007). In both the study sites, infected cases were significantly higher in the urban suburbs. The dominating urban features were high housing density, 1940–1960 period housing, and moderate vegetation cover (Ruiz et al., 2007). However, the study *only* explored the association of infected human cases. The occurrence of WNV infection among mosquitoes is a necessary precursor for human infection and therefore it is important for effective intervention strategies to explore whether the spatial variability of association of mosquito infection follows the same pattern as found in human illness. This study fills this research gap. The objectives are as follows: (1) analyzing the association of urban environmental features that facilitated the viral activities of WNV infection in the TCMA from 2002 to 2007 and (2) comparing the spatial association between WNV infected mosquito pools and human cases with heterogeneous urban features.

The ongoing urbanization in the seven county metropolitan area, including counties of Anoka, Hennepin, Ramsey, Dakota, Scott, Carver and Washington, has profound implications to shape the environmental and socioeconomic characteristics of the region (Fig. 7.1). This 7,700 km^2 seven-county area is the economic hub of a multistate region. Home to 2.8 million people, and forecasted to top 3.5 million by 2020, it is also a major center of sprawl (Ghosh and Manson, 2008). The rapid expansion of rural areas into urban, suburban, and exurb agglomerations, buffered from others by undeveloped land, is an ideal setting for examining the association of urban features and disease transmission. The WNV infection first appeared in the

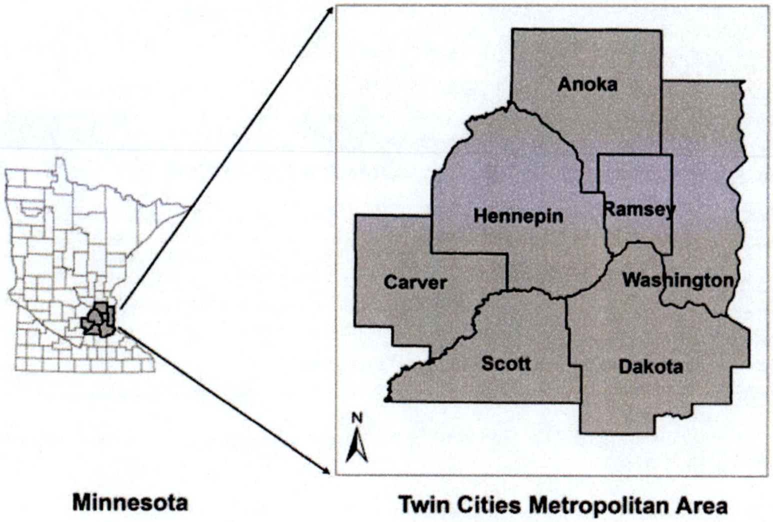

Fig. 7.1 The study: Twin Cities Metropolitan Area of Minnesota

TCMA in 2002. In 2003, clusters of WNV incidences were found in the Twin Cities of Minneapolis and Saint Paul and its surrounding suburban areas. Again in 2006, there was evidence of strong spatial clustering with 480 infected dead birds and approximately 1500 mosquitoes in 90 WNV-positive mosquito pools. These pools are mosquito traps designed to collect mosquitoes of different species for viral analysis on a weekly basis (MMCD, 2004). MMCD has three types of collection traps for mosquitoes, which are distributed throughout the metropolitan area. The trap types are CO2 traps, gravid traps, and sweep nets. The CO2 traps are elevated in the tree canopy and are used for collecting female mosquito samples in their host-seeking phase. The gravid traps, located on the ground are designed to attract female mosquitoes that are seeking oviposition sites (i.e., places to lay eggs). The pattern is shown distinctly in Fig. 7.2.

This chapter is organized as follows. Section 2 briefly discusses the wide range of geospatial applications to address health and environmental issues in urban areas. Section 3 describes the data used, their sources, and how statistics and Geographic Information Science (GIS) techniques are combined in this study. Section 4 interprets the results and finally section 5 provides discussion and conclusion.

7.2 Urban Health and Geospatial Analysis

One could point to a plethora of research studies that have developed and applied geospatial techniques to study environmental health, conservation, neighborhood effects, and health disparities in urban areas (Ghosh and McMaster, 2010). There

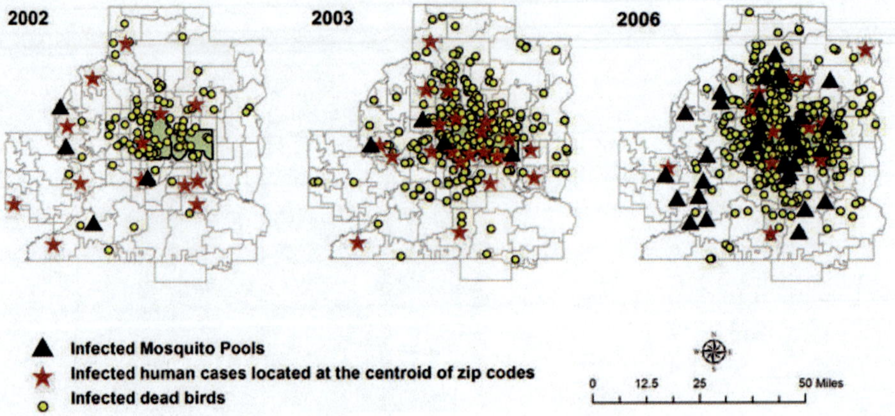

Fig. 7.2 Spatial clustering of West Nile virus occurrences in the urban and suburban areas of Twin Cities Metropolitan Area

have been numerous studies linking urban sprawl, congestion, traffic, urban waste, disturbance of natural areas within a city, and pollution (air, water, and noise) to various health effects. Health outcomes range from mental health (Leventhal and Brooks-Gunn, 2003; Watson et al., 2008; Howell and McFeeters, 2008; Clark et al., 2008; Gary et al., 2007), obesity (Ewing et al., 2003b; Saelens et al., 2003), water related infections (Drechsel et al., 2008; Greenberg et al., 2003; Aramaki et al., 2006), vector-borne diseases (Wartenberg, 1992; CDC, 1997), exposure to radioactive materials (Henriques and Spengler, 1999), lead poisoning (Wartenberg, 1992; CDC, 1997), pedestrian-vehicle accidents (Ewing et al., 2003a), to asthma and other respiratory ailments (Grineski, 2008; Oudinet et al., 2006; Schikowski et al., 2008; Jones et al., 2008; Kyrkilis et al., 2007).

In these studies, geospatial techniques are used in myriad ways. For example, GIS functions are used to identify at-risk population (maps) exposed to radioactive iodine and lead poisoning (Wartenberg, 1992; CDC, 1997). These risk maps are often used by the local health departments to prioritize interventions, while minimizing travel time and expenses in congested urban areas. Recently, in response to the need for spatio-temporal disease diffusion models or risk assessment models, a STIS – a Space Time Information System within a GIS environment – has been developed to visualize and analyze disease rates simultaneously through space and time (G. M. Jacquez and Greiling, 2003; G. M. Jacquez, D. A. Greiling, and A. M. Kaufmann, 2005; Ghosh et al., 2010). There are also several examples where a combination of GIS and Remote Sensing (RS) could prove effective controlling efforts for mosquito-borne infectious diseases such as LaCrosse encephalitis (Kitron, 1997), malaria (Beck et al., 1994), and West Nile Virus (Ruiz et al., 2004; Ozdenerol et al., 2008; Leblond et al., 2007; Zou et al., 2006; Ghosh and Guha, 2010).

Cancer is one such disease where GIS is extensively used to identify clusters (Kulldorff, 1997; Rushton et al., 2004; Lian et al., 2008; Matthews, 2007; Hsu, 2007; Kingsley, 2007; Albert, 2004; G. M. Jacquez and Greiling, 2003; G. M. Jacquez,

D. A. Greiling, and A. M. Kaufmann, 2005; Heineman, 2001; Peleg, 2000; Smith, 1995) and target areas or urban communities in need for screening, education, and testing. Given the wide range of useful applications of geospatial technologies in health research, this study combines statistics with GIS functions to assess the association of WNV infection and urban landscape features.

7.3 Data and Methodology

The incidence data on WNV infected dead birds, mosquito pools, and human cases are obtained from the Minnesota Department of Health and Metropolitan Mosquito Control District. The risk factors associated with the urban morphology of the TCMA area are divided into three categories of environmental, built-environment, and proximity factors. The Table 7.1 summarizes the variables selected for this study

Table 7.1 Description of urban landscape features hypothesized to be associated with West Nile virus transmission

Categories	Risk factors	Sources	References
Incidence data (birds, mosquito pools, and human cases)		MDH, MMCD	
Environmental	Land Cover (14 classes), density of streams/ sq.mile, elevation	NLCD, MnDNR, MetroGIS	Brownstein et al., 2002; (Ruiz et al., 2004)
Built-environment	Density of urban catch basins/sq. mile (dry), density of urban catch basins/sq. mile (wet), density of ditches/sq. mile, housing density/acre, age of houses, density of roads/sq. mile, density of population	MMCD, MnDOT, MPCA, MetroGIS,	MMCD 2004; Ruiz et al., 2004; Huhn et al. 2005; Gibbs et al. 2006; Ruiz et al., 2007
Proximity	Distance to 8 types of wetlands, distance to lakes, distance to open green space, distance to sewers, distance to waste water discharge points, distance to streams, distance to golf courses, distance to trails, distance to bike paths, distance to impaired lakes	MMCD, TLG, MnDNR, MetroGIS	Rappole et al. 2000; Cooke et al. 2006; (Zou et al., 2006)

MDH, Minnesota Department of Health; MMCD, Metropolitan Mosquito Control District; NLCD, National Land Cover Data; MnDNR, Minnesota Department of natural Resources; MnDOT, Minnesota Department of Transportation; MPCA, Minnesota Pollution Control Agency; TLG, The Lawrence Group.

with data sources and related literature. The data, both for incidence and risk factors, were obtained from multiple sources and are aggregated to the zip code level.

The study uses a combination of factorial ecology approach (FA) and geospatial analysis. The FA typically uses Factor Analysis or Principal Component Analysis (PCA) to derive uncorrelated metrics or indices from a set of correlated variables (Berry, 1971). PCA is a powerful multivariate statistical technique that can be used to simplify a dataset by reducing the number of correlated variables into a smaller number of uncorrelated principal components (PCs). The advantages of using PCA are as follows: (1) it reduces the dimensionality of a dataset by retaining PCs which explains the maximum amount of variation, (2) because the PCs are uncorrelated, multicollinearity can be avoided by using the components in place of the original variables, and (3) it is an exploratory tool to identify patterns and relationships among groups of related variables. In this analysis, 40 correlated variables, describing environment and built-environment factors conducive for the transmission of WNV in the TCMA, are reduced to a much smaller number of uncorrelated components. The *Principal Component* function in the "stats" package of R statistical programming language is used for this purpose.

The next important step is to extract the relevant PCs for further analysis. The goal here is to retain the number of components that account for as much variation as possible with the fewest meaningful components. Typically a subset of $k<p$ of components are selected by a three-part process (p = total number of PCs and k = selected PCs). First, the number of PCs is retained by examining the slope of the "scree-plot" (refer to glossary). Second, usually PCs with eigenvalues (refer to glossary) greater than one are selected. Third, through sequential selection, PCs that explain 90% of the cumulative variation of the entire data set are retained. Once the important PCs are retained, factor scores are calculated for each record (zip codes).

Hierarchical agglomerative cluster analysis (refer to glossary), developed within a Geographic Information Science (GISc) environment is conducted with the factor scores as input data. The output of the cluster analysis is the formation of five classes or groups with zip codes showing similar urban landscape characteristics. For the next part, analyzing the association between derived urban classes and WNV infected mosquitoes and human cases; two incidence rates per 100,000 people are calculated. First, the incidence rate of mosquitoes in the WNV infected pools and second, the incidence rate of human cases. A GIS *overlay* operation is used to explore the spatial association of WNV infected mosquito and human incidence rates with the urban classes. Further, in order to assess the differences in the degree to which urban classes affected the transmission, ANOVA is computed on the zip code means of the incidence rates for the different classes. The null hypothesis is H0: *There is no difference of WNV incidence rates among the urban classes.* Since the Levene's test for homogeneity of variance revealed non-constant variances, Brown-Forsythe ANOVA is used instead. This function does not assume normal distribution and homogeneity of variance. Figure 7.3 highlights the important steps of the methodological framework described above.

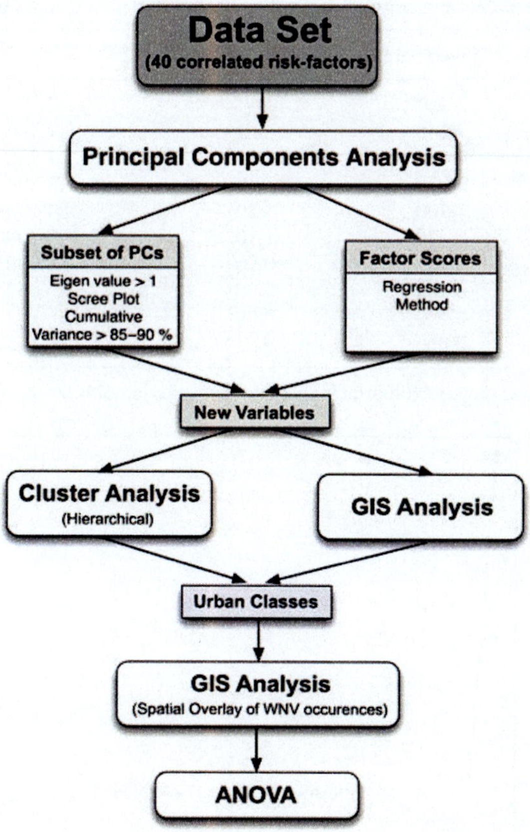

Fig. 7.3 Methodological framework

7.4 Results

Table 7.2 shows eigenvalues and cumulative variation explained by the first 10 PCs. The first five PCs together account for 84% of the total variation. The first PC explains 40% and the second 21%. There is a significant drop in the amount of variance explained from PC1 to PC2. Also, beyond PC5, the eigenvalues of the remaining components drops below one.

The scree plot (Fig. 7.4) also confirms the drop in the eigenvalues beyond PC5. When read left to right across the X-axis, this plot shows a clear separation between PCs with high-explained variance versus low-explained variance at PC5. This point of separation is termed as *elbow*. Thus, based on Table 7.2 and Fig. 7.3, the first five PCs are retained for further analysis. In the next step, factor scores are calculated for each PC, resulting in five new variables: Score1, Score2, Score3, Score4, and Score5. These new uncorrelated variable represent the features of 40 original risk factor variables. Finally, cluster analysis is conducted with these variables to obtain urban landscape classes. Figure 7.5 is a dendogram (refer to glossary), which

Table 7.2 Variance Explained by the selected Principal Components

Principal components	Eigen values	Variation (%)	Cumulative variation (%)
1	12.000	40.40	40.40
2	7.240	21.10	61.500
3	2.620	10.00	71.500
4	2.070	7.30	78.800
5	1.700	5.20	84.000
6	0.470	1.10	85.100
7	0.360	0.90	86.000
8	0.180	0.06	86.060
9	0.120	0.04	86.103
10	0.060	0.01	86.113

Note: The variances for the selected principal components are in bold.

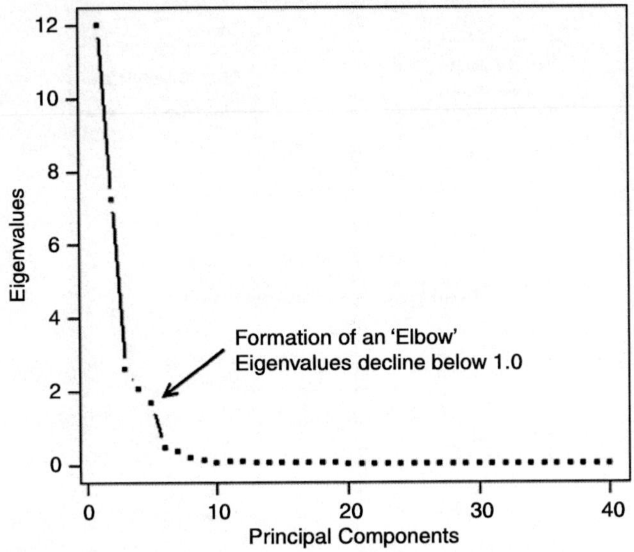

Fig. 7.4 Scree plot showing the eigen values obtained from the principal component analysis with 40 urban variables

is a graphical representation of hierarchical clustering based on similarities and dissimilarities of factor scores obtained from the PCA. The red rectangles show the grouping of zip codes into five groups. Here parsimony is achieved with a small number of clearly defined urban classes.

The five urban landscape classes and their dominant characteristics are as follows:

1. *City, High Density* – high density of urban catch basins, open green space within a distance of 1 mile, high housing density, some 70–75 year-old housing, smaller distance to sewers, and developed high density of development.

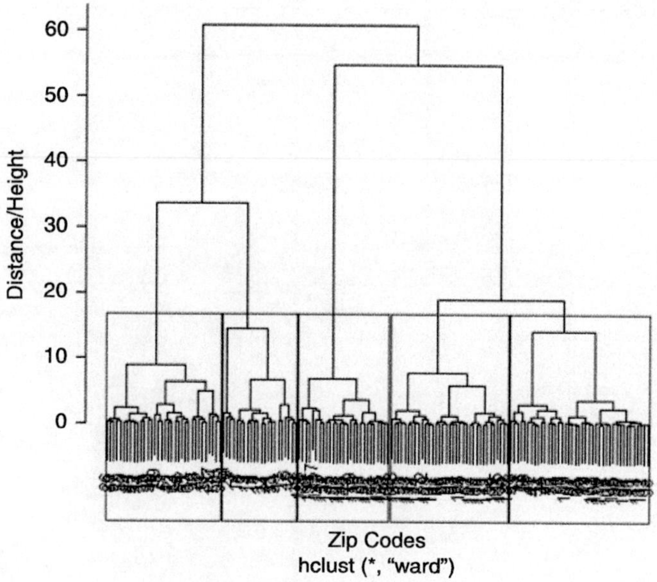

Fig. 7.5 Dendogram showing the hierarchical clustering of zip codes into 5 classes of urban landscape in Twin Cities Metropolitan Area

2. *City, Medium Density* – presence of wetlands like swamps and bogs, high density of urban catch basins, open green space within a distance of 0.5 mile, housing belonging to 1940–1960s period, and developed medium density land cover class.
3. *Suburb* – developed low density land cover class, low housing density, presence of more natural features like lakes, parks, shallow fresh marsh, and diverse vegetation.
4. *Outer Suburb 1* – Recent development houses between 20 and 25 years old, shallow fresh marsh, lakes, and diverse vegetation in the form of shrubs, pastures, and deciduous forest.
5. *Outer Suburb 2* – Agricultural area, pasture, open green space, and deciduous forest land cover.

The spatial distribution of urban classes shows a concentric pattern. The City-High Density class in the center occupies the cities of Minneapolis and Saint Paul. The City-Medium Density encircles the first class and includes cities such as Golden Valley, St. Louis Park, Bloomington, and Richfield. The Suburb class is on the cusp between more urban classes (City-High Density and City-Medium Density) and relatively rural classes (Outer Suburb 1 and Outer Suburb 2). Some of the important small cities and towns classified in the Suburb class are Coon Rapids, Shoreview, and White Bear Lake in the North, and Shakopee, Inner Grove Heights, Rosemount, and Lakeville in the south. From Outer Suburb 1 class one can start to see relatively

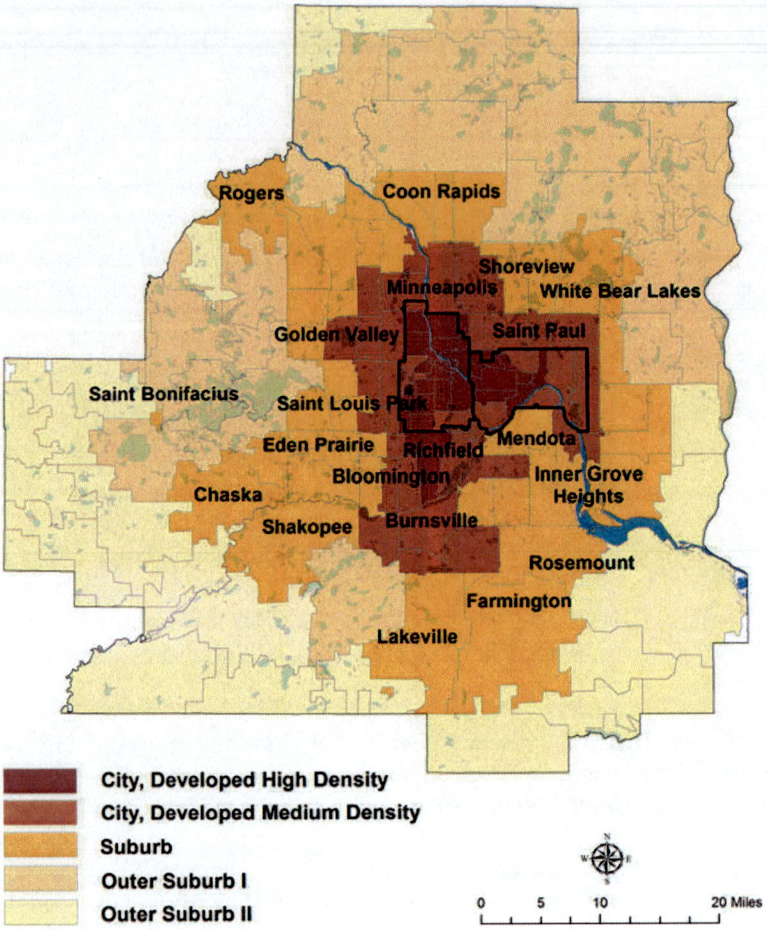

Fig. 7.6 Five urban landscape classes in Twin Cities Metropolitan Area

rural characteristics, which extended further to the Outer Suburb 2. Figure 7.6 shows the spatial distribution of the urban classes.

Overlay of infected dead birds, positive mosquito pools, and human cases on the five urban landscape classes in TCMA depicts a strong spatial association with the City-High density and the City-Medium density classes consistently from 2002 to 2007 (Fig. 7.7). In general the rate of WNV infection decreases with increasing distance from the core urban areas of TCMA.

The highest incidence rate of WNV infected dead birds is found in the City-Medium Density class (1,956.95 per 100,000 people) and the rate of infected dead bird reporting declines significantly in the outer suburb classes (Table 7.3). However, this pattern could be due to urbanization and higher population density. It is possible that in the developed areas, dead birds are easily sighted (and reported)

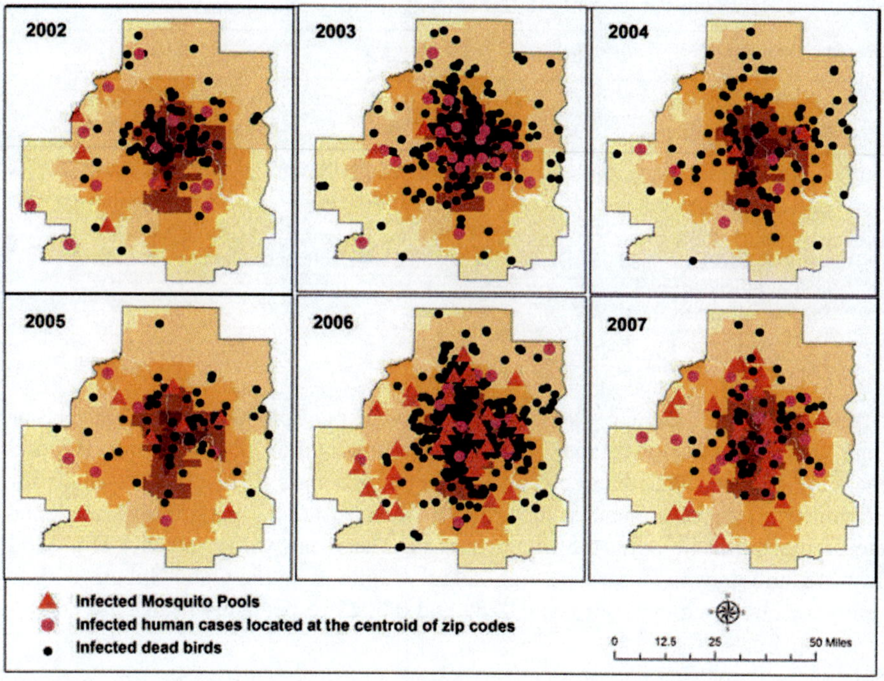

Fig. 7.7 Spatial "overlay" of West Nile virus incidences on the derived urban landscape classes in Twin Cities Metropolitan Area, 2002–2007

Table 7.3 Twin Cities Metropolitan urban classes with West Nile virus case rates among reported infected dead birds, positive mosquito pools, and humans from 2002 to 2006

Urban class	Description	Number of zip codes	Population	WNV dead birds per 100 k	Mosquitoes in WNV infected pool per 100 k	WNV human cases per 100 k
1	City-high density	27	545,267	912.89	88,204.06	9.19
2	City-medium density	36	842,625	1,956.95	66,162.71	11.12
3	Suburban	46	928,789	1,071.54	29,440.40	6.18
4	Outer suburb 1	28	360,382	809.12	10,127.02	2.85
5	Outer suburb 2	22	152,398	706.39	5,085.07	2.91

than less developed areas. Also, in remote rural areas there are fewer people to notice dead birds. In the case of vector population, the City-High density urban class has the highest incidence rate of mosquitoes tested from the infected mosquito pools (88,202.06 per 100,000 people). This rate also declines further as we move outwards from the core cities of Minneapolis and Saint Paul. The Outer Suburb 2 has the lowest vector incidence rate of 5,085.07. Following the similar trend of WNV illness among birds and mosquitoes, the highest incidence rate among humans is reported

Table 7.4 ANOVA results of incidence rate of mosquitoes

Urban classes	Estimate	T-value	P-value	Sig.
City-high density	1,838.61	2.235	0.0269	*
City-medium density	3,035.70	2.907	0.0042	**
Suburb	743.85	0.203	0.8393	
Outer suburb 1	130.50	0.126	0.8999	
Outer suburb 2 +				

F-statistic: 3.535 on 4 and 154 DF, p-value: 0.00863. +, Reference class.
* 95% significance level.
** 99% significance level.

in the City-Medium Density class, with 11.12 cases per 100,000 people. This is four times higher than the lowest rate of 2.85 cases found in the Outer Suburb 2 class and it was almost two times the rate found in the Suburb class (6.18). ANOVA tests for differences in zip code mean values further demonstrate the variation in incidence rate of mosquitoes among the urban classes. The mean values of City-High and City-Medium density classes are significantly higher than the Outer Suburb 2 class (reference class) with p-values as 0.0269 and 0.0042 respectively (Table 7.4).

7.5 Discussion and Conclusion

Utilizing a combination of factorial ecology and geospatial approach, the derived urban landscape classes indicate a variable association with WNV illness in the TCMA. The rate of WNV infection among mosquitoes and humans are highest in the core urban classes of City-High Density and City-Medium Density classes and the rate decreases with increasing distance from the urban core. It is typical to assume that more natural areas found away from urbanized areas would provide more potential habitats for mosquitoes. However, the degree to which an area is natural does not decrease linearly along a transect line outward from the urban core. This generalization is often simplistic. In the City-High and City-Medium density classes, along with significant presence of built area, there are natural areas in the form of lakes, parks, wetlands, golf courses, trails, older residential and commercial buildings, and wedges along old transportation routes, which could very well provide suitable habitats for mosquito breeding. Some of the features of built-environment, such as urban storm water catch basins, construction sites, stock pile of abandoned tires, and swimming pools in the backyards of residential houses, are attractive breeding grounds for mosquitoes. Such composition of urban landscape with natural and man-made features, typically affected by past land use and planning, creates an urban heterogeneous environment suitable for mosquito habitats and thus increase the risk of WNV transmission. This is critical because two out of four WNV carrying vectors, *Culex restuans* and *Culex pipiens,* are predominantly urban mosquitoes.

The specific urban features that contributed to the viral activities of WNV are catch basins, housing density, age of houses, and proximity to lakes, swamps, bogs, and parks. Density of urban catch basins and storm water ponds, built primarily to accumulate polluted urban run-off, emerged both as a predominant feature of urban structure and is strongly associated with WNV occurrences. The flow of water (which would lessen during droughts) and the presence of organic matter in these catch basins would positively affect the breeding conditions for *Culex* mosquitoes (Shaman et al., 2005). Housing characteristics, namely the density of buildings (both residential and commercial) and the age of housing, play an important role in the viral activity of the virus (Ruiz et al., 2007). The age of housing, especially "older houses" approximately 50–60 years old, emerges as an important factor. Plausible explanation for this could be that in lower income neighborhoods older houses are not well maintained. However, future analysis including the interaction effect of income and age of houses could better explain this positive association. Additional variables differentiating residential, commercial, and industrial use, impervious surface, and knowledge of soil characteristics could also be helpful. These variables could provide a better understanding of natural open space and impervious surface, which would be helpful in identification of potential mosquito habitats.

Since, in a typical WNV transmission cycle, infected mosquitoes are a necessary prerequisite for human infections, this study analyzed whether the spatial variability of mosquito infection shows the same patterns as those found in human cases in the TCMA. The GIS overlay shows a strong spatial clustering of WNV infected dead birds, positive mosquito pools, and human cases in the urban/suburban areas (Fig. 7.7). The incidence rate of human cases per 100,000 people is highest in the City-Medium Density class followed by City-High Density and Suburb class. City-High density class shows the highest rate of mosquitoes tested in the infected mosquito pools. The difference in incidence rates for the mosquitoes and humans between the City-High and City-Medium density is not large. In addition, the ANOVA results statistically demonstrated that the means of mosquito incidence rates for zip codes in City-Medium and City-High density urban classes are significantly different than the other classes. Thus, the results ranging from exploratory to confirmatory analysis provide evidences that both the incidence rates of infected mosquitoes and human cases follow similar spatial pattern.

The derived urban landscape classes could further provide a basis for the selection of field sites for MMCD for mosquito traps, collection, and treatment. Finally, this study also contributes to the broader research question in the field of medical and health geography, i.e., how the heterogeneous urban environment affects human health and disease patterns.

7.6 Glossary

Eigenvalue: A vector which, when acted on by a particular linear transformation, produces a scalar multiple of the original vector. In the context of Principal Component and Factor Analysis, eigenvalues are column sums of squared loadings for a factor. It conceptually represents that amount of variance accounted for by a factor.

Scree Plot: A scree diagram plots the variances of principle components on the y-axis against the component number on the x-axis. The term scree refers to the fact that the variance curve resembles the side of a mountain with a scree, or *rock debris*, at the base. When read left to right across the x-axis, this plot shows a clear separation between principle components with high-explained variance versus low-explained variance. Typically the variance curve takes a shape of a half folded arm and the point of separation is termed as "elbow".

Hierarchical Clustering: Cluster Analysis is a process involving grouping a collection of objects (also called observations, individuals, cases, or data rows) into subsets or "clusters", such that those within each cluster are more closely related to one another than objects assigned to different clusters. Central to all of the goals of cluster analysis is the notion of degree of similarity (or dissimilarity) between the individual objects being clustered. There are two major methods of clustering – hierarchical clustering and k-means clustering. In hierarchical clustering the data are not partitioned into a particular cluster in a single step. Instead, a series of partitions takes place, which may run from a single cluster containing all objects to *n* clusters each containing a single object. Hierarchical Clustering is subdivided into *agglomerative* methods, which proceed by series of fusions of the n objects into groups based on the degree of similarity (or dissimilarity).

Dendogram: Dendogram is a branching (tree) diagram representing the steps in a hierarchical cluster analysis. The tree starts with a single cluster (including all the observations) and then branches out into *n* clusters of observation based on the degree of similarity (or dissimilarity) of the observations.

Author Biography

Debarchana Ghosh, Assistant Professor, Kent State University. Dr. Ghosh's research employs an interdisciplinary approach to investigate health and environmental issues. She combines health and medical geography concepts and theories with geospatial technologies such as Geographic Information Science (GIS), modeling, machine learning, and statistics. Specific interests include spatiotemporal modeling of diseases (chronic and infectious), toxic releases, pollution, and their association with unsustainable use of urban and environmental resources, environmental justice, and health disparities.

Her work in the field of health and medical geography started with her Master's thesis titled 'Exposure to Mass Media and its Effect on Mothers Seeking Prenatal Care Services.' For her Ph.D degree, from the University of Minnesota, she applied and developed geospatial techniques to understand the dynamics of West Nile Virus (WNV) in the Twin Cities. Her dissertation is titled 'A Geospatial Analysis of West Nile Virus in the Twin Cities Metropolitan Area of Minnesota.' Currently an Assistant Professor at Kent State University, Dr. Ghosh is working on (1) geospatial Analysis of Psychiatric Mental Health Advanced Practice Nurses in the United States; (2) Analysis of H1N1 cases in the United States and Mexico; (3) Identification of 'Food Deserts' in the Northeast Ohio; and (4) how knowledge of HIV/AIDS is disseminated through intervention programs using social network analysis.

Her research is published in several peer-reviewed journals such as *Computers, Environment and Urban System, Social Computing, Behavioral Modeling, and Prediction, Cartography and Geographic Information Science Journal, Environmental and Planning B, Social Science and Medicine, Geography Compass, The URISA Journal, and Professional Geographer.*

Dr. Ghosh is also an elected member of the Health and Medical Geography Specialty Group of the Association of American Geographers.

References

Albert SM (2004) Cancer screening among older women in a culturally insular community. [Article]. Prevent Med 39(4):649–656 doi:10.1016/j.ypmed.2004.06.016

Aramaki T, Galal M, Hanaki K (2006) Estimation of reduced and increasing health risks by installation of urban wastewater systems. Water Sci Technol 53(9):247–252

Beck LR, Rodrigues MH, Dister SW, Rodrigues AD, Rekmankova E (1994) Remote sensing as a landscape epidemiologic tool to identify villages at high-risk for malaria transmission. Am J Trop Med Hyg 51:271–280

Berry B (1971) City classification handbook: methods and applications. Wiley, New York, NY

Brownstein JS, Rosen H, Purdy D, Miller JR, Merlino M, Mostashari F (2002) Spatial analysis of West Nile virus: rapid risk assessment of an introduced vector-borne zoonosis. Vector Borne Zoonotic Dis 2:157–164

CDC, C. f. D. C. a. P (1997) Screening young children for lead poisoning: Guidence for state and local public health officials, atlanta, ga.

Clark C, Ryan L, Kawachi I, Canner MJ, Berkman L, Wright RJ (2008) Witnessing community violence in residential neighborhoods: a mental health hazard for urban women. J Urban Health Bull N Y Acad Med 85(1):22–38

Drechsel P, Keraita B, Amoah P, Abaidoo RC, Raschid-Sally L, Bahri A (2008) Reducing health risks from wastewater use in urban and peri-urban sub-saharan africa: applying the 2006 who guidelines. Water Sci Technol 57(9):1461–1466

Ewing R, Schieber R, Zeeger C (2003a) Urban sprawl as a risk factor in motor vehicle occupant and pdestrain fatalities. Am J Public Health 93:1541–1545

Ewing R, Schmid T, Killingsworth R, Zlot A, Raudenbush S (2003b) Relationship between urban sprawl and physical activity, obesity, and morbidity. Am J Public Health 18:47–57

Gary TL, Stark SA, LaVeist TA (2007) Neighborhood characteristics and mental health among african americans and whites living in a racially integrated urban community. Health Place 13(2):569–575

Ghosh D, Guha R (2010) Identifying optimal risk factors for prediction and interpretation of west nile virus incidences using genetic algorithm and neural network techniques. Computers, environment and urban systems. *Computers, Environment and Urban Systems, Under minor revisions*

Ghosh D, Manson SM (2008) Robust principle component analysis and geographically weighted regression: urbanization in the twin cites metropolitan area of minnesota. URISA J 20(1):15–25

Ghosh D, Manson SM, McMaster RB (2010, April) Delineating west nile virus transmission cycles at various scales: the nearest neighbor distance-time model. Cartogr Geogr Inf Sci J 37(2): 149–163

Ghosh D, McMaster RB (2010) Chapter 34: Biophysical and human-social applications. In: JD Bossler (ed) Manual of geospatial science and technology, 2/e, vol 2. Taylor and Francis Group LLC, Boca Raton, FL, pp 669–687

Gibbs SEJ, Wimberly MC, Madden M, Masour J, Yabsley MJ, Stallknecht DE (2006) Factors affecting the geographic distribution of West Nile virus in Georgia, USA: 2002-2004. Vector Borne Zoonotic Dis 6(1):73–82

Greenberg M, Mayer H, Miller K, Hordon R, Knee D (2003) Reestablishing public health and land use planning to protect public water supplies. Am J Public Health 93(18):1522–1526

Grineski S (2008) Coping with asthma in the central city: parental experiences with children's health care. J Health Care Poor Underserved 19(1):227–236

Heineman E (2001) Geographic information system for the long island breast cancer study. Epidemiology 12(4):435 [Meeting Abstract]

Henriques WD, Spengler RF (1999) Locations around the hanford nuclear facility where average milk consumption by children in 1945 would have resulted in an estimated median iodine-131 dose to the thyroid of 10 rad or higher, washington. J Public Health Manag Pract 5(2):35–36

Howell E, McFeeters J (2008) Children's mental health care: differences by race/ethnicity in urban/rural areas. J Health Care Poor Underserved 19(1):237–247

Hsu CE (2007) A spatial-temporal approach to surveillance of prostate cancer [Article]. J Natl Med Assoc 99(1):72–87

Huhn G, Austin C, Langkop C, Kelly K, Lutch R, Lampman R et al (2005) The emergence of West Nile virus during a large outbreak in Illinois in 2002. Am J Trop Med Hyg 72:768–776

Jacquez GM, Greiling DA, Kaufmann. AM (2005) Design and implementation space – time intelligence system for disease surveillance. J Geogr Syst 7(1):7–23

Jacquez GM, Greiling DA (2003) Geographic boundaries in breast, lung and colorectal cancers in relation to exposure to air toxics in long island, New York. Int J Health Geogr 17(2:4):1–22

Jones AYM, Lam PKW, Gohel MDI (2008) Respiratory health of road-side vendors in a large industrialized city. Environ Sci Pollut Res 15(2):150–154

Kingsley BS (2007) An update on cancer cluster activities at the centers for disease. Environ Health Perspect 115(1):165–171 [Article]doi:10.1289/ehp.9021

Kitron U (1997) Spatial analysis of the distribution of lacross encephalitis in illinois, using a geographic information sustem and local and global statistics. Am J Trop Med Hyg 57: 469–475

Kulldorff M (1997) Breast cancer clusters in the northeast united states: a geographic analysis. Am J Epidemiol 146(2):161–170

Kyrkilis G, Chaloulakou A, Kassomenos PA (2007) Development of an aggregate air quality index for an urban mediterranean agglomeration: relation to potential health effects. Environ Int 33(5):670–676

Leblond A, Sandoz A, Lefebvre G, Zeller H, Bicout DJ (2007) Remote sensing based identification of environmental risk factors associated with west nile disease in horses in camargue, france. Prevent Veterinary Med 79(1):20–31 [Article]

Leventhal T, Brooks-Gunn J (2003) Moving to opportunity: An experimental study of neighborhood effects on mental health. Am J Public Health 83:1576–1582

Lian M, Jeffe DB, Schootman M (2008) Racial and geographic differences in mammography screening in st. Louis city: a multilevel study. J Urban Health Bull N Y Acad Med 85(5):677–692

Matthews BA (2007) Colorectal cancer screening among midwestern community-based residents. J Commun Health 32(2):103–120 [Article]doi:10.1007/s10900-006-9038-0

MDH (2003) West nile virus maps and statistics. http://www.health.state.mn.us/divs/idepc/diseases/westnile/

MMCD (2004) 2004 operational review & plans for 2005. Annual report to the technical adviosry board. St. Paul, Minnesota: Metropolitan Mosquito Control District

Oudinet JP, Meline J, Chehmicki W, Sanak M, Magdalena DW, Besancenot JP et al (2006) Towards a multidisciplinary and integrated strategy in the assessment of adverse health effects related to air pollution: the case study of cracow (poland) and asthma. Environ Pollut 143(2):278–284

Ozdenerol E, Bialkowska-Jelinska E, Taff GN (2008) Locating suitable habitats for west nile virus-infected mosquitoes through association of environmental characteristics with infected mosquito locations: a case study in shelby county, tennessee. Int J Health Geogr 7:12

Peleg I (2000) The use of a geographic information system (gis) to identify small. Gastroenterology 118(4):3930 [Meeting Abstract]

Rappole JH, Derrickson SR, Hubalek Z (2000) Migratory birds and spread of West Nile Virus in the western hemisphere. Emerg Infect Dis 6(4):319–328

Ruiz MO, Tedesco C, McTighe TJ, Austin C, Kitron U (2004) Environmental and social determinants of human risk during a west nile virus outbreak in the greater chicago area, 2002. Int J Health Geogr 3(8):2–11

Ruiz MO, Walker ED, Foster ES, Haramis LD, Kitron UD (2007) Association of west nile virus illness and urban landscapes in chicago and detroit. Int J Health Geogr 6(10):1–11

Rushton G, Peleg I, Banerjee I, Smith G, West. M (2004) Analyzing geographic patterns of disease incidence: rates of late-stage colorectal cancer in iowa. J Med Syst 28(3):223–236

Saelens B, Sallis J, Black J, Chen D (2003) Neighborhood-based difference in physical acitivity: An environment scale evaluation. Am J Public Health 93:1552–1558

Schikowski T, Sugiri D, Reimann V, Pesch B, Ranft U, Kramer U (2008) Contribution of smoking and air pollution exposure in urban areas to social differences in respiratory health. BMC Public Health 8:179

Shaman J, Day JF, Stieglitz M (2005) Drought-induced amplification and epidemic transmission of west nile virus in southern florida. J Med Entomol 42(2):134–141

Smith TJ (1995) Differences in initial treatment patterns and outcomes of lung cancer. [Article]. Lung Cancer 13(3):235–252

Wartenberg D (1992) Screening for lead exposure using geographic informaiton system. Environ Res 59:310–317

Watson JM, Logan HL, Tomar SL (2008) The influence of active coping and perceived stress on health disparities in a multi-ethnic low income sample. BMC Public Health 8:41

Zou L, Miller SN, Schmidtmann ET (2006) Mosquito larval habitat mapping using remote sensing and gis: implications of coalbed methane development and west nile virus. J Med Entomol 43(5):1034–1041 [Article]

Chapter 8
The Health Impacts of Brownfields in Charlotte, NC: A Spatial Approach

Junfeng Wang

Abstract Brownfield redevelopment in Charlotte, North Carolina has been a success in terms of leveraging private investment, increasing tax bases, and creating job opportunities. Little is known, however, about the potential health impacts of these brownfield sites in the city. This research intends to fill this gap by examining the effect of brownfield sites on neighborhood Low Birth Weight (LBW) rate. The health impact is measured as a function of proximity to brownfield sites, inactive hazardous site density, population's economic status, and the community's socio-economic attributes. The analyses show that being close to brownfield sites is not significantly related to having a higher rate of Low Birth Weight, but the density of brownfields in the census block group is related to a higher LBW rate. The Geographically Weighted Regression (GWR) model reveals that there is a considerable spatial variation in the strength of the health impacts. The local health department has indicated lack of capacity to examine the variation in community health status. The findings of this study can serve as the starting point for local health professionals to identify communities that are impacted by brownfields the most, and therefore more actively participate in brownfield redevelopment.

Keywords Brownfield · Low birth weight · Geographically weighted regression · Buffer · Census block group

8.1 Introduction

Brownfields are "real properties, the expansion, redevelopment, or reuse of which may be complicated by the presence or potential presence of a hazardous substance, pollutant, or contaminant" (Public Law 107–118; H.R. 2869). The EPA estimates that there are more than 450,000 brownfields in the US, most of which are the result of past industrial activities that included the release of hazardous substances to the

---

J. Wang (✉)
Department of Political Science, University of North Carolina at Pembroke, Pembroke, NC 28372, USA
e-mail: jwang86@uis.edu

J.A. Maantay, S. McLafferty (eds.), *Geospatial Analysis of Environmental Health*, Geotechnologies and the Environment 4, DOI 10.1007/978-94-007-0329-2_8,

soil or water on the property. Many of these sites no longer support active industrial or commercial uses and have been abandoned.

Brownfields can be found in most US cities that were centers of industrial activity in the nineteenth century, but whose economies have largely changed in recent years to service and financial sectors. The numbers, sizes, and contamination severity of brownfields in a city depend on its industrial structure. For instance, after the Civil War, the City of Charlotte became a cotton processing center and a railroad hub for the Piedmont region of the Carolinas. In the 1970s, Charlotte started emerging as a banking and finance center. Closed textile facilities became important brownfield redevelopment targets.

A variety of toxicants have been found in brownfields such as heavy metals, solvents, arsenic, polycyclic aromatic hydrocarbons, plasticizers, and insecticides. EPA divides toxicants into two categories: carcinogens and non-carcinogens. Carcinogens are toxicants that are known or suspected to cause cancer. The US EPA Region 9's preliminary remediation goal (PRG) Table provides a reference of chemical concentrations that correspond to fixed levels of risks in soil, air, and water (i.e. a one-in-one million (10^{-6}) cancer risk). Each state then develops its own remediation goals according to the EPA standards.

The exposure risk is the chemical concentration found on site divided by the PRG reference dose for that particular chemical and then multiples by 10^{-6}. For multiple toxicants, simply add the risk for each chemical. Table 8.1 displays the carcinogens found in Charlotte brownfield sites that are above the EPA remediation standards. The last column shows the potential cancer risk that is associated with particular sites.

In October 1996, Charlotte received its first Brownfields Assessment Demonstration Pilot grant of $200,000 from the EPA. The former industrial area known as the South End and the nearby economically depressed Wilmore neighborhood were selected for the pilot program. The Brookhills neighborhood, another depressed area contiguous with Wilmore, was later added to the area to be redeveloped.

Once a blighted area, thousands of people currently are working in the South End. This area has attracted $800 million new development investment (New Ventures, 2006). The great success of the South End redevelopment made Charlotte a national example of brownfields cleanup and reuse.

In May 1999, EPA awarded Charlotte a $500,000 Brownfields Cleanup Revolving Loan Fund (BCRLF) grant. In March 2000, Charlotte received a supplemental assistance grant of $100,000 from EPA for its Brownfields Assessment Demonstration Pilot. With these two grants and a City appropriation of $140,000, the Charlotte City Council approved the creation of the City's Brownfield Assessment Program and started offering environmental assessment and clean up assistance throughout Charlotte's entire distressed geographic areas.

Charlotte's brownfield program has demonstrated its effectiveness in terms of its impact on the local tax base, employment, and other socio-economic measures (Bacot and O'dell, 2006; Schwarz and Hanning, 2007; Chilton, 2007). What is not clear concerning the development of the Charlotte Brownfield Program is whether

Table 8.1 Carcinogens cancer risk calculation table

Charlotte Brownfield site	Soil contaminants	Concentration (mg/kg)	EPA standard (mg/kg)	Ratio	Cancer risk	Site cancer risk
ABC engraves	Hexavalent chromium	180	30	6.00	6.0E–06	6.0E–06
Alpha mills	Arsenic	42.4	0.39	108.72	1.1E–04	1.1E–03
	Chromium VI	474	30	15.80	1.6E–05	
	Tetrachloroethylene	78	0.48	162.50	1.6E–04	
	Trichloroethylene	11	0.053	207.55	2.1E–04	
	1,4-Dichlorobenzene	42	3.4	12.35	1.2E–05	
	Indeno(1,2,3-cd)pyrene	17	0.62	27.42	2.7E–05	
	Dibenz(a,h)anthracene	5	0.062	80.65	8.1E–05	
	Benzo(k)fluoranthene	22	6.2	3.55	3.5E–06	
	Benzo(b)fluoranthene	32	0.62	51.61	5.2E–05	
	Benzo(a)pyrene	23	0.062	370.97	3.7E–04	
	Benzo(a)anthracene	24	0.62	38.71	3.9E–05	
American cyanamid	Arsenic	17	0.39	43.59	4.4E–05	4.4E–05
Camden square	Trichloroethylene	0.52	0.053	9.81	9.8E–06	7.6E–05
	Benzo(a)pyrene	1.6	0.062	25.81	2.6E–05	
	Benzo(b)fluoranthene	1.5	0.62	2.42	2.4E–06	
	Dibenz(a,h)anthracene	1	0.062	16.13	1.6E–05	
	Arsenic	8.47	0.39	21.72	2.2E–05	
Cherokee oil	Arsenic	234	0.39	600.00	6.0E–04	6.0E–04
Dynatech industries	Arsenic	3.4	0.39	8.72	8.7E–06	8.7E–06

Table 8.1 (continued)

Charlotte Brownfield site	Soil contaminants	Concentration (mg/kg)	EPA standard (mg/kg)	Ratio	Cancer risk	Site cancer risk
Hamilton property	Arsenic	4	0.39	10.26	1.0E–05	2.1E–05
	Methylene chloride	94	9.1	10.33	1.0E–05	
Home depot	PCB-1254	4.1	0.22	18.64	1.9E–05	1.9E–05
Midtown	Trichloroethylene	0.059	0.053	1.11	1.1E–06	1.1E–06
Rea asphalt plant	Arsenic	12	0.39	30.77	3.1E–05	3.1E–05
Rusak property	Hexavalent chromium	1000	30	33.33	3.3E–05	3.3E–05
Sonoco flexible packaging facility	Hexavalent chromium	47.8	30	1.59	1.6E–06	1.6E–06
Terrell machine	Cis-1,3-dichloropropene	28	0.78	35.90	3.6E–05	2.2E–03
	Methylene chloride	98	9.1	10.77	1.1E–05	
	Tetrachloroethylene	430	0.48	895.83	9.0E–04	
	Trichloroethylene	66	0.053	1245.28	1.2E–03	

Note: The chemical concentration was from Brownfield agreements, and the cleanup standard was from NC Inactive Hazardous Sits Branch Health-Based Soil Remediation Goals and EPA Region 9 PRG InterCalc Tables.

public health was considered in these policy decisions. Using low birth weight (LBW) rates as an example of potential adverse health outcomes, this study fills the gap by examining the association between brownfield and nearby neighborhoods LBW rates.

8.2 Exposure to Hazardous Waste Sites and Low Birth Weight

Two politically influential and well-cited works by the General Accounting Office (GAO, 1983) and the United Church of Christ (UCC, 1987) found that a high proportion of minority and poor lived in communities with large number of hazardous waste sites. However, the extent to which these hazardous waste sites affect public health in surrounding communities remains uncertain.

Multiple studies have found that low birth weight was significantly elevated in the exposure zones closest to hazardous sites (Vianna and Polan, 1984; Goldman et al., 1985; Goldberg et al., 1995; Baibergenova et al., 2003). Other studies have indicated that the magnitude of the effect was either in the range of birth weight reduction due to smoking during pregnancy (Berry and Bove, 1997), or disappeared after taking into account risk factors such as smoking, chronic disease, and young maternal age (Oliveira et al., 2002).

Socio-economic and demographic variables such as education, race/ethnicity, and income have been studied extensively in epidemiological research, and these factors could explain much of the observed differences in health status (Jolley et al., 1992; Sexton et al., 1993; Adler and Ostrove, 1999). These studies generally concluded that disadvantaged and minority groups who live in areas with high levels of pollution suffer higher rates of disease and death, even after controlling for social class and ethnicity/race.

8.3 Data and Analytical Method

Before the year 2000, only a few brownfield sites were redeveloped in Charlotte. This study assesses the relationship between the health outcomes, measured by census block group LBW rates in the year 2000, and brownfields which have not been redeveloped yet. Data were collected from the North Carolina State Center for Health Statistics, the North Carolina Department of Environment and Natural Resources, the Charlotte Economic Development Office, the US Census, and the Charlotte Chamber of Commerce.

The study is conducted at the census block group level. Previous studies that examined the relationships between toxic sites and public health typically were conducted at the census tract or zip code levels. By focusing on a smaller geographic unit, this study revealed more detailed and accurate information about the potential associations among health problems, socio-economic status, and brownfields.

8.3.1 Study Area

Charlotte is the largest city in North Carolina and is the county seat of Mecklenburg County. There are 373 census block groups in Mecklenburg County. Using the "Clip" function in GIS, whereby the boundaries of one feature are clipped as with a "Cookie-cutter" by the boundaries of another feature, the 329 block groups within the Charlotte city boundary were selected. The selection criterion was that if the centroid of the block group is located within the city boundary, then the block group is considered in the analysis.

8.3.2 Dependent Variables

In this study, low birth weight infants are those live singleton infants with a birth weight less than 2,500 g. Low birth weight rate is then defined as the number of low birth weight infants divided by the total live singleton infants. Figure 8.1 shows that

Fig. 8.1 Low birth weight rate, Charlotte, 2000

LBW rates in block groups located in the west of the city were more likely to be higher than other areas in the year 2000.

Birth weight is one of the strongest predictors of the risk of infant mortality (Institute of Medicine, 1985). If the risk of mortality were same for black and white newborns, over 60% of the deaths of black infants, or about 5,000 deaths in the US, would be averted each year (Sastry and Hussey, 2003). The impacts of low birth weight extend well beyond infant survival. Studies have found that significant associations between birth weight and physiological, developmental, reproductive outcomes, and chronic diseases (Avchen et al., 2001; Rich-Edwards et al. 1997; Reichman, 2005).

8.3.3 Independent Variables

A common weakness in epidemiologic studies is the lack of a measure of the direct exposure to toxicants. When exposure monitoring is unavailable, whether and to what extent chemicals from hazardous waste sites reach the host is largely unknown. One approach is to estimate the exposure based on the presence of hazardous sites. For example, Baibergenova et al. (2003) defined the exposed groups as residing in a zip code containing a PCB site, and comparison groups were defined as residing in a zip code that did not have a PCB site. More commonly, the distance from mother's residence to landfill is used as the measure for exposure (Goldberg et al., 1995; Berry and Bove, 1997; Oliveira et al., 2002). These studies have employed Geographic Information System (GIS) and quantitative techniques to assess the association between hazardous waste sites and adverse reproductive outcomes. The limitations of these methods to estimate exposure are discussed fully in Chakraborty and Maantay's Chapter 5 of this volume.

This study considers brownfields that have enrolled in the Charlotte Brownfield Program. Figure 8.2 shows the geographic locations of Charlotte brownfield sites. These brownfield sites are concentrated in areas around the city center and along highway I-77. This concentration reflects the old land use pattern in Charlotte, as these areas are the old business corridors, and also because the city intentionally encourages redevelopment in these areas. From the very beginning of the Charlotte Brownfield Program, the Economic Development Office, which administrates the program, has put economic success as the top priority. The South End area has the most market potential and was redeveloped first. Then the city decided to expand the Brownfield Program to all sites within the Charlotte Business Corridor Revitalization Geography (Fig. 8.3).

The impact of brownfields is first measured in brownfield density, which is the size of total acres of brownfields in a census block group dividing by the land size of the census block group. To capture the impact of nearby brownfields outside the census block group, a 0.5 mile buffer was created for each brownfield site (Fig. 8.3).

The selection of 0.5 mile is based on the assumption that brownfields outside this buffer have negligible health impacts (Lybarger et al., 1998). Each block group could contain parts of multiple buffers. The number of buffers that completely or

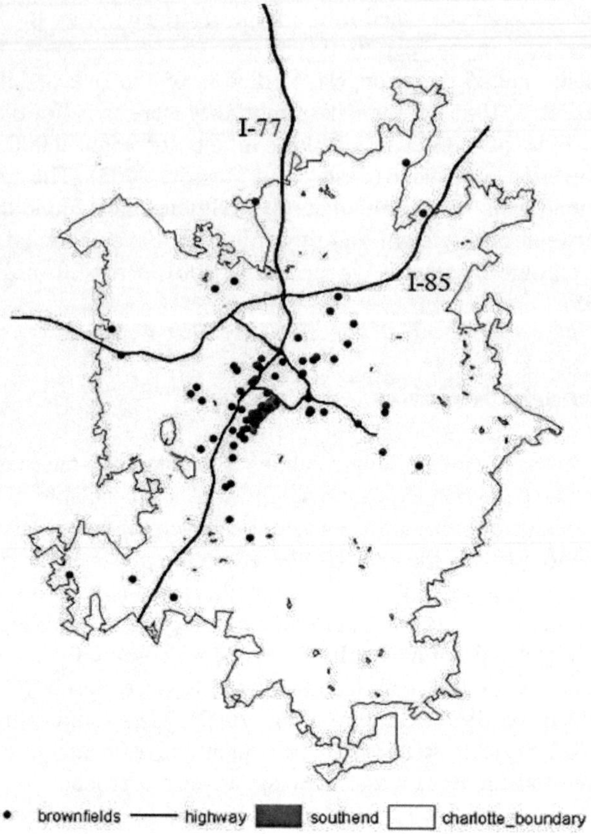

Fig. 8.2 Charlotte brownfield sites

partially overlap a block group is used as one indicator for the proximity to brownfields. Regardless the percentage of the block group that falls into the 0.5 mile buffer from the brownfield, as long as the block group intersects with the buffer, this block group is assumed to be affected by the brownfield.

The North Carolina Division of Waste Management administrates the Inactive Hazardous Sites Program. Inactive Hazardous Sites include facilities in a variety of property types where a hazardous substance was released. These facilities can be either abandoned facilities or ones still in operation. The term "inactive" refers to the fact that cleanup was inactive at the time of program enactment. Usually developers of inactive hazardous sites are owners, while developers of brownfields are not. The reason is that owners of brownfields often do not qualify for government subsidy if they contributed to the pollution. In some sense, inactive hazardous sites are potential brownfields but administrated by a different government agency. This study includes these inactive hazardous sites in the model. The variable is measured as the acreage of inactive hazardous sites divided by census block group land acreage.

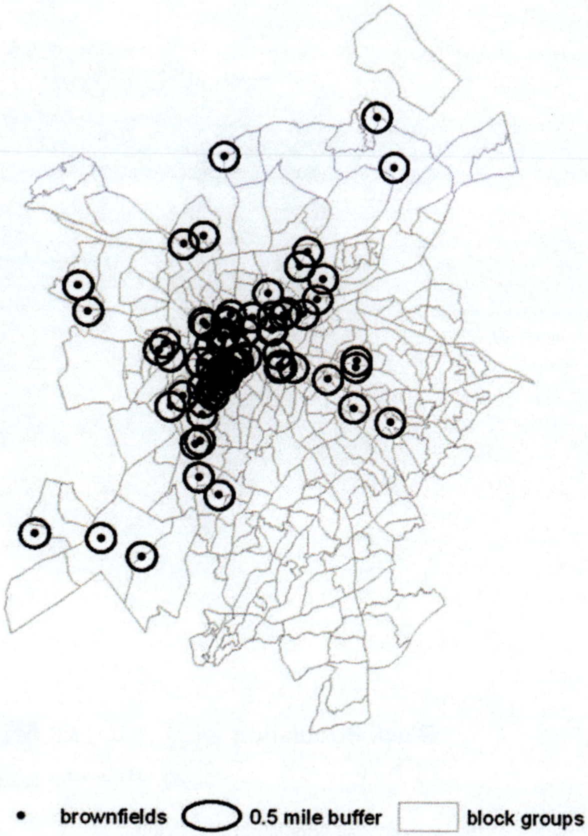

Fig. 8.3 One half mile buffers around brownfields

Olden (1998) found that persons with lower socio-economic status are more likely to live in economically distressed areas and work in more hazardous occupations. Since a primary goal of the Charlotte Brownfield Program is to act as an incentive for redevelopment in distressed areas, the following variables which indicate the level of economic distress in block groups were examined:

Percentage of the population that is African-American
Percentage of people aged 25 and above with less than a high school degree
Percentage of the population living in poverty
Median household income

Figure 8.4 shows a high racial segregation pattern in Charlotte. African-Americans are more likely to live in the western and northern parts of the city.

Median household income shows a similar cluster pattern as in African-American population distribution. There are more poor communities in the western and the northern parts of the city than in the southern areas (Fig. 8.5). There are a few block groups with extremely high unemployment rates around the central city (Fig. 8.6). The west and the northeast sides of the city have relatively higher unemployment rate than other areas.

Fig. 8.4 African-American population distribution

8.3.4 Methods

Ordinary least squares regression (OLS) and geographically weighted regression (GWR) are commonly used to examine the correlations between the presence of hazardous sites and health outcomes. OLS assumes spatial stationarity of the relationship between the dependent variable and the independent variables. It generates a single regression coefficient for each independent variable, and is considered a global regression. GWR, on the other hand, examines the spatial heterogeneities in the relationship, and is considered a local regression. At each data point, it generates a local regression coefficient for each independent variable.

Although GWR has been widely applied in various studies, this technique is prone to the multicollinearity problem. The regression model may not be stable as it depends on the joint-distribution of exogenous variables. The local regression coefficients can be correlated when actually they are not. If two or more exogenous variables are highly correlated, the situation will get even worse. Researchers may misinterpret the meaning of coefficients and reach inaccurate conclusions (Wheeler and Tiefelsdorf, 2005; Griffith, 2008; Ogneva-Himmelberger, Pearsall, and Rakshit, 2009).

Fig. 8.5 Median household income distribution

The binary correlation table shows that the "percentage of African-Americans," "percentage of people aged 25 and above with less than a high school degree," and "percentage of population living in poverty" have correlation coefficients that are close to or larger than 0.7 (Table 8.2).

This indicates a possible multicollinearity problem among these three variables. Tolerance is a measure of collinearity reported in statistics software SPSS. A small tolerance value indicates a multicollearity problem. The Variance Inflation Factor (VIF) measures the impact of collinearity among the variables in a regression model. A larger VIF value confirms the multicollinearity problem. A collinearity statistic reveals that the tolerance value of the variable percentage of people aged 25 and above with less than a high school degree is smaller than 0.3 and its VIF value is above 4. To be on the conservative side, this study uses tolerance value of 0.3 and VIF value of 4 as thresholds in identifying the multicollinearity problem. This problem is solved by only including the variable "percentage of African-Americans," "unemployment rate" and "median household income" in both the OLS and the GWR models.

8.4 Findings

The OLS model explains about 47% of the variance in census block group low birth weight rates. Table 8.3 indicates that census block groups that contain high densities of brownfields have a statistically significant association with low birth

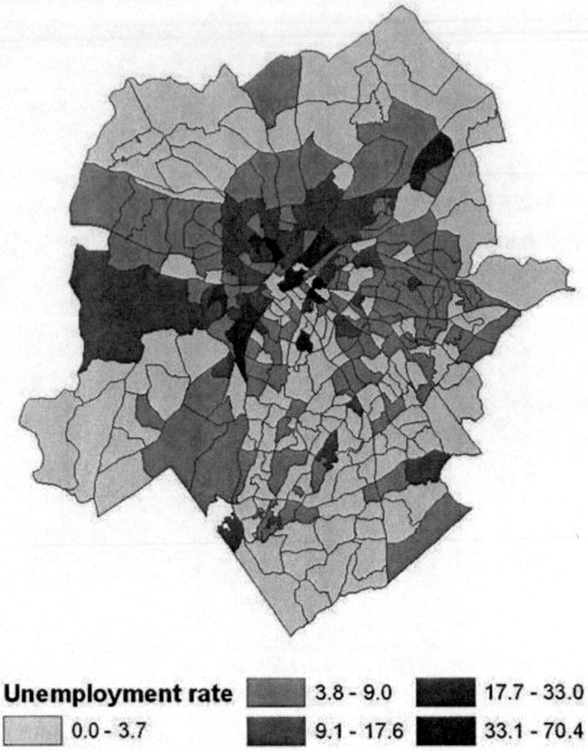

Fig. 8.6 Unemployment rate distribution

Table 8.2 Summary of binary correlation

	Black	Education	Unemployment	Income
Black	1	0.784[a]	0.494[a]	0.689[a]
Education	0.784[a]	1	0.454[a]	0.724[a]
Unemployment	0.494[a]	0.454[a]	1	0.587[a]
Income	0.689[a]	0.724[a]	0.587[a]	1

$N = 329$.
[a]Correlation is significant at the 0.01 level (2-tailed).

weight rates. However, being close to multiple brownfields, measured by the number of brownfield buffers, is not associated with LBW rates. Consistent with previous research, being African-American is an important determinant of low birth weight.

Local Indicators of Spatial Association (LISA) is a spatial analysis technique that gives an indication of the extent of the spatial clustering of similar values around an observation. This study uses GeoDa to conduct this analysis.

Table 8.3 Summary of OLS $N=329$ census block groups

	Coefficients	*P*-value
Intercept	4.77	0.000
Brownfield density	0.19	0.025
Brownfield buffer	−0.11	0.039
Inactive sites density	0.03	0.709
Percent of African-Americans	0.06	0.000
Percent of unemployment rate	0.12	0.000
Median household income	0.00	0.052
Adjusted R^2	0.47	

There are five categories in a LISA cluster map legend:

- Not significant (Areas that are not significant at default significance level of 0.05)
- High-High (High values surrounded by high values)
- Low-Low (Low values surrounded by low values)
- Low-High (Low values surrounded by high values)
- High-Low (High values surrounded by low values).

In the LBW case, high-high and high-low areas may indicate potential problems. In Fig. 8.7, the west/northwest cluster indicates that LBW rates in these block groups are high, and their nearby block groups have similar situation.

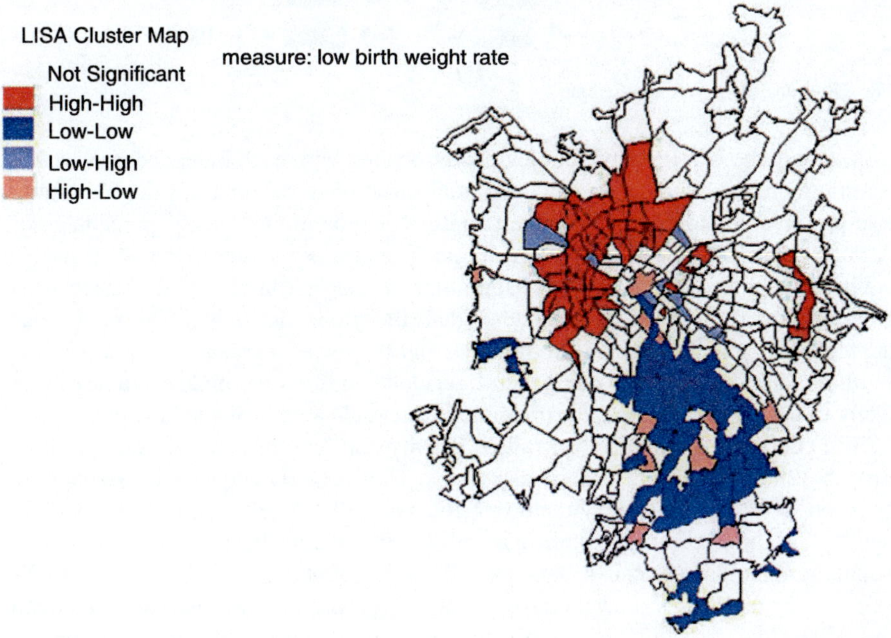

Fig. 8.7 Low birth weight Lisa cluster map

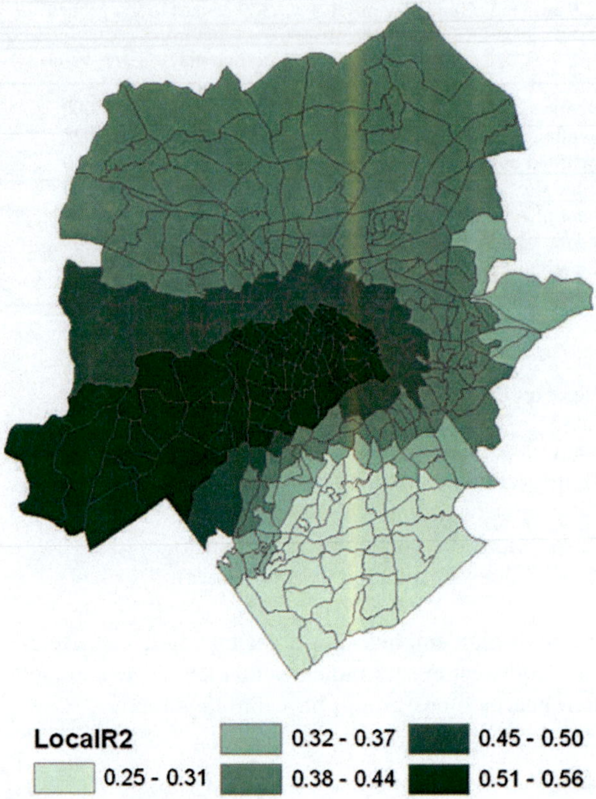

Fig. 8.8 Local R^2 GWR coefficients

Mapping the GWR coefficient pattern associated with each independent variable is a common procedure to display the spatial variation of the relationship. This study also presents the coefficients for the variables of brownfield density, percentage of African-American, and Unemployment rate. The local R^2 values in the GWR model are positive in all 329 block groups and range up to 56%. In one third of the census block groups the R^2 values are higher than the global R^2 value. The model has the highest explanatory power in the west, and the lowest explanatory power in the south of Charlotte (Fig. 8.8). In general, residents in the west of Charlotte are more likely to be African-Americans, unemployed, and close to brownfield sites.

The coefficients between brownfield density and low birth weight rate are positive in 328 out of 329 census block groups. This strongly suggests an association between LBW rate and brownfield density. The GWR coefficients for brownfield density are the highest in eastern part of the city, where there is a large Hispanic population. Unlike African-Americans, Hispanic populations tend to show a LBW rate similar to Whites, and therefore relatively lower than the African-American rate. Therefore, brownfield density becomes the strongest explanatory variable in the eastern of Charlotte than in the rest of the city (Fig. 8.9).

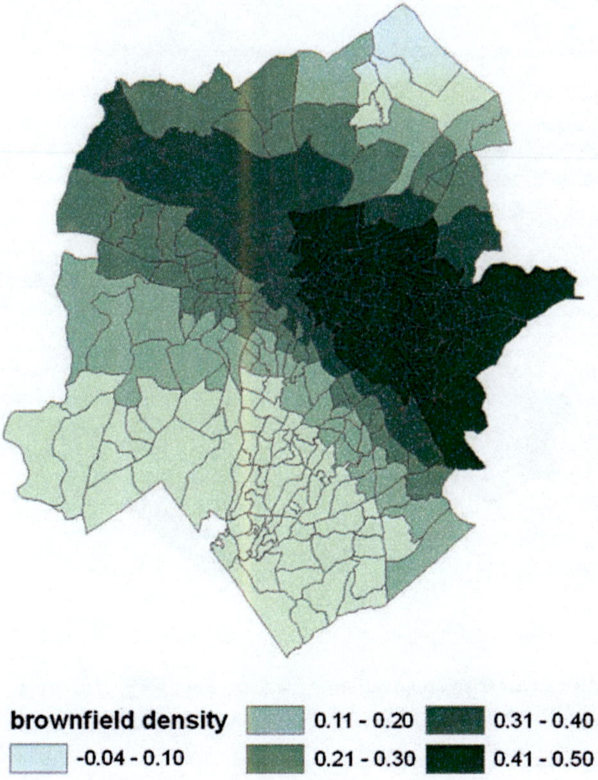

Fig. 8.9 Brownfield density GWR coefficients

The percentage of African-American population in a census block group is another strong indicator of low birth weight rate. Fewer of the brownfields included in this study are in the southern part of the city as compared to other parts of the city. There, African-American population becomes the most important indicator of LBW rate. From the south to the north in the city, African-American population increases. However, the brownfield density also increases. Therefore the association between percentage of African-American population and LBW rate weakens (Fig. 8.10).

The unemployment rate has a positive local R^2 value in all 329 census block groups as the Percent African-Americans, but has a very different pattern of explanatory power (Fig. 8.11). It has the highest explanatory power in areas with low unemployment rate.

8.5 Discussion and Conclusion

This study employed both the OLS and GWR tools to examine the relationship between neighborhood low birth weight rates and the existence of brownfields.

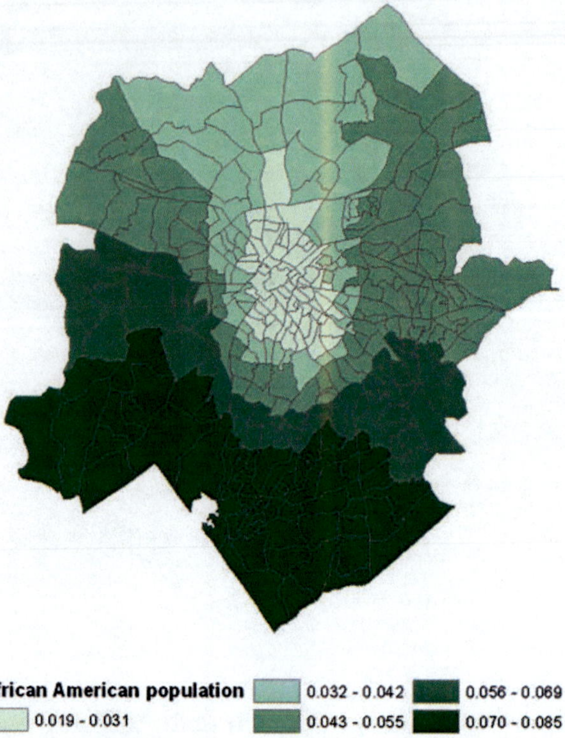

Fig. 8.10 Percent African-American GWR coefficients

The GWR model explained more variance in the dependent variable in one third of the census block groups.

The results show that there is a positive relationship between low birth weight and brownfield density, the percentage of African-Americans, and the unemployment rate at the census block group level. The maps of GWR coefficients help to clarify the spatial variation of these relationships.

Public health was not considered in the city's decision to subsidize brownfield projects. One of the reasons was the lack of information on the association between brownfields and public health. At the early stage of brownfield redevelopment, Mecklenburg County Health Department staff was consulted about the targeted areas of the South End and Wilmore. They attended initial community meetings to respond to any possible health concerns, but public health never came up as an issue. This does not mean that Charlotte has no areas where public health indicators show high rates of health problems. The Mecklenburg Health Department has identified "the crescent area" as having the greatest health concerns. This area spans across the city from the western part of the city over the middle of uptown to the northeast areas. The health department has not conducted a formal study to determine the "why" behind the "what" regarding certain rates

Fig. 8.11 Unemployment rate GWR coefficients

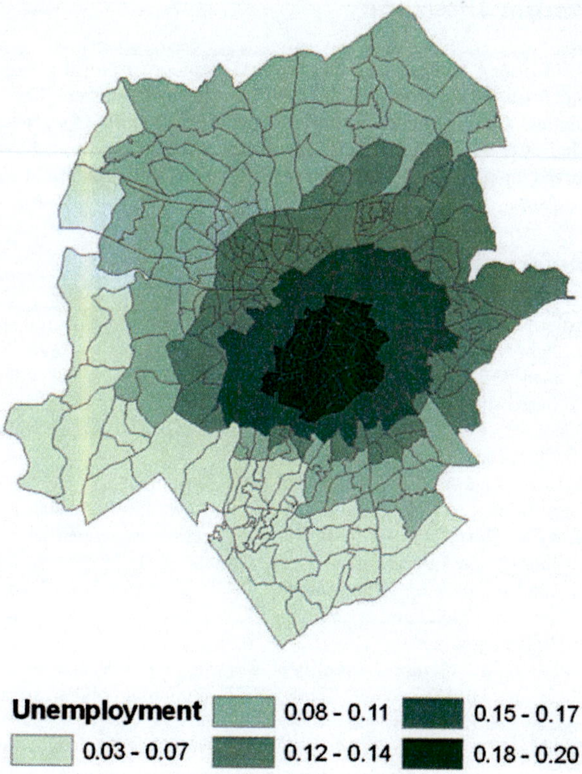

of disease or public health measures in the these neighborhoods (Zimmerman, 2007). Among the possible explanations of the health problems in the crescent are: (1) the lower economic status groups that reside in these areas tend to display unfavorable health behaviors such as poor diet, obesity, drinking, smoking and drug use; (2) the population has limited access to health care and public health services (Zimmerman, 2007); and (3) toxic sites in the area exacerbate the effects of race, class, and ethnicity, as well as other factors that might make these populations more vulnerable to adverse health outcomes, such as residential segregation.

Using multiple data sources, GIS analysis, and various regression tools, this study found that brownfields were more likely to be located in low socio-economic status neighborhoods. These neighborhoods also had above-average low birth weight rates. This indicates that although policymakers and developers did not intentionally include public health outcomes as an important factor in choosing brownfield sites, brownfield cleanup and reuse probably has had positive impacts for social and environmental justice.

Author Biography

Dr. Junfeng Wang graduated from University of North Carolina at Charlotte with a doctoral degree in Public Policy in 2008. The Ph.D. program emphasizes quantitative analytical methods and Geographic Information System (GIS) skills. As a graduate research assistant at UNC Charlotte, Dr. Wang has served on several public health studies for 2 years. The last project she served on at UNC Charlotte was an EPA sponsored brownfield study.

References

Adler N, Ostrove JM (1999) Socioeconomic status and health: what we know and what we don't. Ann N Y Acad Sci 896(1):3–15

Avchen RN, Scott KG, Mason CA (2001) Birth weight and school-age disabilities: A population-based study. Am J Epidemiol. 154:895–901

Bacot H, O'dell C (2006) Establishing indicators to evaluate brownfield redevelopment. Econ Dev Q 20(2):142–161

Baibergenova A, Kudyakov R, Zdeb M, Carpenter DO (2003) Low birth weight and residential proximity to PCB-contaminated waste sites. Environ Health Persp 111(10):1352–1357

Berry M, Bove F (1997) Birth weight reduction associated with residence near a hazardous waste landfill. Environ Health Persp 105(8):856–861

General Accounting Office (GAO) (1983) Sitting of hazardous waste landfills and their correlation with racial and economic status of surrounding communities. General Accounting Office, Washington, DC

Goldberg MS, Goulet L, Riberdy H, Bonvalot Y (1995) Low birth weight and preterm births among infants born to women living near a municipal solid waste landfill site in Montreal, Quebec. Environ Res. 69:37–50

Goldman LR, Paigen B, Magnant MM, Highland JH (1985) Low birth weight, prematurity and birth defects in children living near the hazardous waste site, Love Cancal. Hazard Wastes and Hazard Mater 2(2):209–223

Griffith D (2008) Spatial-filtering-based contributions to a critique of geographically weighted regression (GWR). Environ Plann A 40(11):2751–2769

Jolley D, Jarman B, Elliott P (1992) Socio-economic confounding. In: P Elliott, J Cuzick, D English, R Stern (eds) Geographical and environmental epidemiology, 1st edn. Oxford University Press, Oxford

Lybarger JA, Lee R, Vogt DP, Perhac RM, Spengler RF, Brown DR (1998) Medical costs and lost productivity from health conditions at volatile organic compound-contaminated Superfund sites. Environ Res 79(1):9–19

Ogneva-Himmelberger Y, Pearsall H, Rakshit R (2009) Concrete evidence & geographically weighted regression: A regional analysis of wealth and the land cover in Massachusetts. Appl Geogr. 29:478–487

Olden K (1998) The complex interaction of poverty, pollution, health status. The Scientist 7

Oliveira LM, Stein N, Sanseverino MTV, Vargas VM, Fachel JM, Schuler L (2002) Reproductive outcomes in an area adjacent to a petrochemical plant in southern Brazil. Rev Saude Publica 36(1):81–87

Reichman NE (2005) Low birth weight and school readiness. Future Child 15(1):91–116

Rich-Edwards JW, Stampfer MJ, Manson JE, Rosner B, Hankinson SE, Colditz GA, Willett WC, Hennekens CH (1997) Birthweight and the risk of cardiovascular disease in adult women. Brit Med J 315(7105):396–400

Sastry N, Hussey JM (2003) An Investigation of racial and ethnic disparities in birth weight in Chicago neighborhoods. Demography 40(4):701–725

Sexton K, Gong H Jr., Bailar JC, Ford JG, Gold DR, Lambert WE et al (1993) Air pollution health risks: do class and race matter? Toxicol Ind Health 9(5):843–878

United Church of Christ (UCC) (1987) Toxic wastes and race in the United States: A national report on the racial and socio-economic characteristics with hazardous waste sites. United Church of Christ, Commission for Racial Justice, New York, NY

Vianna NJ, Polan AK (1984) Incidence of low birth weight among love canal residents. Science 226(4679):1217–1219

Wheeler D, Tiefelsdorf M (2005) Multicollinearity and correlation among local regression coefficients in geographically weighted regression. J Geogr Syst. 7:161–187

Zimmerman S (2007) Personal correspondence with Mecklenburg County Health Department epidemiology specialist Sara Zimmerman

Chapter 9
Regional Environmental Patterns of Diarrheal Disease in Bangladesh: A Spatial Analytical and Multilevel Approach

Elisabeth D. Root and Michael Emch

Abstract This study investigates diarrheal disease distributions in Bangladesh using nationally representative household-level survey data integrated with land type maps that characterize flood inundation levels. The spatial distribution of childhood diarrhea is mapped throughout the country and diarrhea rates are stratified by individual, household, and regional-level variables. This study describes national-level trends by integrating spatially-referenced household characteristics and regional-level information on water and sanitation. The world saw dramatic improvements in water availability during the 1980s, designated "International Water Supply and Sanitation Decade" by the UN. Nevertheless, reductions in diarrheal morbidity in much of the developing world have been modest, possibly because of a lack of sufficient parallel improvements in sanitation and hygiene (Levine et al., Lancet 2(7976):86–89, 1976; Esrey et al., Bull World Health Organ 63:757–772, 1985; Hoque et al., Bull World Health Organ 74: 431–437, 1996). In a recent meta-analysis of 64 studies, Fewtrell and Colford (Health, Nutrition, Population Discussion Paper, World Bank, Washington DC, http://www1.worldbank.org/hnp/Pubs_Discussion/Fewtrell&ColfordJuly2004.pdf, 2004) found that water supply, water quality, hygiene, and sanitation programs all reduce diarrheal disease mortality and morbidity. However, multiple interventions did not reduce diarrheal disease any more than approaches that involved only one intervention. This suggests that a better fundamental understanding of the relationship between water, sanitation, household characteristics and diarrheal disease is needed to optimize future interventions. This study begins to investigate these relationships using data at different collected at multiple scales.

Keywords Bangladesh · Diarrheal disease · Multilevel modeling · Spatial analysis · Environment

E.D. Root (✉)
Department of Geography, University of Colorado, Boulder, CO, USA
e-mail: roote@colorado.edu

J.A. Maantay, S. McLafferty (eds.), *Geospatial Analysis of Environmental Health*, Geotechnologies and the Environment 4, DOI 10.1007/978-94-007-0329-2_9,

9.1 Background

Diarrheal diseases cause more deaths among children under five than any other disease in Bangladesh (Hoque and Hoque, 1994). Over the past 2 decades, millions of rural households across Bangladesh, a country with a per capita GNP of US$400, have installed private tube wells in the hope of improving their health. This US$500 million effort was the response of international organizations, including UNICEF, the Bangladesh government and many NGOs, in order to reduce infant mortality by encouraging a switch from surface water to groundwater, which is generally less contaminated with human pathogens. Because limited resources must be used wisely in countries such as Bangladesh, it is necessary to identify risk factors so preventative health programs can focus on specific interventions. Assessing risk for diarrheal disease requires knowledge of the complex and dynamic interaction of biological, socioeconomic, behavioral, and environmental factors. The objective of this study is to advance such knowledge in the context of rural Bangladesh. Specifically, the study identifies the variables related to diarrheal disease risk and analyzes the spatial patterns of diarrhea.

In Bangladesh, diarrheal diseases are caused by many disease agents (Table 9.1). A detailed study of the etiological burden was conducted in a rural area of Bangladesh and included estimates of the diarrheal disease burden in the hospital and in the community (Baqui et al., 1991, 1992). Hospitalized cases represent more severe diarrheal disease overall compared to community based disease burden values. Cholera, which is a disease caused by the bacterium Vibrio cholerae 01, is the most common agent leading to diarrheal hospitalization followed by shigella which is bacterial dysentery. In contrast, diarrhea in the community is mostly caused by

Table 9.1 Diarrheal etiological agents in rural Bangladesh

Disease agent	Percentage detected in community	Percentage detected in hospital
Vibrio cholerae 01	0.4	39
Vibrio cholerae non 01	2.9	3
Shigella	8.6	11
Enterotoxigenic Esherichia coli	12.2	14
Campylobacter	17.6	11
Salmonella	0.1	1
Enteroadhesive Esherichia coli	34.3	–
Enteropathogenic Esherichia coli	13.5	–
Aeromonas	2	–
Pleismonas	0.1	–
Rotavirus	4.3	–
Entaemoeba histolytica	0.4	2
Giardia lamblia	2.2	2
Cryptosporidium	1.9	–
No agent detected	42.1	–

–, Lab test not done.

different forms of Esherichia coli. In an ongoing study by one of the authors (Emch, unpublished) it was found that in rural Bangladesh shallow tubewells (<100 ft) are contaminated with Esherichia coli more than half of the time and more often during the monsoon season. Thus, it is hypothesized in the present study that areas more prone to flooding will have higher diarrheal disease rates.

9.2 Study Data

The primary source of data for this study is the Bangladesh Demographic and Health Survey (BDHS) collected in 1999/2000 (NIPORT et al., 2001). The BDHS relies on a two-stage sample, stratified by urban/rural status, and selected using the Integrated Multi-Purpose master Sample (IMPS) which was created on the basis of 1991 census data by the Bangladesh Bureau of Statistics. The master sample consists of 500 primary sampling units (PSUs), selected with sampling probability proportional to the population of the census enumeration area (NIPORT et al., 2001). A total of 341 PSUs were used for the 1999–2000 BDHS (99 in urban and 242 in rural areas). A systematic sample of 30 households was selected from each primary sampling unit and all ever-married women age 10–49 interviewed. Response rates were very high (96.9%), and the final sample consisted of 9,854 households from which 10,544 women were interviewed.

The Women's Questionnaire was used to collect information from eligible women. Women were asked various questions on health and personal/household characteristic including the health of children under the age of five within their households. The question used to determine prevalence of diarrhea among children was whether a given child under the age of 6 years "Had diarrhea recently?" in the last 24 hours, last week, or longer periods. The BDHS also asked questions related to: breastfeeding and weaning practices; household specific characteristics such as household size, source of drinking water, existence of toilets and condition of housing, and; maternal characteristics such as age, education and literacy. Sample weights were generated and normalized such that the weighted number of cases is identical to the unweighted number of cases when using the full dataset with no selection. Sample weights were used to weight all tabulations examined in this study. Also recorded is the geographic cluster (PSU or census enumeration area) to which the household belongs, enabling researchers to link survey data to outside data sources using geographic location. We obtained the geographic coordinates (latitude and longitude) for the centroid of each geographic cluster and linked survey data to these clusters using the unique cluster identification code assigned to each household.

The data used in this study consisted of a record for each child for which data was reported by the mother. A value of 0 was assigned if the mother did not report that the child "Had diarrhea recently?" while a value of 1 was assigned if a child had diarrhea in the "last 24 hours" or "last 2 weeks". Children for whom data was missing were removed from the dataset. Each child was also assigned household variables and maternal characteristics including: the number of children under 5

Table 9.2 Desriptive characteristics of the sample

Variable	Children with diarrhea ($n = 361$)	Children without diarrhea ($n = 4,775$)
Number of children 5 years or younger in the household (mean/SD)	1.7 (0.92)	1.6 (0.95)
Latrine type		
septic tank	5 (1.3)	186 (3.9)
pit/open/hanging	264 (73.3)	3,477 (72.8)
no facility	92 (25.4)	1,112 (23.3)
Currently breastfeeding	254 (70.6)	2,881 (60.3)
Child given water	341 (94.5)	4,254 (89.1)
Land type		
highland	157 (43.7)	2216 (46.4)
medium highland	164 (45.5)	2216 (46.4)
lowland	39 (10.8)	343 (7.2)

in the household, total number of household members, household drinking water source (piped water, tubewell or surface water source), type of latrine facility (septic tank/toilet, pit/open/hanging latrine or no facility), housing material (natural, rudimentary or bring/cement/tin walls), maternal literacy (reads easily or with difficulty vs. cannot read), whether or not the child was still breastfeeding (yes vs. no), whether or not the child is given water to drink (yes vs. no). Table 9.2 displays descriptive statistics for the study.

The spatial data are derived from the Bangladesh National Database Project (NDP), a geographic information system (GIS) developed to assist in planning and managing spatial data for the Flood Action Plan (FAP) in Bangladesh (ISPAN, 1995). The NDP provided geographic boundary files for districts as well as flood inundation land type, which displays "normal" flood depths during the monsoon season. Using elevation and soil type data, land is classified into several categories: non-flood (highland), medium flood (medium highland and medium lowland), deep flood and very deep flood (lowland) and urban areas. The coordinates of the geographic cluster data obtained from the BDHS were used to assign the flood category to each household used in this study. Figure 9.1 illustrates the different geographic data layers used.

9.3 Methods

Methods include a combination of GIS analyses and statistical modeling. A GIS database was created to integrate the BDHS data with flood maps, to visualize diarrheal distributions, and multilevel models were used to measure associations between these distributions. We first assigned each geographic BDHS cluster a corresponding land type/flood classification code using overlay operations with Hawth's Analysis Tools in ArcGIS 9.2 (Beyer, 2004). Hawth's Tools is a free

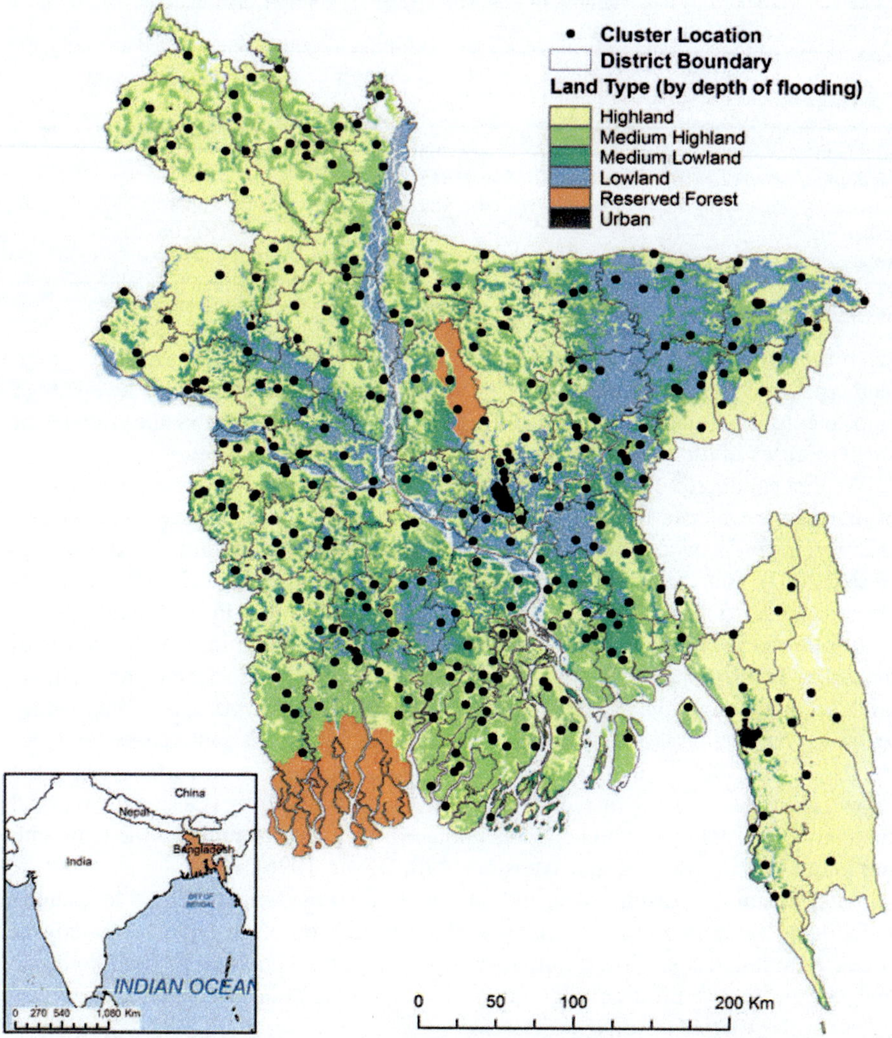

Fig. 9.1 Cluster locations, district boundaries and land type (by flood classification)

extension for ESRI's ArcGIS (specifically ArcMap). It is designed to perform spatial analysis and functions that cannot be conveniently accomplished with out-of-the-box ArcGIS. Overlay operations involve the placement of one map layer (clusters) on top of a second map layer (land type) to create a new map layer that is some combination of the two. Hawth's Tools has specific raster-vector overlay operations not available in ArcMap whereby attributes of a raster dataset can be assigned to observations in a vector dataset. This allowed us to integrate environmental data with the BDHS survey data and assign each household to a land type category. Households assigned to a cluster surrounded by a non-flood area (highlands) were

Table 9.3 Number of cases, number of children 5 years or younger, and diarrhea rate by land use type

Land use type	No. of diarrhea cases	No. of children 5 or under	Rate per 1,000 children
Highland	157	2,408	65.2
Medium	167	2,415	69.2
Lowland	39	389	100.3
Urban	10	98	102.0
Total	373	5,310	70.2

given this land type value for subsequent regression analyses. We chose to group land types into four categories based on similarities in flood inundation levels: high land, medium highland, lowland, and urban land. Table 9.3 shows the number of clusters, cases of diarrhea and children under 5 by land type category.

We first conducted a data visualization step to qualitatively assess spatial patterns of diarrhea prevalence in Bangladesh. Several maps of diarrhea rates were created at different levels of spatial aggregation and using different methods in ArcGIS 9.2 (ESRI, 2007). Diarrhea data were aggregated to the district level and rates (per 1,000 children under 5 years) displayed using district boundaries. Proportional symbol maps allowed us to display the magnitude of diarrheal events in combination with land type. Finally, we created a diarrheal disease prevalence surface using an interpolation technique called kriging. Spatial interpolation is a process of estimating unknown data values for specific locations using the known data values for other points. In this study, we created a continuous diarrheal disease surface using the known diarrheal rates for each discrete cluster location. Kriging is a specific method for spatial interpolation which creates unbiased weighted averages of the data with minimum variance (Oliver and Webster, 1990; Oliver, 1996).

After examining results from the data visualization step, we chose to exclude individuals living in an urban land type classification from our regression analyses because these areas are located in a variety of land type areas (e.g., both highland and lowland), have different methods for coping with flood events and, therefore, different risk factors for diarrheal disease.

To estimate the risk of a diarrheal event associated with land type, maximum likelihood estimates of odds ratios (OR) and 95% confidence intervals (CI) were calculated from multilevel logistic regression models (Diez Roux, 2001; Raudenbush and Bryk, 2002). Children living within the different land use classifications are subject to intraclass correlation, which could lead to overestimation of the significance of the risk factor. Intraclass correlation occurs when individual-level observations are not independent because they share some factor, in this case a specific land use type. Thus, children living in the same land use type will be more similar (probably due to shared risk factors) than children living in different land use types. To account for intraclass correlation, a multilevel model with SAS PROC GLIMMIX procedure (SAS Institute, 2007) was used, which permits the incorporation of random effect terms (e.g., land type). These random effect terms are a method of "controlling for" possible intraclass correlation. Multilevel models combine individual-level (level-1) and area-level (level-2) predictors into one regression equation (a "full" or "combined" model). For this study level-1 predictors include child-specific and maternal characteristics derived from the BDHS and the level-2 predictor is the land type associated with each cluster. We estimated multilevel logistic regression models with a fixed slope value for each predictor variable and random intercepts for each land type. Probability of diarrhea occurrence can be evaluated with the following equation:

$$\log\left(\frac{\Pi_{ij}}{1-\Pi_{ij}}\right) = \beta_0 + \beta_1 X_{ij} + u_{0i} + e_{ij}$$

where:

$\Pi_{ij} = \Pr(Y_{ij} = 1)$, the probability of the *j*th child (level-1) nested within the *i*th land type (level-2) having a diarrheal event
$\beta_{0,}\ \beta_{\ 1}$ are level-2 coefficients (also called fixed effects)
X_{ij} represents a vector of variables that are level 1 predictors of the outcome variable (maternal and child characteristics)
e_{ij} is a level-1 random effect (error term)
u_{0j} is a level-2 random effect (error terms)

The combined influence of individual- and area-level indicators was examined to determine whether risk for children living in a lowland or flood-prone region varied even after controlling for individual characteristics. Considered as potential confounders were number of children under 5 in the household, household drinking water source (piped water, tubewell or surface water source), type of latrine facility (septic tank/toilet, pit/open/hanging latrine or no facility), breastfeeding status (still breastfeeding vs. not breastfeeding) and whether the child was given water to drink (yes vs. no). Only those covariates that showed a significant risk associated with diarrhea during univariate and multivariate analyses were included in the area-level analysis. Univariate analysis, logistic regression and multilevel modeling were conducted in SAS version 9.2 software (SAS Institute, 2008).

9.4 Results

Figure 9.2 shows diarrhea rates for children 5 years or younger aggregated by districts while Fig. 9.3 shows the rate for each geographic cluster. There appears to be significant spatial variation in diarrhea rates with some locations experiencing no diarrhea and several districts reporting more than 300 cases per 1,000 children. Figure 9.3 shows no clear correlation between high diarrhea rates and land type, though a qualitative assessment of the lowland areas near the two major rivers suggests rates may be higher in these areas. The most affected regions appear to lie in the north central area of the country. Figure 9.4, the interpolated prevalence surface, confirms this spatial pattern, and suggests that areas of high risk mirror the major rivers in Bangladesh.

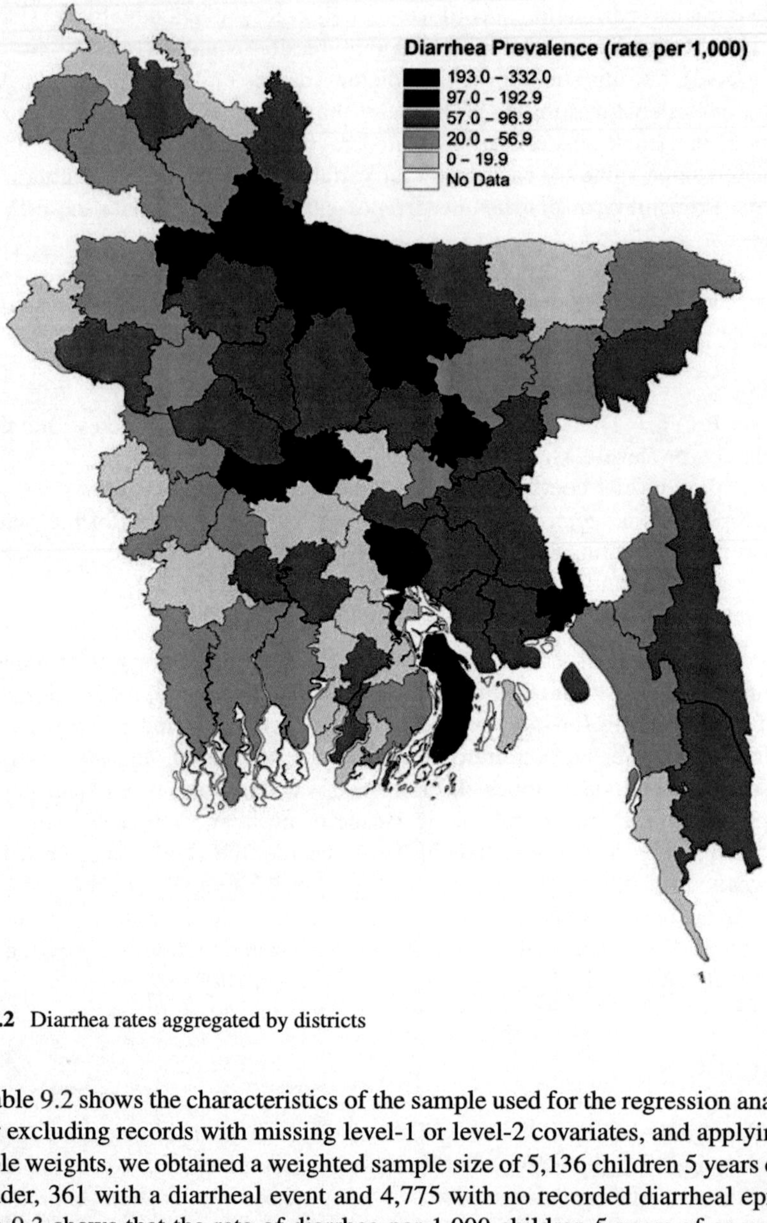

Fig. 9.2 Diarrhea rates aggregated by districts

Table 9.2 shows the characteristics of the sample used for the regression analysis. After excluding records with missing level-1 or level-2 covariates, and applying the sample weights, we obtained a weighted sample size of 5,136 children 5 years of age or under, 361 with a diarrheal event and 4,775 with no recorded diarrheal episode. Table 9.3 shows that the rate of diarrhea per 1,000 children 5 years of or younger varied by land use type, with highland areas reporting a rate of 65.2 and lowland areas reporting a rate of 100.3.

Table 9.4 displays the results of the logistic and multilevel logistic regression models. Model 1 examines the odds of a diarrheal event controlling for individual-level factors only. The odds of diarrheal disease increased with the number of children in the household. Children living in households using a latrine with a septic

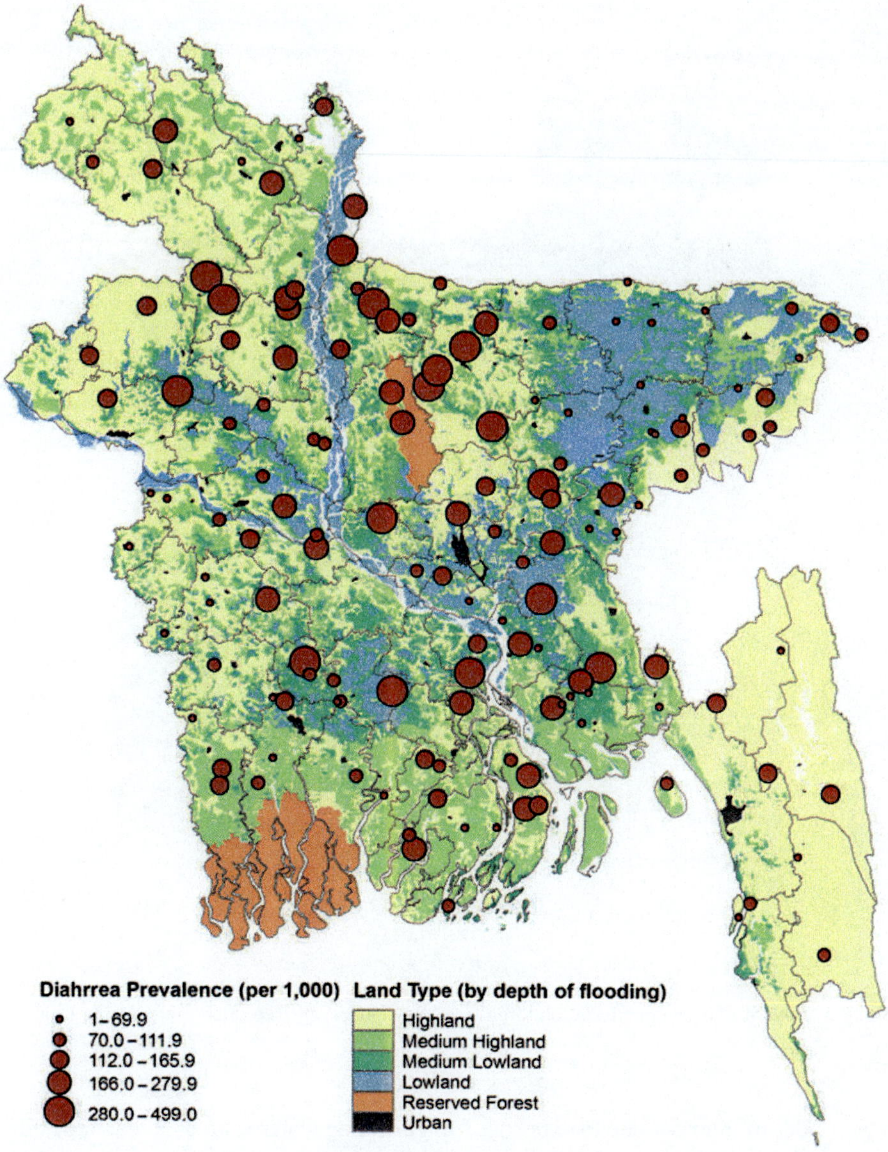

Fig. 9.3 Proportional symbol map showing diarrhea prevalence and land type

tank had the lowest odds of diarrheal disease followed by households using a pit, open or hanging latrine. Children with no latrine facility present had the greatest odds of a diarrheal event. Breastfeeding was also significantly associated with a higher risk of a diarrheal event, though not in the expected direction, while children who received water to drink had a higher odds of diarrhea even than children who only breastfeed. While children who receive drinking water can be expected to have

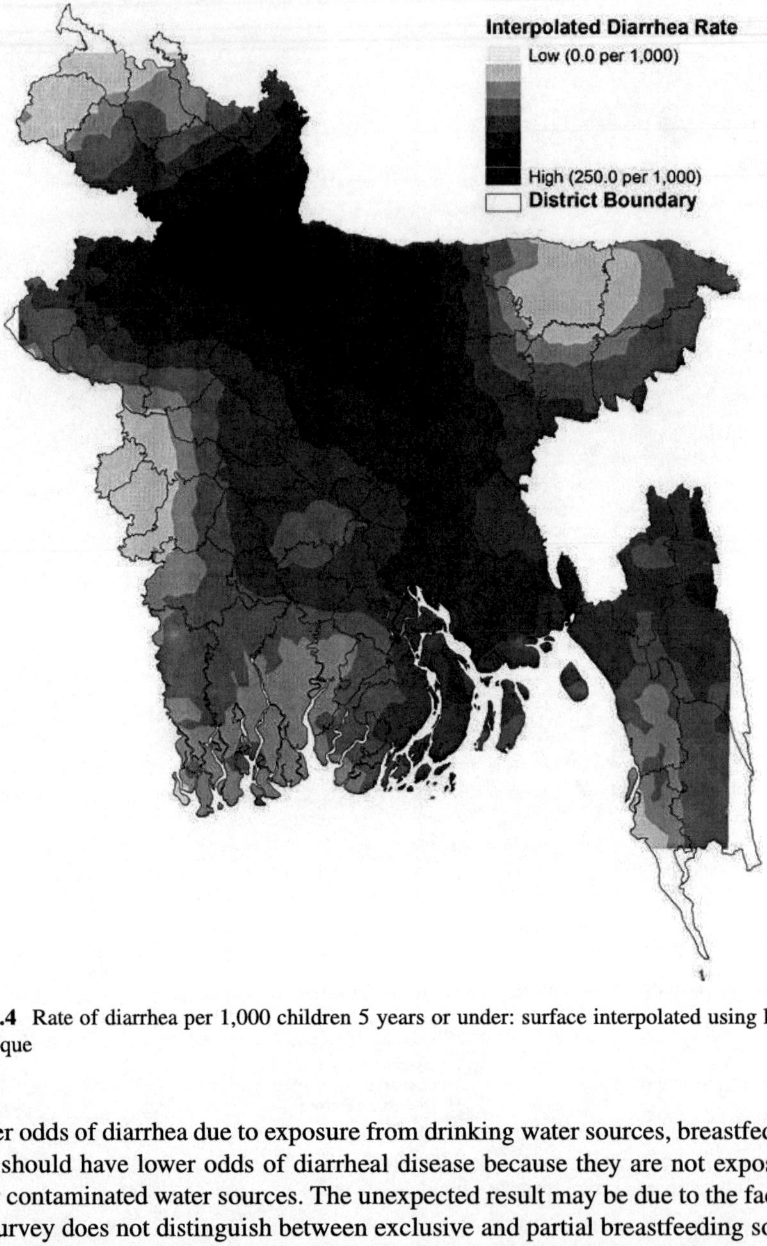

Fig. 9.4 Rate of diarrhea per 1,000 children 5 years or under: surface interpolated using kriging technique

higher odds of diarrhea due to exposure from drinking water sources, breastfed children should have lower odds of diarrheal disease because they are not exposed to other contaminated water sources. The unexpected result may be due to the fact that the survey does not distinguish between exclusive and partial breastfeeding so children who are reportedly being breastfeed also receive other foods or liquids which may be subject to contamination. Model 2 examines the odds of a diarrheal even for each land type, but not considering any individual-level factors. Children living in lowland areas had significantly higher odds of a diarrheal event when compared to children living in a highland area. Children living in the midland areas, however, did not show increased risk for a diarrheal event.

Table 9.4 Logistic and multilevel logistic regression model results

Covariate	Model 1	Model 2	Model 3
Individual-level (level 1) variables			
Number of children 5 years or younger in the household	1.17 (1.04–1.33)		1.17 (1.03–1.32)
Latrine type			
septic tank	0.32 (0.13–0.83)		0.32 (0.13–0.84)
pit/open/hanging	0.94 (0.74–1.21)		0.93 (0.72–1.19)
no facility	1.0		1.0
Currently breastfeeding	1.81 (1.42–2.30)		1.82 (1.43–2.31)
Child given water	2.63 (1.64–4.20)		2.66 (1.66–4.25)
Area-level (level 2) variables			
Land type			
Highland		1.0	
Medium high-/lowland		1.03 (0.83–1.29)	
Lowland		1.56 (1.08–2.25)	
Random effects			
Between-land type variation			0.04 (0.06)

Model 3 examines the odds of a diarrheal event using a random intercept model. Intercepts were allowed to vary by land type and the odds of a diarrheal event controlled for individual-level predictors. Allowing intercepts to vary by land type changed the odds ratios very little. The random effects represent the residual between-land type variation in the outcome after accounting for the individual-level covariates. These can also be interpreted as the residual within land type clustering of the outcome. The estimated variance of the land type intercepts was 0.04 with a standard error of 0.06. In this case, the residual between land type variance is not statistically significant, which is not surprising since the odds of diarrhea for individuals living in medium high-/lowland areas was not significantly different from individuals living in highland areas. The fixed individual-level effects more strongly influence diarrhea outcomes that land type.

Table 9.5 lists, for each combination of land types, the estimated relative risk and associated p-values calculated from the multilevel logistic regression models. The relative risk is a measure of risk due to one land type classification relative to each

Table 9.5 Relative risk and 95% confidence intervals of a diarrheal event by land type

Land type	Relative risk (95% CI)	*p*-value
Highland vs. medium high-/lowland	0.96 (0.78–1.19)	0.708
Highland vs. lowland	0.72 (0.53–0.99)	0.044
Medium high-/lowland vs. lowland	0.75 (0.54–1.03)	0.077

of the other land type classifications. These results suggest that children living in highland areas had 28% less risk of a diarrheal event relative to children living in a lowland area. Children living in a highland and medium high-/lowland area had similar risk for diarrheal event, suggesting these two areas have similar patterns of diarrheal events. Children living in medium high-/lowland areas had approximately 25% less risk of a diarrheal event than children in lowland areas, though this was only borderline significant. Overall, these results suggest that children living in lowland areas have greater risk of a diarrheal event than children living in the other two land types.

9.5 Conclusions

There is much regional variation in diarrheal disease in Bangladesh. Maps reveal that the highest rates of childhood diarrhea are in the northern region of Bangladesh near the Brahmaputra River (Figs. 9.1–9.4). Households with latrines that have septic tanks have significantly lower diarrheal disease rates, suggesting that good sanitation is one of the most important factors in lowering the risk of diarrhea among children. Giving young children water to drink increases their risk of diarrheal disease, probably because the water is contaminated with a variety of pathogens that cause diarrhea. This is particularly problematic because children do not have immunity against diarrheal pathogens while adults do. This is both because children have not been exposed to the agents and their immune systems are not fully developed yet. Children living in lowland areas are more prone to diarrheal diseases than those living in upland areas. This study indicates a 28% increase in risk of diarrhea among children living in a lowland area relative to children living in a highland area. This is likely due to flooding which inundates sanitation systems and also infiltrates the shallow aquifer thus spreading diarrheal pathogens and contaminating tubewell water sources.

Overall, the effects of land use type (level-2 effects) on the risk of childhood diarrheal disease are smaller than the impact of individual and family characteristics, but this does not preclude land use from playing an important role in disease transmission in Bangladesh. Direct behaviors and interactions that place children at risk for contracting diarrheal disease are expected to have the strongest influence, which the results clearly indicate. But area-level factors, such as land use, can also impact child health, though they are often modified by individual-level behaviors. The fact that children living in lowland areas do experience a significant increase in risk of diarrhea when compared to children living in highland areas does indicate some protective effect of living outside of a flood zone.

The geographic approach used in this study reveals that when controlling for individual and household-level variables known to be related to diarrheal disease, there are still regional differences in childhood diarrhea rates which may be related to environmental factors such as flood levels. Using spatial visualization methods, such as maps and kriging, along with multilevel statistical models can reveal more information than traditional univariate data analysis and regression models. This study used a GIS primarily to integrate different data sets based on common geography thus allowing for the inclusion of map data in multilevel models. The GIS was also used to map and visualize the spatial patterns of the disease. Large datasets including the BDHS often include spatial coordinates collected with global positioning system receivers. This study provides not only empirical results about risk of diarrheal disease in Bangladesh but can also serve as an example of how to conduct this type of spatial study using household-level point data along with digital map data in a multilevel framework.

Author Biographies

Elisabeth Root, Ph.D., is an Assistant Professor of Geography and Faculty Research Associate at the Institute of Behavioral Science at the University of Colorado at Boulder. Her research explores geographical patterns of health and disease using quantitative spatial methodologies to understand the socioeconomic and environmental determinants of human health. She is particularly interested in the complex interactions between demographic, socioeconomic and environmental factors that influence human health and how we can quantify these factors and interactions to better understand health outcomes and the impact of health interventions. Dr. Root combines traditional epidemiological study design, complex spatial statistical methods as well as Geographic Information Systems to explore spatio-temporal patterns of chronic and communicable diseases. These methods are also used to understand how the spatial distribution of health and socioeconomic interventions affect health behaviors, social and economic well being and health outcomes. For more information, see Dr. Root's website at: http://spot.colorado.edu/~roote/.

Michael Emch, Ph.D., is Professor of Geography, Fellow at the Carolina Population Center, and Adjunct Professor of Epidemiology at the University of North Carolina at Chapel Hill. His expertise is in infectious disease ecology, neighborhood determinants of health, and geographic information science applications of public health. He leads the Spatial Health Research Group which conducts research that explores spatio-temporal patterns of disease, primarily infectious diseases of the developing world. Disease patterns are studied using a holistic approach by investigating the role of natural, social, and built environments in disease occurrence in different places and populations. Diverse statistical and spatial analytical methods are informed by theory from the fields of medical geography, epidemiology, ecology, and others. These theories and methods are used to examine diverse topics such as the role of population-environment drivers in viral evolution, how social connectivity contributes to disease incidence, and using environmental indicators to predict disease outbreaks. For more information see the Spatial Health Research Group website at http://www.unc.edu/depts/geog/spatialhealthgroup/

References

Baqui AH, Sack RB, Black RE, Haider K, Hossain A, Alim ARMA, Yunus M, Chowdhury HR, Siddique AK (1992) Enteropathogens associated with acute and persistent diarrhea in Bangladeshi children <5 years of age. J Infect Dis 166:792–796

Baqui AH, Yunus M, Zaman K, Mitra AK, Hossain KMB (1991) Surveillance of patients attending a rural diarrhoea treatment center in Bangladesh. Trop Geogr Med 43:17–22

Beyer HL (2004) Hawth's analysis tools for ArcGIS. Available at http://www.spatialecology.com/htools

Diez Roux AV (2001) Investigating neighborhood and area effects on health. Am J Public Health 91(11):1783–1789

Environmental Systems Research Institute, Inc. (ESRI) (2007) ArcView GIS Version 9.2. Redlands, CA

Esrey SA, Feachem RG, Hughes JM (1985) Interventions for the control of diarrhoeal diseases among young children; improving water supplies and excreta disposal facilities. Bull World Health Organ 63:757–772

Fewtrell L, Colford JM Jr (2004) Water sanitation and hygiene: interventions and diarrhoea – A systematic review and meta-analysis, Health, Nutrition, Population Discussion Paper, World Bank. – http://www1.worldbank.org/hnp/Pubs_Discussion/Fewtrell&ColfordJuly2004.pdf

Hoque BA, Hoque MM (1994) Environment and health. In: Rahman AA, Huq S, Haider R, Jansen EG (eds) Environment and development in Bangladesh. The University Press Limited, Dhaka, Bangladesh, p 359

Hoque BA, Juncker T, Sack RB, Ali M, Aziz KM (1996) Sustainability of a water, sanitation and hygiene education project in rural Bangladesh: a 5-year follow-up. Bull World Health Organ 74:431–437

Irrigation Support Project for Asian and the Near East (ISPAN) (1995) Bangladesh national level GIS database. Arlington, VA

Levine RJ, Khan MR, D'Souza S, Nalin DR (1976) Failure of sanitary wells to protect against cholera and other diarrhoeas in Bangladesh. Lancet 2(7976):86–89

National Institute of Population Research and Training (NIPORT), Mitra and Associates (MA), and ORC Macro (ORCM) (2001) Bangladesh demographic and health survey 1999–2000. Dhaka, Bangladesh and Calverton, Maryland [USA]: National Institute of Population Research and Training, Mitra and Associates, and ORC Macro

Oliver MA (1996) Kriging: a method of estimation for environmental and rare disease data. Geol Soc Lond Spec Publ 113:245–254

Oliver MA, Webster R (1990) Kriging: a method of interpolation for geographical information systems. Int J Geogr Inf Syst 4:313–332

Raudenbush SW, Bryk AS (2002) Hierarchical linear models: applications and data analysis methods, 2 Sage Publications, Thousand Oaks, CA

SAS Institute (2007) The GLIMMIX procedure. Available at: http://support.sas.com/rnd/app/papers/glimmix.pdf

SAS Institute (2008) SAS Software Version 9.2. Cary, NC

Chapter 10
Developing a Supermarket Need Index

Laura Smith, Chris Goranson, Jodi Bryon, Bonnie Kerker, and Cathy Nonas

Abstract The New York City Department of City Planning with assistance from the New York City Department of Health and Mental Hygiene developed a supermarket need index to determine the areas in the city with the highest levels of diet-related diseases and largest populations with limited opportunities to purchase fresh foods. The index was created using Geographic Information Systems to measure the need for supermarkets based on high population density, low access to a car at the household level, low household incomes, high rates of diabetes, high rates of obesity, low consumption of fresh fruits and vegetables, low share of fresh food retail, and capacity for new stores. The resulting index identified areas of acute need for additional full-line grocery stores, encompassing portions of the city where approximately three million New Yorkers reside.

Keywords Food environment · Built environment · Food planning · Health policy

10.1 Background

In many neighborhoods across the country, access to healthy foods is limited. The USDA Economic Research Service estimates that 23.5 million people nationwide live in low-income communities called food deserts with little or no access to a supermarket or large grocery store within one mile of their home (USDA, 2009). These neighborhoods suffer from higher rates of obesity, diabetes, and other diet-related diseases. Studies have shown that when people have access to supermarkets that offer healthy food, they tend to be less obese (Morland et al., 2006). Furthermore, these neighborhoods do not experience the additional benefits that supermarkets may provide, including economic vitality and increased job opportunities.

L. Smith (✉)
NYC Department of City Planning, New York, NY, USA
e-mail: lsmith@planning.nyc.gov

J.A. Maantay, S. McLafferty (eds.), *Geospatial Analysis of Environmental Health*, Geotechnologies and the Environment 4, DOI 10.1007/978-94-007-0329-2_10,

In 2004, the Food Trust in Pennsylvania formed a successful public/private partnership to offer flexible loans and grants to supermarkets that opened in underserved areas. In 2007, with the help of a small start-up grant from New York City's (NYC) Office of the Mayor under Mayor Bloomberg, the Food Trust began organizing a similar initiative in New York that included a revolving loan fund and zoning and tax incentives for supermarkets that developed new markets in low-income communities. Recently this initiative received national recognition. The federal government has allocated over $400 million for the Healthy Food Financing Initiative, which will establish grocery stores and other healthy food retailers in underserved urban and rural communities across America.

In early 2007, the New York City Department of City Planning (DCP) was asked by the Mayor's Office to explore issues related to supermarket need in the city, in response to growing concerns over a shortage of supermarkets. The study, conducted with assistance from the New York City Food Policy Task Force, the NYC Economic Development Corporation (NYCEDC) and the NYC Department of Health and Mental Hygiene (DOHMH), showed a widespread shortage of supermarkets and neighborhood grocery stores in the city. The analysis identified the areas with the greatest level of need for fresh food purveyors based on neighborhoods with the highest levels of diet-related diseases and largest populations with limited opportunities to purchase fresh foods, and determined that approximately three million New Yorkers live in high need areas.

The supermarket needs index used data on a variety of food retail, community health and socioeconomic variables to assess levels of need for full line grocery stores in small, walkable neighborhoods of NYC. However, many of the original data sets comprising the index were collected for larger administrative boundaries and statistical geographic entities at varying spatial resolutions. For example, some data were available at the census block or census tract level, while other data were aggregated to a ZIP code. Still others were aggregated to NYC Community Districts or United Hospital Fund (UHF) neighborhoods. As a result, the project depended on the ability to evaluate a number of variables across these different geographies. Since many of these areas overlap and cannot be cleanly aggregated into one another, this posed a particular problem. For instance, NYC Community Districts do not align cleanly within UHF neighborhoods or vice versa. If they did, one approach would be to simply aggregate all datasets up to the largest coincident boundary. One downside to this approach is that it creates much more generalized data for smaller areas, and the pockets of high or low density areas scattered within ZIP codes are averaged out as a result.

Geographic Information Systems (GIS) provided the tools necessary to compare different datasets with different spatial resolutions, and eventually derive an index that was representative of all the variables combined. This ability was critical to developing the supermarket needs index.

The goal of this chapter is to illustrate how a real-world analysis used varying types of information to collectively say something about newly created geographical regions. Because of this methodology, the participating agencies were able to assess access to supermarkets by neighborhood, and identify barriers to access and opportunities for encouraging the development of supermarkets in underserved areas. As a result of these findings, the NYC DCP proposed a zoning text amendment to increase the number of full-line supermarkets and neighborhood grocery stores in some of the city's highest need areas. The city continues a multi-agency effort to address additional need for fresh food retailers throughout the city, focusing on the areas identified through the DCP's supermarket needs index.

10.2 Literature Review

The influence of the built environment on food systems and dietary patterns is complex. More accurate metrics that reflect this complexity are needed to guide community and regional food-related economic development planning (American Planning Association, 2007). In response, this study created a Supermarket Needs Index (SNI) to weigh a number of diverse factors in order to identify high supermarket need areas.

Current research demonstrates a number of trends that highlight the urgent need for more supermarkets to locate in low-income inner-city areas across the country. A 1995 study conducted by the Food Marketing Policy Center at the University of Connecticut confirmed the existence of an "Urban Grocery Store Gap" throughout the country (Cotterill and Franklin, 1995). The study analyzed the country's top twenty-one major metropolitan areas and showed that ZIP codes with high percentages of low-income residents contained fewer supermarkets per capita than wealthier areas. In addition, the study showed that these areas also have the lowest car ownership rates (Cotteril and Franklin, 1995).

The lack of accessible supermarkets in low-income neighborhoods is a contributing factor to high rates of obesity and diabetes, in addition to other health related issues (Gallagher, 2006). Research has shown that a poor diet is a key contributor to obesity (DHHS, 2001) and that a supermarket carries a wider variety of fresh produce than any other food store type (Morland and Filomena, 2007). Specifically, one study found that, holding education and income constant, as grocery store access decreases, obesity increases (Gallagher, 2006). Further evidence of the link between a healthy diet and accessible supermarkets is documented in three individual studies where respondents reported an increase in fresh fruit and vegetable consumption once it became easily accessible (Moreland et al., 2002; Wrigley et al., 2002; Floumoy and Treuhaft, 2005).

Despite the well documented lack of supermarkets in many inner-city low-income neighborhoods, there is evidence of a large unrealized market potential in these areas. In a 1995 analysis completed by Public Voice for Food and Health Policy, it was estimated that as much as \$1 billion in Food Stamp Program purchasing power is lost annually as a result of people not having access to competitively-priced stores in their neighborhoods (Weinberg, 1995). In addition, certain supermarket chains, for example, Pathmark and Super Stop and Shop, have found that their highest grossing stores are in low-income communities (The Food

Trust, 2004). Furthermore, data from a Pathmark that opened in Harlem in 1999 show that this supermarket "met or exceeded industry averages in almost every category" and, that as of 2005, this store had one of the largest produce departments in New York City (Turcsik, 1999). There are a number of documented advantages to locating in inner-cities such as: density of purchasing power, limited competition, and available labor force (Floumoy and Treuhaft, 2005). One of the main reasons current market analyses fail to capture this market potential is because they tend to rely on average income as opposed to the high aggregate purchasing power that exists in dense urban areas (Floumoy and Treuhaft, 2005). Accordingly, current research points to an information gap between supermarket operators and actual market conditions (Floumoy and Treuhaft, 2005; Pothukuchi, 2005).

This study is presented as an example of how a raster-based index can account for the variability in data. This study is the first city-wide comprehensive policy to provide a combination of zoning and financial incentives to address the urban grocery gap.

10.3 Supermarket Need Index

The Supermarket Need Index (SNI), developed by the NYC DCP, is a multi-criteria index that reflects both the health status of local populations and the economic and geographic barriers they face in acquiring fresh food. Using GIS and geospatial analysis, the index identifies areas in the city where large populations with limited opportunities to purchase fresh food also have the highest levels of diet-related diseases. Specifically, the index measures the need for supermarkets based on the following factors: high population density, low access to a car at the household level, low household incomes, high rates of diabetes, high rates of obesity, low consumption of fresh fruits and vegetables, low share of fresh food retail, and lack of stores.

10.3.1 Methodology: Data Selection

As has been highlighted, the Supermarket Need Index was progressive in its consideration of multiple health- and land-use related variables to determine areas of New York City with the greatest need for additional fresh food retail. The process of identifying the appropriate data to incorporate into the index involved consulting with multiple city agencies, but final decisions were left to the NYC DCP. This collaborative effort to acquire the desired data, and the processes of converting multiple data formats from a variety of sources, are where this chapter seeks to be most instructive.

Data used to develop the SNI existed at several incongruent geographies, which posed a particular problem. Some variables – population, household income, access to a car – came from the 2000 Census at either the census block or census tract level.

Data on the proportion of fresh food retail to all food retail came from the 2007 ZIP Business Patterns. ZIP code areas are commonly used as geographic zones because they attach a more specific geography to a location than a state or county, but do not require an individual providing an actual street address. However, ZIP codes are challenging to use since they are defined on the basis of US mail carrier routes (USPS, 2010). As such, ZIP codes as represented on maps are often generalized collections of similar routes (Manifold System, 2010). In GIS, this would mean that the carrier routes are best represented as line features, not polygons, or areas. Since there is a desire to collect information by ZIP Code, the lines are transformed into areas, but as a result lose some definition. Thus, data are assigned to areas that provide relatively good approximations for location, but are not perfect. The ZIP code data included information on the number of food retail establishments and each establishment's appropriate standard industry classification code. Using this code, the determined the type of establishments most likely to sell fresh food, and NYC DCP then calculated the ratio of numbers of stores likely to sell fresh foods to the total of all food retailers by ZIP code. These data were then joined to the ZIP code boundary shapefiles, available through the 2000 Census Tiger website.

To incorporate poverty data, census tract boundaries, which were downloaded through the 2000 Census Tiger file, were merged with data compiled by the NYC DCP that identified Community Development Block Grant (CDBG) eligibility. The CDBG designation identifies the percent of those living in households with incomes below 80% of the median household income, which was $47,100 for a 4-person household in 2000 (NYC DCP, 2010). Tracts where at least 51% of residents are defined as low- or moderate- income based on this income limit are designated as CDBG eligible, and receive CDBG funds. A tract's CDBG eligibility status was used as a proxy for the Supermarket Need Index's determination of poverty among census tracts citywide.

Census block group boundaries were downloaded through the 2000 Census Tiger file website and were loaded with data on the number of households without access to a car and on the population density of each block group (number of people per acre) from Census 2000 American Factfinder.

In contrast, data from the NYC DOHMH files came from the Community Health Survey (CHS) and were geographically much larger than census blocks, census tracts, or ZIP code areas. The survey samples the population using the United Hospital Fund's 42 (UHF) neighborhood designations, which are aggregations of contiguous ZIP codes. For use on the CHS, these neighborhoods have been modified slightly for the addition of new ZIP codes since UHF's initial definitions (NYC DOHMH, 2010). UHF Neighborhoods are used because most people know their own ZIP code, so it is easier to collect this information on a self-reported survey than other geographic boundaries.

The CHS provides a wide range of estimates of chronic disease and behavioral risk factors collected annually through a telephone survey of approximately 10,000 New Yorkers (NYC DOHMH, 2010). Among the questions asked during the survey were questions that assess fruit and vegetable consumption, obesity and diabetes. For example, in 2004 respondents were asked, "How many total servings of fruit and/or vegetables did you eat yesterday?" Choices included "none", "1–4" or "5 or more" (NYC DOHMH, 2010). When age-adjusted, these estimates contribute to a picture of fresh fruit and vegetable consumption in the city. The NYC DCP received the UHF neighborhood shapefiles with these three data points pre-loaded into the attribute tables.

The NYC DCP and DOHMH agreed that these fruit and vegetable consumption and diabetes and obesity rates provided useful guidance in identifying populations in the greatest need of healthy food intervention. While numerous other health variables could have been incorporated into the index as indicators of poor diet, the three chosen sufficiently highlight high need areas for the purpose of the NYC DCP's policy goals. The NYC DOHMH looks more closely at diet-related diseases and food consumption for their own intervention and outreach activities.

10.3.2 Methodology: Shapefile Creation, Hot Spot Analysis, Trade Area Determination

As has been highlighted, the Supermarket Need Index was progressive in its consideration of multiple health- and land-use related variables to determine areas of New York City with the greatest need for additional fresh food retail. The process of identifying the appropriate data to incorporate into the index involved consulting with multiple city agencies. The collaborative effort to acquire the desired data, and the processes of converting multiple data formats from a variety of sources, is where this chapter seeks to be most instructive.

The first step in creating the index was to prepare the data. For two of the variables, Getis-Ord Gi* hot spot analyses were run in order to identify spatial clusters so that the variables could be easily incorporated in the SNI. The two variables were: number of households without access to a car, and number of people per acre within each census block group. A distance threshold of 1,320 ft, or ¼ mile was used. The NYC DCP defined this ¼ mile distance as being the general distance a resident is willing to walk to access daily goods and services. This distance thus approximates a person's neighborhood. The resulting hot spot renderings highlighted neighborhoods where there are clusters of residents without access to a car, and where there are clusters of high population density. Block groups with fewer than 100 residents were excluded from the analysis, so that non-residential block groups did not influence the identification of hot spots.

Estimates for trade areas illustrating the possible market reach for supermarkets were constructed using the Thiessen Polygon tool provided in the ArcInfo analysis toolbox. The input points were existing supermarkets, as provided by the NY State Department of Agriculture and Markets. Each resulting polygon contained only one point (supermarket), and bounded an area where "any location within a polygon is closer to its associated point than to the point of any other polygon" (ESRI, 2010). The attribute table for the Thiessen polygons contained all of the data for the individual grocery stores around which the polygons were created – so each polygon had a store name and store size associated with it.

The population living within each supermarket trade area was determined by converting census block group polygons into centroid points, with each block group population tied to its centroid point. This was done to avoid errors during the spatial join. These block group points were then spatially joined with the trade area polygons, so that the populations associated with each block group were joined with the supermarket trade areas that contained their centroid. Block group populations were summed by Theissen polygon, providing an estimate of the total population within each supermarket trade area.

10.3.3 Methodology: Raster Analysis and Index Calculation

These eight shapefiles, representing vector data, were then converted into raster layers using the Features to Raster tool in Spatial Analyst. Each raster layer consisted of a series of cells, or pixels, with each cell representing a $611' \times 611'$ area on the map. Furthermore, each raster cell included some measure of the layer – an actual integer score assigned to each pixel based on the value of the earlier vector layer. For example, the raster cell might indicate a population density, while another might be the percent of persons within a particular area who were diabetic.

In each case, the data were assigned a weight to represent the perceived influence of the variable, as determined through discussion internally at the NYC DCP, and in consultation with partners at NYC DOHMH. For each variable, the pixel values were reclassified on an integer scale from 0 to 3. For example, diabetes rates were reclassified from their original percentages, so that a value of 0 represented an area where the diabetes rate was at or below the citywide average, and a 1 represented an area where diabetes rates were above the citywide average. This reclassification methodology provided a convenient way to aggregate a pixel's scores on a number of different variables and come up with a representative total score for that specific pixel, on the map. Furthermore, each variable could be weighted in the final calculation based on its perceived influence

The raster layers, shown below, were used in creating an index score that represents the need for and access to supermarkets in each square (pixel) in the map. The index represents the sum of the integer scores for that pixel of the eight variables described above.

Index formula:

SNI = [POP] + [CAR] + [LOWINC] + [DIAB] + [OBES] + [5FRV] + [FFRET] + [CAP]

POP = Population density (high density ranks higher)

CAR = Access to a car (low access ranks higher)

LOWINC = Low income, defined by CDBG eligibility (eligible tracts ranks higher)

DIAB = Diabetes rates by UHF (high rates rank higher)

OBES = Obesity rates by UHF (high rates rank higher)
5FRV = Daily consumption of fresh fruits and vegetables (low rates rank higher)
FFRET = Share of fresh food retail (low rates rank higher)
CAP = Population capacity for new stores (more people served by less supermarket square footage ranks higher)

The combined weighted values at every pixel produced an overall index of need, categorized on a three-tiered scale from less to moderate to high. Pixels with a combined score of 0–5 were determined to have Minimal Need; those with a score of 6–8 were classified as having Moderate Need; 9–12 were classified as being in High Need of additional fresh food retail. This categorization was based on natural breaks in the data when divided into three groups.

Table 10.1 – SNI index variable weights below illustrates how variables were weighted.

Health data (obesity, diabetes, fruit/vegetable consumption) were weighted 0 or 1, depending on whether the pixel had a value below the citywide average for diet-related diseases (1) and whether the consumption of fruits and vegetables among residents in each UHF was above the citywide average (1).

For the food retail variable, raster cells with values at or below the citywide average (where the share of fresh food retail to all food retail was at or lower than average) were weighted 0; those with values above the citywide average share were weighted 2.

Neighborhood clusters of car access were weighted with values of 0 or 2. Cells where the Getis Ord Gi* z score indicated no apparent spatial concentration of residents without access to a car were weighted 0, and cells where the Getis Ord Gi* z score indicated a high concentration of residents lacking access to a car was weighted 2. The assumption was that spatially-clustered neighborhoods where populations lack access to a car are less likely to have cars available for driving to the grocery store.

Neighborhood clusters of population density were weighted in thirds. Pixels where the Getis Ord Gi* z score indicated no apparent concentration of population density were weighted 0, those with a moderate population density spatial association were weighted 1, and pixels with a z score indicating a strong spatial concentration of population density were weighted 2.

Pixels representing areas with low incomes as defined by CDBG funding were weighted 3, while those in tracts with higher income (ineligible for funding) were weighted 0. Since NYC DCP believes that poverty and income play a large role in a family's ability to conveniently access fresh and healthy foods, areas eligible for CDBG funding were weighted heavily.

Pixels from the individual supermarket Thiessen polygons, representing supermarket coverage per population size, were also scored. The supermarket database, received from the New York State Department of Agriculture and Markets, included the store sizes of all food retailers in the City. The Urban Land Institute's Shopping

Table 10.1 Weighting of variables in Supermarket Need Index

Variable	Geography	Weighted value for raster cell			
		0	1	2	3
Percentage of population who were obese by neighborhood[a]	United hospital fund (UHF) neighborhood	At or below citywide average	Above citywide average		
Percentage of population been told having diabetes by neighborhood	United hospital fund (UHF) neighborhood	At or below citywide average	Above citywide average		
Fruit/vegetable consumption	United hospital fund (UHF) neighborhood	At or below citywide average	Above citywide average		
Share of fresh food retail to all food retail	Zip code	Above citywide average		At or below citywide average	
Percent of households with no car	Getis Ord1320 neighborhood clusters	No clustering of low car ownership		Neighborhood clusters of low car ownership	
Population within trade area that cannot be accommodated by existing store in trade area (at 10 K pop/30 K pop ratio)	Thiessen polygon defined trade areas for all supermarkets	Over capacity	Population capacity for up to one additional store	Population capacity for between one and two additional stores	Population capacity for more than two additional stores
Neighborhood clusters of population density	Getis Ord1320 neighborhood clusters	No neighborhood clustering of population density	Some clustering of population density	Strong clustering of neighborhood population density	
CDBG eligibility	Census tract	No			Yes

[a]Obesity, diabetes and fruit vegetable consumption variables from Epiquery: NYC Interactive Health Data System – Community Health Survey 2004. Body Mass Index (BMI) is calculated based on respondents' self-reported weight and height. A BMI between 25.0 and 29.9 is classified as overweight and a BMI of 30 or greater is classified as obese. Diabetes: The exact survey question was: Have you ever been told by a doctor that you have diabetes? Fruit/vegetable consumption: The exact survey question was: How many total servings of fruit and/or vegetables did you eat yesterday? Age-adjusted estimates.

Center Handbook (Casazza, 1985) helped NYC DCP to approximate a supermarket industry trade area standard of 30,000 ft^2 of supermarket square footage for 10,000 people. This ratio took into account the general trade area captured in a neighborhood shopping center developed in the Northeast United States and factored in New York City's densely developed neighborhoods. The ratio helped NYC DCP to approximate how well served a neighborhood was in supermarket square footage and the extent to which capacity for additional supermarket square footage existed. A trade area that was over capacity in its ratio of supermarket square footage to population was weighted 0; a trade area with the capacity for up to one additional 30,000 ft^2 store was weighted 1; a trade area with capacity for up to two additional stores was weighted 2; and a trade area with a capacity for more than two 30,000 ft^2 stores was weighted 3. The 30,000 ft^2 store model was used to simplify the index results; New York City could attract several smaller stores and still achieve the goal of reaching the 30,000 ft^2 per 10,000 people ratio.

Land use and economic related variables (share of fresh food retail to all food retail; percent of households with no car; population within trade area that cannot be accommodated by an existing store in the trade area; neighborhood clusters of population density; and CDBG eligibility) had a higher maximum possible weight than the diet-related variables (percent of population with diabetes; percent of population that is obese; percent of population that did not consume a fruit of vegetable the day before), which were scored no higher than 1. This was done for two reasons: since diet-related diseases (in this case diabetes and obesity) are often correlated, assigning a lower weight to these variables guarded against the double-counting or overemphasis of these variables. Emphasizing zoning and land use data in the SNI produced findings that enabled the NYC DCP to focus on the areas where it can have the greatest influence, neighborhoods with the greatest physical barriers to fresh food access. Future collaboration efforts may incorporate additional health- and land-use variables as participating agencies see fit. The NYC DOHMH, for example, might have a greater interest in developing an index that will illuminate neighborhoods with specific health considerations, or focus specifically on areas with large vulnerable populations, like children or the elderly.

10.4 Results

The resulting total index score provided a measure that took into account the values of each of the separate variables. Combined raster cells with a total score of 0–5 were determined to have Minimal Need; those with a score of 6–8 were classified as having Moderate Need; 9–12 were classified as being in High Need of additional fresh food retail. The raster cells that scored highest were identified as areas that had residents with low access to fresh food, low consumption of fresh food, pervasive health problems, low income, few existing stores but a population capacity to support new ones, high population density and low rates of car ownership. The highest need areas had near-maximum scores for all eight variables. Figure 10.1 also shows the boundaries of the NYC DOHMH's District Public Health Offices (DPHOs),

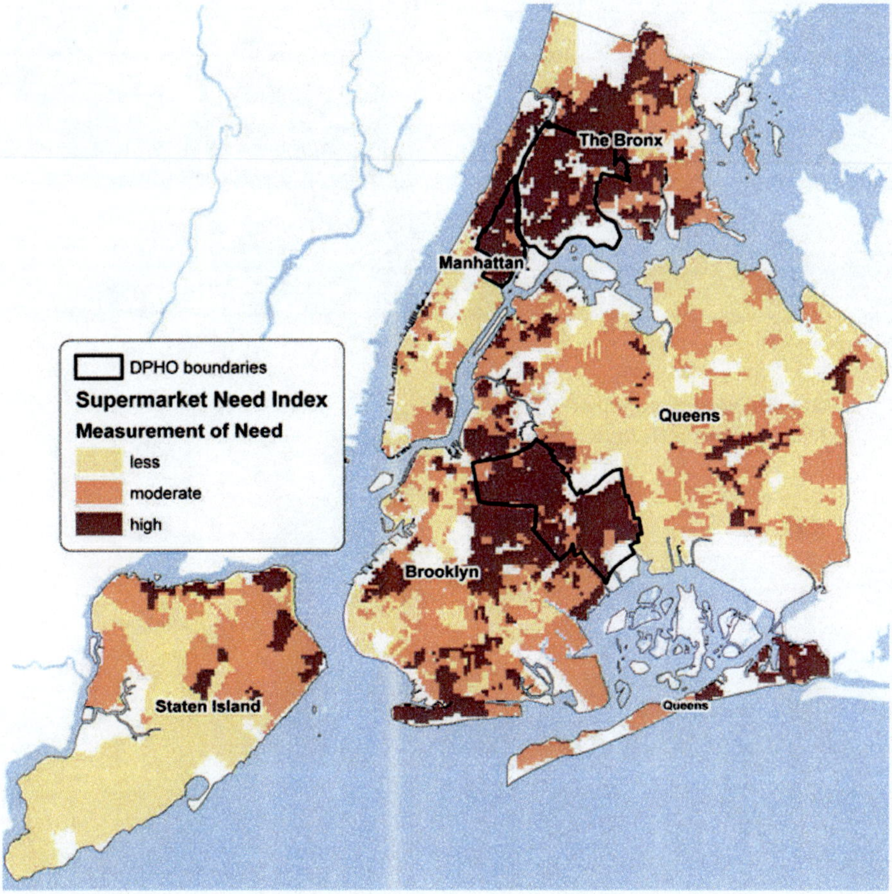

Fig. 10.1 Supermarket Need Index with District Public Health Office boundaries

communities that receive targeted programming to reduce health inequalities across New York City due to their high levels of poverty, disease and premature death. The high need areas encompass the DPHOs and thereby reinforce the necessity for coordinated efforts to create conditions that will enable New Yorkers to make healthier choices to lead healthier lives. Figure 10.2 illustrates the final Supermarket Need Index against the original United Hospital Fund boundaries.

10.5 What We Learned and What We Could Do Differently

The NYC DCP developed this index and the weighting methodology based on the agency's understanding of supermarket need, and the agency's ability to influence policy around land use and the built environment. Data from the NYC DOHMH

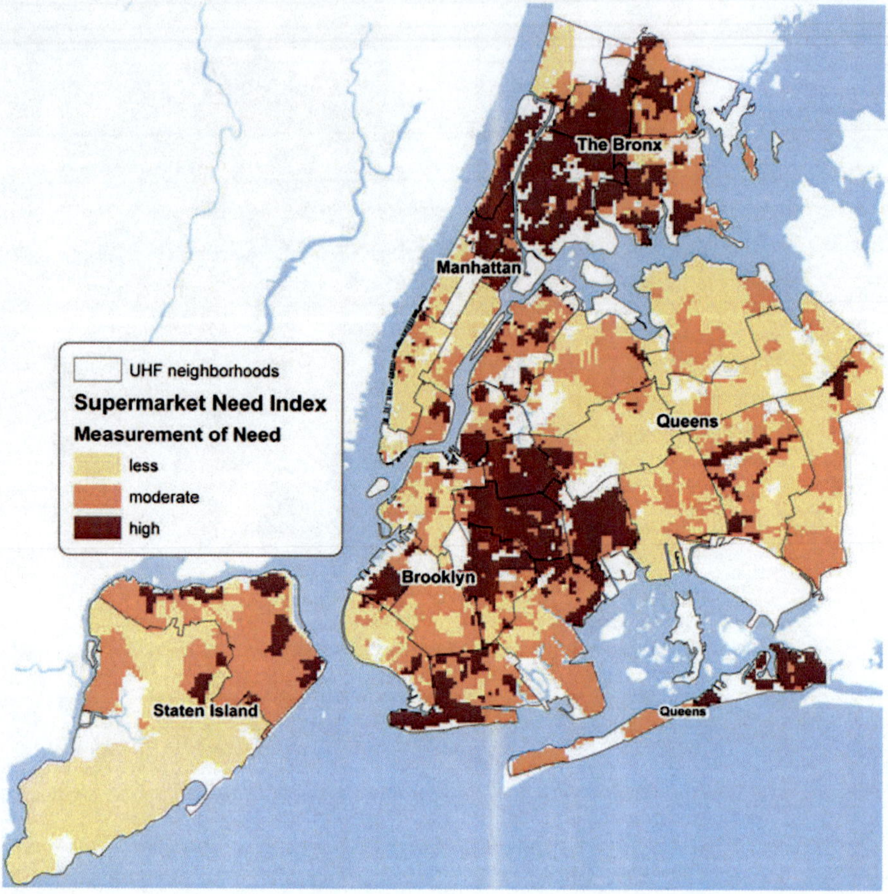

Fig. 10.2 Supermarket Need Index with United Hospital Fund neighborhoods

provided an additional and valuable perspective on the relationship between health, diet-related diseases, and the built environment. As previously discussed, however, land use variables were given more weight in the NYC DCP index because this agency's policies can only influence land use, and by extension, the retail landscape.

Some specific lessons were learned through the development of the Supermarket Need Index. The use of $611' \times 611'$ rasters yielded results at a geographic level that was somewhat arbitrary; a recalculation of the rasters at a square footage size that is more meaningful might be preferable. For example, $528' \times 528'$ rasters could be generated, yielding an area that is 1/10th mile long by 1/10th mile wide, about 2 Manhattan blocks – a distance that is familiar to most New York City residents. By doing this, no background knowledge of the study and no experience with raster data would be needed to measure distances using the cell size or identify meaningful geographies portrayed on the final map. Having each raster cell represent an area that people can generally visualize and mentally map could help the SNI reach a wider audience and would allow people to translate areas on a map into distances they are familiar with in their own neighborhoods.

There were also some limitations to the data used to develop the index. More time and/or specific resources might have allowed the NYC DCP to utilize a different methodology for determining the trade areas for existing stores. For example, the current methodology assumes all residents shop at the store that is physically closest to their homes. This does not take into account highways or natural barriers that might compel a pedestrian shopper to visit the store that is on the more walkable path, rather than a store that is geographically closer but less accessible. Nor does it take into account one's perception of or preference for particular areas or stores. The methodology also did not recognize quality differences among food retailers that might compel someone to travel to a "destination" grocery store, nor the fact that many residents may pick up groceries near their place of employment, school, etc. That said, because city policy aims to increase and improve fresh food retail at the neighborhood level, and because most residents do rely on the grocer closest to their home to meet their daily food needs, identifying neighborhoods with limited options for full-line grocery shopping was considered a valuable finding. Also, because the index considered automobile access in its calculation of high need, there is an understanding that transportation options might be more limited in areas identified as being in high need of additional stores.

For these reasons, trade area definitions and the determination of the capacity for additional food retail square footage were based on an assumption that people shop exclusively at the supermarket closest to where they live and are not drawn to stores outside of their neighborhood. Making fresh food retail available within walking distance from residents is a priority in the city's dense, pedestrian-oriented neighborhoods.

10.6 Future Applications

Rather than considering the results of the data calculations and resulting index as a definitive measure of the need for additional supermarkets, city agencies might want to consider the index as a tool for policy development. Incorporating land use variables and weighting them more heavily than other data allowed the NYC DCP to focus more closely on elements of supermarket access that the agency can directly influence. An index that gives additional weight to health variables, childhood indicators, economic variables, and other factors might be a possibility for agencies dealing with non-land use related aspects of fresh food access and health policy. The use of multiple such indices may lead to the identification of further policy alternatives that could contribute to addressing this issue.

To help community leaders identify the food deserts in their area, USDA recently launched a Food Environment Atlas (www.ers.usda.gov/FoodAtlas/). This new online tool allows for the identification of counties where, for example, more than

40% of residents have low incomes and live more than one mile from a grocery store.

Although this type of mapping is a start, many additional elements must be defined before this initiative can accurately pinpoint neighborhoods with the greatest need for new markets. For New York, one aspect of the learning curve was recognizing and adapting to the differences in the urban environments of Philadelphia and New York City. For example, in New York City there are fewer large, convertible spaces for supermarkets to occupy, a greater number of neighborhoods, and fewer cars. Therefore, it was incumbent upon those working in the city to rethink the mapping of this project: Hence, the supermarket index.

Appendix: Data Sources and Tools Used

Data were collected from several sources, including the NYC Department of City Planning, the US Census, The NYC Department of Health and Mental Hygiene, the Bureau of Labor Statistics, and the NY State Department of Agriculture and Markets. Using ArcGIS 9.2 with Spatial Analyst, this information was mapped to reveal diet-related data and land-use conditions across New York City. The boundaries of variables were non-coterminous in some instances, but generally captured similar demographics.

Data and shapefiles came from city and state agencies:

Supermarkets database – New York State Department of Agriculture and Markets
Obesity and diabetes – NYC Department of Health and Mental Hygiene
Consumption of fresh fruits and vegetables – NYC Department of Health and Mental Hygiene
Share of fresh food retailers to all food retailers – Bureau of Labor Statistics
Population density – US Census 2000
Car ownership – US Census 2000
CDBG eligibility – NYC Department of City Planning
Geographic shapefiles – NYC Department of City Planning

ESRI ArcGIS 9.2 was used to create the index; generally, the ArcView license with extensions contained the tools needed for this process. An ArcInfo license was required for the Thiessen Polygon analysis.

The following ArcToolbox tools were used:
For specific variable calculations:

Getis Ord analysis: Geoprocessing tool reference > Spatial Statistics toolbox > Mapping Clusters toolset > Tools – Getis Ord: Use to identify population density and car ownership hot spots, to isolate areas of truly high density and high/low car ownership and not just capture, for example, a single dense building among a more suburban landscape

Thiessen Polygon creation: Geoprocessing tool reference > Analysis toolbox > Proximity toolset > Tools – Thiessen Polygons: Used to develop approximate trade areas for existing supermarkets in our database – City Planning wanted to capture the areas closest to each existing store and estimate the population living closer to that store than to any other store, in order to determine how well existing stores are able to serve their neighborhood populations.

Feature to Point conversion: Geoprocessing tool reference > Data Management toolbox > Features toolset > Tools – Feature to point: Used to convert block group polygons (and the population data associated with these polygons) into points, allowing City Planning to capture the residential population within each supermarket trade area.

For general analysis:

Conversion to Raster
Spatial Analysis Raster Reclassification
Raster Calculator

Obesity, diabetes and fruit vegetable consumption variables from Epiquery: NYC Interactive Health Data System – Community Health Survey 2004. Body Mass Index (BMI) is calculated based on respondents' self-reported weight and height. A BMI between 25.0 and 29.9 is classified as overweight and a BMI of 30 or greater is classified as obese. Diabetes: The exact survey question was: Have you ever been told by a doctor that you have diabetes? Fruit/vegetable consumption: The exact survey question was: How many total servings of fruit and/or vegetables did you eat yesterday? Age-adjusted estimates.

Author Biographies

Laura Smith, MS, is a City Planner with the New York City Department of City Planning's Housing, Economic and Infrastructure Planning division. Since joining the Department of City Planning in 2007, she has worked on the city's Food Retail Expansion to Support Health (FRESH) program and other initiatives to support commercial development in underserved neighborhoods. Laura holds a Masters of Arts in Geography from the George Washington University.

Chris Goranson, MGIS, GISP, is the Director of the GIS Center of Excellence within the Bureau of Epidemiology Services, NYC Department of Health and Mental Hygiene. The GIS Center is tasked with providing centralized resources, training, consulting, map production and geographic information analysis support to the department. Prior to joining the department in 2004, Chris worked for a professional services firm, supporting various federal client GIS projects including those for the EPA, FHWA, GSA and USGS. Chris teaches a GIS course in public health as adjunct faculty at CUNY's School of Public Health, and Mt. Sinai School of Medicine.

Jodi Bryon, MUP, is a City Planner in the Housing, Economic and Infrastructure Planning Division at the NYC Department of City Planning, where she worked as project manager on developing the

Food Retail Expansion to Support Health program. Jodi received a Master of Urban Planning degree from the University at Buffalo, The State University of New York.

Bonnie Kerker, Ph.D., is the Assistant Commissioner of the Bureau of Epidemiology Services at the NYC Department of Health and Mental Hygiene. Dr. Kerker holds a Ph.D. in Health Policy and Epidemiology and an MPH in Health Policy from Yale University School of Medicine.

Cathy Nonas, MS, RD, CDE, directs the New York City Health Department's Physical Activity and Nutrition program. Trained as a clinical dietitian, she has spent years working to combat Type 2 diabetes, both as a researcher at the federally funded Obesity Research Center and as an assistant clinical professor at Mt. Sinai School of Medicine. Since joining the Health Department two years ago, Nonas has worked on policy initiatives to increase physical activity for young children and improve access to healthy foods in underserved neighborhoods. She helped develop New York City's calorie-posting regulation, and has been instrumental in the NYC Green Cart initiative, an effort to expand the number of mobile food vendors selling fresh fruits and vegetables.

References

American Planning Association Policy Guide on Community and Regional Food Planning (May, 2007) CDA International, Ltd. Manifold System 8.00 User Guide. Zip Codes are Not Areas. August 6, 2010. http://www.manifold.net/doc/zip_codes_are_not_areas.htm

Casazza J (1985) Shopping center development handbook, 2nd edn. Urban Land Institute, Washington, DC

Cotterill RW, Franklin AW (April, 1995) The Urban Grocery Store Gap. Food Marketing Policy Center, University of Connecticut, Storrs, CT

Environmental Systems Research Institute, Inc. ArcGIS Desktop 9.2 Help (August 6, 2010). http://webhelp.esri.com

Flournoy R, Treuhaft. S (2005) Healthy food, healthy communities: improving access and opportunities through food retailing. The California Endowment, Woodland Hills, CA

Manifold.net. Manifold ® System 8.00 (August 6, 2010). http://www.manifold.net/doc/manifold.htm#zip_codes_are_not_areas.htm

Mari Gallagher Research and Consulting Group (2006) Examining the impact of food deserts on public health in Chicago. Mari Gallagher Research and Consulting Group, Chicago, IL

Morland K, Diezroux, A, Wing S (2006) Supermarkets, other food stores, and obesity the atherosclerosis risk in communities study. Am J Prev Med 30(4):333–339

Morland K, Wing S, Diezroux. A (2002) The contextual effect of the local food environment on residents' diets: the atherosclerosis risk in communities study. Am J Public Health 92(11):1761–1768

Morland K, Filomena. S (2007) Disparities in the availability of fruits and vegetables between racially segregated urban neighbourhoods. Public Health Nutr 10:12

New York City Department of City Planning (2000) Census Tract – Community Development Block Grant Eligibility Report (Revised). (August 6, 2010). http://www.nyc.gov/html/dcp/html/resource/censustractreport.shtml

New York City Department of Health and Mental Hygiene (August 6, 2010) Community Health Survey: Methodology. http://www.nyc.gov/html/doh/html/survey/chs-methods.shtml

New York City Department of Health and Mental Hygiene (August 6, 2010) Community Health Survey: Survey Data on the Health of New Yorkers. http://www.nyc.gov/html/doh/html/survey/survey.shtml

New York City Department of Health and Mental Hygiene (August 6, 2010) Epiquery: NYC Interactive Health Data System – Community Health Survey 2004. http://nyc.gov/health/epiquery

Pothukuchi K (2005) Attracting supermarkets to inner-city neighborhoods: economic development outside the box. Econ Dev Q 19:232–244

The Food Trust, Farmer's Market Program Evaluation 2004. Retrieved from http://www.thefoodtrust.org/catalog/resource.detail.php?product_id=68

The Food Trust, Philadelphia City Council Calls for Hearings on Supermarket Development to be Held in May, 2004. Retrieved from http://www.thefoodtrust.org/catalog/resource.detail.php?product_id=60

Turcsik R (1999) Taking root: a suburban-style Pathmark blossoms in New York's Harlem. Prog Grocer 78:(7)

U.S. Department of Health and Human Services. (2001) The surgeon general's call to action to prevent and decrease overweight and obesity. U.S.G.P.O. Retrieved from http://www.surgeongeneral.gov/topics/obesity/calltoaction/CalltoAction.pdf on August 13, 2010

USDA Economic Research Service (June, 2009) Access to Affordable and Nutritious Food: Measuring and Understanding Food Deserts and Their Consequences. Report to Congress. http://www.ers.usda.gov/Publications/AP/AP036/AP036.pdf

United States Postal Service (August 6, 2010) What is a Carrier Route. http://www.usps.com/faqs/zipZIPlookup-faqs.htm?WT.z_zip4link=FAQ

Weinberg Z (1995) No place to shop: the lack of supermarkets in low-income neighborhoods. Public Voice for Food and Health policy, Washington, DC

Wrigley N, Warm D, Margetts. B (2003) Deprivation, diet, and food-retail access: findings from the Leeds 'Food Deserts' study. Environ Plann A 35(1):151–188

Chapter 11
Asthma, Air Quality and Environmental Justice in Louisville, Kentucky

Carol Hanchette, Jong-Hyung Lee, and Tim E. Aldrich

Abstract Many analyses have demonstrated that environmental hazards tend to be concentrated in areas with higher numbers of low-income populations and people of color. We used geographic information science (GISc) and statistical analyses to examine issues of air quality and asthma occurrence among urban children in Louisville, Kentucky. The results of our analyses indicate that there is a well-defined spatial cluster of high rates of childhood asthma hospitalizations in western Louisville, an area of the city that is notorious for its poor air quality and the poor economic and physical health of its residents. Analyses also confirmed a strong seasonal pattern to asthma, with a fall peak. The multi-factorial etiology of asthma makes it difficult to pinpoint specific triggers for acute asthma episodes. Analyses of EPA criteria pollutants and volatile organic compounds from local air monitoring sites showed very little correlation with hospital admissions, although acetone, acrylonitrile and chloroform manifested similar seasonal patterns. In order to address the environmental justice concerns of disproportionate siting vs. minority move-in, we used GISc to examine patterns of residential mobility in western Louisville over a 60-year period. The polluting industries in western Louisville's "Rubbertown" preceded the local in-migration of African-Americans, the majority of which took place from 1960 to 1970. While the increasing African-American presence in the community has resulted in a community with greater social cohesion over time and successful community-based initiatives to reduce air toxics emissions have been implemented, significant health disparities in western Louisville must continue to be addressed.

Keywords Asthma · Environmental justice · Louisville Kentucky · Air pollution · Seasonality

---

C. Hanchette (✉)
Department of Geography and Geosciences, University of Louisville, Louisville, KY 40292, USA
e-mail: carol.hanchette@louisville.edu

J.A. Maantay, S. McLafferty (eds.), *Geospatial Analysis of Environmental Health*, Geotechnologies and the Environment 4, DOI 10.1007/978-94-007-0329-2_11,

11.1 Introduction

Many analyses have demonstrated that environmental hazards, such as air pollution and contamination from toxic wastes, are not equitably distributed. They tend to be concentrated in areas with higher numbers of low-income populations, people of color, and less-advantaged social classes. In this book chapter, we aim to examine issues of air quality and asthma occurrence among urban children in Louisville, Jefferson County, Kentucky, with a particular focus on the western industrial area of "Rubbertown," an area that is notorious for its poor air quality and the poor economic and physical health of its residents. The fate and transport of polluting substances, residential mobility and economic/political processes may all play a role in western Louisville's high asthma rates. All of these factors are inherently spatial, and geographic information science (GISc) provides an ideal approach for analyzing their impacts on health.

11.2 Background

11.2.1 Asthma Epidemiology

Asthma is one of the most common serious chronic diseases of childhood and is the third leading diagnosis for hospitalization among children (Akinbami et al., 2009, ALA, 2008). For the latter portion of the twentieth century, asthma was a vague diagnosis, representing a mosaic of conditions characterized by respiratory deficiency. However, it has gradually emerged as a distinct disease entity. It is not only an inflammatory process, as may characterize bronchitis or other respiratory infections; nor is it simply bronchial constriction, as may occur with allergic reactions. In fact, asthma is comprised of both bronchial inflammation and constriction. It is a reactive airway condition induced by irritation of the bronchial tree as illustrated by inhalation of fine particles, or breathing during high ambient temperatures (Helms, 2000). While the exact causes of asthma are hard to pinpoint, known triggers include dust mites, molds, cockroaches, pet dander, secondhand smoke, ozone and particulate matter, and leaf mold, particularly *alternaria* and *cladosporium* (Alp et al., 2001; Arbes et al., 2003; Lwebuga-Mukasa et al., 2004)

Unfortunately, no population-based data on asthma prevalence in the US exist. Most national and state statistics are based on information from the National Health Interview Survey, the National Survey of Children's Health, the Behavioral Risk Factor Surveillance System (BRFSS), the National Ambulatory Medical Care Survey, and asthma hospital discharges. In 2006, nearly 23 million people in the United States had asthma. Over 6.8 million of them were children, a number that represents over 9% of US children (Akimbami et al., 2009, ALA, 2008). Asthma prevalence rates are highest in the 5–17 year age group, at 106.3 per 1,000 population (ALA, 2008). In children, prevalence is higher in boys (ALA, 2008; Bjornson and Mitchell, 2000); in adults, females have higher rates (ALA, 2008). In general, rates are higher among people of color and in families with lower incomes (ALA, 2008; Aligne et al., 2000).

The estimated prevalence of asthma in children varies geographically, with high prevalence rates in states in the industrial northeastern US, including Kentucky (ALA, 2009). Year 2007 data from Kentucky's most recent Asthma Surveillance Report indicate that asthma prevalence for children 11 years and younger was 10.6%. The rates were higher for middle and high school students, at 13.6 and 11.8%, respectively. Rates for Kentucky's African-Americans were higher for all age groups, with the largest disparity among high school students – 22.4% for blacks, 11.3% for whites (Nunn et al., 2009). Geographically, asthma prevalence is highest in the southeastern portion of Kentucky (Appalachia). However, the Allergy and Asthma Foundation of America included Louisville (located in Jefferson County, north central Kentucky) in the top 100 worst cities in America for people with asthma, with a ranking of 53. Louisville ranked number three in the nation for spring allergies (AAFA, 2009a, b).

Information about asthma morbidity is also available from hospital discharge databases. In 2006, the US hospital discharge rate for asthma as a primary diagnosis was 149 per 100,000 (ALA, 2009). In 2007, the overall age-adjusted rate for Kentucky was 145.7 per 100,000 and the highest age-specific rate was for the age group 0–4 at 424 per 100,000 (Nunn et al., 2009).

Unfortunately, hospital discharge data do not always provide a good estimate of asthma prevalence. Patients who lack good preventive care and/or health insurance may be more likely to seek primary care in hospital emergency rooms. They may also live farther from a primary care physician (Jones et al., 2004). However, for many local areas, such as Louisville, hospital records often provide the only available source of data. Our analysis uses hospital discharge data to examine geographical and social inequalities in asthma in Louisville and to explore the determinants of geographic variations.

11.2.2 Asthma in Louisville: Recent Studies

Recent studies of asthma in Louisville have indicated that there are strong spatial and seasonal variations. Using hospital discharge data from 2000 to 2001, Jones et al. (2004) reported a band of zip codes with high pediatric asthma hospitalization rates extending from northwestern Louisville to the central portion of Jefferson County, with the highest rates in northwestern Louisville. This band corresponded with a spatial distribution of poverty, low educational attainment and African-American population. A similar geographic pattern was mapped by the Louisville Metro Housing Coalition (2005), using hospital admissions data from 2000 to 2003. Jones et al. (2004) illustrated that the locations of pediatric practices were mutually exclusive of the residential areas in Louisville that had the greatest numbers of emergency department admissions.

Asthma hospitalizations vary seasonally as well as spatially. Statewide, the highest percentage of Kentucky hospitalizations occur in winter (29.5%) and fall

(27.5%) (Nunn et al., 2009). The pattern appears to be slightly different in Louisville where Lee (2006) documented a strong, statistically significant seasonal pattern to childhood asthma hospital admissions, using 2003–2004 data: a summer trough followed by a strong fall peak.

The spatial patterns noted by Jones et al. (2004) and the Louisville Metro Housing Coalition (2005) identify high rates of pediatric asthma in areas with high poverty and high concentrations of African-American populations. Because the northwestern corner of Jefferson County is an area known for its poor air quality, a major environmental justice concern is whether environmental exposures are associated with a spatial concentration of poor health outcomes such as asthma in nearby residential areas.

11.3 Spatial Analysis of Hospital Discharge Data, 2005–2008

Since the Jones et al. (2004) study, successful community-based initiatives to reduce air toxics emissions have been implemented. The company responsible for the largest 1,3-butadiene emissions has installed technology to significantly reduce them. The Louisville Metro Air Pollution Control Board has implemented a series of regulations that go above and beyond EPA air quality measures. This program, the Strategic Toxic Air Reduction (STAR) Program, was approved in 2005 and requires nearly 200 companies that emit polluting chemicals to develop plans to reduce emissions. The primary purpose of our analysis is to utilize GISc and spatial analyses to examine patterns of asthma hospitalizations since the implementation of the STAR program.

11.3.1 Hospital Discharge Data

After obtaining Institutional Review Board approval from the Kentucky Cabinet for Health and Family Services (CHFS), we obtained hospital discharge records from Kentucky's Office of Health Policy for all Jefferson County residents with a primary diagnosis of asthma (diagnosis codes 493.0–493.99) for years 2005–2008. These data consisted of 5,347 total hospitalizations with 1,856 (34.7%) hospitalizations of children ages 0–19. Each record was stripped of personal identifiers and included admission and discharge dates, patient age and gender, zip code of residence, diagnosis codes, referral sources, treatment costs, and insurance types. While the dataset included fields for race and ethnicity, these were rarely populated. All records are for inpatient hospitalizations. CHFS did not collect outpatient data during the study period. Each record represents a single hospital admission/discharge, thus the time unit of analysis is daily.

Table 11.1 shows demographic characteristics of Jefferson County children hospitalized for asthma. A comparison of this table with a similar table in the Jones et al. (2004) study reveals interesting differences. First, a much larger percentage of hospitalizations occurred among children (34.7 now vs. 12.7% in 2000–2001).

Table 11.1 Demographic characteristics of children hospitalized for asthma, Jefferson County, Kentucky, 2005–2008

Characteristic	Number of cases	Percent of cases[a]
Gender		
Male	1,078	58.1
Female	778	41.9
Age in years		
0–4	930	50.1
5–9	495	26.7
10–14	261	14.1
15–19	170	9.2
Referral source		
Physician	389	21.0
Emergency department	1,456	78.4
Other	11	0.6
Insurance type		
Commercial	452	24.4
Federal program	1,245	67.1
Other	159	8.6
Total	1,856	100

[a]Due to rounding, percentages may not add up to 100.

The gender breakdown in 2005–2008 is nearly the inverse of 2000–2001. The age distribution of cases has changed as well. While the largest percent of hospitalizations was still in the 0–4 age group, the percent drops off steadily with age. In 2000–2001, the percentage in the 15–19 year age group was only slightly lower than that in the 0–4 year age group (38.4 vs. 37.2%), whereas in 2005–2008, it was substantially lower. This may be due to the fact that smoking rates among Kentucky high school students declined dramatically from 2000 to 2006 (Jones, 2007). Unfortunately, rates increased slightly in 2008 (Tooms, 2009). The referral source in the two studies is a near mirror image, with only 34.3% of cases referred from emergency departments in 2000–2001. Another striking difference is the flip in insurance type. In 2000–2001, commercial insurance accounted for 71.1% of cases. Now, the predominant source is federally-funded programs, including Louisville's Medicaid managed care program, Passport.

Over the 4-year period, the hospitalization rate per 100,000 was 248.70. Jefferson County intercensal estimates were used as population denominators (US Census Bureau, 2010). There was an overall trend for increasing rates over time (Table 11.2).

11.3.2 Spatial Analysis

We computed hospital discharge rates for all Jefferson County mappable zip codes ($n = 36$), using 2007 age/race/sex population estimates from GeoLytics (2007) as

Table 11.2 Hospital discharge rates per 100,000, children 0–19

Year	No. of discharges	Population	Rate per 100,000
2005	424	184,564	229.73
2006	421	185,620	226.81
2007	484	186,963	258.87
2008	527	189,122	278.66
All years	1,856	746,269	248.70

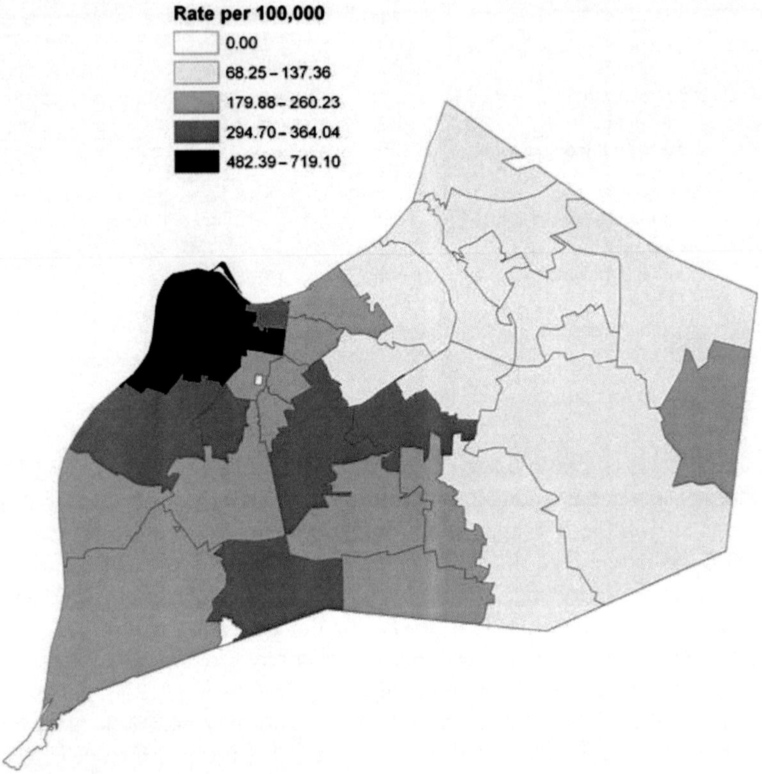

Fig. 11.1 Childhood asthma hospital discharge rates, Jefferson County ZIP codes, 2005–2008. Map source: ESRI Data (ESRI, Redlands, CA)

the denominator. We made the decision to use US Postal Service (USPS) boundaries as opposed to the Census Bureau's Zip Code Tabulation Areas (ZCTAs) because the asthma database used the former. ArcGIS 9.3 (ESRI, Redlands, CA) was used for mapping and GIS analysis. Figure 11.1 shows the spatial distribution of asthma discharge rates by zip code, using a Jenks natural breaks classification.

The spatial distribution of hospital discharge rates is striking. Two patterns stand out. The first is the concentration of high rates in the northwesternmost zip codes of the county, in the vicinity of the Rubbertown chemical/industrial complex. The second is the divide between the eastern part of the county (higher income, mostly white population) and the western portion.

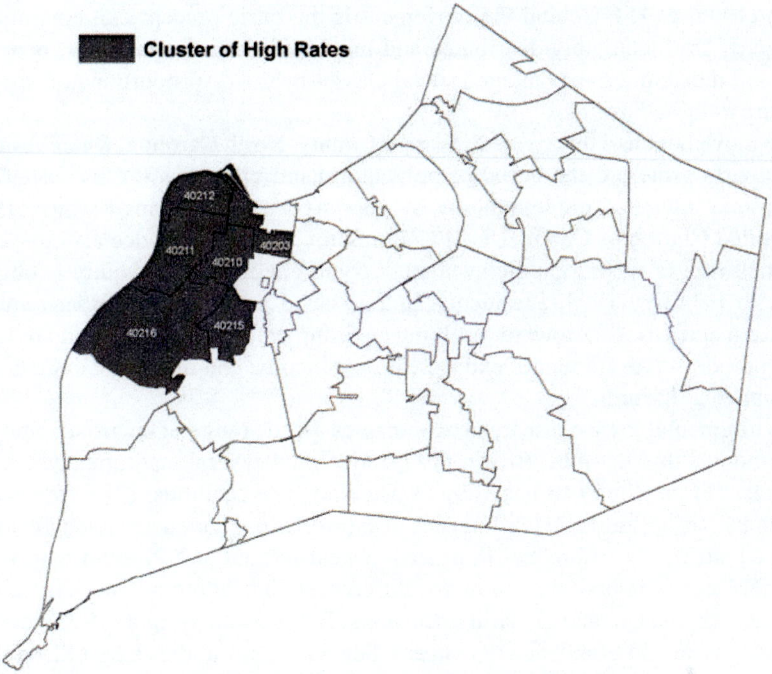

Fig. 11.2 Cluster of high hospitalization rates, northwestern Jefferson County

Figure 11.1 should be interpreted with some caution, however, due to small numbers in the rate numerator. Six of the zip codes had fewer than 10 discharges. For that reason, we used the local index of spatial autocorrelation (LISA), GeoDa 0.9.5-i software, to determine whether there were any significant clusters of high rates (Anselin, 2003). Due to the instability of the rates, we used the Empirical Bayes LISA option, with queen's contiguity and a spatial lag of 1. The result in shown in Fig. 11.2. The rates are highly clustered (Moran's $I = 0.5961$) and significant at $p = 0.01$. A cluster of high rates is comprised of six zip codes in the Rubbertown area: 40203, 40210, 40211, 40212, 40215 and 40216. A cluster of low rates exists in the eastern portion of the county, but is not shown in Fig. 11.2.

The remainder of our analyses, i.e. an examination of seasonal trends and the relationship of asthma hospital admission rates to air pollution, focuses on children residing in the Rubbertown area, the Jefferson County asthma "hot spot." This region of Louisville (hitherto referred to as the "West End"), is an area with many environmental and social justice issues. It is notorious for the poor quality of its air and the poor economic and physical health of its residents.

11.4 Environmental Justice: Disproportionate Siting Versus Minority Move-In

Environmental justice, also referred to as "environmental racism," or "environmental equity," was termed the "fastest growing social movement in recent years" by

Brown (1995, p. 15). Behind the movement is the basic precept that environmental hazards are inequitably distributed and that low income populations, people of color, and those in less-advantaged social classes bear a disproportionate burden of exposure to these hazards.

The movement had its origins in Warren County, North Carolina, where residents and activists protested the burial of polychlorinated biphenyls (PCBs) in 1978 in a rural area inhabited predominantly by poor African-Americans (Brown, 1995). The United Church of Christ (UCC) Commission on Racial Justice also played an important role in raising public awareness about environmental equity (Coughlin, 1996). In February 1994, President Clinton issued an Executive Order requiring federal agencies to work toward resolving environmental justice issues and reducing the disproportionate amount of exposure of low income and minority populations to environmental hazards.

Environmental justice research encompasses a wide range of environmental concerns and health outcomes. Brown (1995) identified several environmental justice concerns: (1) proximity to hazardous waste sites and facilities, (2) differences in exposure to air pollution, (3) differences in exposure to other environmental hazards (e.g. TRI sites), (4) differences in procedures carried out at NPL (and other) sites, (5) differences in regulatory action, (6) differences in health outcomes due to environmental exposures, and (7) siting decisions. Bullard and Wright (1993) included pesticides as an additional area of concern. Several recent studies have utilized GISc to examine asthma within an environmental/social justice framework (Corburn et al., 2006; Maantay, 2007).

Environmental justice issues can result from the disproportionate siting of hazardous facilities in minority/low-income neighborhoods or from changing residential patterns, i.e. minority move-in to these neighborhoods, or both (Pastor et al., 2001). Brown (1995) refers to these processes as a "causal" argument, with the deliberate placement of environmental hazards in poor and minority neighborhoods, or a "drift" argument, where poor and minority populations moved into an area with environmental hazards. In the case of the latter, i.e. minority move-in, it is important to seek an understanding of the historical, economic and political processes that have resulted in population mobility and their contribution to environmental and social justice issues. Hanchette (2008) has demonstrated this using lead poisoning as an example. In this section, we examine the primary sources of Louisville's air pollution and the neighborhood composition of western Louisville.

Louisville's West End is notorious for its industrial pollution. The West End is home to Rubbertown, a chemical/industrial complex that started with a Standard Oil of Kentucky refinery in 1918. During World War II, the US was cut off from 90% of its natural rubber supply and synthetic rubber production (and associated chemicals) was established in Rubbertown, hence its name (Kleber, 2001). Western Louisville is home to the bulk of the county's Toxic Release Inventory Sites. Figure 11.3 shows Rubbertown's location in western Louisville and its proximity to the cluster of zip codes comprising Louisville's asthma hot spot.

Louisville's environmental justice issues stem predominantly from minority move-in, which Pastor et al. (2001) note is a market-driven trade-off: increased

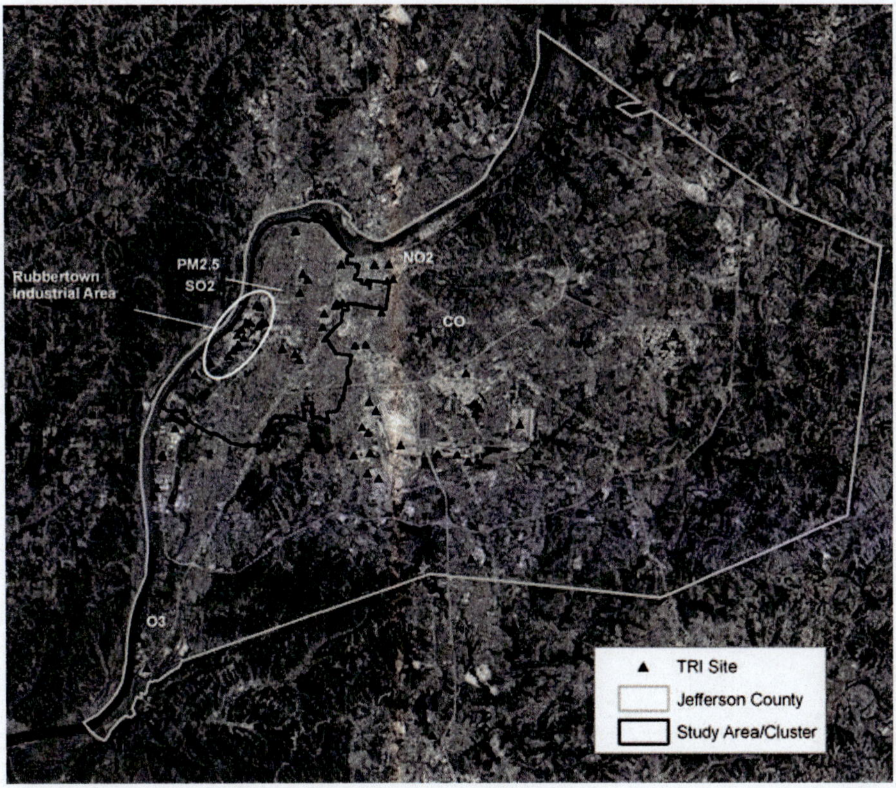

Fig. 11.3 The Rubbertown chemical/industrial area, adjacent to the Ohio River, is circled in *white*. The study area is outlined in *black*. Downtown Louisville is to the east. To the west of the Ohio River are the bisected hills of southern Indiana. Topographically, Louisville is in a basin, and polluted air is often trapped by temperature inversions. Location of criteria pollutant monitors is indicated by pollutant. Map source(s): ESRI, Redlands, CA

neighborhood health risks for better housing. In the 1940s and 1950s, many West End neighborhoods were inhabited by middle- to upper-income whites. Areas close to the river are characterized by distinctive craftsman-style homes and a park designed by Frederick Law Olmstead. The past 60 years have seen dramatic neighborhood changes in racial makeup, the most significant of which were described by Anderson (1980). Today, with the exception of the Portland community in the north, the area is largely African-American (Fig. 11.4).

Prior to the twentieth century, Louisville's African-American residents were dispersed throughout the city. By the mid-1900s, there were several identifiable clusters of neighborhoods with African-American majorities. These are shown in Fig. 11.4; shaded areas represent census tracts with an African-American population over 50%. The most dramatic changes took place from 1960 to 1970, with "white flight" to the eastern suburbs and neighboring Oldham County. Anderson (1980) noted that in this decade, over 40,000 people were involved in a massive property exchange that occurred over in nine census tracts. Although the changes from 1970 to 2000

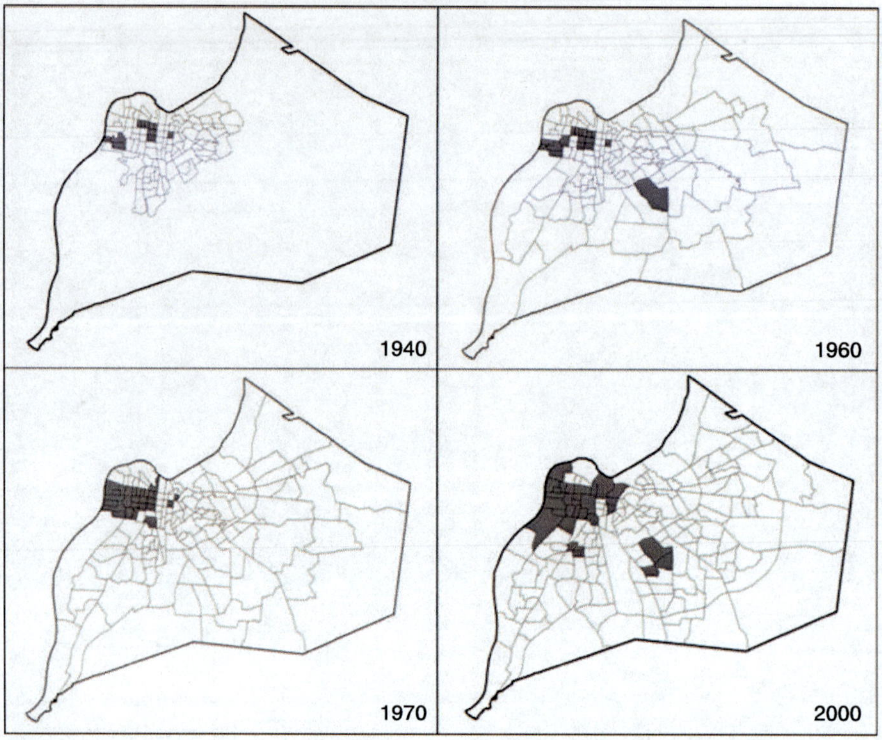

Fig. 11.4 Change in major African-American residential areas, Louisville, Kentucky, 1940–2000. Shading represents areas with an African-American population greater than 50%. Source: National Historic Geographic Information System (2009); US Census (1990); US Census (2000)

are less dramatic by decade, the year 2000 map in Fig. 11.4 shows the heavy concentration of the African-American population in the West End. This area is also predominantly poor, as indicated by the map in Fig. 11.5. The greatest spatial concentration of poverty is located in the West End, which supports the minority move-in hypothesis. Over several decades, higher-income white populations were replaced by lower-income populations of color.

For over two decades, residents of the West End have expressed concern over air quality and exposure to dangerous chemicals, including 1,3-butadiene. The Agency for Toxic Substances and Disease Registry (ATSDR) published a report on the area in 1992, with inconclusive results due to a paucity of air monitoring data. In response to citizen concerns, the Jefferson County Division of Environmental Health and Protection established the West Jefferson County Community Task Force in 1999, a citizens group that identified environmental concerns in western Louisville neighborhoods. Funding was obtained for an intensive air toxics monitoring network, with ten sites in the West End and two control sites located farther away (Fig. 11.6). Samples of over 50 volatile organic chemicals (VOCs) were monitored from 2000 to 2005 to determine whether residents were being exposed to toxic air pollutants that posed risks to human health. This project was a collaboration among

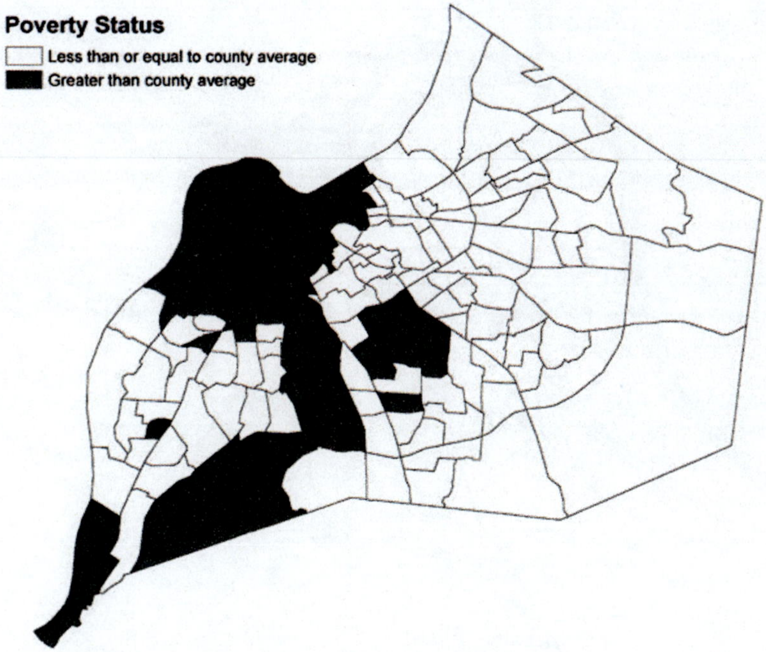

Fig. 11.5 Poverty by census tract. The shaded areas represent tracts with a poverty rate higher than the county average of 12%. Source: US Census (2000)

the Louisville Metro Air Pollution Control District, the University of Louisville, the Commonwealth of Kentucky, and the West Jefferson County Community Task Force.

The resulting West Louisville Air Toxics Study, carried out by an independent research organization, documented high concentrations of harmful air toxics, including carcinogens, in Louisville's West End (Sciences International, Inc. 2006). As a result of the study, the Louisville Metro Air Pollution Control District implemented the Strategic Toxic Air Reduction (STAR) program. Ongoing monitoring of air toxics is being carried out at six of the original sites by the Air Quality Lab at the University of Louisville.

11.5 West Louisville Study Area

The asthma hot spot identified by mapping and subsequent spatial autocorrelation analysis consists of six Louisville zip codes. During the 2005–2008 study period, there were 832 asthma hospitalizations for children ages 0–19 in the study area. The characteristics of these children differ from the county-wide study population (Table 11.1) in the following ways. A lower percentage of children are in the 0–4 year age group (45.1 vs. 50.1%), indicating a slight weight towards school-aged children. Emergency department referral sources are higher, at 82.2%. Finally,

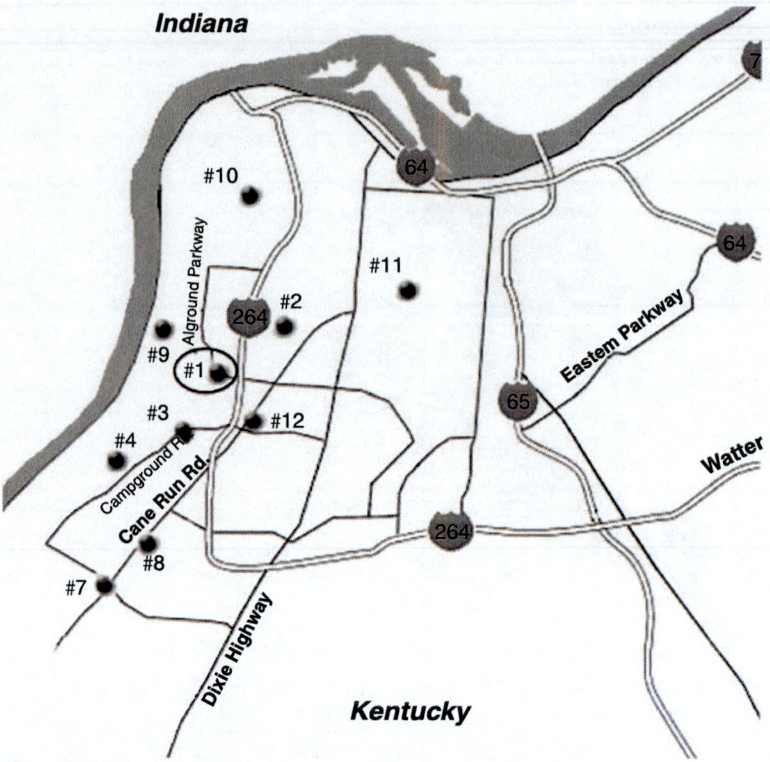

Fig. 11.6 West Louisville air toxics monitoring sites. Source: West Jefferson County Community Task Force (2009). The Firearms Training monitoring site is *circled*

a higher percentage of children are on federal health insurance programs (81.7 vs. 67.1%).

11.5.1 Seasonal Trends in Asthma Hospitalizations

In order to understand the potential relationship between asthma hospitalizations and polluted air, we first examined seasonal trends in asthma rates and air pollution measurements. All dates used in these analyses were hospital admission dates (as opposed to discharge dates).

Evidence of seasonal patterns has been reported in previous studies (Chen et al., 2000; Hashimoto et al., 2004; Ivey et al., 2001, 2003; Lee, 2006; McConnell et al., 2002). One study took place in Louisville. Using weekly asthma inpatient and outpatient data from Norton/Kosair hospitals, Lee (2006) reported a strong seasonal trend in Louisville, with high rates in fall and low rates in summer. In an attempt to explain seasonal patterns, Lee examined correlations between asthma hospitalizations and the following: temperature, humidity, precipitation, wind speed, wind direction, particulate matter ($P.M_{2.5}$), carbon monoxide, and sulfur dioxide. Most correlations were low or negative, with the exception of precipitation ($R = 0.28$) and $P.M_{2.5}$ ($R = 0.32$).

Our air pollution analysis attempts to address the following questions: (1) do seasonal peaks in Jefferson County (specifically West End) asthma hospitalizations persist through time and, if so, (2) are they related to air pollutants not analyzed by Lee? These would include ozone and several chemicals and volatile organic compounds (VOCs), the latter emitted by industries in western Louisville.

Figure 11.7 shows the distribution of asthma hospitalizations in the study area, by year and season. Seasons were delineated according to solstice and equinox dates for each year. Four-year totals are shown in Table 11.3. It is obvious that the highest number of admissions consistently occurs in the fall.

We used a one-way ANOVA procedure with a square-root transformation of counts, to determine whether there was a significant difference in asthma admissions by season. Our results indicate that there was, $F(3{,}592) = 5.623, p = 0.01$. Post-hoc tests, Tukey's honestly significant difference and least significant difference (LSD),

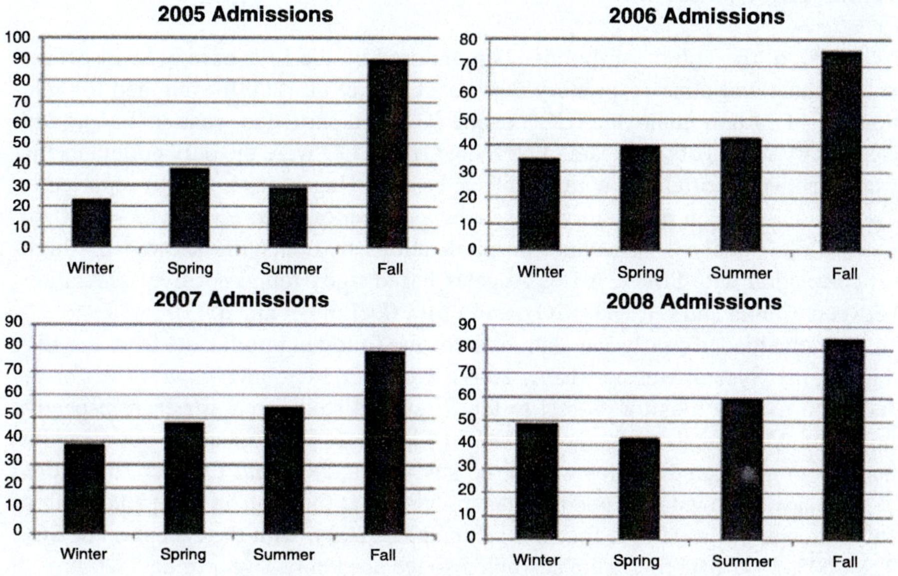

Fig. 11.7 Graphs of asthma admissions, by season, 2005–2008

Table 11.3 Number and percent of asthma hospitalizations by season, all years combined

Year/season	Number	Percent
Winter	146	17.5
Spring	169	20.3
Summer	187	22.5
Fall	330	39.7
Total	832	100.0

indicated that the largest difference was between Fall vs. all other seasons. While our findings support Lee's earlier (2006) results, the summer trough noted by Lee is not present after 2005.

A closer examination of the timing of asthma hospitalizations revealed that the "fall peak" doesn't actually begin in fall. Asthma rates jump in mid-August, which corresponds closely with the start date of Louisville's public school year. This is especially obvious when weekly rates are examined, as they were in an earlier study. We obtained academic calendars from Jefferson County Public Schools for the study years 2005–2008. We categorized all days during the 4-year period using a binary system (0 = not in school, 1 = in school) which accounted for summers, weekends, and all school holidays, including teacher work days. In-school and out-of-school rates were calculated for each year, using hospital admissions as the numerator and school vs. non-school days as the denominator. Table 11.4 shows the categorical rate differences. For all years, in-school rates were higher than out-of-school rates.

11.5.2 Air Pollutants

Nearly all of the criteria air pollutants monitored by the EPA have been associated with asthma and other respiratory diseases. Chen et al. (2000) examined the relationship of carbon monoxide (CO), ozone (O_3) and particulate matter (PM_{10}) with elementary school absenteeism in Nevada. O_3 and CO were the only pollutants that were positively correlated with the absentee rate. Hagen et al. (2000) reported an association between respiratory disease hospitalizations and $PM_{2.5}$ ($RR = 1.038$). Lee's (2006) analysis of asthma hospitalizations in Louisville showed the highest correlation with $PM_{2.5}$. A Los Angeles-based study found positive associations between asthma and O_3, NO_2, SO_2, and PM_{10} (Delfino et al., 2003).

We downloaded yearly raw data files for the following pollutants from the EPA Air Quality System website (EPA, 2009): CO, NO_2, O_3, $PM_{2.5}$ and SO_2, for the Jefferson County monitor closest to the West End study area, for each respective pollutant. Monitor locations are shown in Fig. 11.3. Unfortunately, not all pollutants were monitored at sites within the study area. NO_2 and CO were monitored at sites east of downtown, and O_3 was monitored at the Watson Lane site south of Rubbertown. We used data for all 4 years, 2005–2008, with the exception of SO_2. The 2005–2007 data files were corrupted so we used only 1 year of data, 2008.

Table 11.4 Hospital admission rates by school calendar

Year	In school rate	Out of school rate
2005	0.8851	0.4712
2006	0.75	0.5078
2007	0.8994	0.5484
2008	0.9881	0.6263
All years	0.8802	0.5391

Most pollutants were monitored continuously, with 1-h averages recorded. The exception was $PM_{2.5}$ which was recorded as a daily 24-h average. Ozone is only monitored in Louisville from March 1 to October 31. For each continuously-monitored pollutant, we analyzed daily and seasonal average and maximum values.

We also examined six chemical compounds monitored at the Lousville Police Firearms Training Site (Site #1, Fig. 11.6), which is located in the heart of the study area: 1,3-butadiene, acetone, acrylonitrile, benzene, chloroform and toluene. The sampling frequency for acetone, benzene and toluene was every 12 days. The frequency for the others varied, and was lower. We did not use readings that did not meet the Minimum Quantitation Limit (MQL) of the chemical. Three of the compounds – chloroform, 1,3 butadiene, and acrylonitrile – were selected because of their recognized capability to serve as respiratory irritants. A fourth compound, formaldehyde, is perhaps the most well established respiratory irritant, but there was a dearth of monitoring data for it in Louisville (Liteplo and Meek, 2003). Chloroform has an exceptionally low vapor pressure and is well linked to bronchial constriction. 1,3 butadiene is particularly associated with rubber production, a prominent portion of the Louisville industry. Acrylonitrile is less studied than the other three respiratory antagonists, but it is regarded as a pulmonary carcinogen (Palmer et al., 2002; Schneider and Freeman, 2001).

In addition, we selected three "marker" chemicals for exposure to possible "mixtures" of respiratory irritants (Cassee et al., 1996). The light aromatic hydrocarbon compounds of toluene, benzene, and acetone were defined in this role. From early times with studies of off-site migration of hazardous substances, there has been a practice of identifying "markers" that would be readily identified and that might serve to describe the distribution of myriad and unspecified compounds (Kim et al., 1980). These three agents are widely prevalent with the industrial processes in Louisville, and are actively reported in monitoring studies. Again, it not the asthma-linked effect of these compounds (although they have very low vapor pressures and trigger inhalation sensitivity), rather it is their value for these studies to describe a "general" exposure to respiratory hazards within Louisville. Hagen et al. (2000) reported the strongest associations with benzene, toluene and formaldehyde in their study of respiratory diseases. Positive associations between childhood asthma and benzene, formaldehyde, acetone, 1,3-butadiene and toluene (among others) were reported in a Los Angeles study (Delfino et al., 2003).

Figure 11.8 shows seasonal averages and maximums for the five criteria pollutants and six chemical compounds. Few of the pollutants have a fall peak. Ozone, which is commonly regarded as an asthma trigger, is highest in spring and summer, with a high summer maximum. Particulate matter ($PM_{2.5}$) is highest in summer, and sulfur dioxide has a spring average high with a winter maximum high. Nitrogen oxide is the only criteria pollutant with a high fall average; however its maximum is lowest in fall. Four of the chemical compounds show fall highs: acetone, acrylonitrile, benzene (maximum only) and chloroform.

We computed bivariate correlations for all criteria pollutants and chemical compounds. Several studies have indicated that models with air pollution concentrations

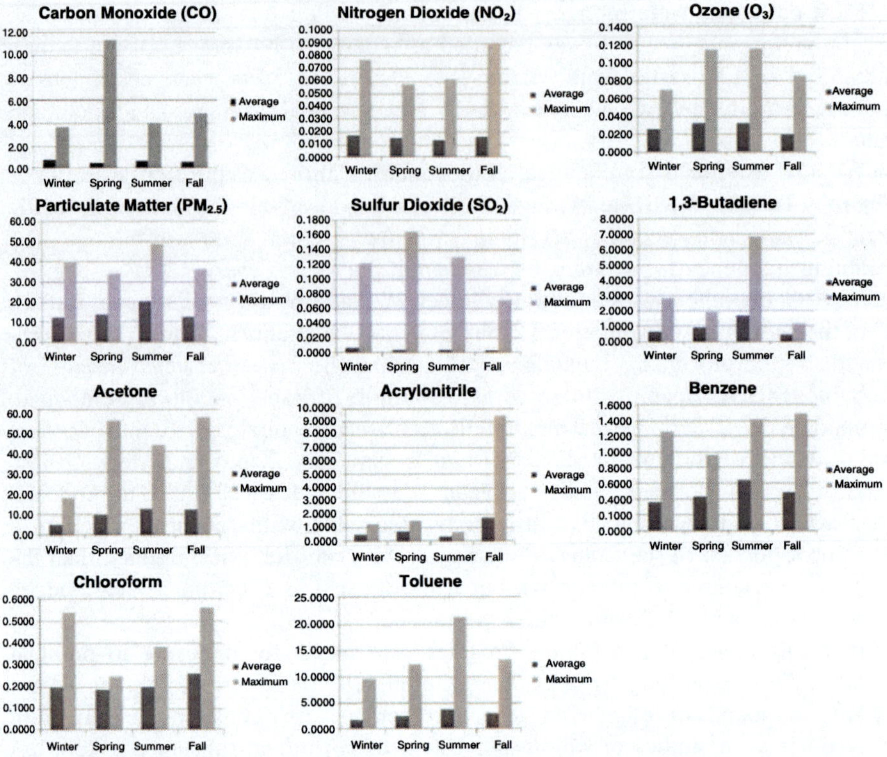

Fig. 11.8 Graphs of seasonal averages and maximums for air pollutants

at an exposure lag of zero (i.e. same day) have stronger associations with asthma and respiratory diseases than models with a time lag (Delfino et al., 2003; Hagen et al., 2000). For this reason, we correlated same-day pollution measures (exposure lag 0) with daily hospital admission counts. Distributions of criteria pollutants were normal; chemical compounds and hospital admissions were left-skewed. The former were normalized with log transformations, but many dates had zero hospital

Table 11.5 Correlation of chemical compounds with daily asthma rates

Pollutant/VOC	Sampling frequency	No. of records	Days with cases	R
1,3-Butadiene	Varied	54	24	−0.134
Acetone	12 days	108	48	0.034
Acrylonitrile	Varied	34	19	0.199
Benzene	12 days	106	48	−0.099
Chloroform	Varied	41	20	−0.130
Toluene	12 days	107	47	0.030

admission counts. Correlation coefficients are Spearman's rank. All correlations with criteria pollutants (daily average and maximums) were slightly negative and are not shown here. Correlations between asthma admissions and the six VOCs were all negative or near zero, with the exception of acrylonitrile ($R = 0.199$, Table 11.5). In interpreting these results, it should be noted that lack of data aggregation (i.e. daily counts vs. weekly or monthly rates) may be partly responsible for low coefficients.

11.6 Discussion and Conclusions

The results of our analyses indicate that there is a well-defined spatial pattern of high rates of childhood asthma hospitalizations in western Louisville. Analyses also confirmed a strong seasonal pattern to asthma for all study years, with a fall peak. The seasonality of asthma is well-documented, across different climates (Crighton et al., 2001; Langley-Turnbaugh et al., 2005). The spatial cluster of high rates in western Louisville does not appear to have diminished with the implementation of the STAR program. However, the program's compliance deadline is nearly 3 years away. The multi-factorial etiology of asthma makes it difficult to pinpoint specific triggers for acute asthma episodes, and it is difficult to tie criteria pollutants and specific chemical compounds to asthma hospitalizations. Some of this may be due to spatial location of EPA monitoring sites and the time interval (minimum 12 days) between chemical compound samples. The higher admission rates while school is in session raises some interesting questions. Future research should include an examination of pollutant measurements at school sites (Spira-Cohen et al., 2010) as well as traffic analysis in western Louisville.

In order to address the environmental justice concerns of disproportionate siting vs. minority move-in, we used GISc to examine patterns of residential mobility in western Louisville over a 60-year period. The polluting industries in western Louisville's "Rubbertown" preceded the local in-migration of African-Americans, the majority of which took place from 1960 to 1970. Although used in a different context, Pastor et al. (2001) referred to this rapid shift in population as "ethnic churning" – a time of racial/ethnic tensions and lowered social capital. While the increasing African-American presence has resulted in a community with greater social cohesion over time and successful community-based initiatives to reduce air toxics emissions have been implemented, significant health disparities in western Louisville must continue to be addressed.

The use of geographic information science was critical to our analysis. We used it to map asthma rates in Jefferson County and identify a statistically significant cluster of high rates, which we then designated as our study area. GISc allowed us to perform visual analyses of population changes over time and was used to select appropriate monitoring sites for our air pollution and VOC analyses. Because geographic information system software contains both spatial and non-spatial data, most of our asthma and air pollution data were analyzed in a GIS environment.

Author Biographies

Dr. Carol Hanchette is an Associate Professor in the Department of Geography and Geosciences at the University of Louisville, with an associate appointment in the University's School of Public Health and Information Sciences. Her teaching and research foci are medical geography, spatial analysis and the use of geographic information systems (GIS) for health analyses. She has published several papers reporting on the use of geographic methods in environmental epidemiology and public health.

Jong-Hyung Lee is currently affiliated with the Department of Physics, Chungbuk National University, Cheongju, Korea. He received his B.S. in Geography from the University of Louisville, where he examined the impact of air quality and atmospheric conditions on childhood asthma. In 2009, he received his M.S. in Geography from the University of Illinois at Urbana-Champaign.

Dr. Tim E. Aldrich is an Associate Professor in the College of Public Health, Department of Biostatistics and Epidemiology, East Tennessee State University. From 2002 to 2010, he also served as Adjunct Associate Professor in the University of Louisville's School of Public Health and Information Sciences. Dr. Aldrich has a wide range of research interests in epidemiology and environmental health and has published extensively.

References

Akinbami LJ, Moorman JE, Garbe PL, Sondik EJ (2009) Status of childhood asthma in the United States, 1980–2007. Pediatrics 123:S131–S145

Aligne AC, Auinger P, Byrd RS, Weitzman M (2000) Risk factors for pediatric asthma: contributions of poverty, race, and urban residence. Am J Respir Crit Care Med 162(3 Pt 1):873–877

Allergy and Asthma Foundation of America (AAFA) (2009a) Spring allergy capitals, 2009. Retrieved on February 9, 2010 from: http://www.aafa.org/display.cfm?id=9&sub=33

Allergy and Asthma Foundation of America (AAFA) (2009b) 2009 asthma capitals: the most challenging places to live with asthma. Retrieved on May 25, 2009 from http://www.aafa.org/pdfs/2009ACPublicList.PDF

Alp H, Yu BH, Grant EN, Rao V, Moy JN (2001) Cockroach allergy appears early in life in inner-city children with recurrent wheezing. Ann Allergy Asthma Immunol 86(1):51–54

American Lung Association (2008) Lung disease data. Retrieved on February 9, 2010 at http://www.lungusa.org/about-us/publications/

American Lung Association (November, 2009) Epidemiology & statistics unit, research and program services. Trends in Asthma Morbidity and Mortality

Anderson JL (1980) Changing patterns of Louisville's black residential areas. In: Bierman DE (ed) An introduction to the Louisville region: selected essays. University of Louisville, Louisville, KY, 39–48

Anselin L 2003 GeoDaTM 0.9 user's guide. Spatial analysis laboratory, department of agricultural and consumer economics, University of Illinois, Urbana-Champaign and Center for Spatially Integrated Social Science. Urbana, IL. Retrieved on February 21, 2010 from: http://geodacenter.asu.edu/pdf/geodaGA.pdf

Arbes SJ, Cohn RD, Yin M, Muilenburg ML, Burge HA et al (2003) House dust mite allergen in U.S. beds: results from the first national survey of lead and allergens in housing. J Allergy Clin Immunol 111:408–414

Bjornson CL, Mitchell I (2000) Gender differences in asthma in childhood and adolescence. J Gender Specific Med 3(8):57–61

Brown P (1995) Race, class and environmental health: a review and systematization of the literature. Environ Res 69:15–30

Bullard RD, Wright BH (1993) Environmental justice for all: community perspectives on health and research needs. Toxicol Ind Health 9:821–841

Cassee FR, Groten JP, Feron VJ (1996) Changes in the nasal epithelium of rats exposed by inhalation to mixtures of formaldehyde, acetaldehyde, and acrolein. Fundam Appl Toxicol 29(2):208–218

Chen L, Jennison BL, Yang W, Omaye ST (2000) Elementary school absenteeism and air pollution. Inhal Toxicol 12:997–1016

Corburn J, Osleeb J, Porter M (2006) Urban asthma and the neighborhood environment in New York City. Health Place 12:167–179

Coughlin SS (1996) Environmental justice: the role of epidemiology in protecting unempowered communities from environmental hazards. Sci Total Environ 184:67–76

Crighton EJ, Mamdani MM, Upshur REG (2001). A population-based time series analysis of asthma hospitalizations in Ontario, Canada: 1988 to 2000. BMC Health Sciences Research 1:7. Retrieved February 21, 2010 from: http://www.biomedcentral.com/1472-6963/1/7

Delfino RJ, Gong H Jr, Linn WS, Pellizzari ED, Hu Y (2003) Asthma symptoms in Hispanic children and daily ambient exposures to toxic and criteria air pollutants. Environ Health Persp 111(4):647–656

EPA (Environmental Protection Agency) (2009) Air quality system data download. Retrived February 27, 2009 from: http://www.epa.gov/ttn/airs/airsaqs/detaildata/downloadaqsdata.htm

Geolytics (2007) Estimates Professional: Estimates and Projections data set for 2007–2012

Hagen JA, Nafstad P, Skrondal A, Bjorkly S, Magnus P (2000) Associations between outdoor air pollutants and hospitalization for respiratory diseases. Epidemiology 11(2):136–140

Hanchette CL (2008) The political ecology of lead poisoning in Eastern North Carolina. Health Place 14(2):209–216

Hashimoto M, Fukuda T, Shimizu T, Watanabe S, Watanuki S, Eto Y, Urashima M (2004) Influence of climate factors on emergency visits for childhood asthma attack. Pediatr Int 46:48–52

Helms PJ (2000) Issues and unmet needs in pediatric asthma. Pediatr Pulmonol 30(2):159–165

Ivey MA, Simeon DT, Juman S, Hassanally R, Williams K, Monteil MA (2001) Associations between climate variables and asthma visits to accident and emergency facilities in Trinidad, West Indies. Allergol Int 50:29–33

Ivey MA, Simeon DT, Monteil MA (2003) Climatic variables are associated with seasonal acute asthma admissions to accident and emergency room facilities in Trinidad, West Indies. Clin Exp Allergy 33:1526–1530

Jones KL (2007) Kentucky youth tobacco survey 2006. Kentucky Department for Public Health, Tobacco Prevention and Cessation Program, Frankfort, KY

Jones VF, Lawson P, Robson G, Buchanan B, Aldrich T (2004) The use of spatial statistics to identify asthma risk factors in an urban community. Pediatr Asthma Allergy Immunol 17:3–13

Kim CS, Narang R, Richards A et al (1980) Love Canal: chemical contamination and migration. Proceedings of National Conference on Management of Uncontrolled Hazardous Waste Sites, U.S. EPA, pp 212–219

Kleber JE (2001) The encyclopedia of Louisville. The University Press of Kentucky, Lexington, KY

Langley-Turnbaugh SJ, Gordon NR, Lambert T (2005) Airborne particulates and asthma: a Maine case study. Toxicol Ind Health 21:72–95

Lee J-H (2006) Pediatric asthma admissions and seasonal variables in Jefferson County, Kentucky. Senior undergraduate thesis. Department of Geography and Geosciences, University of Louisville, Louisville, KY

Liteplo RG, Meek ME (2003) Inhaled formaldehyde: exposure estimation, hazard characterization, and exposure-response analysis. J Toxicol Environ Health B 6(1):85–114

Louisville Metro Housing Coalition March 2005 Out of breath: childhood asthma, poverty and housing. Available at: http://www.metropolitanhousing.org/index.cfm?article_id=63

Lwebuga-Mukasa JS, Oyana TJ, Wydro P (2004) Risk factors for asthma prevalence and chronic respiratory illnesses among residents of different neighbourhoods in Buffalo, New York. J Epidemiol Commun Health 58:951–957

Maantay J (2007) Asthma and air pollution in the Bronx: methodological and data considerations in using GIS for environmental justice and health research. Health Place 13:32–56

McConnell M, Rob D, Berhane K, Gilliland F, London J, Stephanie I, Talat G, James W, Avol E (2002) Asthma in exercising children exposed to ozone: a cohort study. Lancet 359:386–391

National Historic Geographic Information System (2009) Retrieved May 24, 2009 from: http://www.nhgis.org/

Nunn K, Hudson N, Robeson S (2009) Kentucky asthma surveillance report 2009. Kentucky Department for Public Health, Frankfort, KY

Palmer LJ, Silverman ES, Weiss ST, Drazen JM (2002) Pharmacogenetics of asthma. Am J Respir Crit Care Med 165(7):861–866

Pastor M Jr, Sadd J, Hipp J (2001) Which came first? Toxic facilities, minority move-in, and environmental justice. J Urban Aff 23(1):1–21

Schneider D, Freeman N (2001) Children's environmental health risks: a state-of-the-art conference. Arch Environ Health 56(2):103–110

Sciences International, Inc (2006) Final report. West Louisville air toxics study risk assessment. Prepared for Louisville Metro Air Pollution Control District. Alexandria, VA. Retrieved May 24, 2009 from: http://www.louisvilleky.gov/NR/rdonlyres/49F54DC0-8153-4355-8FE3-73F3741FEDF6/0/WLATS20012005Final.pdf

Spira-Cohen A, Chen LC, Kendall M, Sheesley R, Thurston GD (2010) Personal exposures to traffic-related particle pollution among children with asthma in the South Bronx, NY. J Expo Sci Environ Epidemiol 20:446–456

Tooms MR (2009) Kentucky youth tobacco survey 2008. Kentucky Department for Public Health, Tobacco Prevention and Cessation Program, Frankfort, KY

U.S. Census Bureau (1990) Summary tape file 1 (STF1). Retrieved May 24, 2009 from: http://factfinder.census.gov

U.S. Census Bureau (2000) Summary file 1 (SF1). Retrieved May 24, 2009 from: http://factfinder.census.gov

U.S. Census Bureau (2010) Population estimates. Retrieved February 9, 2010 from: http://www.census.gov/popest/datasets.html

West Jefferson County Community Task Force (2009) Air toxics monitoring sites. Retrieved May 30, 2009 from: http://www.wjcctf.org/air/monitor_sites.php

Chapter 12
The Impact of Changes in Municipal Solid Waste Disposal Laws on Proximity to Environmental Hazards: A Case Study of Connecticut

Ellen K. Cromley

Abstract Environmental policies affect proximity to environmental hazards. In the late 1980s, the State of Connecticut implemented mandatory recycling laws to improve management of municipal solid waste. At that time, more than 80% of the State's 169 towns disposed of trash within their own borders. The regulatory change redirected flows of waste to transfer stations and trash-to-energy plants. To assess changes in the proximity to hazards associated with this shift, the origins and destinations of solid waste shipment flows are modeled using data for 2008. Ton-weighted distances to disposal are estimated for 2008 and compared to the distances if solid waste had continued to be disposed of within towns. Changes in municipal solid waste management in Connecticut have differentially impacted local communities. Residents of the Town of Hartford, particularly low-income minority residents in the North End, have been affected by the operation of municipal solid waste management facilities in Hartford, which receive waste from almost half the towns in the state. The implementation of environmental policies intended to improve municipal solid waste disposal at the state level adversely affected proximity to environmental hazards in selected communities and the abilities of local communities to improve environmental quality in their own jurisdictions.

Keywords Municipal solid waste · Trash-to-energy · Recycling · Waste flows · Weighted distance

12.1 Background

12.1.1 Flows of Municipal Solid Waste

This case study assesses changes in proximity to environmental hazards associated with changes in municipal solid waste disposal practices in Connecticut. A number

---

E.K. Cromley (✉)
Department of Community Medicine and Health Care, University of Connecticut School of Medicine, Farmington, CT 06030-6325, USA
e-mail: cromley@charter.net

J.A. Maantay, S. McLafferty (eds.), *Geospatial Analysis of Environmental Health*, Geotechnologies and the Environment 4, DOI 10.1007/978-94-007-0329-2_12,

of chapters in this book are concerned with the measurement of distances to environmental hazards in studies modeling environmental exposure and patterns of health and disease. These hazards are often viewed as fixed in terms of their locations and zones of impact. In reality, however, the geography of environmental hazards is dynamic.

The research presented in this case study is an initial attempt to look at how we can develop GIS-based methods for investigating the changing geography of environmental hazards. Municipal solid waste disposal is an appropriate subject for study using these methods. Municipal solid waste (MSW), what we commonly think of as "trash," is the stream of material collected through community sanitation programs. It is generated by households, businesses, and organizations. Industrial waste and medical waste are generally not included in MSW. Paper and yard waste, such as leaves and clippings, account for a sizeable share of the municipal waste stream in the US (Environmental Protection Agency, 2009).

The US Environmental Protection Agency's municipal solid waste hierarchy ranks methods for dealing with MSW (Environmental Protection Agency, 2008). Source reduction, also referred to as waste prevention, is the most environmentally sound method. It involves reducing the amount and toxicity of materials entering the solid waste stream by changing the design, manufacture, packaging, and use of materials. Recycling, including composting, is next in the hierarchy. In the US, about 33% of MSW generated annually is recycled (Environmental Protection Agency, 2009). At the bottom of the waste hierarchy is disposal, which includes dumping in landfills and combustion in incinerators. Some combustion facilities use the heat from incineration to generate electricity, and this energy recovery provides some benefits over incineration of MSW alone (Michaels, 2007).

In practice, communities use a mix of methods to manage solid waste. There have been important shifts in these methods over time, in response to changing regulations at the federal and state levels. Connecticut has made a significant shift to combustion with energy recovery along with recycling over the last 20 years.

12.1.2 The Connecticut Context

In 1989, the State of Connecticut enacted legislation requiring mandatory recycling of certain types of municipal solid waste effective January 1, 1991, with a goal of recycling 25% of the state's solid waste stream by that date (Connecticut Department of Environmental Protection, 2005). Connecticut is one of only a handful of states with mandatory recycling, although there are a number of counties and municipalities which have adopted mandatory recycling. This initiative was taken for several reasons: many municipal landfills were reaching their permitted capacities; there were new federal regulatory standards for landfills which many existing landfills did not meet; and there was concern that some landfills posed a contamination threat to groundwater that served as a source of public drinking water. Much of the drinking water in Connecticut is taken from groundwater.

To meet this goal, nine items were designated for mandatory recycling rather than disposal. These items include: glass food and beverage containers, used motor oil, vehicle (lead-acid) batteries, scrap metal, corrugated cardboard, newspaper, metal food and beverage containers, leaves, white office paper (private residences exempt). On May 1, 1996, nickel-cadmium batteries were added to the list of mandatory recyclables. Instead of being recycled curbside, these batteries are recycled at retailers, businesses, municipalities and other sites though a take-back program sponsored by the battery manufacturers. As of October, 1998, grass clippings were banned from solid waste disposal facilities. In 1993, Connecticut's General Assembly passed legislation which, among other provisions, raised the state's recycling/source reduction goal to 40% by the year 2000. To help achieve this goal, many municipalities have added additional items to their programs including plastic resins #1 and #2, magazines and junk mail, and even textiles. Connecticut's recycling/source reduction rate was about 30% for fiscal year 2005. The rate does not include redeemable deposit containers (Connecticut is a bottle bill state), auto scrap, or certain other commercial recyclables.

Along with the effort to increase recycling, Connecticut also took steps to develop plants for combustion with energy recovery, trash-to-energy plants referred to as "Resource Recovery Facilities." The Connecticut General Assembly passed the law (Public Act 73-549) establishing the Connecticut Resources Recovery Authority (CRRA) in 1973 and Bridgeport, Connecticut, was selected as its first site for a regional trash-to-energy project (Connecticut Resources Recovery Authority, 2009). After a lengthy development effort including malfunctions and bankruptcy, the Bridgeport plant went into operation in 1988. Hartford was selected as the site of the second plant and the plant began operation in 1987. In 2008, Connecticut had 6 trash-to-energy plants in operation, about 7% of the total number of plants operating across 25 states in the US (Michaels, 2007). Trash-to-energy plants are more common in many European countries, but many states and municipalities in the US have been reluctant to embrace the technology (Rosenthal, 2010).

As a consequence of changes in federal and state regulations, the landscape of municipal solid waste disposal changed dramatically. Town landfills were closed and in some cases became transfer stations for recyclables (Fig. 12.1). Towns are the basic units of local government in Connecticut and other New England states. Connecticut has 169 towns which cover the entire state; there are no unincorporated areas in Connecticut. As the town landfills were closed, intermediate processing centers and transfer stations for un-recycled municipal solid waste developed (Fig. 12.2). Trash-to-energy plants were built (Fig. 12.3), and ash landfills were created to handle the ash produced by incineration (Fig. 12.4). Connecticut's experience over the last 20 years in shifting from landfills to recycling and trash-to-energy provides useful information for communities considering making a similar shift. The experience in Connecticut illustrates the consequences of these shifts on proximity to hazards. The impacts of these changes are geographically uneven. GIS-based methods can be used to improve our understanding of the impacts by revealing differences across places.

Fig. 12.1 A town landfill, now closed, is operated by the town as a transfer station for recyclables. Individuals who are town residents can drop off paper, plastics, and other material for recycling

Fig. 12.2 This company provides a wide range of recycling and waste management services. It is an intermediate processing center for recyclables and it is also a transfer station for municipal solid waste collected in towns and transferred to trash-to-energy plants

Fig. 12.3 The mid-Connecticut resource recovery facility developed on the site of an existing utility plant on the banks of the Connecticut River. The facility operates continuously and can burn coal if solid waste is not available

Fig. 12.4 The Hartford landfill is on a site in North Hartford next to the Connecticut River which was opened by the City of Hartford in 1940 for use as a town landfill. After the Connecticut resources recovery authority leased the landfill in 1982, the site was developed into an 80-acre municipal solid waste landfill and a 16-acre ash landfill. The landfill accepted its final deliveries on December 31, 2008. Disposal of 10 million cubic yards of solid waste on the site resulted in the 138-ft high hill seen in this photograph

12.2 Methods

12.2.1 Data

The Connecticut Department of Environmental Protection publishes data on solid waste and recycling. Data on tons of un-recycled municipal solid waste generated in each town and disposed of within the state in FY 2008 were obtained from a series of these reports Connecticut Department of Environmental Protection, (2009a, b, c). The 2008 data were the most recent data available at the time of this research. Data on flows out of state, by-pass waste not processed at trash-to-energy facilities but sent elsewhere for disposal, and other non-processible wastes are not included in this study.

A GIS application was developed and data layers were created modeling the origins and destination of MSW flows in the state. These GIS data layers included town centroids, transfer stations, resource recovery facilities, active landfills, and ash landfills in 2008 (Fig. 12.5). Next, GIS line data layers were created using straight lines to connect the origins and destinations of solid waste flows. Three sets of flows were modeled: flows directly from towns to active landfills and to resource recovery facilities (Fig. 12.6), indirect flows from towns through transfer stations to active landfills and to resource recovery facilities (Fig. 12.7), and flows of ash from resource recovery facilities to the two active ash landfills (Fig. 12.8). The number of tons of MSW associated with a particular flow from an origin to a destination was added as an attribute of each line segment.

12.2.2 Methods

Understanding the impact of solid waste management practices requires us to understand how solid waste is transported from the places where it is generated to the places where it is transshipped or incinerated, and to the places where it is ultimately disposed in landfills. In order to assess the impact of the changing state policies on municipal solid waste disposal, tons of MSW were weighted based on the distances they were moved in the state.

First, the distance-weighted tons of MSW that would have resulted if the 2008 waste had been disposed of within the town where the waste originated was estimated by multiplying the number of tons generated in the town by a distance equal to one-half the radius of a circle with the same area as the town. This provides an estimate of in-town movement of solid waste from households and businesses to a hypothetical in-town disposal site such as a town-operated landfill of the sort closed following the introduction of mandatory recycling.

Next, the actual distance-weighted tons of MSW in FY 2008 were estimated by multiplying the number of tons shipped over each of the links by the straight-line distance of the link from flow origin to flow destination. Data on the actual routes of shipment were not available so network distances could not be used. The difference

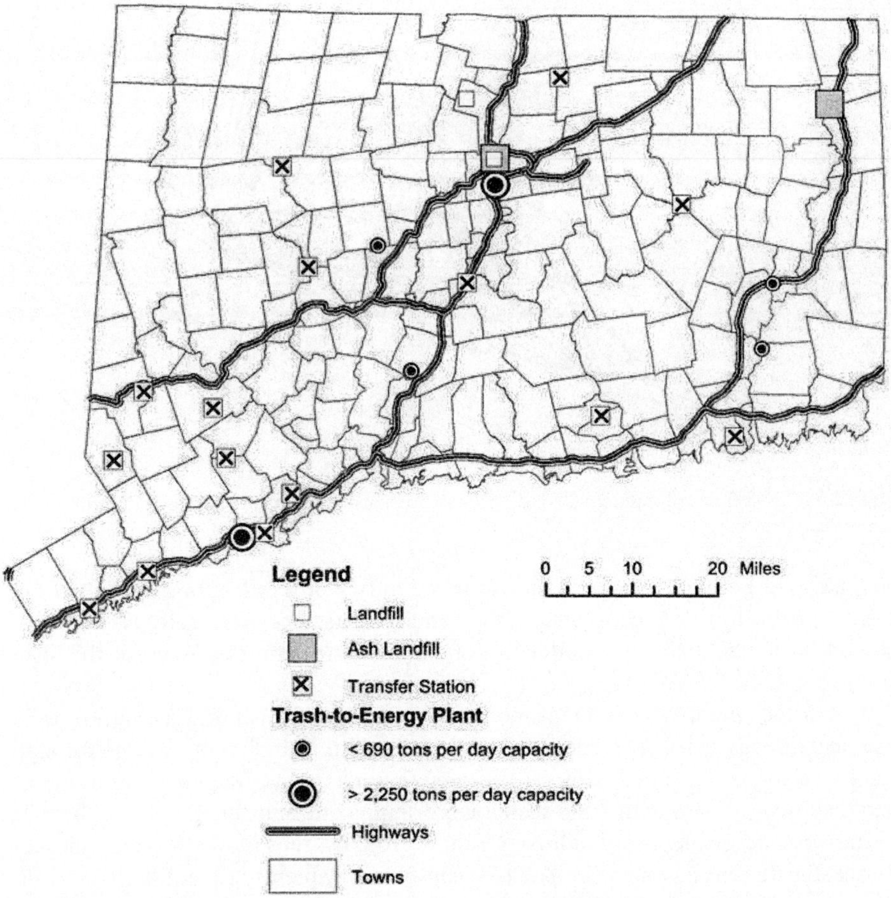

Fig. 12.5 Facilities handling municipal solid waste generated in Connecticut's 169 towns include landfills, ash landfills, transfer stations, and trash-to-energy plants. These facilities are located near major highways in most cases

in distance-weighted tons of MSW indicates the distance effects of the change in state municipal solid waste disposal policy.

Finally, the distance-weighted tons of MSW ending up in each town are mapped for both the hypothetical flows and for the actual flows in FY 2008. The populations in towns receiving most of the waste are compared to the population in the state as a whole.

12.3 Results

A total of 2,366,908 tons of MSW were generated in towns and disposed of in-state in 2008. The ton-distance for hypothetical in-town disposal was estimated to

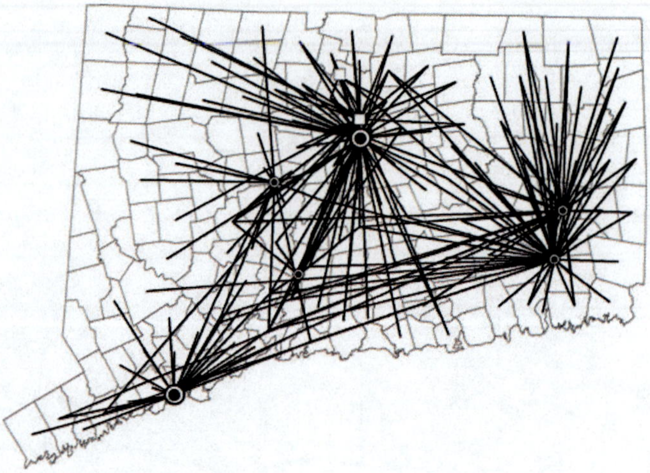

Fig. 12.6 Direct flows of municipal solid waste to trash-to-energy plants and landfills from town centroids

be 6,802,911 because most trash would travel only 1 or 2 miles for disposal within a town. In this hypothetical scenario, the pattern of distance-weighted tons of MSW ending up in each town is similar to population distribution by town in the state (Fig. 12.9a).

Under the actual system, the same number of tons of trash originating in towns was sent directly to trash-to-energy plants and landfills or indirectly to them through transfer stations. As noted, there were 2,366,908 tons of MSW generated in towns and disposed of in-state in 2008 which were shipped to transfer stations or directly to trash-to-energy plants or landfills. Of these, 769,094 tons of MSW were first sent to transfer stations and then shipped to trash-to-energy plants or landfills. The trash-to-energy plants shipped 489,060 tons of ash to one of two active ash landfills for a total of 3,625,062 tons of materials moved. The actual ton-distance was 57,065,989 (about 8 times greater than the total for hypothetical in-town disposal) because trash was shipped over much longer distances.

The pattern of distance-weighted tons of MSW ending up in each town under the actual patterns of MSW flow in FY 2008 shows that waste disposal is much more concentrated under the system of recycling and combustion in trash-to-energy plants. Only 24 of the state's 169 towns received MSW from other towns (Fig. 12.9b). The maps use the same class intervals to aid comparison, except for the minimum and maximum amounts which differ for the two distributions.

In terms of community impacts, Hartford, the state capital, has been the most severely affected by the shift in policy over the last 20 years. In FY 2008, the Town of Hartford had located within its borders one of the two large-capacity trash-to-energy plants in the state, a landfill, and one of the two ash landfills in the state. It also had an intermediate processing facility for recyclables. Eighty-five towns and transfer stations located around the state shipped 1,004,296 tons of MSW to the Hartford Landfill and/or the Hartford trash-to-energy plant in FY 2008

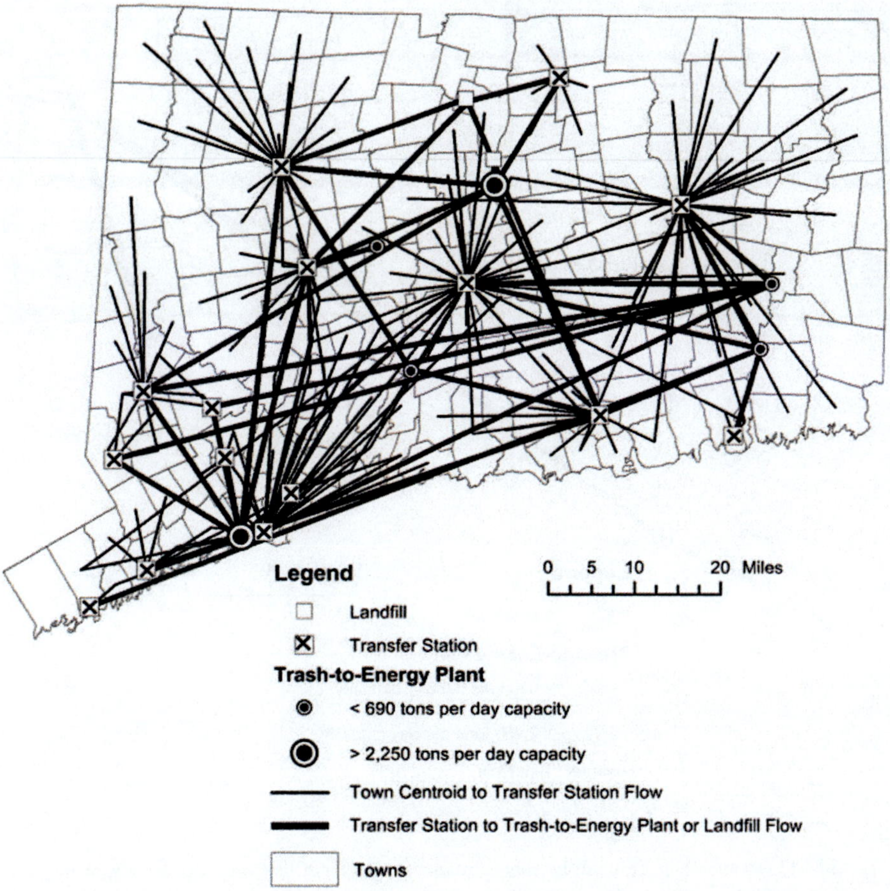

Fig. 12.7 Indirect flows of municipal solid waste from town centroids to transfer stations and from transfer stations to trash-to-energy plants and landfills

(Figs. 12.6 and 12.7). In addition, 155,640 tons of ash were deposited in its ash landfill (Fig. 12.8). This means that 32% of all tons of MSW and ash moved in the state ended up in Hartford.

Hartford's population fell from 139,739 in 1990 to 121,067 in 2009 (Connecticut Department of Economic and Community Development, 2010), a decline of 13% during the period when the trash-to-energy plant was operating. During this period, the population of Hartford County increased about 3% and the population of Connecticut increased 6%. Hartford's median household income in 2009 was \$30,379 compared to \$64,189 in Hartford County and \$68,055 in the state. More than 30% of the town's population lived in poverty in 1999. About 62% of Hartford's population is minority compared to 29% in Hartford County and 21% in the state. The existing landfill and power plant coupled with Hartford's central location in the state, good highway access, and relatively large population are factors that made

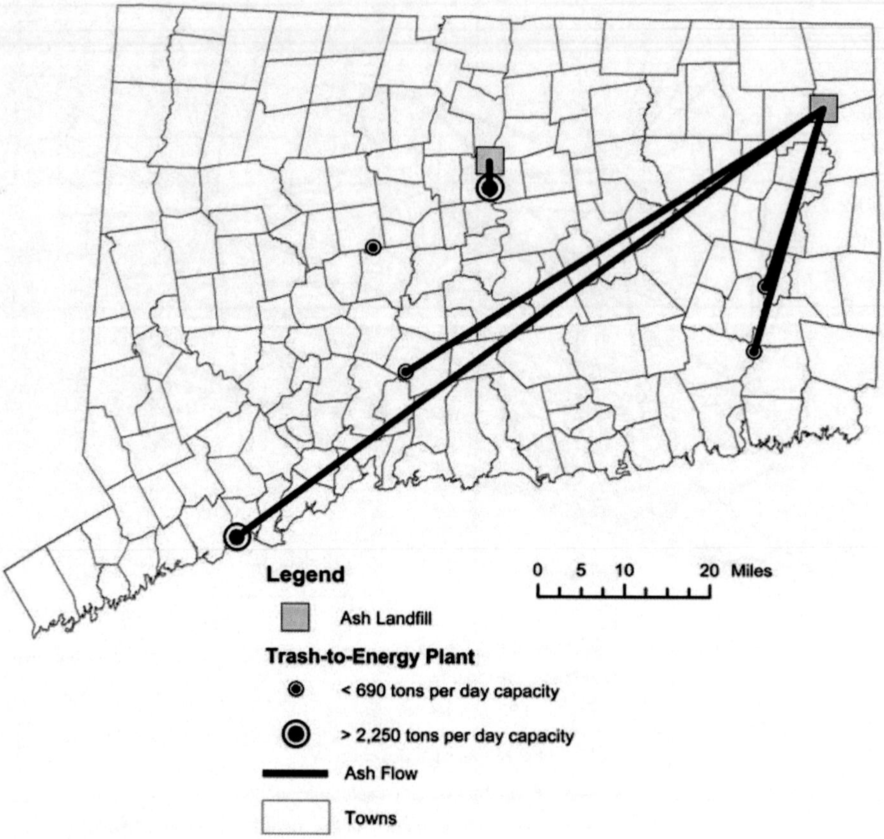

Fig. 12.8 Flows of ash from trash-to-energy plants to ash landfills. Most ash is sent to the ash landfill in Putnam, Connecticut. The trash-to-energy plant located in Bristol sends it ash out of state to a landfill in New York State

Hartford a reasonable choice for locating MSW management facilities there. The degree to which Hartford became the destination for MSW solid waste and recyclables from so many other communities in the state, however, has had a major impact on the town.

In 1994, a Hartford-based community organization petitioned the Agency for Toxic Substances and Disease Registry to examine the impact of the Hartford Landfill on the health of residents (Agency for Toxic Substances and Disease Registry, 1998). The Hartford Landfill site includes an 80-acre municipal solid waste landfill and a 16-acre ash landfill (Fig. 12.4). In 2008, 25 towns across the state sent MSW for disposal to the Hartford Landfill. At the time of the study conducted in 1994, approximately 10,000 people lived within 1 mile of the landfill. The report concluded that, because of the presence of hydrogen sulfide, residents who were sensitive to unpleasant odors might experience episodes of nausea or headaches as a result of the short-term maximum peaks in hydrogen sulfide concentration.

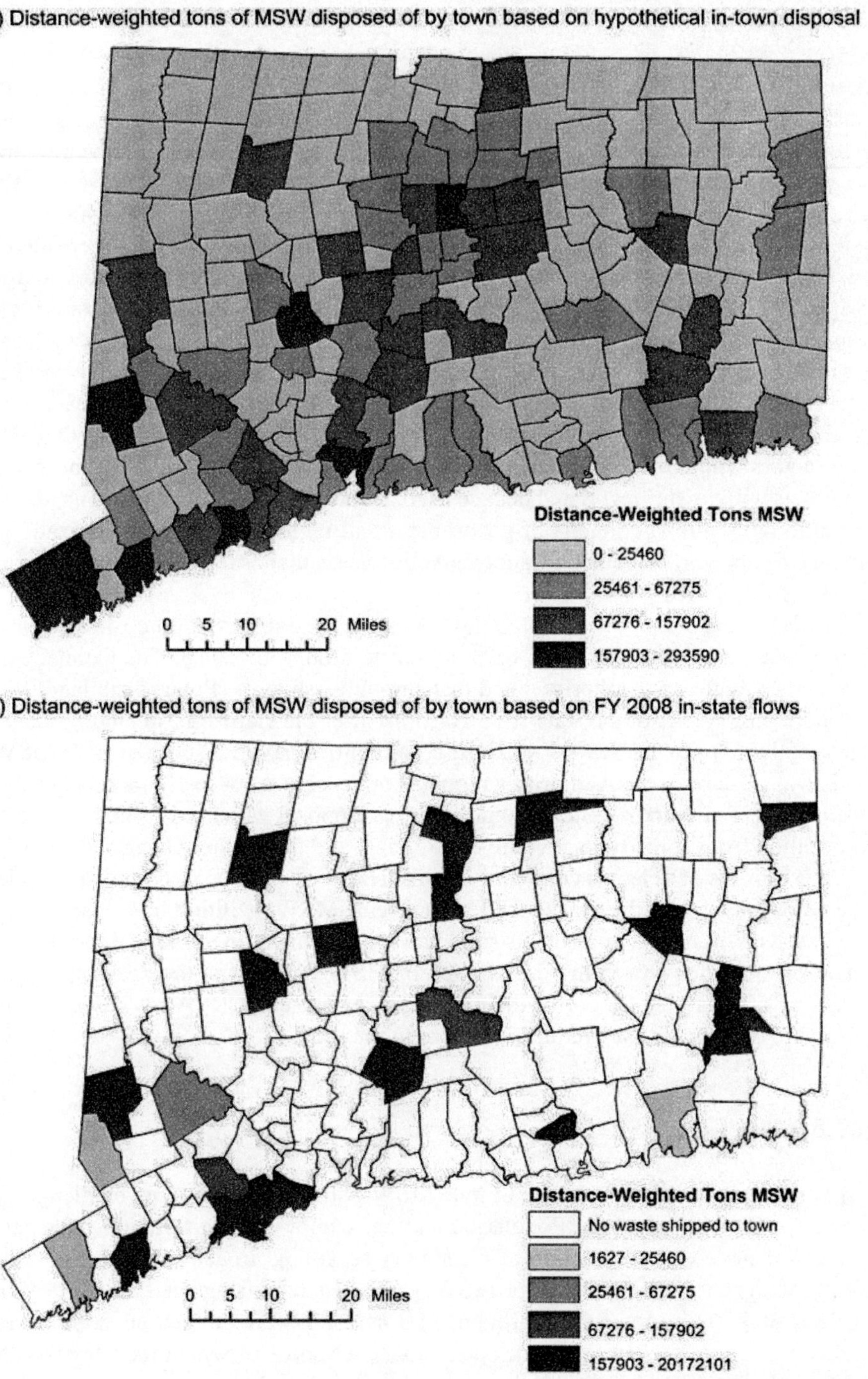

Fig. 12.9 The top figure shows the distance-weighted tons of MSW by town if the actual tons of MSW generated by each town in FY 2008 had been disposed of in-town. The bottom figure shows the distance-weighted tons of MSW by town based on actual flows of MSW in FY 2008. Some towns which generate large amount of MSW have no waste disposed of in-town

12.4 Discussion

State environmental policies have radically re-structured the geography of waste disposal in Connecticut. The scale of facilities has increased and MSW management is more concentrated in a handful of towns in the state. The shift to recycling and incineration of un-recycled solid waste emphasizes two intermediate levels in the EPA solid waste hierarchy. Recycling ranks higher than disposal in the hierarchy, but not as high as source reduction. Incineration with energy recovery is considered by some to be superior to disposal by incineration without energy recovery and to disposal in landfills. However, the capital costs of trash-to-energy plants are high, especially when bonding and operating costs are considered. Operation of these plants has created new and powerful quasi-private corporate entities with a need to secure waste streams for several decades to cover their costs. The 2008 data used in this study are the last that will be reported by towns. Beginning in 2009, the contractors who transport the waste will be responsible for reporting. As the initial contracts with towns negotiated decades ago are ending in 2012, some towns, working through regional councils of government and regional planning organizations, are trying to work together to address solid waste disposal needs (Leavenworth, 2010; Smith, 2010).

Transporting waste out of communities increases transportation costs and environmental costs associated with trucking waste around the state. The Connecticut Resources Recovery Authority noted that shipping ash to the Putnam ash landfill in the northeast corner of the state from Hartford after the Hartford ash landfill closed added $9 (15%) to the $69-per-ton cost communities pay to dispose of their MSW. This was a factor in the Authority's proposal to develop a 90- to 100-acre landfill in the town of Franklin in eastern Connecticut, a proposal which was defeated due to opposition from some state government officials and the public (Owens, 2009).

The impacts on selected communities, like Hartford, have been significant. Under the new system, with the shift to a few, large-scale MSW facilities, it is clear that not every town will have to manage its solid waste in the town. In this situation, the goal of towns becomes preventing the location of MSW facilities in their towns, despite the costs of arranging for out-of-town disposal. At present, CRRA has suspended its efforts to develop an ash landfill anywhere in the state.

12.5 Conclusion

Almost 20 years after the change in state MSW policy in Connecticut, challenges in dealing with MSW remain. Population and per capita waste generation rates have increased since the introduction of mandatory recycling. In addition, rates of recycling of the increased per capita MSW generated have stagnated since 1995 at around 30% (Connecticut Department of Environmental Protection, 2006), well below recycling rates in many European countries some of which recycle 60–70% of MSW (Eurostat, 2010) and in some cities in the US such as San Francisco which recycles 72% of its waste (Barringer, 2008). It is interesting to speculate whether removal of waste has contributed to complacency about source reduction and recycling in towns that ship their waste to other locations in the state.

Some advocates of combustion with power generation point to the costs of transporting MSW to out-of-state landfills as a reason for developing trash-to-energy facilities. Although Connecticut has made trash-to-energy a centerpiece of its MSW policies for the last two decades, Connecticut's reliance on out-of-state disposal is growing. From 1994 to 2004, out-of-state disposal of Connecticut generated MSW increased from 27,000 tons per year to 327,000 tons per year (Connecticut Department of Environmental Protection, 2006). Connecticut sends municipal solid waste to New York, Massachusetts, Ohio, and Pennsylvania, and it is beginning to receive solid waste from some other states, especially Rhode Island.

Trash-to-energy plants may be valuable facilities in the municipal solid waste disposal system after high rates of source reduction and recycling have been achieved, if there is a sufficient stream of remaining waste in a localized area to support the capital investment in the plant, and if disposal of ash waste can be managed without shipping the waste long distances. There is a great need for better modeling of solid waste disposal facilities, flows, transport costs, and community impacts. Spatial analyses, aided by geographic information systems, can play a role in this endeavor, if data are publicly available. These analyses would make it clear who pays and who benefits in the complex geography of solid waste management.

Author Biography

Ellen K. Cromley is an Assistant Clinical Professor in the Department of Community Medicine, in the University of Connecticut's School of Medicine. She completed a Ph.D. in medical geography at the University of Kentucky, and was a Professor in the Department of Geography at the University of Connecticut. Ellen is a co-author with Sara McLafferty of *GIS and Public Health*, Guilford Press.

References

Agency for Toxic Substances & Disease Registry (1998) Health Consultation Hartford Landfill, Hartford County, Connecticut, CERCLIS No. CTD983870692. Retrieved from www.atsdr.cdc.gov/HAC/pha/hartford/har.html on June 27, 2010

Barringer F (2008) A city committed to recycling is ready for more. The New York Times. Wednesday, May 7, Late Edition-Final, Section A, p 23

Connecticut Department of Economic and Community Development (2010) Town profiles January 2010. Retrieved from www.cerc.com/TownProfiles/ on June 27, 2010

Connecticut Department of Environmental Protection (2005) Recycling in Connecticut. Retrieved from www.ct.gov/dep/cwp/view.asp?A=2714&Q=324892 on June 27, 2010

Connecticut Department of Environmental Protection (2006) State of Connecticut state solid waste management plan, amended December, 2006. State of Connecticut, Hartford, CT

Connecticut Department of Environmental Protection, Bureau of Material Management & Compliance Assurance Recycling Office (2009a) First Destination (Multi-Town Transfer Station or Landfill or Resource Recovery Facility) of Municipal Solid Waste (MSW) Generated within the Borders of Connecticut Municipalities and Disposed in FY2008. Retrieved from www.ct.gov/dep/lib/dep/reduce_reuse_recycle/data/first_destination_of_msw_disposed_by_ct_munis_fy2008.pdf on April 8, 2010

Connecticut Department of Environmental Protection, Bureau of Material Management & Compliance Assurance Recycling Office (2009b) Municipal Solid Waste (MSW) Reported Received by Connecticut Resource Recovery Facilities and Landfills FY2008. Retrieved from www.ct.gov/dep/lib/dep/reduce_reuse_recycle/data/msw_received_by_ct_rrfs_and_lfs_fy2008.pdf on April 8, 2010

Connecticut Department of Environmental Protection, Bureau of Material Management & Compliance Assurance Recycling Office (2009c) Solid Waste Reported Transferred Out from Connecticut Municipal Solid Waste Resource Recovery Facilities (RRFs – Waste-to-Energy Facilities). Retrieved from www.ct.gov/dep/lib/dep/reduce_reuse_recycle/data/solid_waste_reported_sent_out_from_ct_rrfs_fy2008.pdf on April 8, 2010

Connecticut Resources Recovery Authority (2009) Timeline. Retrieved from www.crra.org/pages/timeline.htm on June 27, 2010

Environmental Protection Agency (2008) Wastes – Ask a Question. What is the municipal solid waste (MSW) hierarchy? Retrieved from waste.custhelp.com/cgi-bin/waste.cfg/php/enduser/std_adp.php?p_faqid_1040 on June 27, 2010

Environmental Protection Agency (2009) Municipal solid waste generation, recycling, and disposal in the United States: Facts and Figures for 2008. Environmental Protection Agency, EPA-530-F–009–021, Washington, DC

Eurostat (2010) Environment in the EU27 40% of Municipal Solid Waste Recycled or Composted in 2008. Eurostate News Release 43/2010. Retrieved from epp.eurostat.ec.europa.eu/cache/ITY_PUBLIC/8-19032010-AP/EN/8-19032010-AP-EN.PDF on July 11, 2010

Leavenworth, J (2010) Trash Switch Considered. The Hartford Courant. Wednesday, April 28, Final 5 Edition, CTNOW, p B6

Michaels, T (2007) The 2007 IWSA directory of waste-to-energy plants. Integrated Waste Services Association, Washington, DC

Owens, D CRRA Drops Plan for Ash Dump in Franklin. The Hartford Courant. Friday, August 28, Final 5 Edition, Section A, p A1

Rosenthal, E (2010) Europe Finds Clean Fuel in Trash; U.S. Sits Back. The New York Times. Tuesday, April 13, Late Edition-Final, Section A, p 1

Smith, L (2010) Town Joins Search for Bids. The Hartford Courant. Tuesday, June 15, Final 5 Edition, CTNOW, p B10

Chapter 13
Global Geographies of Environmental Injustice and Health: A Case Study of Illegal Hazardous Waste Dumping in Côte d'Ivoire

Florence M. Margai and Fatoumata B. Barry

Abstract Global environmental injustice, the unfair distribution of hazardous activities and materials in disadvantaged communities, is increasingly evident in the African continent through transboundary pollution and illegal disposal of hazardous wastes. Studies are needed to uncover the underlying factors that account for these trends and the detrimental health effects in the host communities. This chapter examines a recent incident involving the disposal of hazardous wastes in Abidjan, Cote d'Ivoire. Specifically, in August 2006, hazardous wastes consisting of mercaptans, hydrogen sulfide, phenols, and hydrocarbons were dumped illegally in seventeen locations resulting in approximately fifteen deaths and thousands of injuries. The chapter examines the circumstances under which the incident occurred and the communities that were most affected by the incident. Atmospheric dispersion models are used to delineate the plume footprints of the hazardous releases in Abidjan. The generalized risk zones are then integrated into a Geographic Information Systems (GIS) to assess the environmental impacts of exposure and the demographic profile of residents within these high risk zones.

Keywords Global environmental justice · Environmental health · Hazardous wastes · Transboundary pollution · Atmospheric dispersion modeling · GIS

13.1 Introduction

Nearly three decades since the emergence of the environmental justice (EJ) movement in North Carolina, the story of environmental pollution and injustice continues to unfold within and beyond the borders of the United States to include new narratives of resource exploitation, toxic contamination, and transnational pollution. Scientific accounts show that these environmental hazards are increasingly global in scope, though their attendant impacts on residents remain highly contextualized and disproportionately distributed by race, ethnicity and class (Westra and Lawson,

F.M. Margai (✉)
Department of Geography, State University of New York, Binghamton, NY, USA
e-mail: margai@binghamton.edu

J.A. Maantay, S. McLafferty (eds.), *Geospatial Analysis of Environmental Health*, Geotechnologies and the Environment 4, DOI 10.1007/978-94-007-0329-2_13,

2001; Adeola, 2001; Bullard, 2005; Betsill et al., 2006; Grineski and Collins, 2008; Myers, 2008; Steady, 2009). In both rich and poor economies, including those known to be egalitarian societies such as Sweden, researchers have found that residents in the low income, working class, minority and/or indigenous communities remain by far, the ones that are most impacted and unfairly harmed by these environmental hazards (Chaix et al., 2006; Fairburn et al., 2009; Adeola, 2009; Higgs and Langford, 2009). They are also the least likely to benefit from environmental remediation, risk compensation, or legislative efforts designed to redress the problems.

As global accounts of EJ struggles unfold, the emerging literature remains highly focused on the plight of residents in the wealthy industrialized nations, the Global North (Su et al., 2009; Chakraborty, 2009; Crowder and Downey, 2010; Landrigan et al., 2010; Martuzzi et al., 2010). More recently however, there has been a definitive shift toward the examination of these concerns within the Global South, notably in African and Latin American countries (Bullard, 2005; Betsill et al., 2006; McCaffrey, 2008; Myers, 2008; Carruthers, 2008; Phalane and Steady, 2009; Barry, 2010). Some scholars now contend, and rightly so, that in these countries, the struggles for environmental justice existed all along, albeit in disparate forms of consciousness, practice, and mobilization, and therefore unrecognized in the Western World (Carruthers, 2008; Myers, 2008). These recent efforts by various scholars and activists are designed to bring attention to these concerns, and more importantly, deepen our understanding of the root causes and the emerging health risks facing the residents in the host communities.

The research described in this chapter falls within the realm of these latest studies, seeking to explore the emerging geographies of global environmental injustices through GIS and spatial analytical methods that delineate the exposure risks and potential health effects among vulnerable populations. The overall objectives are two-fold. The first is to conceptualize the global dynamics of EJ by examining the underlying mechanisms and drivers behind the expansion and cross-border transfer of polluting industries and products into poor countries. Using transboundary hazardous flows as an example, a two-way conceptualization is offered to explain the reasons for industrial flight in the Global North, and the emergence of pollution havens in the Global South. The second objective is to showcase the use of geospatial methodologies to assess the differential patterns of population exposure to these hazards. The EJ literature offers many examples of robust GIS/statistical techniques for validating EJ concerns, but these have been traditionally employed in settings within the developed world (Mennis, 2002; Harner et al., 2002; Mennis and Jordan, 2005; Fairburn et al., 2009). In this chapter, we hope to apply a few of these approaches in a developing country by examining the environmental health risks arising from an illegal hazardous waste disposal incident in Abidjan, Côte d'Ivoire. Of particular interest is the Toxic Demographic Difference Index (TDDI), one of many EJ indices proposed by Harner et al. (2002), and the atmospheric dispersion modeling approach previously used by Chakraborty and Armstrong (1996) and Margai (2001). Using these approaches, we will examine the spatial extent and variability of the chemical exposure risks resulting from this incident. Atmospheric dispersion models will be used to generate the toxic plume footprints and identify the demographic profile of the residents that are most likely to suffer adverse health consequences from the incident.

13.2 The Global Expansion of EJ: The Case of Hazardous Industries, Products and Wastes

Environmental health injustices may be defined as the disproportionate and unfair distribution of hazardous substances and other environmental hazards in disadvantaged communities with crippling impacts on the health of the residents (Bullard, 2005; Maantay, 2002; Landrigan et al., 2010). In this new millennium, often dubbed the era of corporate hegemony and globalization, the spatial dimensions of these injustices are inherently global in scope, and more readily manifested in myriad forms that reflect the operations and practices of transnational corporations (TNCs). Alongside the usual documentation of toxic waste inequities in industrialized countries, EJ studies now include many examples of corporate activities that significantly impact the living environments of the poor and indigenous groups in developing regions notably in Latin America and sub-Saharan Africa (Westra and Lawson, 2001; Adeola, 2001; Betsill et al., 2006; Steady, 2009). Accounts of corporate environmental injustice range from oil, gas and mineral resource exploitation or "megamining" operations, to deforestation and use of harmful pesticides in agribusinesses, hazardous waste shipments, and overall threats to communal property rights, land use, and traditional lifestyles (Westra and Lawson, 2001; Bullard, 2005; Bury, 2007; Steady, 2009). Of interest in this chapter are the health threats that arise from hazardous materials and electronic wastes originating from developed countries. The characteristics and trends in the production and movement of these hazards are briefly described below.

13.3 Trends in the Production and International Shipment of Hazardous Materials and Wastes

The characterization of hazardous materials vary by chemical composition and geographic origin, however, these are generally regarded as substances that are toxic, flammable, explosive, or corrosive, and can cause significant adverse health effects when exposed to humans, animals or the environment. Estimates show that approximately 300–500 million tons of hazardous wastes are generated globally each year and nearly 90% of these wastes are produced in the wealthy industrialized countries (UNEP, 2006a). Though most of these wastes are handled internally in these countries, about 2% is exported to other countries. According to Kocasoy (2003), the leading global exporters of hazardous waste are all located in the Global North and include: Germany, the Netherlands, USA, Canada, Switzerland, Austria, Sweden, Norway, Italy, Spain, France, Denmark, Finland, Portugal and Great Britain. Between 1993 and 2001, transnational shipments of wastes increased from 2 million tons to more than 8.5 million tons (UNEP, 2006). Further, since the early 1980s, increasingly large amounts of these wastes have been sent to poorer nations for disposal (Lipman, 2002; Clapp, 2001; Gbadegesin, 2001).

Electronic wastes (e-wastes) have also been added to the mix of international shipments. Globally, it is estimated that 20–25 million tons are produced each year, primarily in Europe, the US and Australia (Robinson, 2009). India, China and many African countries have been the most actively involved in the importation of these wastes. Scientific studies show that these products contain harmful contaminants including Polychlorinated Biphenyls (PCBs), Polybrominated Diphenyl Ethers (PBDEs), dioxins and furan (PCDD/Ff) and heavy metals particularly antimony, lead, mercury, cadmium, and nickel (Wong et al., 2007; Robinson, 2009; Frazzoli et al., 2010).

Concerns about the emergent health risks associated with the international transfer of these hazardous materials previously led to a number of multilateral environmental agreements including the Basel Convention (ratified in 1992), and the Bamako Convention, which became effective in 1998 (UNEP, 1992; Basel Convention, 2007). However, serious doubts remain over the effectiveness of these international treaties for a number of reasons. First, official reports of the shipments are often incomplete, and may under-represent the true volume of the hazardous flows between the countries. Further, to circumvent environmental regulations, some of these materials, particularly the e-wastes, may be labeled as reusable or recyclable products since they sometimes contain valuable metals such as copper and platinum. It is evident that most of the shipments that arrive in developing countries are misclassified, often containing products that are obsolete, and completely unusable (Schmidt, 2004; The Basel Action Network, 2007). Additionally, these low income countries may not have the proper tools, technologies or expertise to efficiently reprocess or recover these materials. Workers rely instead on crude recycling practices such as dissolving the e-products in strong acids, or burning different parts to reclaim the materials. Unused materials are later disposed through open burning, at local dumpsites, or abandoned along roadsides, endangering the lives of the local residents and ecosystems.

13.4 Conceptualizing Transboundary Industrial Operations and Waste Flows

When examining the root causes behind the expansion of toxic operations and related waste flows into developing countries, many scholars contend that these activities are directly linked to political and economic globalization (Frey, 2003; Asante-Duah, 1992; Asante-Duah and Nagy, 2001; Gbadegesin, 2001; Adeola, 2001; McCurdy, 2001; Lipman, 2002). Over the years, previous barriers to the international flow of goods and services have been removed through the enactment of trade liberalization policies and the formation of regional and supranational organizations. These changes have resulted in new economic structures, and greater interdependence between countries. TNCs have willingly responded to these changes by leveraging their economies of scale to reduce costs while expanding

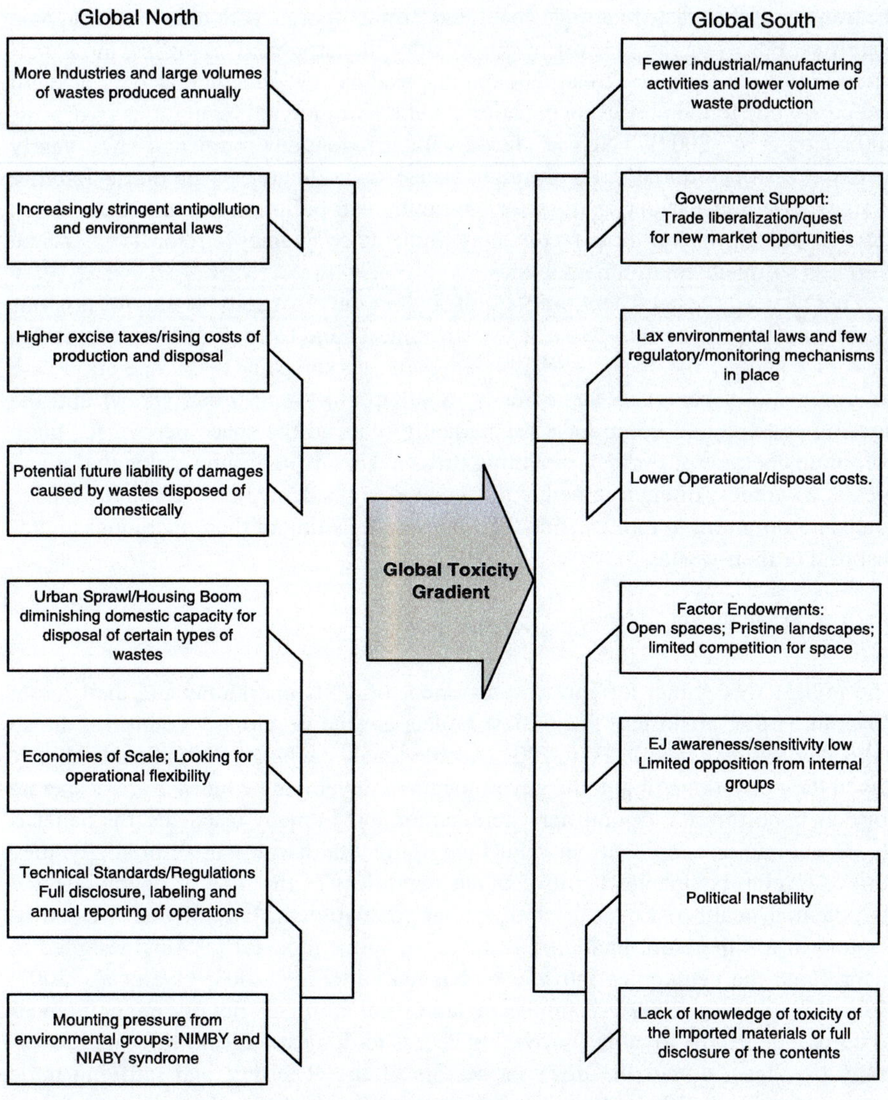

Fig. 13.1 Root causes and drivers of global environmental injustices

their global assets. Drawing on these discussions in the literature, the conceptualization offered in Fig. 13.1 specifies the core drivers and mechanisms that account for these toxic flows to low income countries. These are characterized here as the push and pull factors.

13.4.1 Push Factors of Hazardous Flows

The push factors are affiliated with the hazardous operations and waste products generated in the source regions. These factors contribute to "industrial flight", and/or the transfer of the waste materials generated from corporate activities to low income countries. For example, in most of the wealthy industrialized countries, the acquisition of permits for waste treatment and disposal has become increasingly difficult and costly due to the prevailing economic and environmental regulations (Kocasoy, 2003; Coe et al., 2007). TNCs are faced with stringent environmental laws, yearly increases in operational costs through excise taxes, increases in future liability charges, or compensatory charges for environmental pollution in the industrialized nations. One or more of these factors may force the companies to relocate elsewhere with less stringent environmental laws.

There is also the persistent problem of finding the most suitable site for noxious facilities, often characterized as Locally Unwanted Land Uses (LULUs). Companies have to convince the local residents and their workers that these operations are environmentally benign. Recent trends in suburbanization, urban sprawl and the housing construction boom have dramatically reduced the space needed for these corporate operations. Further, mounting pressures from environmental activists and greater awareness of environmental justice among residents in these countries have forced corporations to look beyond their borders for siting of their operations and/or disposal of their wastes.

13.4.2 Pull Factors of Hazardous Flows

The pull factors attract the inward movement of TNC operations and their waste flows into other countries. The easiest targets are the developing countries, the so called "*pollution havens*" (Frey, 2003; Asante-Duah and Nagy, 2001; Steady, 2009). Given their economic plight, the governments of developing countries are desperate for new opportunities to jumpstart their economies. In many instances, the belief is that economic development outweighs environmental harm. Not surprisingly, they turn to foreign companies for investment regardless of the type of operations and its potential health risks on the indigenous populations. Many of these scenarios abound in the literature including instances of governmental reforms designed to incorporate the economies into the global market (Frey, 2003; Coe et al., 2007; Bury, 2007; Steady, 2009). Examples include eliminating restrictions on remittances of corporate profits, royalties, providing access to domestic credit, offering tax stability packages, privatizing different sectors of the economy, and ratifying trade agreements (Bury, 2007). Countries offering such incentives often become attractive destination points for waste handlers or companies seeking to relocate their base operations away from more expensive or environmentally conscious locations.

There are also lower operational costs in low income countries. These may originate from the natural endowments of these countries such as the availability of land, resources, and characteristics of the local population (who are perhaps less cognizant or aware of the EJ issues). For example, while treating or disposing of a hazardous waste material is estimated to cost an average of \$2,000–\$3,000 per ton in a wealthy industrialized country, through illicit negotiations, it could be a meager \$2.50 a ton for the same material in an African country (Gbadegesin, 2001). Not surprisingly, there have been many scandals of illicit hazardous waste exports in the region (Gbadegesin, 2001; Kocasoy, 2003; Bullard, 2005; Steady, 2009). Political instability also accounts for the importation of hazardous materials, as previously demonstrated in Somalia (Hussein, 2001) and Guinea-Bissau (Clapp, 2001). Foreign companies and governments take advantage of situations during which they are least likely to face governmental resistance, public opposition, or large expense in their negotiations to relocate or transfer their waste products.

Finally, as noted earlier, many of the hazardous wastes transactions involve the lack of full disclosure of the true content of the wastes (Morton, 1996). The receiving countries may not be fully aware of the toxicological properties of the wastes that are being imported either because the transactions are made between unscrupulous individuals or the wastes are mislabeled to disguise the true contents. Further, the receiving countries may not have the waste handling facilities or the technical expertise to evaluate the contents of the waste shipment. Without the right equipment and technical expertise to test, recycle, treat or dispose of these wastes the end result is improper disposal through open burning, or dumping on residential properties.

Overall, Fig. 13.1 shows that the international movement of transnational operations and wastes are driven in part by the conditions in the source areas, and those within the destination areas. Organizing these factors in a conceptual framework allows for a better understanding of the motives and mechanisms underlying these flows. The prevailing conditions, or the observed differences in physical, political, economic, demographic and environmental characteristics between the source and the destination countries contribute to the geographic expansion of these toxic operations (Castleman and Navarro, 1987; Kocasoy, 2003; Coe et al., 2007; Steady, 2009). The observed differential is characterized here as the *global toxicity gradient*, such that the greater the difference between the countries, the greater the likelihood of transfer of toxic operations between countries, thus producing new geographies of global environmental injustice. TNCs take advantage of these differentials by implementing new global strategies that enable the international outsourcing of some or all of their production processes to low income countries. We proceed next to the discussion of the case study completed in Cote d'Ivoire.

13.5 A Geographic Case Study of Illegal Dumping of Hazardous Wastes in Cote d'Ivoire

Cote d'Ivoire is a low income developing country of roughly 21 million people, residing along the Gulf of Guinea, in West Africa. In August 2006, hazardous mixtures of mercaptans, hydrogen sulfide, phenols, and hydrocarbons originating from the Netherlands, were dumped illegally around the port city of Abidjan, resulting in several deaths and injuries. The political environment and economic circumstances leading up to this incident, underscore many of the points noted above.

13.5.1 Conditions Preceding the Hazardous Waste Incident

Historically, France had a great presence in Cote d'Ivoire from 1647 until 1960 when it gained independence. Thereafter, the country went through a long period of relative political stability and economic growth under the leadership of their first president, Félix Houphouët-Boigny. This peaceful period ended in 1993 following his death. Successive presidents were criticized for being corrupt, and inept in the management of the country's finances. Rebel factions and coup attempts repeatedly destabilized their governments notably in 1999 and 2002. A related area of concern, politically, was linked to the issue of citizenship, "Ivorianness", or one's status as an "Ivoirite" (Skogseth, 2006). This issue came up because of the significant influx of immigrants resulting from ongoing conflicts in other parts of West Africa. In 2006, proof of citizenship became a hotly debated issue especially for the 3.5 million people without the proper credentials. The courts were ordered to examine the plight of these residents, but the proceedings were interrupted by groups opposed to the hearings. Violence ensued and this created the threat of another coup.

13.5.2 Hazardous Waste Incident

It was against the backdrop of immigrant xenophobia, political and economic instability that the hazardous wastes incident occurred. On August 19th 2006, the Probo Koala, a Panamanian registered ship arrived at the Abidjan port. Though registered in Panama, the ship (owned by a Greek company), had been chartered by Trafigura, the world's third largest independent oil trader with global assets and employees in over forty countries. With its base operations in Amsterdam and London, most of the company's oil storage and delivery facilities, were located elsewhere, in Africa, Central and South America (Kao and Bosley, 2009).

Trafigura contends that the ship's contents were "slops" or wastewater from the washing of ship tanks. However, detailed news accounts released since the initiation of this study show that the hazardous materials originated from "coker gasoline", a type of fuel that contains large amounts sulphur and silica (Duckett, 2009). This fuel was being produced by Pemex, a Mexican based company. Due to some operational difficulties and the lack of storage to store the excess gasoline, Pemex sold about 84,000 tons of this gasoline to Trafigura. The product was trucked to Brownsville, Texas, and subsequently loaded on board the Probo Koala. The ship later anchored in Gibraltar, and on board, it is widely believed that Trafigura embarked on an experimental process to strip the sulphurous products off the coker gasoline. The end product, Naptha, was reportedly resold for a profit. Unfortunately, this experimental process also generated about 500 tons of harmful waste materials that could not be readily discarded.

The Probo Koala initially transported the wastes to Amsterdam, and there they would have been charged 500,000 Euros to properly treat and dispose the materials. After several inquiries, they successfully negotiated a deal with a subcontractor in Ivory Coast to dispose of the wastes for a significantly lower price of 18,500 Euros. Upon arrival in Abidjan, the first tanker from the Probo Koala disposed of its initial load of toxic wastes at a local waste dump located in Akouedu, the Southern part of the city. Over the course of the next several days, tankers disposed the rest of the wastes across 16 other sites around Abidjan. Subsequent investigations by UNEP led to the identification of all but two of these locations.

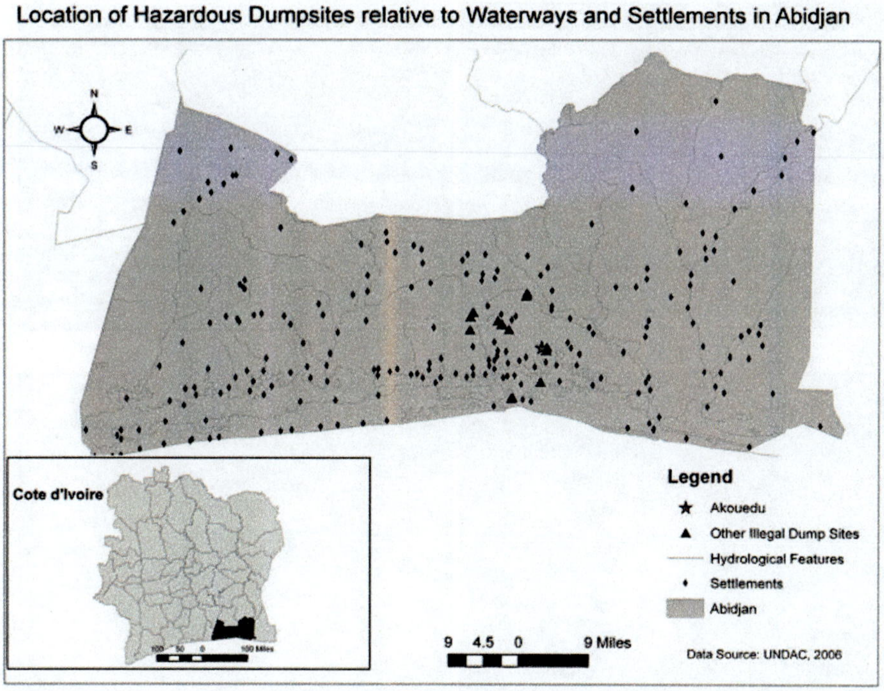

Fig. 13.2 Location of the illegal dumpsites in Abidjan

Figure 13.2 shows the distribution of these sites relative to the hydrological features and residential settlement in the city. Aside from Akouedu which is a primary landfill, field evaluation of these sites shows that they are either near open vegetated or forested areas dispersed around the city, or along roadsides, creeks, sewage systems and lagoons (See Fig. 13.3a through d for a sampling of these sites).

13.6 Geographic Data Sources and Preprocessing

Demographic Data: In this study, we used the 2005 Cote d'Ivoire data acquired through the Demographic and Health Survey (DHS), a standardized survey that is administered periodically by ICF Macro in over 50 developing countries (DHS, 2005). For Cote d'Ivoire, this large national database consisted of 9,686 records of individuals interviewed during the survey period. Using a multistage probability sampling approach, this nationally representative dataset was compiled from 253 population clusters, or census enumeration areas. Geographically referenced data, including the longitude, latitude and altitudinal information, were collected for each cluster enabling the integration of the data into a GIS. Within each cluster, households were randomly selected, and adults (women of reproductive ages 15–49 and men aged 15–59 years) within those households were subsequently interviewed.

Fig. 13.3 Sample pictures of illegal dumpsites

The average size of the sample or "take" per cluster varied from about 30 to 40 households in rural areas, and 20–25 households in urbanized locations. Of the 253 clusters, about 26% (66) of the sample with 1,708 individuals were from Abidjan, where the hazardous incident occurred.

The national DHS data was first brought into the *SPSS 16.0* statistical software, and following the preprocessing of the data, including the application of sampling weights, several variables were selected for inclusion in the study: locale (rural/urban), ethnicity, religiosity, age, education, housing characteristics (ownership, length of stay), employment, and access to basic amenities such as electricity, piped water, sanitation, radio, and a vehicle. Also included was a wealth index, a proxy measure of socio-economic status (SES) computed by DHS.

The analysis of ethnicity required a few extra steps given the diverse composition of the Ivorian population. There are over sixty ethnic groups in the country and these are often organized into four major clusters based on their linguistic commonalities and other cultural attributes: Mande, Gur, Krou and the Kwa (Skogseth, 2006). In this study, we chose to focus on the Kwa and Krou ethnicities for two reasons. First, there are regional differences in the distribution of the four ethnic clusters. The Kwas and Krous reside primarily in the subtropical areas of the country, particularly along the coastal areas of the south where the hazardous incident occurred. The Mande and Gur ethnicities were not included because they tend to reside mostly in the Savannah regions of the North, and therefore least likely to be impacted by the incident. A second reason for choosing the Kwa and Krou ethnicities is that they both represent the population groups of interest in terms of majority/minority relations. In this study, as with most EJ applications, the goal was to examine the distribution of environmental risks among residents that belong to either the majority or minority population groups. The Kwa cluster represents the ethnic majority in the country comprising about 44% of the population (Skogeth, 2006). This cluster is characterized by subgroups such as Ajoukrou, Abbey, Aboure, Alladian, Abidji, Abron, and Agni people. As part of the preprocessing of the data therefore, all individuals who self-identified as belonging to the any of these subgroups were classified as Kwa, the ethnic majority. At the other end of the ethnic spectrum is the Krou cluster representing the ethnic minority, and constituting only about 12% of the Ivorian population. This ethnic cluster is characterized by the Ahizi along with subgroups such as the Bakwe and Didi people. Those who self-identified with these subgroups were placed in the Krou cluster, and designated as ethnic minorities.

In this study, we also considered religiosity which is another notable source of group marginalization, conflict, and environmental inequity in many world regions. Preliminary review of the literature on Ivorian religiosity suggested a mixture of religious beliefs that included Islam, Christianity, and traditional animistic views (Skogeth, 2006). The religious differences also tended to conform to the ethnic regional boundaries noted earlier with Moslems residing mostly in the northeastern/Sahelian reaches of the country, and Christians (mostly Catholics) residing within the urbanized areas and along coastal regions, the original points of contact with European missionaries. Over the years however, these distinctive religious patterns have been altered by rural-urban migration and group intermixing. Though, both the Kwas and Krous are more likely to be Christians than their counterparts in the North, a crosstabulation of these two groups using the national DHS data showed significant differences in religious composition. Of the roughly 43.7% who self-identified as Moslems, only about 36.3% belonged to the Kwa majority, and the others were ethnic minorities (Table 13.1). The national sample also revealed that approximately 37% of the population self-reported as Christians of which 52% were Catholics, 24% Protestants, and the rest belonged to other denominations. Overall, based on the data, there appeared to be a fairly equitable distribution of both Christians (37%) and Moslems (44%) within the general population but with observable differences in religious beliefs among the major ethnicities.

Hazardous Waste Data: The details regarding the incident were garnered in 2006 by contacting officials in the United Nations Institute for Training and Research (UNITAR), and the Operational Satellite Applications Program (UNOSAT). These agencies were the lead coordinators in the emergency response to the incident. The data acquired consisted of the geographic coordinates of the disposal sites, the type and amount of chemicals released, and official injury statistics.

Table 13.1 Crosstabulation of national DHS respondents by ethnicity and religiosity

			KWA		
			Other	KWA majority	Total
Religion	Catholic	Count	153	1,696	1,849
		% within KWA	9.1	21.2	19.1
		% of total	1.6	17.5	19.1
	Protestant	Count	32	825	857
		% within KWA	1.9	10.3	8.9
		% of total	0.3	8.5	8.9
	Other Christian religions	Count	77	796	873
		% within KWA	4.6	10.0	9.0
		% of total	0.8	8.2	9.0
	Moslem	Count	1,322	2,902	4,224
		% within KWA	79.0	36.3	43.7
		% of total	13.7	30.0	43.7
	Tribal religions	Count	84	1,694	1,778
		% within KWA	5.0	21.2	18.4
		% of total	0.9	17.5	18.4
	Other religions	Count	5	84	89
		% within KWA	0.3	1.1	0.9
		% of total	0.1	0.9	0.9
Total		Count	1,673	7,997	9,670
		% within KWA	100.0	100.0	100.0
		% of total	17.3	82.7	100.0

Religion * KWA Ethnicity Crosstabulation using National DHS sample.
Pearson's Chi square = 1,042.8(df 5,1); $p<0.0001$.

The United Nations Disaster Assessment and Coordination (UNDAC) team was specifically in charge of investigating the causes and effects of the hazardous disposal. They confirmed that roughly 528 tons of hazardous wastes were dispensed from the ship. Also prior to the Probo Koala's departure to Cote d'Ivoire, samples of the ship's content had been tested in Amsterdam. These results along with the field samples taken from dumpsites in Cote d'Ivoire confirmed that the wastes consisted primarily of the following chemicals: hydrogen sulfide, mercaptans, phenols, hydrocarbons (a mixture of olefins, naphtenes, parrafins, and aromatics). All of these chemicals were known to be harmful to human health following exposure through various environmental routes and pathways. According to UNDAC however, groundwater wells were distant from the polluted sites and therefore, no immediate health risks were expected from the drinking water sources. The primary concern was with atmospheric dispersal of the pollutants and contamination through inhalation.

Additional data regarding the toxicity of these chemicals were obtained from a report compiled for Trafigura, after the incident had occurred. This report corroborated most of the environmental health concerns identified by the UN agencies. Further, the analysts categorized the chemicals into three groups: (i) those that are harmful only on close contact; (ii) those that are volatile and may achieve atmospheric concentrations that are harmful at some distance; and (iii) those that are harmful only on close contact, but may degrade further into substances that are volatile and cause harm at some distance (Minton, 2006).

Since the incident, the injury statistics have fluctuated over time, however, the original records show that by September 18th 2006, health care professionals had treated over 44,000 people (UNEP, 2006). A total of 15 were believed to have died shortly after exposure to the contaminants (Agence France-Presse, 2007). UNDAC reported that future health problems should not be expected because the chemicals found in the wastes generally had short periods of toxicity, and less likely to be persistent or bioaccumulative. However, the nauseous odors can still be smelled around some of the dumping sites 3 years after the incident (Koffi, 2009).

13.7 Analytical Procedures and Results

As stated earlier, our primary goal in this case study was to use GIS-based methods to evaluate the spatial distribution of the illegal dumpsites and determine the likelihood of disproportionate exposures to these hazardous releases based on the demographic attributes of the residents. From an EJ perspective, one could argue that the entire country was unfairly targeted by the illicit activities of the TNC. However, we were primarily interested in learning more about the local geographies of injustice, in particular the demographic characteristics of the neighborhoods and residents that were most impacted by this incident. Were these likely to be the neighborhoods of poor families, of religious or ethnic minority groups? Were these recent immigrants, unemployed, or residents with low educational attainment? Was their profile generally consistent with what we know about other EJ communities around the world? Addressing these questions was important not only for profiling global EJ populations with significant environmental exposures, but also planning for a more extensive environmental epidemiological study. The geospatial analysis was performed in a four stage sequence as described below.

13.7.1 Geo-Demographic Analysis Using Data Generated from Sampled Points

The first step involved the analysis of the demographic data to generate maps depicting the distributional patterns of the population. The DHS data, as noted earlier, were collected from 253 residential clusters of which 66 of these sampled communities were from the city of Abidjan, providing a good representation of the study area. Using SPSS, we first obtained the univariate estimates of the variables. The relevant information was then integrated into ESRI's ArcGIS 9.3.

Within the GIS, the DHS data were mappable at multiple spatial scales including the cluster level, and administrative areal levels 2 and 1. We started the analysis first at the cluster level, and then aggregated upwards to administrative level 2, the smallest areal unit available. Since the demographic data were based on sampling points across the country, the ordinary kriging approach was used to generate predictive maps for each of the demographic indicators. For each variable, we first examined the data distribution, and variograms, before deciding on the best model for spatial interpolation. Figure 13.4a to b are examples of maps generated from the geostatistical analysis of two of these variables, religiosity (Moslems) and Education. The maps reveal the national distribution of these variables at administrative level 2, and the insets show the more localized distributions within Abidjan.

13.7.2 Delineation of Chemical Impact Zones

The next step involved the delineation of impact zones or "footprints" of the chemical releases within which residents were at highest risk of suffering adverse health consequences. For this, we used *ALOHA 5.4.1.2* (Areal Location of Hazardous Atmospheres), a modeling program developed by the US. NOAA (National Oceanic and Atmospheric Administration) and EPA (Environmental Protection Agency). The program calibrates the threat zones of hazardous releases based on the toxicological properties of the chemical, the amount and location of release, and the atmospheric conditions at the time of an incident. The footprints that are generated are transportable into GIS packages.

Using the information supplied by UNDAC, the dumpsites were spatially referenced in the modeling environment. As with most risk assessment studies, the analytical emphasis in the ALOHA program was based on the worst case scenarios. So, of all the chemicals detected at the disposal sites, we chose to model Hydrogen Sulfide and Mercaptans (methyl and ethyl), these being the most volatile chemicals in the wastes. These substances were also identified as the ones producing the most harmful atmospheric effects during the incident (UNEP, 2006).

An estimate of 7.75 tons of hydrogen sulfide (H_2S) deposited per site produced a footprint with a 2.2-mile radius of excess risk (Immediately Dangerous to Life or Health: IDLH), and an overall risk zone stretching up to 6 miles (See Fig. 13.5a). According to the ATSDR database (accessed in 2007), exposures to lower concentrations of this chemical often result in irritation of the throat, nose and eyes and respiratory problems. Inhaling or ingesting hydrogen sulfide at high levels may cause coma and death. These health effects were corroborated in the Minton report, which characterized H_2S as a corrosive and highly toxic gas with negative health effects at levels as low as 20 ppm (parts per million). At levels between 250 and 500 ppm, pulmonary edema can occur and exposures beyond these levels result in breathing difficulties, loss of consciousness and death. All of these prognoses are consistent with the health problems observed among some residents at the time of the incident in Abidjan.

For the mercaptans, an estimated amount of 7.75 tons of methyl mercaptan per site led to a 1.4 mile zone of excess risk (See Fig. 13.5b). Mercaptans have very

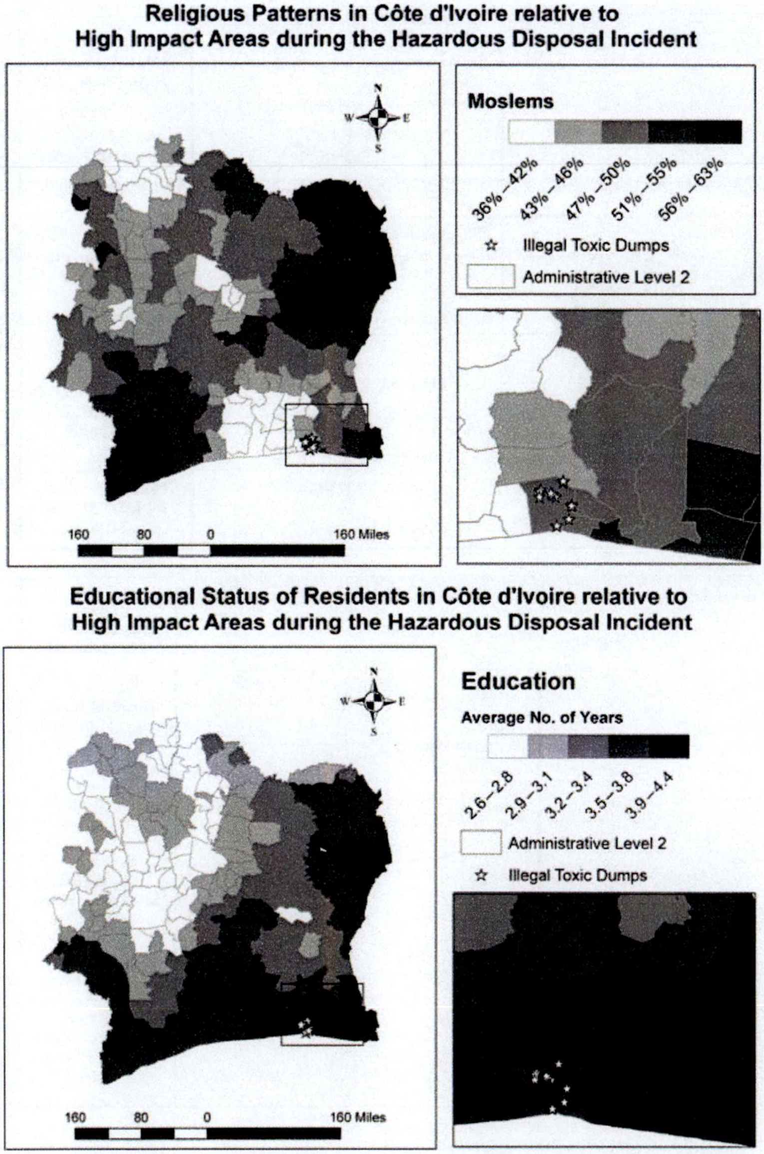

Fig. 13.4 Predictive maps of selected demographic indicators in Cote d'Ivoire

strong, unpleasant odors and are also known to result in nausea, headaches, and breathing difficulties. ATSDR reports show that high levels of prolonged exposure to these chemicals cause anemia, coma, and death (ATSDR, 2006).

Beyond the two chemicals noted above, additional health risks reported during the Abidjan incident included skin burns and ulcerations. These are believed to have been caused by direct or close contact with some of the other chemicals such as

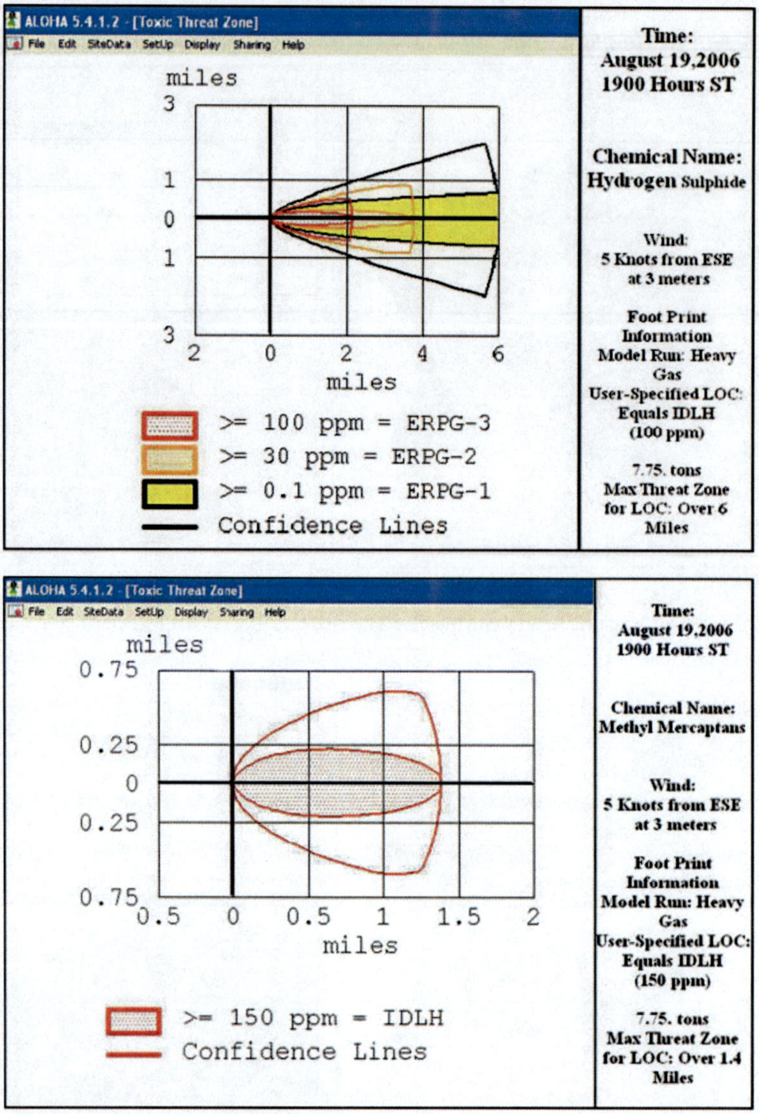

Fig. 13.5 Footprint of hazardous chemicals released during the 2006 incident in Abidjan. **a** Footprint of the hydrogen sulfide gas release. **b** Footprint of the methyl mercaptan gas release

sodium hydroxides, and the byproducts of the reaction between hydrogen sulfides and sodium hydroxides (Minton, 2006).

Figure 13.6 shows the integration of the footprint generated for hydrogen sulfide at one of the hazardous sites. Three zones are shown, at 0.1 ppm, 30 ppm, 100 ppm along with their corresponding confidence intervals. The risk zone at 100 ppm, approximately 2.2 times from the toxic sites, is the IDLH area deemed

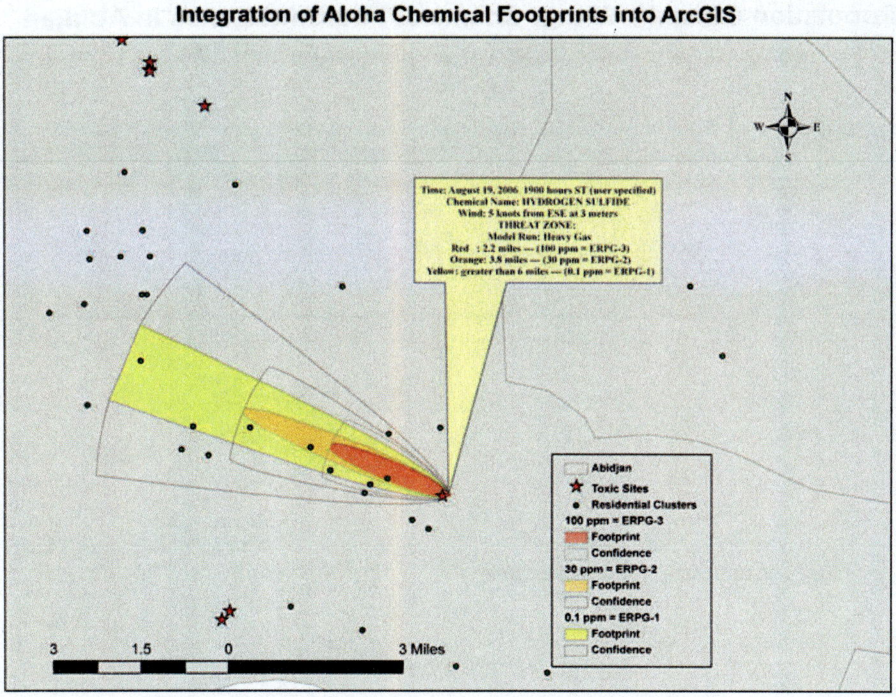

Fig. 13.6 Integration of Aloha chemical footprints into ArcGIS

to be the most dangerous to life and health. For the purposes of this study, we chose this 2.2 mile radius as the generalized risk zone around each dumpsite, the highest threat zone over which the most harmful effects of these chemicals were likely to be felt.

13.7.3 Linking the Threat Zones to the Demographic Data Layers

The third stage of the analysis involved a detailed evaluation of the communities that were within the threat zones delineated in the preceding step. To create a demographic profile of these risk zones, a spatial query was performed in ArcGIS to identify all residential clusters that fell within or part of the 2.2 mile risk zone of each dump site. Nineteen of the city's 66 residential clusters in the sampled DHS database fell into this buffer (Fig. 13.7). These areas were designated as high impact areas and all others were characterized as low impact areas. The demographic characteristics of the residents (religiosity, ethnicity, age, education, employment, wealth, housing composition and length of stay) of these buffer zones were then extracted and exported into SPSS for statistical analysis.

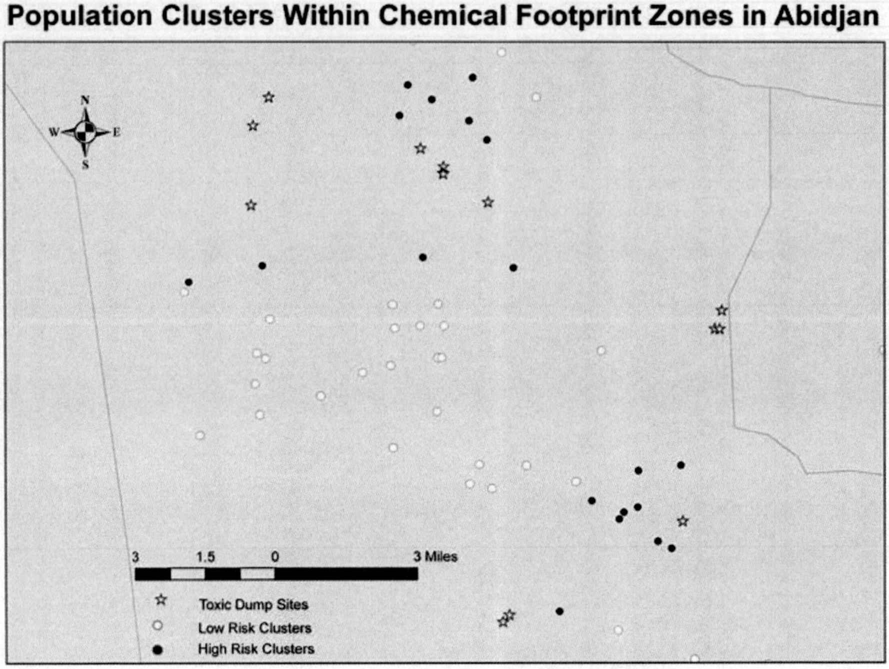

Fig. 13.7 Location of the population clusters within the areas of excessive risk in Abidjan

13.7.4 Statistical Comparison of Threat Zones

The final stage of the analysis involved a computation of univariate measures, the calibration of Toxic Demographic Difference Index (TDDI) for each variable, followed by multivariate analysis using stepwise logistic regression. The univariate statistics generated for the Abidjan sample are reported in Table 13.2. The majority of the residents are Kwa, and surprisingly, no foreigners were identified in the DHS database. About 42% of the respondents have no formal education and about a third fall into the middle and lower rankings of the wealth index. The proportion of Christians and Moslems is fairly even in the sample taken in Abidjan and appears to be nationally representative. The average age of respondents is 27.8 years, and they live with an average of 8 members in their households, both characteristics fairly typical of many African communities.

The TDDI is a variant of the independent samples t-test that compares the mean differences among variables measured across two different samples. This procedure was applied to the DHS sample extracted for Abidjan which consisted of 796 individuals residing within the 19 clusters identified as high risk areas and 912 individuals within the 47 clusters identified as low risk areas. The results of the TDDI are reported in Table 13.3. Those who faced the greatest risk of exposure to the chemical hazard were long term residents and therefore less likely to be immigrants (Table 13.3). A greater proportion of the residents in the high risk zones

Table 13.2 Frequency distribution of selected attributes of residents in Abidjan

Categorical variable	$N = 1{,}708$ (%)
Hazardous risk zones	
Low	53.4
High	46.6
Kwa	80.9
Ahizi	19.0
Foreigners	0
Employed	65.7
Religion	
Christians	44.4
Moslems	42.9
Other	12.8
Education	
No education	41.9
Incomplete primary	19.2
Complete primary	3.3
Incomplete secondary	26.0
Complete secondary	1.3
Higher	8.4
Wealth index	
Poorest	10.7
Poorer	11.2
Middle	13.5
Richer	24.2
Richest	40.3

were Moslems as well. The analysis also revealed that the housing conditions in the high risk zones were substandard with residents less likely to have access to piped water, electricity and other basic amenities. When compared to the low risk zones, the household sizes were larger, and also likely to have more children under the age of five.

Further analysis using logistic regression analysis involved the use of the risk zone as a binary outcome (dependent variable). The odds of being exposed to the hazardous chemicals were expressed as a function of the following independent variables: wealth index, education, religiosity, ethnicity, employment, length of stay in the community, and access to piped water, flush toilet, and electricity (See Table 13.4). A stepwise procedure was employed to minimize the likelihood of multicollinearity among the variables. The model derived at the end of the iterative procedure consisted of six variables: education, wealth, access to piped water, flush toilet, household size, ethnicity and length of stay in the community (Table 13.4).

The exposure risk model was highly significant with a Nagelkerke R^2 of 0.584. This measure belongs to a family of statistical tests that are commonly used to assess the overall fit of regression models by examining the collective contribution of the independent variables in explaining or predicting the outcome (dependent) variable. In logistic regression analysis, variants of this test include Hosmer and Lemeshow's

Table 13.3 Toxic demographic difference indices for hazardous risk zones

	Riskzone	*N*	Mean	Std. deviation	*T* test	TDDI (1-*p*)
Current age – respondent	Low risk	912	27.55	8.637	–1.306	0.806
	High risk	796	28.12	9.404		
Length of stay	Low risk	874	9.2197	11.03404	–12.131*	1.00
	High risk	740	17.6662	15.99013		
Kwa	Low risk	912	0.8114	0.39140	0.255	0.202
	High risk	796	0.8065	0.39526		
Ahizi	Low risk	912	0.1886	0.39140	–0.124	0.099
	High risk	796	0.1910	0.39330		
Working	Low risk	912	0.6009	0.48999	–5.351*	1.00
	High risk	796	0.7224	0.44812		
Moslem	Low risk	912	0.3443	0.47540	–7.631	1.00
	High risk	796	0.5251	0.49968		
PPwater	Low risk	912	0.9167	0.27654	18.255*	1.00
	High risk	796	0.5540	0.49739		
Electric	Low risk	912	0.8914	0.31125	21.355*	1.00
	High risk	796	0.4548	0.49826		
No. of household members	Low risk	912	7.93	4.664	–4.154*	1.00
	High risk	796	8.97	5.498		
No. of children 5 and under	Low risk	912	0.99	1.071	–7.211*	1.00
	High risk	796	1.48	1.672		

*Significant at $p < 0.01$.

Table 13.4 Stepwise logistic regression of exposure risk to hazardous chemical releases

Variables in the equation	*B*	S.E.	Wald	df	Sig.	Exp(*B*)
Education			8.830	3	0.032	
higher education (reference)						
No education	0.702	0.243	8.320	1	0.004	2.017
Primary	0.606	0.248	5.958	1	0.015	1.833
Secondary	0.440	0.233	3.559	1	0.059	1.553
No. of house members	0.043	0.013	10.996	1	0.001	1.044
Access to piped water	−1.914	0.191	100.254	1	0.000	0.147
Access to flush toilet	−1.162	0.177	43.197	1	0.000	0.313
Wealth index			61.716	4	0.000	
richest group (reference)						
Poorest	21.151	2777.017	0.000	1	0.994	1.533E9
Poorer	2.879	0.373	59.677	1	0.000	17.805
Middle	0.304	0.212	2.047	1	0.153	1.355
Richer	0.169	0.174	0.940	1	0.332	1.184
Length of stay	0.036	0.006	41.269	1	0.000	1.037
KWA ethnic group	−0.904	0.200	20.425	1	0.000	0.405
Model summary						
Final step	–2 log likelihood 1,308.431		Cox and snell R^2 0.438		Nagelkerke R^2 0.584	

R^2, and Cox and Snell's R^2, all of which typically compare the baseline regression model (without the independent variables) to the final model (obtained by including the relevant predictor variables). The interpretation of the Nagelkerke R^2, and along with the related measures is consistent with the interpretation used for the traditional R^2 statistic in ordinary regression analysis, such that a value close to 0 indicates a poor model fit, whereas a value close to 1 suggests a near perfect fit. Thus, in this study, one can conclude that almost 60% of the spatial variability in exposure risk to the chemicals can be explained or predicted by the independent variables that were included in the model. Judging from the regression coefficients $\{B\}$ and related odds ratios $\{EXP\ (B)\}$, the strongest predictor was the wealth index. Those who were poor in the city faced the greatest risk of exposure to these chemicals. The odds ratio also showed that residents with little or no formal education, were twice as likely to be exposed. The odds declined with increasing education but education remained a significant predictor overall. When examining the characteristics of the household environment, residents without the basic amenities such as piped water and modern sanitation facilities were the most likely to suffer from this tragedy. The odds were only slightly higher (4%) for those who had lived in these communities longer and lived in large units with several members. Finally, the Kwa majority group faced a slightly lower risk of exposure to the toxic chemicals when compared to the other minority groups; and unlike the TDDI results, religiosity was not a significant factor in this multivariate model, a finding that is probably linked to the stepwise procedure used to create a parsimonious model with only a subset of variables that account for the most variance in exposure risk.

13.8 Discussion and Conclusions: EJ Lessons Learned from This Study

This study has sought to document the root causes underlying cross-border transfers of hazardous operations and products, and the emerging health risks in low income countries. Using the case study from Cote d'Ivoire, the results underscore many of the salient themes associated with the emerging global geographies of EJ: (i) the involvement of transnational companies; (ii) the international mobility of the wastes and related products; (iii) the leveraging of global economies of scale by profiting off of the reprocessing and sale of naptha, while embarking on cost-saving measures to dispose of the harmful waste products in a developing country; (iv) the selection of a country that was politically and economically unstable at the time of the negotiation; (v) the illicit nature of the operations involving an incompetent local subcontractor, along with the lack of full disclosure of the true composition of the wastes; and (vi) the significant health impacts on residents in the local communities. In this case study, at least four countries were involved, the most important one being the Netherlands, the headquarters of the TNC.

The use of GIS and statistically based methodologies to uncover the local geographies of EJ was perhaps the most beneficial in this study. The ALOHA dispersion models helped define the general radius of the threat zone based on the atmospheric conditions and the toxicological properties of the most dangerous chemicals. Subsequent analysis using the GIS buffers, the TDDI and logistic regression models produced results that were fairly consistent with other EJ communities around the world. Even though an entire country was targeted for the waste delivery, the localized effects of the hazard were most evident in the low income neighborhoods. Ethnic minorities and long term residents with low educational attainment and limited access to basic amenities faced greater risks of exposure than their counterparts. Further, even though the country had a fairly balanced distribution of Christians and Moslems, the TDDI results showed that Moslems were more likely to be exposed to the hazardous releases, though the variable failed to enter the final multivariate model derived from logistic regression analysis.

The study further illustrates the functionality of geographic tools and methodologies in the analytical evaluation of global EJ concerns. To our knowledge, most studies of global EJ have so far been theoretical in scope, and few have attempted to apply these tools to validate the claims. This research offers a structured approach toward addressing these issues by integrating GIS, atmospheric dispersion modeling and statistical methods. The results however must be interpreted as only a first step toward a more detailed epidemiological investigation that requires long term biomonitoring of the affected groups, environmental remediation of the polluted sites, and the implementation of environmental regulations that would prevent such incidents from recurring. It is only through these collective efforts that true environmental justice can be achieved.

Author Biographies

Florence M. Margai, Ph.D., is a Professor and Director of Graduate Studies in Geography at Binghamton University, New York. She received her doctorate degree in Geography with a concentration in Environmental Health Analysis from Kent State University, Ohio. Dr. Margai's research interests revolve around the use of geographic methodologies such as GIS and Geostatistics in the mapping and assessment of environmental health risks and negative health outcomes. She studies the distribution of global environmental health hazards, emerging and re-emergent infectious diseases (ERIDs), contaminated environments and health risks, health disparities among vulnerable populations. Her past and current research projects include studies of Malaria Morbidity and Treatment Approaches in Sierra Leone, Food Insecurity and Childhood Health in Burkina Faso, Toxic Exposures in Cote D'Ivoire, and the linkages between Lead Poisoning and Learning Disabilities in US cities. She has worked with several non-profit organizations in the U.S. and Africa to assist with the geographic targeting of vulnerable population groups for disease intervention and health promotional campaigns. Examples include the INDEPTH Network (Ghana), the Food Bank of the Southern Tier, MALAMED (Sierra Leone), and BioMedware (Ann Arbor, Michigan). Dr. Margai has authored and co-edited three books and more than 30 refereed publications.

Fatoumata B. Barry has a Dual Master's Degree in International Affairs and Natural Resources/Sustainable Development from the American University in Washington D.C. and the University for Peace in San José, Costa Rica, May 2010. She also holds a Bachelor of Arts Degree in Philosophy, Politics, and Law with a Minor in Africana Studies from Binghamton University.

Her work includes field research on environmental injustices from the oil industry in the Niger Delta region of Nigeria, specifically researching the health effects of gas flares and oil spills in host communities. She has also worked with the American Lung Association National Headquarters in Washington D.C., the Niger Delta Professionals for Development in Nigeria and the International Fund for Agricultural Development in Italy.

References

Adeola FO (2001) Environmental injustice and human rights abuse: the states, MNCs, and repression of minority groups in the world system. Hum Ecol Rev 8(1):39–59

Adeola FO (2009) From colonialism to internal colonialism and crude socioenvironmental injustice: anatomy of violent conflicts in the Niger delta of Nigeria. In: Steady FC (ed). Environmental justice in the new millennium global perspectives on race, ethnicity and human rights. Palgrave Macmillan, New York, NY

Agence France-Presse (February 17, 2007) Death toll from ICoast pollution rises to 15. http://www.reliefweb.int/rw/RWB.NSF/db900SID/YAOI-6YK47K?OpenDocument&rc=1&emid=AC-2006-000134-CIV. Last Accessed July 7, 2007

Agency for Toxic Substances and Disease Registry (July, 2006) Toxicological Profile: Hydrogen Sulfide. http://www.atsdr.cdc.gov/toxprofiles/tp114.html. Last Accessed July 10, 2007

Agency for Toxic Substances and Disease Registry (July, 2006) Toxicological Profile: Methyl Mercaptans. http://www.atsdr.cdc.gov/toxprofiles/tp139.html. Last Accessed July 10, 2007

Asante-Duah D, Kofi D, Saccomanno FF, Shortreed JH (1992) The hazardous waste trade: can it be controlled? Environ Sci Technol 26:1684–1693

Asante-Duah K, Nagy IV (2001) A paradigm of international environmental law: the case for controlling the transboundary movements of hazardous wastes. Environ Manag 27(6):779–786

Barry FB (2010) Environmental injustices: conflict and health hazards in the Niger Delta. A research paper submitted in partial fulfillment of Dual Master's Degrees in international affairs and natural resources and sustainable development (American University) 73 pp

Betsill MM, Hochstetler K, Stevis D (eds) (2006) International environmental politics. Palgrave Macmillan, New York, NY

Bullard RD (2005) The quest for environmental justice. Human rights and the politics of pollution. Sierra Club Books, San Francisco, CA

Bury J (2007) Mining and migration in the Peruvian Andes. Prof Geogr 58(3):378–389

Carruthers DV (2008) Environmental justice in Latin America. Problems, promises and practice. MIT Press, Cambridge, MA

Castleman BI, Navarro. V (1987) International mobility of hazardous products, industries and wastes. Ann Rev Public Health 8:1–19

Chaix B, Gustafsson S, Jerrett M, Kristersson H, Lithman T, Boalt A (2006) Children's exposure to nitrogen dioxide in Sweden: investigating environmental injustice in an egalitarian country. J Epidemiol Comm Health 60:234–241

Chakraborty J (2009) Automobiles, air toxics, and adverse health risks: environmental inequities in Tampa Bay, Florida. Ann Assoc Am Geogr 99(4):674–697

Chakraborty J, Armstrong MP (1996) Using geographic plume analysis to assess community vulnerability to hazardous accidents. Comput Environ Urban Syst 19(5/6):341–356

Clapp J (2001) Toxic exports: the transfer of hazardous wastes from rich to poor countries. Cornell University Press, Ithaca, NY

Coe NM, Johns J, Ward. K (2007) Mapping the globalization of the temporary staffing industry. Prof Geogr 59(4):503–520

Crowder K, Downey. L (2010) Interneighborhood migration, race, and environmental hazards: modeling microlevel processes of environmental inequality. Am J Sociol 115(4):1110–1149

Demographic and Health Surveys (2005) Cote d'Ivoire. MEASURE DHS. ICF Macro, Calverton, MD

Duckett A (2009) Trafigura story breaks. The Chemical Engineer October 13, 2009

Fairburn J, Butler B, Smith G (2009) Environmental justice in South Yorkshire: locating social deprivation and poor environments using multiple indicators. Local Environ 14(2):139–154

Frazzoli C, Orisakwe OE, Dragone R, Mantovani A (2010) Diagnostic health risk assessment of electronic waste on the general population in developing countries' scenarios. Environ Pollut 149(2):131–140

Frey RS (2003) The transfer of core-based hazardous production processes to the export processing zones of the periphery: the maquiladora centers of northern Mexico. In: Kick E, Jorgenson A (eds) Globalization and the environment. J World Syst Res 9(2):317–354

Gbadegesin S (2001) Multinational corporations, developed nations, and environmental racism: toxic waste, exploration, and eco-catastrophe. In: Westra L, Lawson BE (eds) Faces of environmental racism: confronting issues of global justice. Rowman & Littlefield Publishers, Inc., Lanham, MD, pp 187–202

Grineski SE, Collins. T (2008) Exploring patterns of environmental injustice in the global south: Maquiladoras in Ciudad Juaŕez, Mexico. Popul Environ 29:247–270

Harner J, Warner K, Pierce J, Huber. T (2002) Urban environmental justice indices. Prof Geogr 54(3):318–331

Higgs G, Langford M (2009) GIscience, environmental justice, and estimating populations at risk: the case of landfills in Wales. Appl Geogr 29:63–79

Hussein AM (2001) Environmental degradation and environmental racism. In: Westra L, Lawson B (eds) Faces of environmental racism: confronting issues of global justice. Rowman & Littlefield Publishers, Inc., Oxford, pp 203–227

Kao I, Bosley C (2009) Factbox: Trafigura, 3rd largest Independent oil trader. http://www.reuters.com/article/idUSTRE58G38R20090917. Last Accessed April 10, 2010

Kocasoy G (2003) Environmental exploitation of hazardous wastes by developing countries. J Environ Prot Ecol 4(3):587–593

Koffi C. (September 19, 2009). I.Coast Toxic dump 'still claiming lives'. http://www.physorg.com/news172591777.html. Last Accessed July 13, 2010

Landrigan PJ, Rauh VA, Galvez. MP (2010) Environmental justice and the health of children. Mount Sinai J Med 77(2):178–187

Lipman Z (2002) A dirty dilemma: hazardous waste trade. Harv Int Rev 23(4):68

Maantay J (2002) Mapping environmental injustices: pitfalls and potential of geographic information systems in assessing environmental health and equity. Environ Health Persp 110: 161–171

Margai FL (2001) Health risks and environmental inequity: a geographical analysis of accidental releases of hazardous materials. Prof Geogr 53(3):422–434

Martuzzi M, Mitis F, Forastiere F (2010) Inequalities, inequities, environmental justice in waste management and health. Eur J Public Health 20(1):21–26

McCaffrey KT (2008) The struggle for environmental justice in Vieques, Puerto Rico. In: Carruthers DV (ed) Environmental justice in Latin America: problems, promise and practice. The MIT Press, Cambridge, MA

McCurdy H (2001) Africville: environmental racism. In: Westra L, Lawson B (eds) Faces of environmental racism: confronting issues of global. Rowman & Littlefield Publishers, Inc., Oxford, pp 95–112

Mennis J (2002) Using geographic information systems to create and analyze statistical surfaces of population and risk for environmental justice analysis. Soc Sci Q 83(1):281–297

Mennis JL, Jordan L (2005) The distribution of environmental equity: exploring the spatial nonstationarity in multivariate models of air toxic releases. Ann Assoc Am Geogr 95(2):249–268

Mintin J (2006) Caustic Tank Washings, Abidjan, Cote d'Ivoire. Report submitted to Trafigura. http://image.guardian.co.uk/sys-iles/Guardian/documents/2009/10/16/mintonreport.pdf. Last Accessed June 20, 2010

Morton SC (1996) Hazardous waste trade http://www.web.net/~smorton/waste_trade.html. Last Accessed March 31, 2008

Myers GA (2008) Sustainable development and environmental justice in Africa cities. Geogr Compass 2:1–14

Phalane ME, Steady. FC (2009) Nuclear energy, hazardous waste, health and environmental justice in South Africa. In: Steady FC (ed) Environmental justice in the new millennium global perspectives on race, ethnicity and human rights. Palgrave Macmillan, New York, NY

Robinson BH (2009) E-waste: an assessment of global production and environmental impacts. Sci Total Environ 408(2):183–191

Schmidt CW (2004) Crimes: Earth's expense. Environ Health Persp 112:A97–A103

Skogseth G (2006) Cote d'Ivoire: Ethnicity, Ivoirite and Conflict. LandInfo, 1–35

Steady FC (2009) Environmental justice cross culturally: theory and praxis in the African Diaspora and in Africa. In: Steady FC (ed) Environmental justice in the new millennium global perspectives on race, ethnicity and human rights. Palgrave Macmillan, New York, NY

Su JG, Morello-Frosch R, Jesdale BM, Kyle AD, Shamasunder. B, Jerrett. M (2009) An index for assessing demographic inequalities in cumulative environmental hazards with application to Los Angeles, California. Environ Sci Technol 43(20):7626–7634

The Basel Action Network (2007) E-Waste Nigeria. http://www.ban.org/ban_news/2006/0612_dirty_business.html. Last Accessed July 7, 2007

The Basel Convention (2007) http://www.basel.int/text/documents.html http://www.rachel.org/bulletin/pdf/Rachels_Environment_Health_News_1019.pdf. Last Accessed July 7, 2007

United Nations Environmental Program (UNEP) (1992) Basel convention on the control of transboundary movements of hazardous wastes and their disposal. http://www.basel.int/text/documents.html. Last Accessed July 7, 2007

United Nations Environmental Program (UNEP) (2006a) Tackling illegal trade in ozone-depleting substances and dangerous waste – 'Project Sky Hole Patching'http://www.roap.unep.org/press/NR06-06.html. Last Accessed July 14, 2010

United Nations Environmental Program (UNEP) (2006b) United Nations Disaster and Coordination report. http://www.loe.org/images/070223/undac_civ_11sep.pdf. Last Accessed June 15, 2008

Westra L, Lawson. B (2001) Faces of environmental racism: confronting issues of global. Rowman & Littlefield Publishers, Inc., Oxford, pp 95–112

Wong MH, Wu SC, Deng WJ, Yu XZ, Luo Q, Leung AO, Wong CS, Lusemburg WJ (2007) Export of toxic chemicals – a review of the case of uncontrolled electronic-waste recycling. Environ Pollut 149(2):131–140

Chapter 14
Environment and Health Inequalities of Women in Different Neighbourhoods of Metropolitan Lagos, Nigeria

Immaculata I.C. Nwokoro and Tunde S. Agbola

Abstract Despite all the policies evolved by the various governments in Nigeria to maintain a healthy environment, inequalities in health persist among women in Lagos. This study examines the nature of the relationship between environmental health factors and health status of women in different neighbourhoods of metropolitan Lagos. All the 17 local government areas (LGAs) were selected to achieve 100% representation. Questionnaires (no = 1,150) were administered to randomly selected women aged 18 years and above. A total of 9 Focus Group Discussions (FGDs) were held with women of same age from different neighbourhoods. Data analysis was by descriptive statistics, chi-square tests, one-way ANOVA and logistic regression. GIS was employed to show the spatial variation of health status of women across neighbourhoods. Results show that the mean of environmental diseases experienced by women varied among income neighbourhoods but while the difference in means between the low and medium income groups was highly significant at $p < 0\text{–}5$, that of the medium and high income groups was not. GIS highlighted the high income neighbourhood as having women in the highest health status. The more the access to pipe borne water, the lower the incidence of diarrhea (Wald = 19.125, $p < 0.05$) Also, diarrhea increased with age, irrespective of neighbourhood location. The FGDs identified stress as a major cause of ill health among women across neighbourhood groups. The study identified various neighbourhood environmental factors that affect the health of women. Improved environmental conditions are germane to improving the health status of women in metropolitan Lagos while emphasis is placed on attending to the stressors that affect women's health.

Keywords Environmental health · Health status · Metropolitan Lagos · Neighbourhoods · Inequalities · Women

---

I.I.C. Nwokoro (✉)
Department of Urban and Regional Planning, University of Lagos, Lagos, Nigeria
e-mail: ifunanya66@yahoo.com; inwokoro@unilag.edu.ng

J.A. Maantay, S. McLafferty (eds.), *Geospatial Analysis of Environmental Health*, Geotechnologies and the Environment 4, DOI 10.1007/978-94-007-0329-2_14,

14.1 Introduction

Inequalities in health in society are often a key concern for health policy makers. Stephens et al. (1992) observed that over the last few decades, there have been many studies highlighting the health problems of the urban poor in the cities of Africa, Asia and Latin America. Yet, a review of the international literature reveals that, until recently, the extent of intra-urban differentials in social, environmental and health conditions between groups in cities has been poorly understood (World Health Organization WHO, 1994). Many global summits and international fora, as well as researchers like Wagstaff et al. (1999), and Whitehead (1999) have shown ample evidences pointing to the substantial differences in expectancy and general health conditions between rich and poor people in Europe and in most developing countries like Nigeria. They are also concerned that despite overall economic growth, the health status of women, particularly in developing countries, has not been given adequate attention.

Several major demographic shifts which began after World War II have continued unabated and have even accelerated in some regions. Accordingly, Metropolitan Lagos has experienced a very high rate of urbanization over the last years. The United Nations Commission on Human Settlements (UNCHS), 2001 estimates that by 2015, the population of Lagos will be in excess of 20 million, making it the world's third largest mega-city. Associated with this intense urbanization are observable problems and challenges posed by deteriorating housing conditions, inadequate infrastructure facilities, human and environmental poverty, and declining quality of life among other issues. Stephens (1992) observed that the process of urbanization has differential impacts on a city's environmental, socioeconomic, health and political conditions. She was particularly emphatic about the relationship of the environment to its health consequences.

According to MacArthur and Bonnefoy (1998), environmental health involves those aspects of public health concerned with the factors, circumstances, and conditions in the human surroundings that can exert an influence on health and well-being. Some environmental health factors examined include sources of water, sanitation methods, sources of energy for cooking, and drainage conditions. Usually, vulnerable groups (women, children and, the poor) are most strongly affected (Population Reference Bureau, USA, 2004).

The major concern of this research, therefore, is to unravel the observable health inequalities existing among women of different socio economic groups living in different neighbourhoods. This investigation will lead to the understanding of the nature of environmental health inequalities that exist among women in Metropolitan Lagos. It further investigates if differences between women in health status vary by their social status and also examines the environmental factors that affect women's health in the study area. In addition, the research tries to show spatially the health inequalities across the socio economic neighbourhoods of women in Metropolitan Lagos.

14.2 Context of the Research – Metropolitan Lagos

The study area is located in the South Western part of Nigeria on the narrow coastal plain of the Bight of Benin. The metropolitan area, centrally located within the coastal frontage of Lagos state, comprises seventeen (17) Local Government Areas (LGAs) (Fig. 14.1) while the non metropolitan area comprises 3 LGAs. The wards are classified into low, medium and high income groups. The Lagos metropolitan area comprises 88.7% of Lagos state, a total of 19.87 km^2. Formerly the capital of Nigeria, Lagos is a huge metropolis which originated on islands separated by creeks, such as Lagos Island.

Lagos is the most populous conurbation in Nigeria with 7,937,932 inhabitants at the 2006 census (Federal Government of Nigeria official gazette, 2007). It is currently the second most populous city in Africa, and also estimated to be the second fastest growing city in Africa and the 7th fastest in the world (City Mayors), http://en.wikipedia.org/wiki/Lagos-cite_note-2 immediately following Bamako, the capital city of Mali, Africa. The city is the economic and financial capital of Nigeria. "Such a growing population, especially coupled with weak urban planning and management machinery is one of the main causes of environmental degradation. The growing population and increasing living standards are often accompanied by pressure on existing infrastructural facilities like housing, water, electricity and telecommunication" (State of Lagos Mega city, 2004, p. 40).

The State of Lagos Mega City Report (2004) notes that residential density is as high as 4,000 people per ha in several parts of the city due to its small land area harbouring a large population. It further reveals that 53.3% of poor neighbourhoods in Lagos depend on water from wells, while only 10.5% of same group of people have access to pipe borne water. The rest of these groups of people purchase water from water vendors. Due to the absence of pipe borne water supply in the low income areas of Lagos, 55.5 and 32.5% of residents still use pit and bucket latrines respectively. Only 10% of residents have access to water closet toilet systems. These are very critical issues for health.

It is important to highlight some of the major health indicators of Lagos State. According to Lagos State Government (2002), state health institutions (government general hospitals and health centres) are 30 in number, malaria is the most common disease with about 60% of out patient visit to health centres, Infant mortality rate is 85 deaths per 1,000 live births (LB), Under- five mortality rate is 150 deaths per 1,000 live births, Maternal mortality rate is 650 deaths per 100,000 live births, Immunization coverage is 40% of children and HIV/AIDS prevalence is 6.6%.

Another problem of urbanization in Lagos and indeed the whole of Nigeria is that of urban poverty. Poverty is defined as those living on less than US$1.00 per day. The United Nations Development Programme (UNDP), 2003 found that the average levels of urban poverty in Lagos were 50% for men and 56% for women. One of the major reasons adduced for this is that women in Nigeria find it easier to get jobs in the informal than in the formal sector, most of which are of low incomes and low productivity. This is another concern of this research.

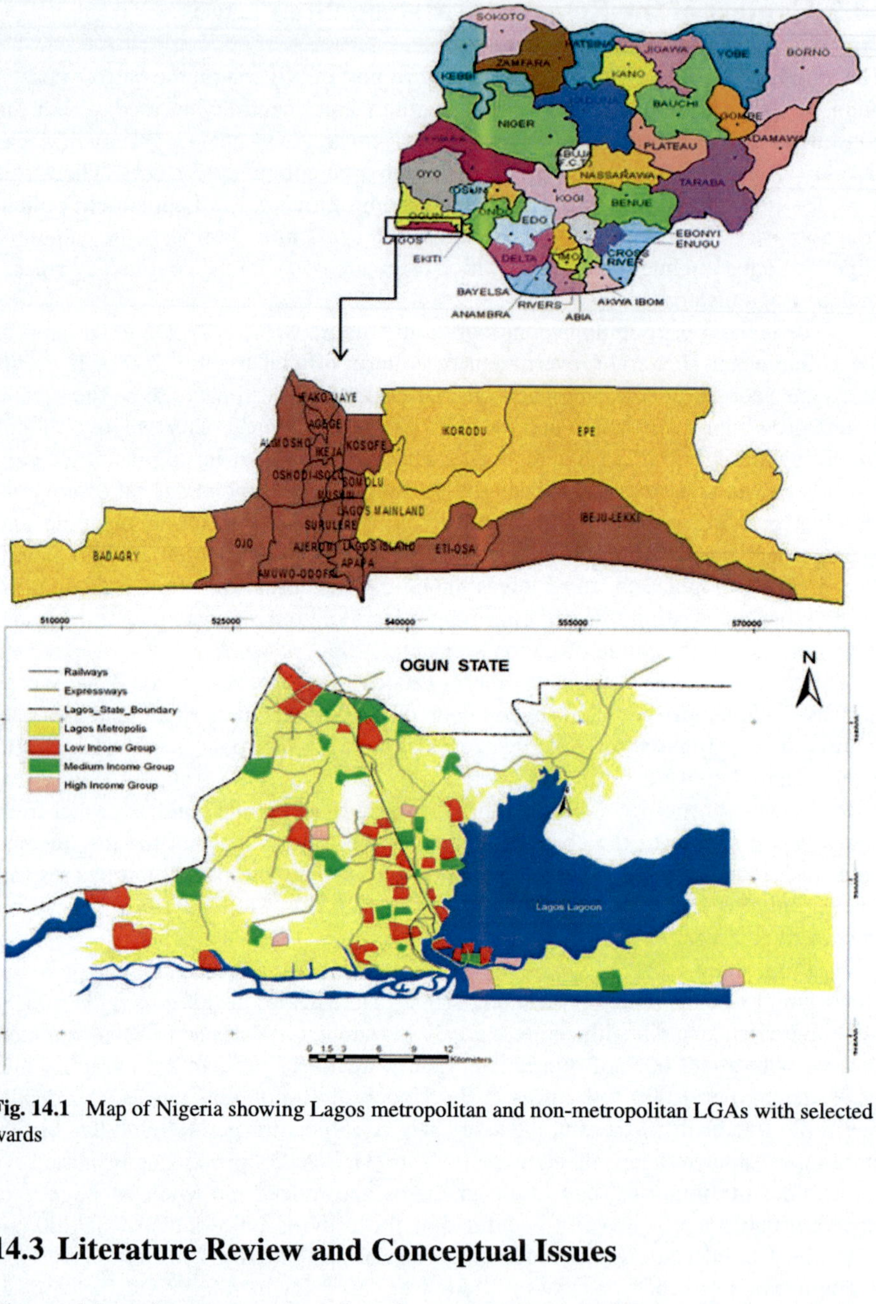

Fig. 14.1 Map of Nigeria showing Lagos metropolitan and non-metropolitan LGAs with selected wards

14.3 Literature Review and Conceptual Issues

The models and concepts adopted as framework for investigating the factors influencing the health of women include "The Health Field Model" (Evans and Stoddart, 1990), "Feminist Political Ecology" (Rocheleau et al., 1996) and the Concept of "Equity and Health" (WHO, 1999; Gwatkin, 2002).

The Health Field Model, proposed by Evans and Stoddart (1990) provides a broad conceptual framework for considering the factors that influence health in a community. It is different from a biomedical model that views health as merely the absence of disease. The health field model, in addition to including functional capacity and well-being as health outcomes, also emphasizes general factors that affect many diseases or the health of large segments of the population rather than specific factors that account for small changes in health at the individual level. The authors note the factors that contribute to differences in heath status as social, environmental, physical, economic, behavioral, and genetic factors. The model established that health outcomes are the product of complex interactions of factors rather than of individual factors operating in isolation.

Although this model provides a good understanding of the research topic in terms of environmental and socio-economic factors that influence health outcomes for large groups of people, it fails to address the gender aspect of this research. Kawachi et al. (1999) define gender as "a social construct regarding culture-bound conventions, roles, and behaviors for, as well as relations between and among, women and men and boys and girls". Annandale and Hunt (2000) observed that gender inequalities in health have been a major area of sociological research interest since the early 1970s. They also stated that "rising to prominence on a wave of interest in the social relations of gender which challenged the empirical, theoretical and methodological core of sociology during the 1970s and early 1980s, the search for an explanation for differences in male and female morbidity and mortality, alongside interest in the relationship between variations in women's social circumstances and their health has been a vital part of feminists attempts to challenge the detrimental effects of patriarchy on women's health." (Annandale and Hunt, 2000, p. 1) According to the authors, by the 1980s, the focus of research changed to women-only samples and works that examined differences in health among women. Still, few of these studies embraced the concept of gender as something other than biological conditions. The research challenge, therefore, is to explore the ways in which gender continues to be an important representation of inequality, while recognizing the diversity of experiences within the category of women.

The aspect of health inequality most relevant to this study is spatial inequalities in health. According to Graham (2000), inequalities between places are marked by inequalities between individuals. However, Macintyre (1997) argues that areas have an effect on spatial inequalities in health, although individual factors are the primary cause. In operational terms, pursuing equity in health can be defined as striving to eliminate disparities in health between more and less advantaged social groups, that is, groups that occupy different positions in a social hierarchy (Gwatkin, 2002). This is useful to this work in terms of different socio-economic groups of women as identified in Metropolitan Lagos. The most important issues in discussing equity are health status, allocation of resources, and access to and utilization of services. Philips (1990) observed that there are apparent disparities in the geographical distribution of health facilities, for example, between regions, between urban and rural areas, between rural areas and within urban areas. This study is related to the disparity within urban areas in distribution of health facilities. The study is also concerned with the type and level of access of women from different socio-economic groups to health care facilities.

Coming from the feminist perspective, another model that appears useful in explaining these relationships is "Feminist Political Ecology" by Rocheleau et al. (1996) which deals with the complex context in which gender interacts with class, race, culture and national identity to shape our experience of and interests in the environment. "While there are several axes of power that may define people's access to resources, their control over their workplace and home environments and their definitions of a healthy environment, we focus on gender as one axis of identity and difference that warrants attention" (Rocheleau et al., 1996, p. 5). The major flaw of this theory to this study is that it has not considered the relationship between the environment of women and health. At this point, we may begin to tease out the relationships between society, place, gender and health and how they play out in the world, particularly as the focus on women's health begins to move away from medicine. From the above debates on the link between gender, environment and inequality in health, it does appear that there is no one single model or concept that encapsulates these complex relationships. Nevertheless, each aspect of this discourse has presented a rich understanding of this research. For a more comprehensive outlook suitable for this research, the Health Field Model, was modified by adding gender and class to the determinants of health and changing some of the arrows to be in double direction as shown in Fig. 14.2.

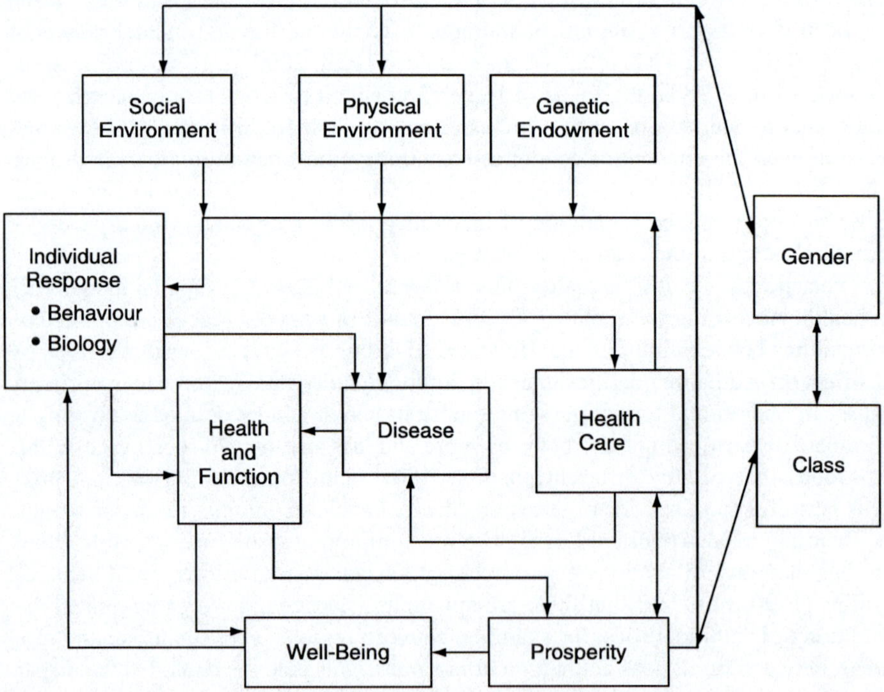

Fig. 14.2 A model of the determinants of health, modified after Evans and Stoddart (1990)

Environmental health practice addresses emerging health risks arising from the pressures that human development places on the environment. World Health Organization (WHO), and other national and international organisations indicate that the interaction between environment and health is far more complex than is commonly understood. As observed by Bradley et al. (1992), around 25–33% of the burden of disease in industrialised countries can be attributed to environmental factors, with the bulk of these affecting children and vulnerable groups especially women. For example, in a 1999 survey, the WHO observed that some 89% of people are concerned about the potential impact of the environment on their health. Some researches confirm the influence of environmental health factors on health. Guerrant et al. (1983) observed that diarrhoea rates at all ages in 297 study participants were statistically significantly higher in poor urban and poor rural areas of Brazil than the non-poor central urban area. They also noted that diarrhoea risks were 2.2 times higher for children in households without pit toilets, compared to those with pit toilets. Also a study from Monrovia, Liberia by Molbak et al. (in Bradley et al., 1992) concludes that water related environmental health problems in urban areas are linked to storage and hygiene factors as well as to safe water variables. Similarly, Surjadi et al. (1993) and McGranahan and Kjellen (1994), in a household survey undertaken in Jakarta found indoor air pollution from cooking, crowding, humidity and poor ventilation to be important factors for predicting respiratory illness, particularly among female householders.

All the above findings suggest that more work need to be done in understanding the impact of the environment on health of different socio economic classes of women, especially, in Metropolitan Lagos.

14.4 Research Methods

Qualitative and quantitative methods of data collection were used in this study. Focus Group Discussions (FGDs) were held with women of age 18 years and above in three local government areas (LGAs) four groups from the low income areas, three from the medium and two from the high. Each group consisted of ten women drawn from different wards, educational, social and professional backgrounds. Also each group was met four times to ascertain consistency of comments. The choice of the number of groups from each local government area was informed by the findings in the literature which showed that women from the low income group are more vulnerable to environmental health problems.

The household questionnaire was structured into eleven sections where information were elicited on variables like age, Income in Naira (N) per month, family size, level of education, occupation, housing conditions, environmental health conditions, nutritional values, health status of the women and policy issues. The study required the classification of the study area into income neighbourhoods as follows; Low income/high density (LI/HD), Medium income/medium density (MI/MD), High income/low density (HI/LD). The income indicators used here are socio-economic characteristics of neighbourhoods for example, type of accommodation (rooming, flat, bungalow, duplex, etc); crowded compounds (accommodating 6 or more households); and monthly income of household (those earning less than the minimum wage of N7,500 per month). The neighbourhoods were then classified according to an urban pathology index (UPI) based on percentage scores on socio economic characteristics as mentioned above. These variables agree with those used as indicators of socio economic status in Urban Ecology Analysis (Berry and Horton, 1970; Timms, 1971). This could not be achieved effectively at the local government level because they are too heterogeneous and large. Some LGAs have some mixed income neighbourhoods. To avoid this problem, the classification was done at the ward level which is a smaller unit with more homogenous characteristics.

All the 17 Metropolitan LGAs were selected to achieve a 100% spatial representation of the study area from where 72 wards were selected for the study. Selection of wards from each local government area was based on proportional representation, both in size and population. Out of the 72 wards, 43 were LI, 22 MI and 7 HI neighbourhoods. At the household level, 1,150 copies of a questionnaire were administered to randomly selected women aged 18 years and above.

Health survey records were obtained from the Lagos State Ministry of Health, LGAs and one hospital within each sampled ward. Maps of Nigeria showing Lagos State in various contextual details, were all transferred from the analogue to digital state using Geographic Information System (GIS). These maps did not only describe the study area but were useful in the spatial analysis of data from the field. The political map of Nigeria (1:100,000) by Federal Surveys (1968) was scanned, geo-referenced and vectorized using ArcGIS 9.2 software by ESRI. The Lagos state map (1:25,000) was also scanned, geo-referenced and digitized. Features such as LGAs boundaries, water bodies and wards (in polygons) were extracted as themes. Roads, railways and state boundaries were digitized as polygons. Attributes tables were then updated to accommodate the questionnaire survey based on ward level. Query builder was used to obtain results shown in Figs. 14.3, 14.4, 14.5, and 14.6.

Data analysis was done using descriptive statistics (tables, graphs, mean), chi-square tests, one-way ANOVA and logistic regression analysis. This study utilised the logistic regression model to determine the risk of occurrence of environmental health diseases using some environmental factors. The data sets for the study which are in binary variables also justify the use of logistic regression. Geographic Information Systems (GIS) was employed to show the spatial variation of health status of women across various income groups.

14.5 Results and Discussions

This section discusses the major findings of the research. The subsections include socio economic conditions, health conditions, neighbourhood environmental conditions of women in metropolitan Lagos as well as analysis of Focus Group Discussions. The major limitation of the GIS research results is that they are based on percentage of total number of respondents instead of the actual number.

14.5.1 Socio Economic Conditions of Women

It is interesting to find a substantial number of women in the low and medium income groups earning above N45,000 per month due to joint earnings from two jobs. There is also a significant difference between the earning capacities of income of the women across the income groups. The results are shown in Tables 14.1 and 14.2. Other socio economic conditions considered in the analysis are age and educational attainment of the women. However, they are discussed along with the health conditions of women in the next sub section to ascertain how they affect their health status. Figure 14.3 also shows the spatial distribution of women that earn less than N15,000 monthly. Most of the women are in the low income group.

Table 14.1 Relationship between (*N*) per month and income classification (%)

Income (*N*)	LI	MI	HI
< 15,000	40.1	18.4	5.0
15,000–30,000	27.2	14.4	7.5
30,001–45,000	8.6	18.4	15.0
> 45,000	24.1	48.9	72.5
Total	100	100	100
	(732)	(305)	(40)

Source: Field work, 2006.
LI, low income, MI, medium income, HI, high income.

Table 14.2 Chi-square test for income (*N*) per month and income classification (%)

	Value	Df	Asymp sig. (2-sided)
Pearson chi square	79.777	6	0.000

Source: Field work, 2006.

14.5.2 The Health Conditions of Women in Metropolitan Lagos

According to Newbold (1999), a major barrier for women in the achievement of the highest attainable standard of health is inequality, both between men and women and among women in different geographical regions, social classes and indigenous and ethnic groups. This section discusses the health conditions and status of women in different income groups. It also examines inequality of health among women of different income groups and investigates the rationale for the differences. The variables considered for this analysis are: disease pattern of women, most frequently experienced diseases, and mortality rate.

Table 14.3 reveals that the most reported disease by the low income women is diarrhoea, while respiratory infection was the most reported by the medium and high income women. The analysis of the secondary based data from hospitals and Lagos

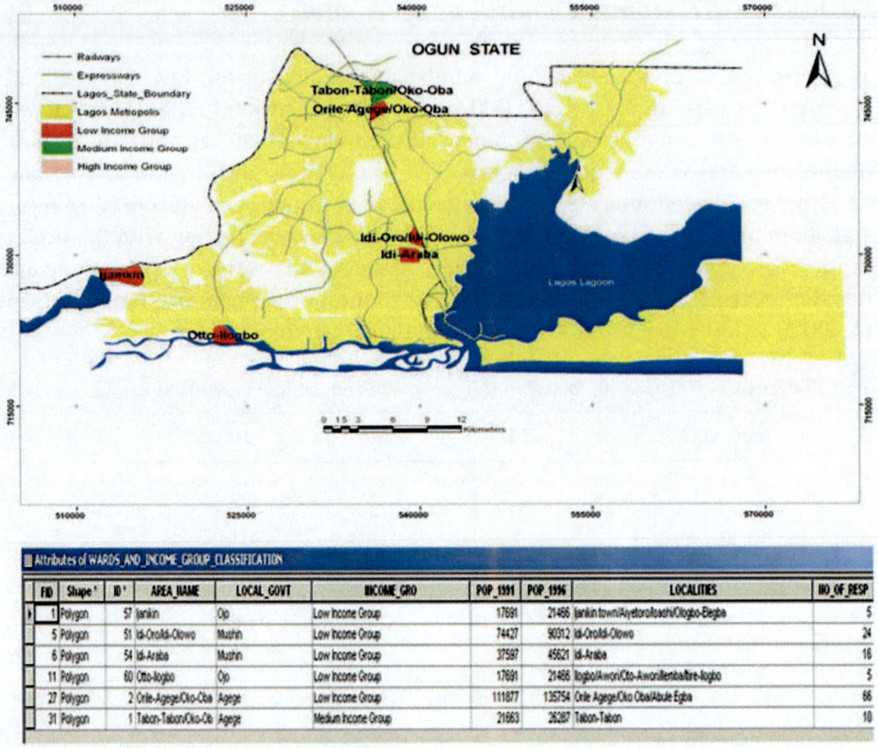

FID	Shape*	ID*	AREA_NAME	LOCAL_GOVT	INCOME_GRO	POP_1991	POP_1996	LOCALITIES	NO_OF_RESP
1	Polygon	57	Ijanikin	Ojo	Low Income Group	17691	21466	Ijanikin town/Aiyetoro/Isashi/Ologbo-Elegba	5
5	Polygon	51	Idi-Oro/Idi-Olowo	Mushin	Low Income Group	74427	90312	Idi-Oro/Idi-Olowo	24
6	Polygon	54	Idi-Araba	Mushin	Low Income Group	37597	45621	Idi-Araba	16
11	Polygon	60	Otto-Ilogbo	Ojo	Low Income Group	17691	21466	Ilogbo/Awori/Oto-Awori/Ilemba/Ire-Ilogbo	5
27	Polygon	2	Orile-Agege/Oko-Oba	Agege	Low Income Group	111877	135754	Orile Agege/Oko Oba/Abule Egba	66
31	Polygon	1	Tabon-Tabon/Oko-Ob	Agege	Medium Income Group	21663	26287	Tabon-Tabon	10

Fig. 14.3 Result of query showing wards with women earning less than N15,000 per month (>= 80%)

Table 14.3 Disease prevalence of women according to income groups in the past 1 year (%)

Type of disease	L1	M1	H1
Malaria	65.4	40.0	62.5
Diarrhoea	74.9	65.6	37.5
Dysentery	58.3	51.5	32.5
Typhoid fever	70.1	71.1	55.0
Cholera	35.8	25.9	20.0
Respiratory infection	58.5	75.4	62.5
Tuberculosis	33.6	14.8	30.0
Others	37.8	43.9	42.5

Source: Authors field work, 2006.
Note: Multiple responses possible.

State Ministry of Health show that the most reported diseases are malaria, respiratory diseases and typhoid fever. These confirm the results from the questionnaire interviews. Figure 14.4 further highlights the spatial analysis of malaria prevalence, that is women that have suffered from this disease at more than four times in the year. The GIS results show that malaria prevalence cuts across all income groups.

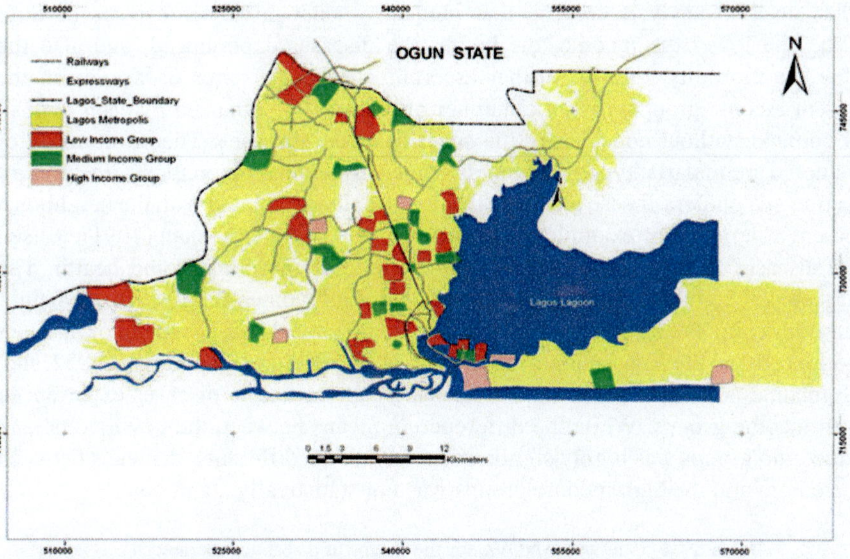

Attributes of WARDS_AND_INCOME_GROUP_CLASSIFICATION

FID	Shape*	ID*	AREA_NAME	LOCAL_GOVT	INCOME_GRO	POP_1991	POP_1996	LOCALITIES	NO_OF_RESP
13	Polygon	44	Idumota	Lagos Island	Low Income Group	3311	4018	Idumota	2
14	Polygon	45	Agbarawu/Obadina	Lagos Island	Low Income Group	21697	26328	Agbarawu	7
15	Polygon	21	Obalende	Eti-Osa	Low Income Group	29716	36058	Obalende	1
16	Polygon	46	Epetedo	Lagos Island	Low Income Group	56495	68553	Epetedo/Oke Popo	13
17	Polygon	33	Old Ifako-Karaole	Ifako-Ijaiye	Low Income Group	48816	59235	Ifako Agege	24
18	Polygon	34	Alakuko/Kollington	Ifako-Ijaiye	Low Income Group	16235	19700	Alakuko	3
19	Polygon	32	Parade-Abule-Egba	Ifako-Ijaiye	Low Income Group	17438	21159	Abule-Egba/Iupeju-Titun	8
20	Polygon	50	Otto/Iddo	Lagos Mainland	Low Income Group	13673	16591	Otto/Ido	3
22	Polygon	61	Bajulaiye	Shomolu	Low Income Group	99948	121279	Bajulaiye/Morocco/Shomolu	53
23	Polygon	62	Abule-Okuta/Ilaje/Barig	Shomolu	Low Income Group	95990	116477	Ifako bariga/Abule Okuta/Ilaje	33
24	Polygon	41	Ketu/Alapere	Kosofe	Low Income Group	111867	135742	Ketu/Alapere	47
25	Polygon	10	Ikotun/Ijegun	Alimosho	Low Income Group	58817	71370	Olayemi/Ikotun/Ijegun	20
26	Polygon	11	Shasha/Akowonjo	Alimosho	Low Income Group	51564	62569	Shasha/Akowonjo	25
27	Polygon	2	Orile-Agege/Oko-Oba	Agege	Low Income Group	111877	135754	Orile Agege/Oko Oba/Abule Egba	66
28	Polygon	3	Isale Oja/Idimangoro	Agege	Low Income Group	30495	37004	Isale Oja/Idimangoro	27
29	Polygon	4	Oyewole/Papa Ashaf	Agege	Low Income Group	122158	148225	Moricas/Papa Ashafa/Oyewole/Alagba/Mulero	66
30	Polygon	31	Ogba-Oke-ira	Ifako-Ijaiye	Medium Income Group	56578	68654	Ijaiye/Oke-ira/Ijaiye-Ogba	19
32	Polygon	37	Adeniyi Jones/Ogba	Ikeja	Medium Income Group	7264	8814	Ogba	1
33	Polygon	7	Ipaja North	Alimosho	Medium Income Group	57035	69208	Ipaja/Mosan/Abesan/Akinogun	21
34	Polygon	69	Ajao Estate	Oshodi/Isolo	High Income Group	11919	14463	Ajao Estate	2
36	Polygon	55	Iba Estate	Ojo	Medium Income Group	7263	8813	Iba	5
37	Polygon	70	Isolo	Oshodi/Isolo	Medium Income Group	44860	54434	Isolo	10
38	Polygon	35	Agidingbi/Omole/Ojodu	Ikeja	Medium Income Group	33950	41196	Ojodu/Agidingbi/Omole/Alode	16
39	Polygon	40	Isheri/Olowo Ira	Kosofe	Medium Income Group	24680	29946	Isheri/Magodo/olowo-Ira/Shangisha	11
40	Polygon	14	Festac II	Amuwo-Odofin	High Income Group	0	0	Festac II	18
41	Polygon	13	Festac I	Amuwo-Odofin	Medium Income Group	0	0	Festac I	18
42	Polygon	36	Onilekere	Ikeja	Medium Income Group	16663	20220	Onilekere/Akowonjo/Mangoro	6
43	Polygon	38	Onigbongbo/Military C	Ikeja	High Income Group	6410	7782	Maryland estate/Onigbongbo	3
44	Polygon	39	Anthony/Mende	Kosofe	Medium Income Group	24638	29896	Mende/Anthony	5
45	Polygon	52	Ilupeju	Mushin	Medium Income Group	45880	55673	Ilupeju/Palmgrove	5
46	Polygon	64	Gbagada Phase I/Oba	Shomolu	Medium Income Group	12024	14590	Gbagada/Obanikoro	6
47	Polygon	48	Iwaya/Onike	Lagos Mainland	Medium Income Group	36410	44181	Iwaya/Onike	4
48	Polygon	63	Akoka/Ilaje	Shomolu	Medium Income Group	23926	29032	Akoka	2
49	Polygon	47	Yaba/Igbobi	Lagos Mainland	Medium Income Group	39007	47332	Fadeyi/Yaba	22
50	Polygon	67	Aguda	Surulere	Medium Income Group	50563	61355	Aguda	22
51	Polygon	68	Adeniran Ogunsanya	Surulere	Medium Income Group	118752	144097	Surulere	64
52	Polygon	43	Onikan/Okesuo	Lagos Island	Medium Income Group	7245	8791	Araromi odo	3
53	Polygon	23	Victoria Island	Eti-Osa	High Income Group	40558	49215	Victoria Island/Lagoon City	18
54	Polygon	24	Ikoyi I	Eti-Osa	High Income Group	34776	42198	Ikoyi	11
55	Polygon	22	Ikesan Housing Estate	Eti-Osa	Medium Income Group	12475	15137	Ikesan/Tepa/Igbara/Maiyegun/Osapa/ Osapa London	21
56	Polygon	25	Lekki I	Ibeju-Lekki	High Income Group	1892	2296	Oso/koro/Ebute Lekki/Lekki	5
57	Polygon	19	Iganmu	Apapa	Medium Income Group	33765	40971	Iganmu	10

Fig. 14.4 Result of query showing malaria prevalence (more than 4 times per year) amongst women in different income groups in Lagos metropolis

The detailed analysis confirm that inequality exist in certain aspects of their health. The lower the income, the higher the diseases experienced, and also the higher the mortality rate. To further ascertain the health status of women in the different income groups, first, the number of diseases experienced by each woman was counted without considering the severity of the diseases. The co-morbidities considered are malaria, typhoid fever, diarrhoea, dysentery, tuberculosis, respiratory infection and cholera. These are those that are usually associated with the neighbourhood environment. For example, Benneh et al. (1993); McGranahan (1991) utilised these diseases in their various works on household environment and health. The next step was to find the mean of all diseases for all women in all the three neighbourhoods using the ANOVA method as shown in Table 14.4. The mean of diseases experienced for the low income group is 4.3434, medium income is 3.8852 and, high income is 3.4250. Also, Table 14.5 shows the means of diseases experienced for all income groups. While the difference in means between the low income and the medium groups was highly significant $p < 0.05$, the difference in means between the medium and the high income groups was not statistically significant.

Table 14.4 One-way ANOVA for the means of diseases experienced

Income group	No. of respondents	Mean of diseases
LI	731	4.3434
MI	305	3.8852
HI	40	3.4250
Total	1,076	4.1794

Table 14.5 Post hoc tests – mean differences

Income group	Income classification	Mean difference	Significance
LI	MI	0.4581[a]	0.001
	HI	0.9184[a]	0.007
MI	LI	–0.4581[a]	0.001
	HI	0.4602	0.307
HI	LI	–0.9184[a]	0.007
	MI	–0.4602	0.307

Source: Authors field work, 2006.
[a]Mean difference is significant at 0.05 level.

The next step was to classify the means as previously discussed to get three classes of women. Those with low disease experience, or good health, medium disease experience or fair health and high disease experience or poor health. To find out the health status of women in Metropolitan Lagos, a cross-tabulation was done for health status and income classification. Results also confirmed that there is no significant difference between the health status of medium and high income women.

One out of every 3 women has poor health status in the low income group while only 10% of women in the high income group have poor health status. However, results from the GIS query in Fig. 14.5 show that women with the poorest health (above 80%) are mostly in the low and medium income groups. Consequently, there is ample evidence that inequality exists in health of the women in Metropolitan Lagos.

Relating the health status of women and their socio economic conditions, results show that women who are in the best health status fall within ages 21 and 40 considered the most reproductive and productive (See Tables 14.6 and 14.7). However, age

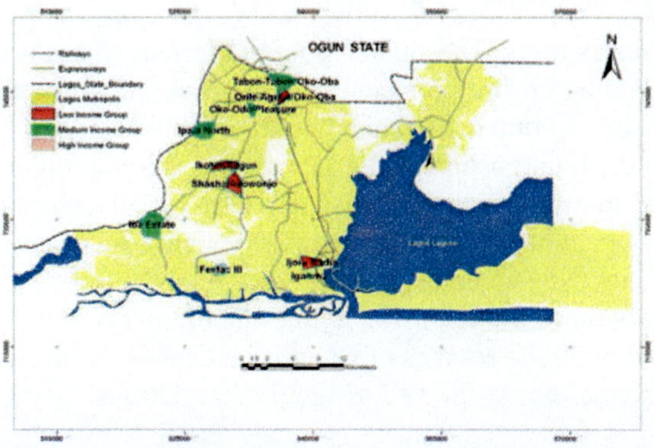

Attributes of WARDS_AND_INCOME_GROUP_CLASSIFICATION

ID	AREA_NAME	LOCAL_GOVT	INCOME_GRO	POP_1991	POP_1996	LOCALITIES	NO_OF_RESP
20	Ijora Badia	Apapa	Low Income Group	55617	71890	Ijora Badia/Oloye	10
10	Ikotun/Ijegun	Alimosho	Low Income Group	58817	71370	Olayemi/Ikotun/Ijegun	20
11	Shasha/Akowonjo	Alimosho	Low Income Group	51564	62589	Shasha/Akowonjo	25
2	Orile-Agege/Oko-Oba	Agege	Low Income Group	111877	135754	Orile Agege/Oko Oba/Abule Egba	66
1	Tabon-Tabon/Oko-Ob	Agege	Medium Income Group	21663	26287	Tabon-Tabon	10
7	Ipaja North	Alimosho	Medium Income Group	57035	69208	Ipaja/Mosan/Abesan/Akinogun	21
8	Oko-Odo/Pleasure	Alimosho	Medium Income Group	110533	134124	Ogbomosho/Oke-Odo/Oke-odo/Agege	1
55	Iba Estate	Ojo	Medium Income Group	7263	8813	Iba	5
14	Festac II	Amuwo-Odofin	High Income Group	0	0	Festac II	18
19	Iganmu	Apapa	Medium Income Group	33765	40971	Iganmu	10

Fig. 14.5 Result of query in GIS showing wards with higher (> 80%) poor health status

Table 14.6 Relationship between the health status and age of all women (%)

Health status	<20	21–30	31–40	41–50	51–60	>60
Good	16.3	21.1	19.1	16.1	6.9	17.2
Fair	64.9	55.0	47.8	48.4	53.4	63.5
Poor	18.8	23.9	33.0	35.5	39.7	19.3
Total	100	100	100	100	100	100

Table 14.7 Chi-square test for health status and age for all women (%)

	Value	Df	Asymp sig. (2-sided)
Pearson chi square	33.829	10	0.000

Source: Authors field work, 2006.

has a significant relationship with the health status of women only in the low and medium income groups, while level of educational attainment shows no significant relationship in any of the income groups.

14.5.3 Neighbourhood Environmental Conditions of Women in Metropolitan Lagos

One of the greatest challenges to environmental health research is the fact that environmental contributors to diseases are multi-faceted and not limited to one specific agent. The environmental health variables measured include, sources of water supply, drainage condition, type of toilet facilities, sources of energy for cooking and, nature of building (Ventilation). The null hypothesis tested in this section is that there is no significant relationship between neighbourhood environmental factors and the health conditions of women. Chi square tests and logistic regression analyses were used to buttress the relationships and establish variations within and between the income groups. A summary of the Chi – square tests for all environmental variables and income classification (sig. <0.05) show very significant associations. This means that the different environmental variables differ between the income groups and consequently the women's choices of living environments vary with income level. For example, out of all the sources of water investigated, the use of well water and water from hawkers at home showed a significant relationship with the health status of all women.

Six major sources of water were used for the study The major sources investigated are stream/pond, well, piped water, borehole, street hawker and other sources (not specified). These were further regrouped into two for the purpose of this discussion. Thus, if stream/pond, well and street hawker, it means a poor source but otherwise good. Results in Table 14.8 reveal that the major source of water used by the low income respondents at home is street hawker which accounts for 77.5%. These are the people who get water for residents in gallons or tins for a fee. The sources of the water are not usually known and the water is usually contaminated before it gets to the final user. Only 37.4% of the respondents use piped water. For medium income respondents, 51.4% use well, 51.8% piped water and 49.8% street

Table 14.8 Relationship between sources of water supply at home and income classification (%)

Source of water	LI	MI	HI
Stream/pond	44.5	33.8	17.5
Well	70.5	54.4	35.0
Piped water	37.4	51.8	87.5
Borehole	25.1	36.4	85.0
Street Hawker	77.5	49.8	42.5
Others	33.6	16.4	22.5

Source: Authors field work, 2006.
Note: Multiple responses possible.

hawker. However, 87.5% of the high income respondents have access to piped water, 85.0% use borehole and 42.5% still patronize street hawkers. 33.1% of women that use well water at home have poor health status. In summary, while piped water and borehole sources are used mostly by the high income respondents at home, well, piped water and street hawkers by the medium income, the low income respondents use mainly the street hawkers and well sources.

Another environmental factor considered was drainage condition. Only 3 wards in the low income neighbourhood had poor drainage facilities as highlighted in Fig. 14.6. Most of the drainage facilities across all income neighbourhoods had fair drainage conditions.

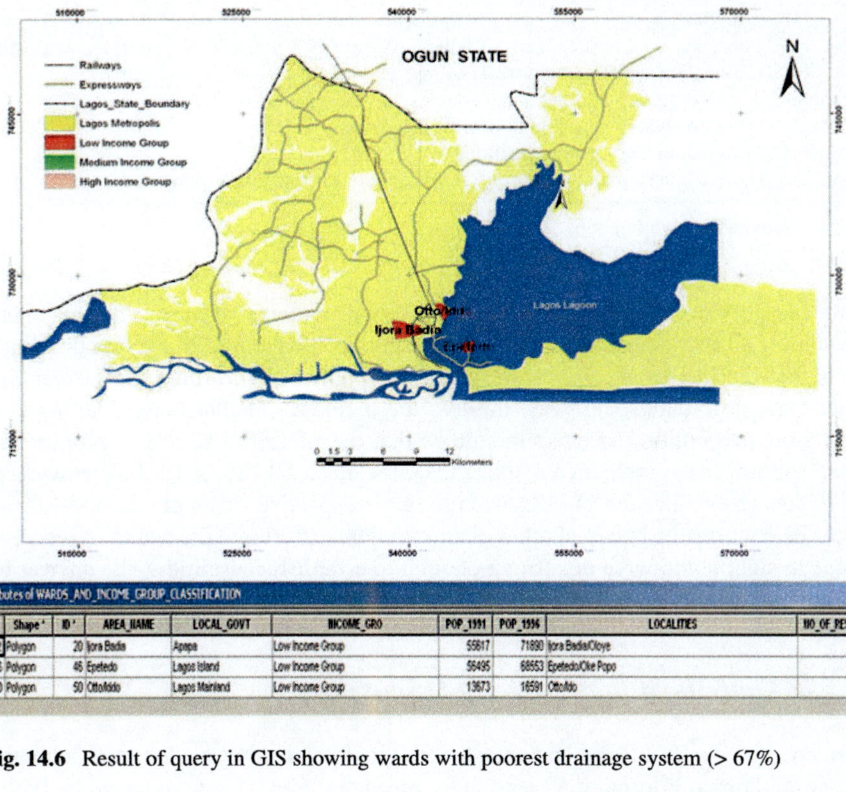

ttributes of WARDS_AND_INCOME_GROUP_CLASSIFICATION

FID	Shape*	ID*	AREA_NAME	LOCAL_GOVT	INCOME_GRO	POP_1991	POP_1996	LOCALITIES	NO_OF_RESP
12	Polygon	20	Ijora Badia	Apapa	Low Income Group	55617	71890	Ijora Badia/Oloye	10
16	Polygon	46	Epetedo	Lagos Island	Low Income Group	56495	68553	Epetedo/Oke Popo	13
20	Polygon	50	Otto/Iddo	Lagos Mainland	Low Income Group	13673	16591	Otto/Ido	3

Fig. 14.6 Result of query in GIS showing wards with poorest drainage system (> 67%)

To ascertain whether diarrhoea depends on any of the environmental factors earlier mentioned, a logistic regression analysis was conducted. The parameter estimates table summarizes the effect of each predictor. The ratio of the coefficient to its standard error, squared, equals the Wald statistic. If the significance level of the Wald statistic is small (less than 0.05) then the parameter is useful to the model. The predictors and coefficient values shown in the last step are used by the procedure to make predictions. The logistic regression results showed in Table 14.9 established

Table 14.9 Logistic regression for all income group diarrhoea as the dependent variable against all water sources, amount spent monthly on water, toilet facilities, method of excreta disposal, access to nearest piped borne water and distance of well from septic tank variables

Variables in the equation

		B	S.E.	Wald	df	Sig.	Exp(*B*)
Step 5(e)	Stream water source	1.351	0.654	4.272	1	0.039	3.861
	Other water sources	0.570	0.218	6.808	1	0.009	1.768
	Access to nearest piped borne water	–0.245	0.056	19.125	1	0.000	0.782
	Toilet facilities	–0.078	0.033	5.727	1	0.017	0.925
	Diarrhoea as a result of poor excreta disposal method	0.110	0.020	31.248	1	0.000	1.116
	Constant	–1.868	0.780	5.738	1	0.017	0.154

a Variable(s) entered on step 1: stream water source
b Variable(s) entered on step 2: other water sources
c Variable(s) entered on step 3: access to nearest piped borne water
d Variable(s) entered on step 4: toilet facilities
e Variable(s) entered on step 5: diarrhea as a result of poor excreta disposal method

Source: Authors field work, 2006.

some causality between diarrhoea and some environmental factors. The most significant for all the income groups is that the poorer the method of excreta disposal, the higher the incidence of diarrhoea. For the high income group, the farther the well from septic tank/soak away, the less the incidence of diarrhoea. The logistic regression also confirmed existing studies that the more the access to pipe borne water, the lower the incidence of diarrhoea, (Wald $= 19.125$, $p < 0.05$). However, the incidence of diarrhea increased with age irrespective of neighbourhood location. The relationship between other environmental factors and the health of women, although significant, were not strong enough to establish causalities. The answer to this emerged from the FGDs.

14.5.4 Analysis of Focus Group Discussions

In an attempt to authenticate and corroborate the information from the questionnaire survey, additional information was sought through the FGD, a time tested and reliable method in healthcare research and investigation. The discussions with diverse groups of women have produced a wealth of information about the ways in which different women perceive the environments they inhabit. The diversity of responses among and within groups suggests that the environmental influences on women's health are far more complex than is generally recognized. Since the research was focused on women, all participants were women. This was to give them the freedom of speech and expression, especially in the discussion of sensitive issues on diseases concerning them.

Perhaps the most interesting result on the health condition of the women that manifested from the FGDs is stress. Women from all the three groups said that stress, in addition to bad environment, affects their health. They further said that although the condition of the environment can affect their health, there are other associated factors of which stress is the most prominent. This is well captured by participant 7, a middle aged woman in the low income group who stated that:

> Stress is a major cause of ill-health for me. In the morning before 5am, I wake up, bath the children and prepare them for school, cook for the family, and then go to market where I hawk. I do not have a house help to assist in all these chores and my husband does not care to help. I get home late to continue caring for the family.

Participants from the medium and high income neighbourhoods shared similar experiences. However, the impact of stress differed in their ability to manage it among the different income groups. The professional woman in the high income group is able to afford the services of a house help which to an extent reduces her stress and consequently less impact on her health. Here lies the inequality in their health status. As corroborated by women in the focus groups, access to healthcare and facilities do not only depend on affordability but also by choice and preferences. Discussions on the socio economic factors did not deviate from the results of the household questionnaire. The low and medium income women report the absence of drainages in their environment and where they exist, are blocked by refuse dumps. However, their report on the type of water supply used agrees with the household questionnaire interview. While the low and medium income women use mainly well and street hawkers, the high income women use mainly piped water in addition to well.

14.6 Conclusions and Policy Statements

The most important socio economic factor in the health status of women is the income earned monthly. This is most apparent in the low and medium income groups. It also does appear that having high income status is a precondition for healthier environments and better health services, given competing demands on resources.

The study identified various neighbourhood environmental factors that affect the health of women. It is important for the state government to ensure the availability of and universal access to safe drinking water and sanitation. This will reduce the amount of money spent buying water from local vendors. It will also ensure that the women drink clean and healthy water, thereby reducing the incidence of diarrhea. It is also necessary to ensure full and equal access to health-care infrastructure and services for women of all social and income groups in metropolitan Lagos Since poverty was identified as one of the major reasons of not getting adequate medical care, programmes aimed at eliminating it should be encouraged. The Lagos state government should pursue social, human development, education and employment policies to eliminate poverty among women in order to reduce their susceptibility to ill health and to improve their health.

The Focus Group Discussions across all income groups identified stress as a serious cause of ill health. This factor should be taken seriously in further planning for women in Metropolitan Lagos. Men should be encouraged to share equally in child care and household work and to provide their share of financial support for their families. This will reduce the disproportionate and increasing burden on women who have multiple roles within the family and the community.

In conclusion, improved environmental conditions are germane to improving the health status of women in metropolitan Lagos while emphasis is placed on attending to the stressors that affect women's health.

14.6.1 Areas for Further Research

This study mainly focused on the health inequalities among women of different neighbourhood income groups. It did not do any comparison with the men. It will be interesting to have a comparative analysis of this kind of study. Such information will be necessary in reducing the perceived health inequalities that exist among men and women. It will also be informative to consider the severity of diseases in analyzing health status, for example, considering diseases that are mild, severe, and deadly or terminal.

Most importantly, the issue of stress as highlighted by women during the focus group discussions should be critically looked into. This is the major area of further research from this study. The reasons for the stress, the nature of the stress and the group of women mostly affected should form some of the research questions.

Acknowledgments The authors sincerely thank Ms Nihinlola Dupe Olayinka, a lecturer in the Department of Surveying and Geoinformatics, University of Lagos, Nigeria for helping with the GIS aspects of this work.

Author Biographies

Immaculata Nwokoro is a First Class Scholar of Urban and Regional Planning from the Rivers State University of Science and Technology, Portharcourt, Nigeria. The other degrees obtained are Masters in Urban and Regional Planning (MURP) and a Ph.D. in Urban and Regional Planning from University of Ibadan then Masters in Geoinformatics (MGIT). She was the best graduating student in all the above mentioned degrees/programmes.

Dr. Nwokoro is a registered and an active member of The Nigerian Institute of Town Planners (NITP) and the Town Planners Registration Council (TOPREC). Other professional associations include Geoinformation Society of Nigeria [GSN], the Institution of Highways and Transportation, England, Environmental Health Society of Nigeria (EHSON), International Human Dimensions Programme (IHDP), Nigerian Chapter and The International Society for Urban Health.

In the last 10 years, she has been involved in teaching, research, administrative and community service in the University of Lagos, Nigeria. Her research interest is Environment and Health, Gender issues and GIS, and has presented papers at different international and local conferences in this area. Dr. Nwokoro made a 2 month study visit to the Department of Geography, Lancaster University, United Kingdom for a collaboration on this research topic, and has published in a number of leading journals.

Tunde S. Agbola is a professor of Urban and Regional Planning in Nigeria's premier University, the University of Ibadan, Ibadan, Nigeria. He had his first degree in Economics from the Ahmadu Bello University, Samaru, Zaria in Nigeria before proceeding to the University of Pennsylvania, Philadelphia, USA for his Master of City Planning (MCP), M.A. and Ph.D. degrees. He returned to Nigeria immediately after in 1983 and has been lecturing at Ibadan since then.

His research interests are in the areas of Housing Development and Planning, Planning Theory and Processes, Urban Violence, Urban Municipal Finance and Environmental Management. He has consulted for local, state and national governments in Nigeria and many international agencies like the UNDP, UNICEF, UN-HABITAT, WHO, UMP and many others. His hobbies include reading, travelling and sight seeing, and photography. He is happily married with children.

References

Annandale E, Hunt K (2000) Gender inequalities in health. Open University Press, Buckingham

Benneh G, Songsore J, Nabila S, Amuzu AT, Tutu KA, Yangyuoru Y (1993) Environmental problems and the urban household in the Greater Accra Metropolitan Area, Ghana Stockholm. Stockholm Environment Institute, Ghana

Berry BJL, Horton FE (1970) Geographic perspective of urban systems. Prentice Hall, Englewood Cliffs, NJ, Chapter 10

Bradley DJ, Stephens C, Harpham T, Cairncross S (1992) A review of environmental health impacts in developing country cities. Urban management and the environment. The World Bank, Washington, DC

City Mayors.com World's fastest growing Cities and Urban areas from 2006–2010. www.citymayors.com

Evans RG, Stoddart GL (1990) Producing health, consuming health care. Soc Sci Med 31:1359

Federal Republic of Nigeria Official Gazette (May 15, 2007) Legal notice on publications of the details of the breakdown of the National and State Provincial Totals, 2006 Census

Graham H (2000) Understanding health inequalities. Open University Press, Buckingham

Guerrant RL, Kirchoff LV, Shields DS (1983) Prospective study of diarrhoeal illnesses in North-Eastern Brazil: patterns of disease, nutritional impact, etiologies and risk factors. J Infect Dis 148:986–997

Gwatkin DR (2002) Who would gain most from efforts to reach the millennium development goals for health: an inquiry into the possibility of progress that fails to reach the poor. The World Bank, Washington, DC

Kawachi I, Kennedy BP, Gupta V, Prothrow-Stith D (1999) Women's status and the health of women and men: a view from the states. Soc Sci Med 48(1):21–32

Lagos State Government (2002) Report of the third lagos state economic summit, Ehingbeti 2002, Ministry of Economic Planning and Budget, Lagos State

MacArthur I, Bonnefoy X (1998) Environmental health services in Europe 2 – Policy Options. WHO Regional Publications, European Series, No.77. WHO Regional Office for Europe, Copenhagen

Macintyre S (1997) What are spatial effects and how can we measure them? In: Dale A (ed) Exploiting national survey and census data: the role of locality and spatial effects, (CCSR occasional paper 12). University of Manchester, Manchester

McGranahan G (1991) Environmental problems and the urban household in the third world countries. Stockolm Environment Institute, Stockolm

McGranahan G, Kjellén M (1994) Household environmental problems in Jakarta. Stockholm Environment Institute, Stockholm

Newbold RR (1999) Women's health and the environment: the next century: advances in Uterine Leiomyoma based article. National Institute of Environmental Health Sciences (NIEHS), USA

Philips DR (1990) Health and health care in the Third world, Chapter 4, Longmans

Population Reference Bureau (2004) Improving the health of the world's poorest people in celebrating 75 years 1929–2004. Washington, DC, USA, Website: www.prb.org. policy brief

Rocheleau D, Thomas- Slayter Band, Wangari E (1996) Feminist political ecology: global issues and local experiences. Routledge, New York, NY

State of Lagos Mega City and other Nigerian Cities Report (2004) Ministry of economic planning and budget. Lagos State, Nigeria

Stephens C (1992) Environment and health in developing countries: an analysis of Iintra-urban mortality differentials using existing data. Health Research Unit, Adabraka Polyclinic, Accra

Surjadi C, Padhmasutra L, Wahyuningsih D (1993) Respiratory diseases of mothers and children and environmental factors among households in Jakarta. Environ Urban 5(2):78

Timms TWG (1971) The urban mosaic. Cambridge University Press, Cambridge

United Nations Commission on Human Settlements (UNCHS) (2001) The state of the World's cities, 2001

United Nations Development Programme (2003) Human development report. UNDP, New York, NY. www.undp.org

Wagstaff A, van Doorslaer E, van der Berg H, Calonge S, Christiansen T, Citoni G (1999) Equity in the finance of health care: some further international comparisons. J Health Eco 18:263–290

Whitehead MM (1999) Where do we stand? Research and policy issues concerning inequalities in health and in healthcare. Acta Oncol 38(1):41–50

World Health Organization (1994) Assessment of fracture risk and its application to screening for post-menopausal women. Technical Paper Series NO 843. WHO Scientific Study Group, Geneva

World Health Organisation (WHO) (1999) Principles and basic concepts of equity and health. Division of Health and Human Development, PAHOWHO, Washington, DC

Chapter 15
Housing Quality and Racial Disparities in Low Birth Weight: A GIS Assessment

Sue C. Grady

Abstract Geospatial technologies such as geographic information systems (GIS) and spatial statistics allow researchers to analyze conceptually meaningful spatial datasets on environmental health topics to target interventions and generate new hypotheses for future research. This environmental health study focuses on the relationship between racial residential segregation, housing quality and value and low birth weight in the city of Flint, Michigan. A GIS was constructed with aerial imagery as the base; and maps of racial residential segregation and tax parcels -i.e., building footprints overlaid to examine the effect of maternal exposure to housing quality and value on low birth weight in these neighborhoods. Geographically weighted regression (GWR) was used to analyze these spatial datasets. The findings from this research showed that substandard and well-maintained housing were dispersed throughout the city of Flint, with a higher density of substandard housing in areas of segregation and concentrated poverty. Pregnant mothers living in well-maintained housing in racially mixed neighborhoods received some protection compared to similar mothers living in housing with minor disrepairs. This protection was not observed for mothers living in well-maintained housing in highly segregated neighborhoods, controlling for the same risk factors. GIS and spatial statistics were essential tools in this environmental health study.

Keywords Environmental health · GIS · Low birth weight · Housing · Neighborhoods · Spatial statistics

15.1 Introduction

Housing is an important social determinant of health and substandard housing is reported to be a major public health problem (Greenberg, 1999; Krieger and

S.C. Grady (✉)
Michigan State University, East Lansing, MI 48824, USA
e-mail: gradys@msu.edu

J.A. Maantay, S. McLafferty (eds.), *Geospatial Analysis of Environmental Health*, Geotechnologies and the Environment 4, DOI 10.1007/978-94-007-0329-2_15,

Higgins, 2002; Shaw, 2004; Breysee et al., 2004; Acevedo-Garcia et al., 2004; Hood, 2005; Lawrence, 2006; Rauh et al., 2008). Housing is shaped by the social and structural context of neighborhood environments, and poor individuals may be more exposed to environmental hazards in their homes than individuals of higher socioeconomic status because of their personal inability to tackle disrepairs in addition to the structural constraints imposed upon them. Over the last three decades, economic restructuring and the decentralization of jobs from the city to the suburbs have affected many US cities especially in the Midwest and Northeast Regions. This trend increased class isolation with high unemployment in the inner cities and economic growth in the suburbs. Racial barriers in social mobility limited the movement of African-Americans out of the inner cities into the suburbs resulting in an increase in class *and* racial isolation, which today is referred to as concentrated poverty. The links between racial residential segregation, concentrated poverty, housing and health is an area of intense investigation.

This study on housing and health takes place in Flint, Michigan an urban area that is highly segregated by race and concentrated poverty. A case study assessing the effect of maternal exposure to substandard housing on low birth weight outcomes is presented. It is hypothesized that (a) the odds of delivering a low birth weight infant will be higher for mothers living in substandard housing than in well-maintained housing; (b) the untoward effect of living in substandard housing will be stronger in racially segregated neighborhoods; and (c) the protective effect of living in well-maintained housing will be stronger in racially mixed neighborhoods. Substandard housing refers to housing that has major or moderate disrepairs. Well-maintained housing refers to housing without disrepairs. Since housing condition is highly correlated with housing value it is assumed that substandard housing will be concentrated in poor neighborhoods and well-maintained housing will be concentrated in higher income neighborhoods, regardless of level of racial residential segregation. Substandard housing located in areas of concentrated poverty and high racial segregation is expected to have the lowest value because of the combined structural constraints imposed on these neighborhoods.

The health outcome studied is low birth weight (LBW) defined as infants born less than 2,500 g. Low birth weight infants are at increased risk of infant mortality or development disorders after birth and later on in life (McCormick, 1985; Ellenberg and Nelson, 1979; Daghistani et al., 2002). Low birth weight is therefore, a critically important public health problem in the US. Research shows that the incidence of LBW is especially among infants born to African-American mothers living in highly segregated US cities (Grady, 2006).

This study also demonstrates the utility of geospatial technologies in environmental health research. Geospatial technologies are becoming widely used in environmental health research and practice because of the ability to integrate and analyze hazard + exposure + health outcome datasets within and across "places". These tools are used in the study to conduct a small area analysis of the association between housing and low birth weight in racially segregated and mixed neighborhoods. A geographic information system (GIS) is constructed in ArcGIS version 9.3 (ESRI, 2010) with aerial imagery as the base, providing a pictorial representation

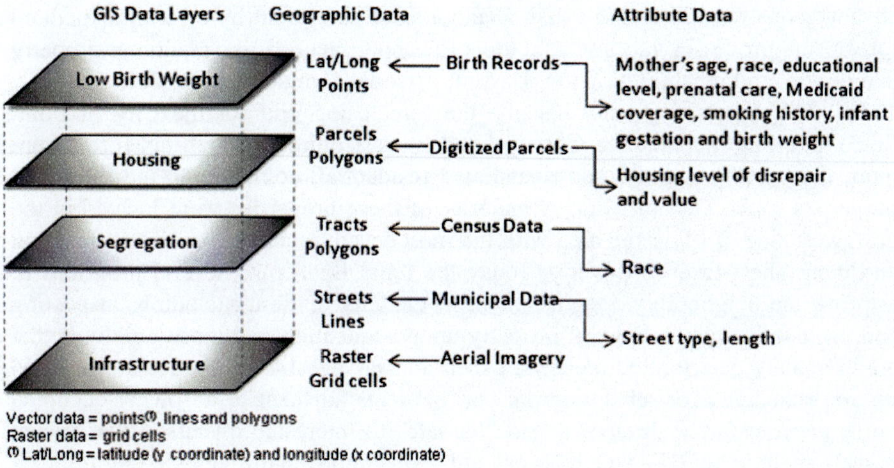

Fig. 15.1 Diagram of GIS and data layers

of Flint's physical and built environment infrastructure. Maps of racial segregation are created at the census tract level and overlaid onto the aerial imagery to display highly segregated and racially mixed neighborhoods. Digitized tax parcels – i.e., digital building footprints for the city of Flint are also added to examine the spatial patterns of housing disrepairs and value within these neighborhoods. Vital statistics birth data are geocoded to the parcels to identify the houses in which mother's lived at the time of their infant's birth. Figure 15.1 displays the GIS constructed for this research and the geographic and attribute datasets input to build the data layers.

Two types of geographic data are used in this study (a) vector data – e.g., geocoded low birth weight records, streets, railroads, census tracts and parcels represented as points, lines and polygons and (b) raster data – i.e., aerial imagery represented as grid cells. Attribute data that describes characteristics of mothers and infants and housing condition and value are joined to the geographic files and mapped to visualize their spatial patterns. These data are further analyzed using spatial regression models. This case study highlights the value of implementing geospatial technologies in environmental health research and practice.

15.2 Flint, Michigan

In an attempt to address substandard housing in inner cities in Michigan, the Michigan legislature passed Public Act 123, the Delinquent Property Tax Foreclosure Act (Legislative Council, State of Michigan, 2009) in 1999 giving counties a choice to take control of foreclosed properties or to leave them to the responsibility of the State (Kildee, 2004; CRC, 1999). Counties that chose to take control of foreclosed properties were given powers to create a county-level "Land Bank" that would manage the foreclosed properties (Kildee, 2005). Land Banks are advantageous for addressing urban abandonment and blight by offering expedited foreclosure processes and financial tools to support actions like planning, property demolition, and rehabilitation (M. Bassett, personal communication, 2007).

In 2002, Genesee County became the foreclosing unit for the City of Flint. Since that time the Genesee County Land Bank acquired through tax foreclosure more than 4,400 blighted and abandoned residential, commercial and industrial properties in the City of Flint. A majority of these properties were located in the northwest side of Flint, the area with the most severe racial residential segregation and the highest level of poverty. Today the Land Bank runs several programs to improve the neighborhood environment in the City of Flint including, inspecting and evaluating each individual property upon acquisition and those buildings that are extremely deteriorated are demolished and salvageable properties are boarded up and scheduled for rehabilitation work. Parcels adjacent to an owner-occupied home are transferred through a "side lot sale" to interested owners. Other vacant parcels are maintained – i.e., grass cut and debris removed through the Land Bank's Clean and Green Program (Genesee Land Bank, 2009).

Housing may have an effect on maternal and infant health through three exposure pathways: quality of life, environmental risks and infectious disease transmission. Indoor environmental quality (IEQ) assessments consist of measuring air temperature, humidity and dampness, air pollutants, disease vectors (e.g., mouse allergens and cockroaches), the presence of overcrowding and broken structures in the home. These hazards may contribute to psychological distress (Morbidity and Mortality Weekly Report, 2001; Hyndman, 1990; Krieger and Higgins, 2002; Rich-Edwards and Grizzard, 2005; Borders et al., 2007), reduced quality of life, acute or chronic lung conditions (Rosenstreich et al., 1997; Dejmek et al., 2000; Choi et al., 2006), viral or bacterial pneumonia (Pearl et al., 1998; Marsh et al., 1999) and/or traumatic injuries (Tinetti et al., 1988; Shenassa et al., 2004). Such social and biological changes may contribute to low birth weight through neuroendocrine, immune, vascular and/or behavioral mechanisms (Valero de Bernabé et al., 2004). This GIS provides a baseline from which to study housing impacts on low birth weight. Data layers will be updated or added in the future to more fully assess the above exposure pathways and changes in maternal and infant health as a result of the Genesee Land Bank's initiative to rehabilitate houses and properties in Flint.

15.3 Data and Methods

15.3.1 Study Area

Flint is located in central Michigan. Aerial imagery of 6-ft resolution – i.e., the grid cell size = 6 × 6 ft obtained for the year 2006 was downloaded from the Michigan Center for Geographic Information – Geographic Data Library's website at http://www.michigan.gov (MCGI, 2010) and input into the GIS to visualize land use and land cover (Fig. 15.2). The city is physically divided by the Flint River that runs northeast to southwest; two Interstate highways, I496 and I69, that also run north-south and east-west; and railroad tracks running in parallel to the highways.

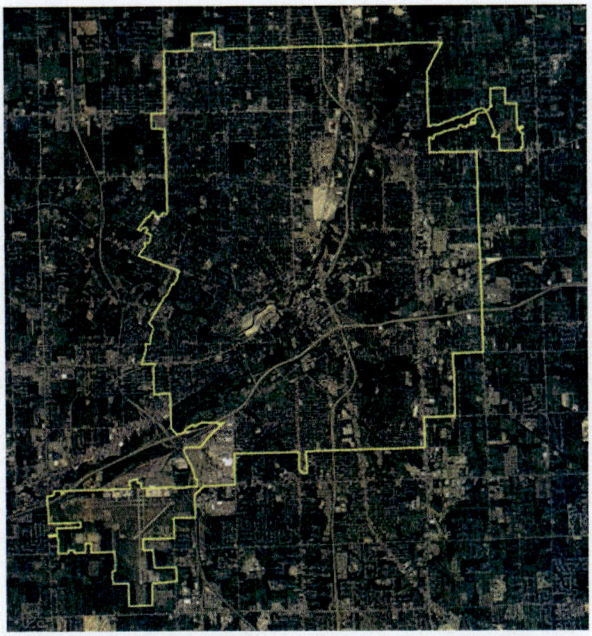

Fig. 15.2 Study area: aerial imagery of Flint, Michigan, 2006

These physical landmarks also act as social barriers limiting the interaction of racial and ethnic groups. African-Americans live primarily north of the Flint River and west of I496 and whites and other racial and ethnic groups live in the remaining parts of the city. While area-level poverty is highly correlated with racial segregation suggestive of concentrated poverty (data not shown), there are also black segregated neighborhoods that are not poor – e.g., western side of Flint and south of the Flint River and poor neighborhoods that are not black segregated – e.g., mid-city east of the Flint River. The spatial isolation of racial groups by socioeconomic status and the high rate of low birth weight infants among African-American mothers in Flint make this an ideal study location for this health geographic research. Figure 15.3 shows the general locations of industry and commercial establishments, vacant and green space in Flint.

15.3.2 Study Population

The study population comprises all live births in the city of Flint between 1995 and 2006 ($n = 27{,}983$). During this time period a majority of mothers were African-American (53.6%) and mothers of other racial groups were reported as White, Asian, American Indian and Hawaiian and Pacific Islanders. In 2003 the estimated population was 120,292 people, which represented a 3.7% decrease in population from year 2000 and an 11.6% decrease in population between 1990 and 2000 (US Bureau of the Census, 2000). Over the last two decades the population of Flint has declined following the elimination of jobs in the auto industry.

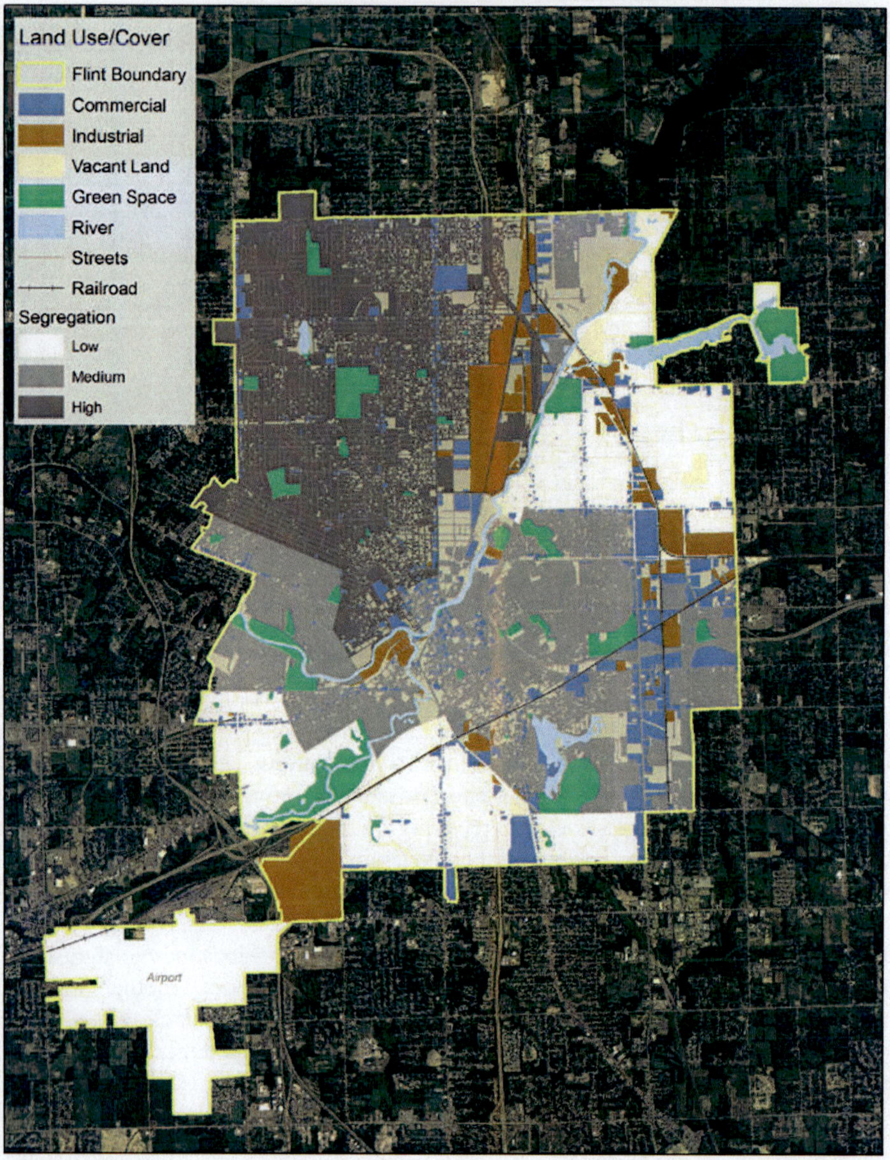

Fig. 15.3 Land use and land cover by level of racial residential segregation in Flint, Michigan 2000

15.3.3 Methods

Parcel data for the city of Flint was used to assess level of housing disrepair and value. Parcel data are digital "building footprints" digitized from tax parcel maps. Attributes of the buildings used in this study were building type – e.g., residential, commercial or industrial and tax assessor value for all single-family dwellings. In 2000 the Genesee County Land Bank and the city of Flint extended their routinely collected data on parcels to include a more in-depth "environmental baseline assessment" (EBA) by driving through neighborhoods and documenting the quality of housing characteristics, such as the condition of sidewalks, foundation, exterior, stairs and roof and the extent of water, fire and smoke disrepairs. These attributes were graded and summed into the disrepair categories "major disrepair" ($n = 98$), "moderate disrepair" ($n = 1{,}312$), "minor disrepair" ($n = 6{,}778$) and "well-maintained" ($n = 4{,}153$). For this analysis houses of major or moderate disrepair were grouped together ($n = 1{,}410$) and referred to as "substandard" housing.

At the neighborhood level racial residential segregation indices were calculated using data on race from the US Bureau of the Census (2000) for the year 2000 at the census tract level. A spatial isolation index developed by Wong (2002) was utilized in this study. The spatial isolation index measures the likelihood that African Americans living in a census tract will come into contact with another racial or ethnic group. The index ranged from 1 = high black isolation to 0 = low black isolation. In this study, 0.7 or above was used to indicate highly segregated neighborhoods and less than 0.7 represented racially mixed neighborhoods (Massey and Denton, 1993). For a complete description of the method used to calculate the spatial isolation index used in this study, please refer to the publication by Dr. Wong (2002).

Vital statistics birth records in the city of Flint were obtained from the Michigan Department of Community Health (MDCH, 2009) for the years 1995 to 2006. The addresses of mothers at the time of their infant's birth were geocoded to the parcel data. Geocoding to parcel data is less common than geocoding to street addresses because parcel data are less available in cities and virtually absent for rural areas. Geocoding to parcel data is more accurate than geocoding to street files because the birth data are matched to the centroid (middle) of the building; whereas, records geocoded to street files are interpolated between street numbers with the x, y coordinates located in reference to the street and not the home. In general geocoding to parcel data results in higher positional accuracy than geocoding to street files but the match rate is lower (Cayo and Talbot, 2003). In this study 23,684 of birth records were automatically or interactively (manually) geocoded to the parcel data resulting in an 85% match rate. Those birth records that did not geocode to parcels ($n = 4{,}299$) were removed from the analysis.

Low birth weight defined as infants born less than 2,500 g was the adverse birth outcome studied. Other maternal and infant characteristics including weeks of gestation (continuous), race (African-American = 1), maternal age (less than 20 years = 1), mother's education (less than high school = 1), prenatal care (no care or no care during first trimester = 1), health insurance (Medicaid = 1), smoking (Yes = 1) and maternal medical risk conditions (Yes = 1) were used as control variables to better understand the effects of substandard (=1) or well-maintained (=1) (base, minor disrepairs = 0) housing on low birth weight outcomes, independent of maternal and infant characteristics.

Spatial regression models were estimated using GeoDa freeware software (Anselin, 2004). The advantage of spatial regression analysis over ordinary least squares regression analysis is that these models incorporate basic diagnostics of spatial autocorrelation (Tobler, 1979; Anselin, 2004). In this study, spatial autocorrelation among mothers of similar race was assumed to be high due to the high levels of racial residential segregation in the City of Flint. The spatial regression analyses were therefore, stratified by highly segregated versus racially mixed neighborhoods to allow for the comparison of low birth weight outcomes of mothers living in substandard or well-maintained housing in these areas. The estimation of spatial regression models in GeoDa is best using a continuous dependent variable; therefore, birth weight in grams was used as a dependent variable.

15.4 Results

15.4.1 Descriptive Analyses

A majority of mothers in Flint lived in general housing (96.2%) and fewer lived in high rise residential buildings (1.5%), mobile homes (2.2%) or commercial buildings converted to residential (<1%). A majority (99.7%) of mothers also rented their homes. The mean value of houses for African-American mothers was substantially less, $23,843 than that for mothers of other racial and ethnic groups, $50,103. Stratified by race and housing condition there was a decrease in almost all maternal risk factors and low birth weight incidence with increasing housing quality; however, racial disparities in these outcomes persisted (Table 15.1). For example, the rate of low birth weight for African-American mothers decreased as the quality of housing improved (range, major-moderate disrepairs = 16.3% to well-maintained = 14.9%) but racial disparities in low birth weight persisted – i.e., African-American mothers living in well-maintained houses (14.9%) were almost twice as likely as other mothers also living in well-maintained housing (7.7%) to have a low birth weight infant. Similar findings were observed for the known risk factors for low birth weight – e.g., young age, low education, lack of prenatal care, Medicaid insurance, smoking during pregnancy and medical risk factors.

Figure 15.4 shows that substandard housing is dispersed throughout the city of Flint but there is a higher density in highly segregated neighborhoods. Well-maintained housing is also dispersed throughout the city. In highly segregated neighborhoods, there is a high concentration of well-maintained housing along the northwest side of Flint.

Figure 15.5 is a map of housing value in the City of Flint. A high density of low value housing – i.e., less than $60,000 is located in highly segregated neighborhoods north and west of the Flint River. There is also a high density of homes of similar value east and south of the river outside of the central business district located near the city-center. Housing of higher value – i.e., $60,000–$120,000 is located along the periphery of the city in highly segregated, moderately segregated

Table 15.1 Descriptive statistics of maternal characteristics by level of house disrepair for African-American and other women living in Flint, Michigan

Housing condition by race	*N*	Low birth weight (%)	Age < 22 years (%)	Education < high school (%)	Lack of prenatal care[a] (%)	Medicaid insurance (%)	Smoking behavior (%)	Medical risks (%)
African-American	12,702	15.5	57.7	34.7	28.0	70.3	15.3	12.3
Major + moderate disrepair	1,410	16.3	64.1	46.1	33.7	80.4	33.4	13.0
Minor disrepair	6,778	15.8	57.7	37.4	29.2	73.8	16.9	12.5
Well maintained	4,153	14.9	55.0	26.7	24.0	60.1	11.3	11.6
Other racial groups	10,982	8.3	47.9	27.5	14.3	52.6	29.2	15.7
Major + moderate disrepair	853	8.6	57.5	45.1	22.3	75.8	35.6	15.7
Minor disrepair	4,826	8.8	50.8	30.5	15.5	57.1	31.8	16.1
Well maintained	4,733	7.7	41.1	19.2	11.2	40.4	23.7	15.3

% = percent of total overall and by level of housing disrepair.

[a]Lack of prenatal care defined as no prenatal care or no prenatal care in the first trimester.

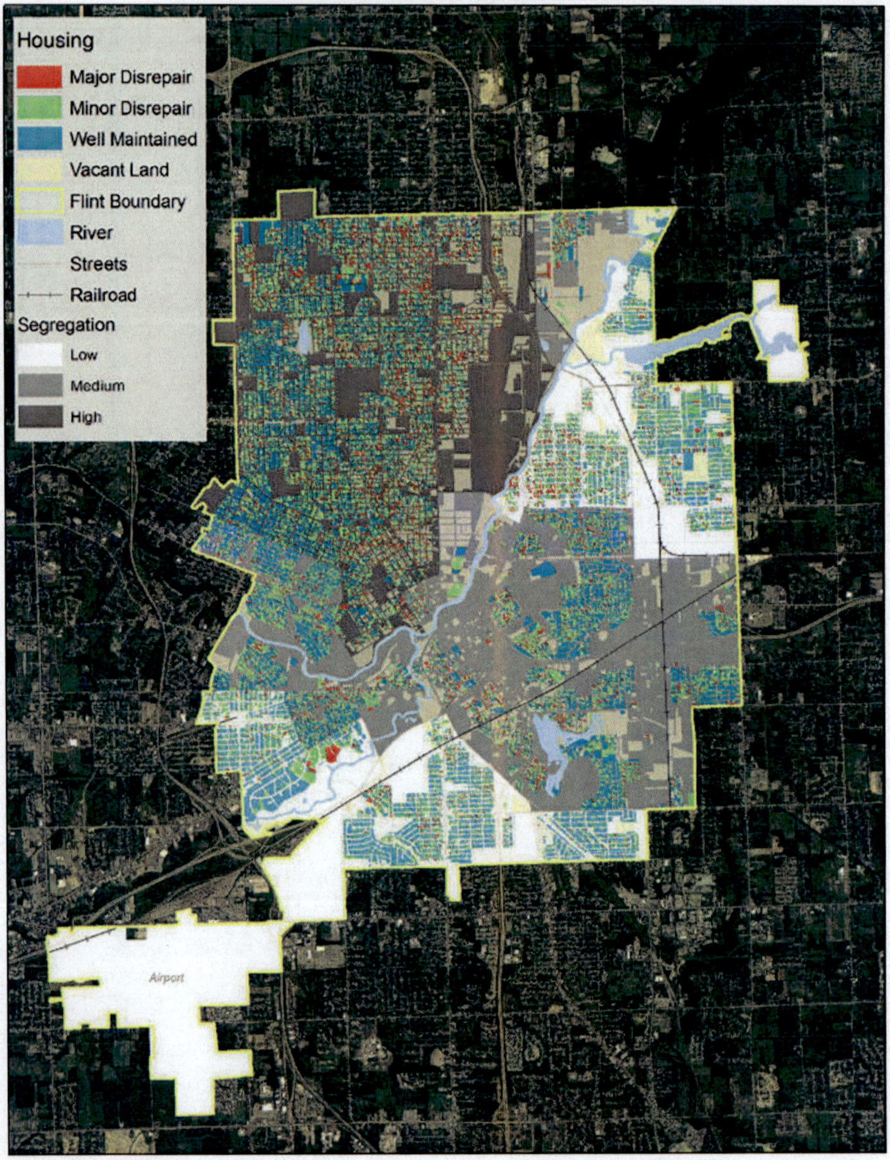

Fig. 15.4 Housing assessment by disrepair status in Flint, Michigan, 2000–2001

and low segregated areas. High valued homes – i.e., greater than $120,000 are primarily located in racially mixed neighborhoods of Flint, especially east of the central business district and in the southwest corner of the city. There are also high value homes in the highly segregated areas of the city but those are dispersed across this area rather than clustered.

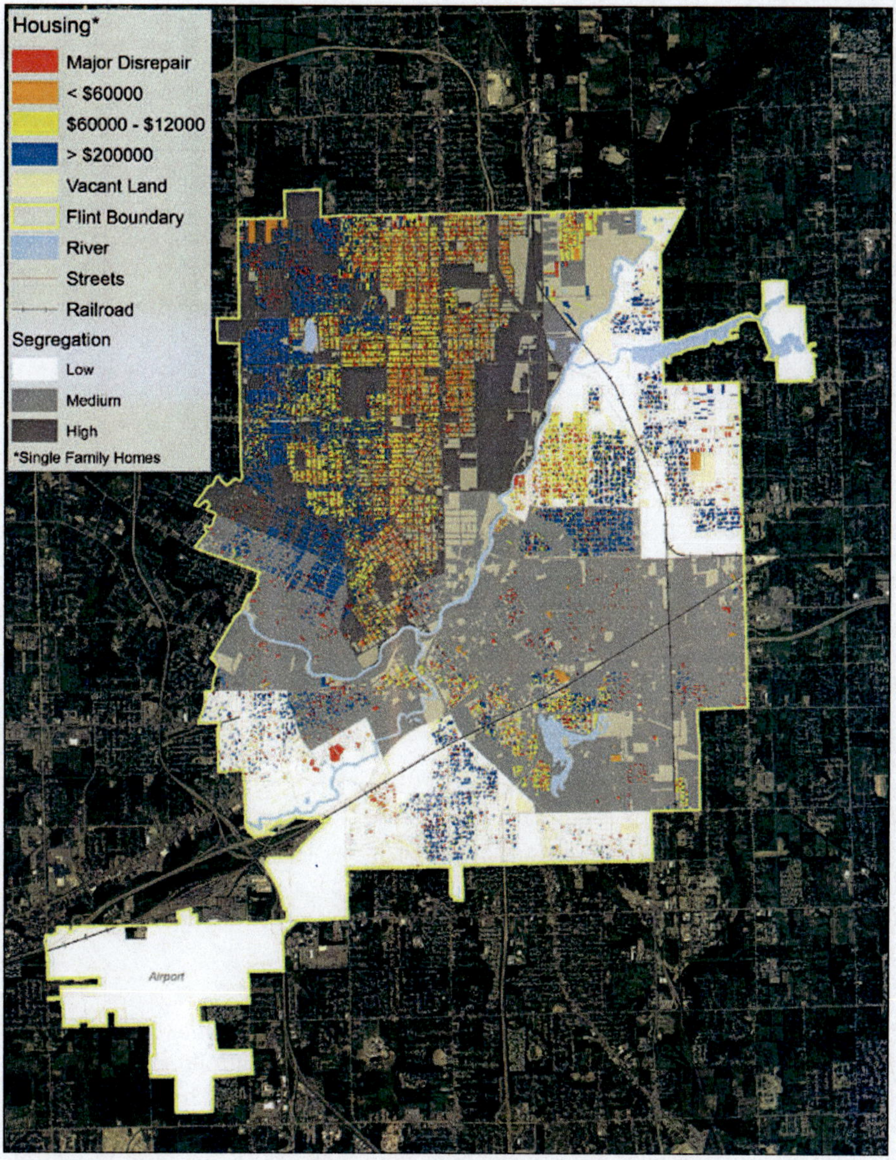

Fig. 15.5 Housing value in Flint, Michigan 2000

15.4.2 Spatial Analyses

The results from the spatial regression models are presented in Tables 15.2 and 15.3. Infants born to mothers in racially mixed neighborhoods were on average 460.0 g heavier than infants of mothers born in racially segregated neighborhoods, controlling for socio-demographic risk factors. Infants of African-American mothers born

Table 15.2 Global spatial regression analysis – infant's born into highly segregated areas of flint controlling for housing major/moderate disrepair and well-maintained[a]

Housing condition by race	*B*	Std error	*t*-statistic	*p*-value
Low birth weight intercept	−1606.39	56.38	−28.48	0.00
Gestation	127.26	1.35	93.83	0.00
African-American	−135.86	18.75	−7.24	0.00
Age < 22 years	−13.39	10.58	−1.26	0.20
Education < high school	−59.77	10.90	−5.48	0.00
No prenatal care	23.07	30.06	0.76	0.44
Medicaid	−29.15	11.26	−2.58	0.00
Smoking	−125.47	13.40	−9.36	0.00
Major/moderate disrepair	−12.49	15.39	−0.81	0.41
Well maintained	−9.21	11.09	−0.83	0.40

Adjusted $R^2 = 0.465$; AIC = 164355.
[a]Including other risk factors presented in Table 15.2.

Table 15.3 Global spatial regression analysis – infant's born into moderate and less segregated areas of flint controlling for housing major/moderate disrepair and well-maintained[a]

Variables	*B*	Std error	*t*-statistic	*p*-value
Low birth weight intercept	−2066.43	62.05	−33.30	0.00
Gestation	140.70	1.58	88.82	0.00
African-American	−188.98	10.95	−17.25	0.00
Age < 22 years	−57.19	9.58	−5.96	0.00
Education < high school	−40.94	11.11	−3.68	0.00
No prenatal care	6.63	39.09	0.16	0.86
Medicaid	−4.44	9.75	−0.45	0.64
Smoking	−185.41	10.61	−17.46	0.00
Major/moderate disrepair	24.85	17.36	1.43	0.15
Well maintained	18.60	9.30	1.99	0.04

Adjusted $R^2 = 0.411$; AIC = 197347.
[a]Including other risk factors presented in Table 15.3.

in highly segregated neighborhoods were on average –53.3 g lighter than African-American infants born in racially mixed neighborhoods, controlling for the same risk factors. Smoking, the most important determinant of low birth weight significantly reduced birth weight by –125.4 g in highly segregated neighborhoods and –185.41 g in racially mixed neighborhoods. Importantly, the birth weight of infants born to mothers living in housing with major/moderate disrepairs was not significantly less than that of infants born to mothers living in housing with minor disrepairs, in both highly segregated and racially mixed neighborhoods. However, in racially mixed neighborhoods, infants born to mothers living in well-maintained housing had significantly increased birth weight (+18.6 g, *p*-value = 0.04) compared to infants born to mothers living in housing with minor disrepairs. This preliminary finding suggests that pregnant mothers living in well-maintained housing in racially mixed neighborhoods receive some protection from potential detrimental exposures compared to similar mothers living in housing with minor disrepairs. This protection was not observed for mothers living in well-maintained housing in highly segregated neighborhoods (–12.4 g, p-value = 0.41), controlling for the same risk factors.

15.5 Discussion

The purpose of this chapter was to demonstrate the utility of geospatial technologies in an environmental health research project. This pilot case study assessed the effects of maternal exposure to housing on low birth weight outcomes in an urban area with high levels of racial residential segregation and concentrated poverty. A GIS was constructed that included data layers to visualize the physical and built environment, the social environment – i.e., racial residential segregation and housing levels of disrepair and value in relation to birth outcomes. Spatial regression analyses were implemented to estimate the effect of substandard or well-maintained housing on birth weight, controlling for maternal and infant risk factors and stratified by highly segregated and racially mixed neighborhoods. Importantly, substandard housing was not found to be significantly associated with reduced birth weight in infants born to mothers living in highly segregated or mixed neighborhoods. This study however, did show preliminary evidence that living in well-maintained housing in racially mixed neighborhoods provided some infant birth weight protection. Why this protective effect was observed in racially mixed neighborhoods and not in highly segregated neighborhoods warrants further investigation.

Future research should more fully describe the characteristics of housing and how housing conditions are shaped by racial residential segregation and concentrated poverty in order to further disentangle housing versus neighborhood effects on low birth weight incidence in these areas. This could be accomplished by, updating and adding additional data layers to the GIS that are "conceptually" relevant and meaningful for this research. For example, the EBA assessment could be expanded to include additional attribute information on indoor environmental quality as well as the socioeconomic status of residents. In this study it was difficult to determine if the protective effect of well-maintained housing on low birth weight was due to improved indoor environmental quality, higher socioeconomic status of residents, higher socioeconomic status of the neighborhood, or a combination of all three conditions. An updated EBA assessment joined to the parcel data would help to answer this question. At the neighborhood level concentrated poverty is characterized by high unemployment resulting in a reduced tax base and reduction in services and amenities – e.g., schools, mass transportation facilities and park maintenance normally found in these locations. Private business owners may also flee to more prosperous areas leaving neighborhoods without amenities to support personal and family health such as grocery stores, pharmacies and other retail – e.g., clothing outlets. Since housing is shaped by the social and structural context of neighborhood environments, adding additional data layers on available resources and amenities to the GIS would provide additional opportunity to explore the pathway(s) by which distal (neighborhood) and proximal (housing) exposures impact maternal health and infant health in Flint.

Qualitative data on resident's perception of their housing (Dunn, 2002) and neighborhood could also be added to the GIS. This approach of acquiring qualitative data for input into a GIS is called "participatory GIS" and is becoming widely used in the field of GIScience. For example, Greenberg's (1999) qualitative assessment of residents living in severely deprived – i.e., blighted and distressed neighborhoods in New Jersey described their experience as "dispiriting, demeaning and profoundly dehumanizing" (Lewis et al., 1973 in Greenberg, 1999) with feelings of "a loss of control" (Greenberg, 1999). The inability to reconcile these structural constraints crime and violence and untoward stress-reducing behavior such a smoking, alcohol and illicit drug use increased and neighborhoods become unsafe. Greenberg (1999) reports a "hierarchy of needs" perceived by residents to improve blighted-distressed neighborhoods beginning with the elimination of crime and removal of severe physical blight followed by the rehabilitation of existing buildings and housing. Residents perceived that these changes would result in a greater sense of safety and security in their neighborhoods and quality of life at home. Future research should therefore, incorporate "participatory GIS" in environmental health studies to further understand how residents perceive the indoor environmental quality of their homes as well as their personal ability/inability – i.e., socioeconomic status to maintain disrepairs. Using GIS, these data could also be mapped to visualize how residents' responses vary across neighborhoods.

15.6 Conclusions

This study represented a small area analysis because it focused on localized variation in hazards and health outcomes within an urban area. Other environmental health projects may require the researcher to examine larger or smaller areas depending on the type of hazard being measured, the pathway(s) of exposure and/or the demographic characteristics of population(s) at risk of acquiring the disease being investigated. Geospatial technologies such as GIS and spatial statistics allow the researcher to incorporate conceptually meaningful datasets – e.g., quantitative and qualitative data on an environmental health topic at a multitude of scales over time – to assess their relationships for planning interventions and to generate new hypotheses for future research. The merger of techniques from the fields of GIScience and public health offers promising new approaches to study and respond to detrimental environmental impacts on health.

Author Biography

Sue C. Grady, Ph.D., MPH, is an Assistant Professor of Geography at Michigan State University. Her research focus is medical (health) geography, human ecology, spatial epidemiology, and health disparity research. She utilizes geographic information systems (GIS) to measure the spatial/geographic distribution of disease and/or risk factors for disease and multilevel modeling to investigate disease etiology. She is currently investigating the reproductive health of African-American women and the risk of adverse birth outcomes and infant mortality associated with living in racially segregated and poor neighborhoods. She has published in geography, epidemiology, public health and urban studies journals.

References

Acevedo-Garcia D, Osypuk TL, Werbel RE (2004) Does housing mobility policy improve health? Housing Policy Debate 15(1):49–98

Anselin L (2004) GeoDa 0.9.5-i. Release Notes. Retrieved on August 2009 from http://geodacenter.asu.edu/software/documentation

Borders A, AE, Grobman WA, Amsden LB, Holl JL (2007) Chronic stress and low birth weight neonates in a low-income population of women. Obstet Gynecol 108(2 Pt 1):331–338

Breysse P, Farr N, Galke W, Lanphear B, Morley R, Bergofsky L (2004) The relationship between housing and health: children at risk. Environ Health Persp 112(15):1583–1588

Cayo MR, Talbot TO (2003) Positional error in automated geocoding of residential addresses. Int J Health Geogr 2:10

Choi H, Jedrychowski W, Spengler J, Camann DE, Whyatt RM (2006) International studies of prenatal exposure to polycyclic aromatic hydrocarbons and fetal growth. Environ Health Persp 114(11):1744–1750

Citizens Research Council of Michigan (CRC) Delinquent Property Taxes as an Impediment to Development in Michigan 1999; Report No. 325. Farmington Hills, MI. Accessed in 2007 from http://www.crcmich.org

Daghistani KJ, Jamal TS, Zakzouk SM (2002) Hearing impairment in low birth weight children. Saudi Med J 23(3):311–313

Dejmek J, Solansky I, Il B, Lenicek J, Sram RJ (2000) The impact of polycylic aromatic hydrocarbons and fine particles on pregnancy outcomes. Environ Health Persp 108: 1159–1164

Dunn JR (2002) Housing and inequalities in health: a study of socioeconomic dimensions of housing and self reported health from a survey of Vancouver residents. J Epidemiol Comm Health 56:671–681

Ellenberg JH, Nelson KB (1979) Birth weight and gestational age in children with cerebral palsy or seizure disorders. Arch Pediatr Adolesc Med 133(10):1044–1048

Environmental Research Systems Institute (2010) Retrieved on September 1, 2009 from http://www.esri.com

Genesee County Land Bank Retrieved on August 2009 from http://www.thelandbank.org/newsroom.asp

Grady SC (2006) Racial disparities in low birthweight and the contribution of residential segregation: a multilevel analysis. Soc Sci Med 63:3013–3029

Greenberg MR (1999) Improving neighborhood quality: a hierarchy of needs. Housing Policy Debate 10(3):601–624

Hood E (2005) Dwelling disparities: how poor housing leads to poor health. Environ Health Persp 113(5):A310–A317

Hyndman SJ (1990) Housing dampness and health amongst British Bengalis in east London. Soc Sci Med 30:131–141

Kildee DT (2004) Reusing forgotten urban land: the Genesee county urban land redevelopment initiative. Housing Facts Findings 6(2):3–5

Kildee DT (2005) Reusing forgotten urban land: the genesee county urban land redevelopment initiative. Housing Facts Findings 6(2):3–5

Krieger J, Higgins DL (2002) Housing and health: time again for public health action. Am J Public Health 92(5):758–768

Lawrence RJ (2006) Housing and health: beyond disciplinary confinement. J Urban Health: Bull N Y Acad Med 83(3):540–549

Legislative Council, State of Michigan. Retrieved on June 2009 from http://www.legislature.mi.gov/(S(0ntq5rn4jwlobp55kveir3yh))/mileg.aspx?page=ExecuteSearch&query=list&isearchfulltext=delinquent%20property%20tax%20foreclosure

Lewis P, Lowenthal D, Tuan Y (1973) Visual blight in America. Association of American Geographers, Washington, DC

Marsh A, Gordon D, Pantazis C, Heslop P (1999) Home sweet home? The impact of poor housing on health. The Policy Press, Bristol

Massey DS, Denton NA (1993) American apartheid. Harvard University Press, Cambridge

McCormick MC (1985) The contribution of low birthweight to infant mortality and childhood morbidity. N Engl J Med 312:82–90

Michigan Center for Geographic Information (MCGI) (2010). http://www.mi.gov/cgi. Retrieved July 2010

Michigan Department of Community Health (2009) Retrieved on March 20, 2009 from http://www.michigan.gov/mdch. Michigan Center for Geographic Information. Retrieved on August, 2009 from http://www.cgi.mi.gov

Mood EW (1993) Fundamentals of healthful housing: their application in the 21st century. In: Burridge R, Ormandy D (eds) Unhealthy Housing: Research, Remedies and Reform. Spon Press, New York, NY, pp 256–282

Morbidity and Mortality Weekly Report (2001) Heat-related deaths-Los Angeles County, California, 1999–2000, and United States, 1979–1998, 50:623–625

Pearl JK, Dickerson J, Li J (1998) Effects of damp and mould in the home on respiratory health: a review of the literature. Allergy 53:120–128

Rauh VA, Landrigan PJ, Claudio L (2008) Housing and health intersection of poverty and environmental exposures. Ann NY Acad Sci 1136:276–288

Rich-Edwards JW, Grizzard TA (2005) Psychosocial stress and neuroendocrine mechanisms in prterm delivery. Am J Obstet Gynecol 192(Suppl 5):S30–S35

Rosenstreich DL, Eggleston P, Kattan M (1997) The role of cockroach allergy and exposure to cockroach allergen in causing morbidity among inner-city children with asthma. N Engl J Med 336:1356–1363

Shaw M (2004) Housing and public health. Ann Rev Public Health 25:397–418

Shenassa ED, Stubbendick A, Brown MJ (2004) Social disparities in housing and related pediatric injury: a multilevel study. Am J Public Health 94(4):633–639

Tinetti ME, Speechley M, Ginter SF (1988) Risk factors for falls among elderly persons living in the community. N Engl J Med 319(26):1701–1707

Tobler W (1979) Cellular geography. In: Gale S, Olsson G (eds) Philosophy in Geography. Reidel, Dordrecht, 379–386

United States Bureau of the Census (2000) Retrieved on September 17, 2009 from http://www.census.gov

Valero de Bernabé J, Soriano T, Albaladejo R, Juarranz M, Calle ME, Martínez D, D-R (2004) Risk factors for low birth weight: a review. Eur J Obstet Gynecol 116:3–15

Wong D, Wong DWS (2002) Modeling local segregation: a spatial interaction approach. Geogr Environ Modelling 6(1):81

Part III
Geospatial Methods in Investigating Environmental Health

Chapter 16
Participatory Mapping as a Component of Operational Malaria Vector Control in Tanzania

Stefan Dongus, Victoria Mwakalinga, Khadija Kannady, Marcel Tanner, and Gerry Killeen

Abstract Global efforts to tackle malaria have gained unprecedented momentum. However, in order to move towards the ambitious goal of eliminating and eventually eradicating malaria, existing tools must be improved and new tools developed. The City of Dar es Salaam, Tanzania, is home to the first operational community-based larviciding programme targeting malaria vectors in modern Africa. In an attempt to optimize the accuracy of the application of larvicides, a participatory mapping and monitoring approach was introduced that includes (1) community-based development of sketch maps of the target areas, and (2) verification of the sketch maps using laminated aerial photographs in the field which are later digitized and analyzed using Geographical Information Systems (GIS). The participatory mapping approach developed enables gap-free coverage of targeted areas with mosquito larval habitat control, and more equal distribution of the workload of field staff. The procedure has been tested, validated and successfully applied in 56 km^2 of the city area. Currently, the approach is being scaled up to an area of about eight times that size, thus covering most of the urban area of Dar es Salaam. The procedure is simple, straightforward, replicable and at relatively low cost. It requires only minimal technical skills and equipment. In the case of Dar es Salaam, the resulting database provides a spatial resolution of administrative boundaries that is almost 50 times higher than that of previously available data. This level of detail can be very useful for a wide range of other purposes rather than merely malaria control, for example implementation of council programmes in a variety of sectors and spatially-explicit analyses for research and evaluation purposes.

Keywords Mapping · Participatory · GIS · Malaria control · Tanzania

---

S. Dongus (✉)
Vector Group, Liverpool School of Tropical Medicine, Liverpool, UK; Ifakara Health Institute, Dar es Salaam, United Republic of Tanzania
e-mail: sdongus@ihi.or.tz

J.A. Maantay, S. McLafferty (eds.), *Geospatial Analysis of Environmental Health*, Geotechnologies and the Environment 4, DOI 10.1007/978-94-007-0329-2_16,

16.1 Background: Malaria

Malaria is one of the most serious health problems the developing world is facing. Despite enormous and diverse control efforts, recent estimates of the World Health Organization indicate that almost one million people die of malaria every year. About 90% of malaria-related deaths occur in sub-Saharan Africa, and most of the victims are children under 5 years of age. Most importantly, the global burden of malaria entails around 45 million DALYs (disability adjusted life years), i.e. 45 million healthy years of life are lost every year due to malaria (WHO, 2009).

Malaria is caused by protozoan parasites, *Plasmodia spp.,* that are transmitted to humans by the bites of infected *Anopheles* mosquitoes. Typical symptoms of the disease are fatigue, headache, dizziness, body pain, chills, nausea and vomiting (Warrell and Gilles, 2002). In African children, the most frequent presentations of malaria are severe anemia and cerebral malaria (Greenwood et al., 2005). This presentation of the disease often leads to coma, and if untreated, to the death of the patient.

The larvae of *Anopheles* mosquitoes generally breed in temporary small ponds, pools and puddles, and in more permanent habitats such as marshes. Most aquatic habitats are freshwater, although some *Anopheles* species also breed in saline waters. Most anophelines avoid polluted waters for breeding (Warrell and Gilles, 2002, p. 70), although this seems to change especially in relatively polluted environments such as urban areas (Sattler et al., 2005). Examples for important breeding sites are shallow open sun-lit pools such as borrow pits, drains, car tracks, hoof prints around ponds and water holes, and pools resulting from the overflow of rivers or left by receding rivers, and rainwater collecting in natural depressions (Gillies and De Meillon, 1968, p. 209).

Global efforts to tackle malaria have gained unprecedented momentum after a paradigm shift from malaria control to malaria eradication following the Gates Malaria Forum in Seattle in October 2007 (Roberts and Enserink, 2007). However, in order to move towards the ambitious goal of eliminating and eventually eradicating malaria, existing tools must be improved and new tools developed. Such tools need to be cost-effective, scalable, and compatible to prevailing health and social systems (Tanner and de Savigny, 2008).

The most important and widely applied malaria control tools available today comprise artemisinin-based combination therapies for treatment, intermittent preventive treatment in pregnancy, rapid and reliable diagnosis of malaria, and vector control by insecticide-treated bednets as well as indoor residual spraying (WHO, 2009). Complementary vector control measures aimed at source reduction that can play an important role in specific settings are biological larviciding (Fillinger et al., 2008; Mukabana et al., 2006) and environmental management (Castro et al., 2009; WHO, 1982, 2004).

Particularly in urban areas, measures aimed at source reduction by larviciding can be a complementary strategy (Castro et al., 2004; Chaki et al., 2009; Fillinger et al., 2008; Geissbühler et al., 2009). Half of the population of Africa will soon live in urban areas (UN, 2008), which offers an opportunity to tackle the malaria problem at its source, as urban breeding sites of malaria transmitting mosquitoes are limited and relatively easy to access.

Maps that are tailored to fulfil the specific needs of field teams as well as their supervisors and programme managers are a prerequisite for successful implementation of such community-based larviciding interventions (Dongus et al., 2007; Fillinger et al., 2008). In areas with limited data availability such as urban sub-Saharan Africa, such maps can be created by participatory use of aerial imagery and simple GIS applications. This chapter describes how an operational malaria vector larviciding programme in the City of Dar es Salaam, Tanzania, is applying this approach as one of its components.

16.2 Community-Based Larviciding of Malaria Vector Mosquitoes in Dar es Salaam, Tanzania

Dar es Salaam, the largest city and de facto capital of Tanzania (Fig. 16.1) with an estimated 2.9 million inhabitants in 2007 (UN, 2008), is home to the first operational community-based larviciding programme in modern Africa, the *Dar es Salaam Urban Malaria Control Programme* (UMCP). The UMCP has been initiated by the Dar es Salaam City Council as a pilot program to develop sustainable and affordable systems for larval control as part of routine municipal services (Castro et al., 2004; Mukabana et al., 2006). Specifically, the UMCP implements the regular application of microbial larvicides (Fig. 16.2) through community-based but vertically managed delivery systems. The current phase of the UMCP launched in March 2004 has achieved a substantial impact on reduction of malaria transmission intensity (Fillinger et al., 2008; Geissbühler et al., 2009).

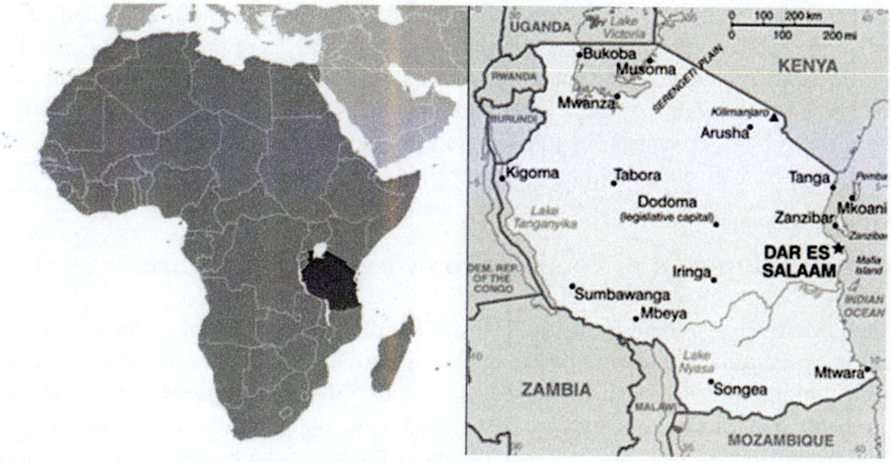

Fig. 16.1 Map of Africa and Tanzania

Fig. 16.2 **a** UMCP field worker searching for *Anopheles* larvae; **b** example for *Anopheles* breeding site; **c** microbial larvicide granules (*Bacillus thuringiensis* var. *israelensis*); **d** UMCP field worker applying larvicide

The UMCP is fully integrated into the existing local administrative system of the Dar es Salaam City Council with the endorsement of the Ministry of Health. It operates on all five administrative levels of the city (listed in hierarchical order): city council, municipalities, wards, neighbourhoods, and more than 3,000 so-called ten-cell-units (TCUs). The main tasks on the four upper levels are project management and supervision, whereas the actual mosquito larval control is organized and implemented at the level of the smallest administrative units, the TCUs. A TCU typically comprises about ten houses, in some cases even more than one hundred. Each TCU is headed by an elected chairperson representing the ruling party.

16.3 Development of Participatory Mapping Procedure

The UMCP aims at identifying and larviciding all breeding sites of malaria vectors in the programme's intervention area. The operational challenges of such a large-scale programme with exhaustive coverage call for very simple implementation protocols that can be executed by community-level staff with minimal education. One of the approaches used is participatory mapping of the target areas. The participatory mapping approach was developed with the aim of enabling complete coverage of targeted communities with larval control by optimizing the quality and spatial coverage of community-derived sketch maps. The overall goal of developing this procedure was to improve programme management systems in a way that makes routine larval surveillance and control truly effective, while enabling the integration of the valuable knowledge of community members. The approach should be easily replicable, adaptable and transferable to any other comparable city in Tanzania or Africa, provided the necessary resources and policy support are available. It should particularly take into account the resource situation and limited availability of maps and remote sensing data in such settings, which cannot be compared to western countries yet. The following paragraphs describe the features of the procedure (for a more detailed description please refer to Dongus et al., 2007).

16.3.1 The Preliminary Sketch Map

As the first activity in the mapping sequence, the UMCP field workers draw preliminary sketch maps (Fig. 16.3a) of their areas of responsibility, with training and support from UMCP staff. The purpose of the sketch maps is to enable the UMCP field workers to assign a unique number to any larval habitat found within a plot, and to enable supervisory staff to identify it unambiguously during spot checks in the field. Features included in the sketch maps are roads, pathways, drains or other landmarks for better orientation, boundaries of the TCUs, and a subdivision of the whole TCU area into individually numbered plots based on regular use or ownership. Attached to every sketch map is a form describing details about each plot such as the house number, the name of the household head, and the number of households. There is one sketch map for each TCU. The sketch maps do not necessarily look like the area itself from the air (Fig. 16.3b), but nevertheless provide good guidance for the UMCP field workers. The system corresponds to the existing administrative boundaries. This makes it easier for the UMCP field workers to orient themselves in the field, as most community members are aware of the number of the TCU or the name of the TCU leader their household is located in, and thus can be asked if in doubt.

16.3.2 Technical Mapping with Aerial Photographs

The next step, referred to as "technical mapping" as opposed to "sketch mapping", entails verifying, correcting and formalizing the preliminary sketch maps in the field by a technical team in collaboration with the UMCP field workers. By using aerial photographs, all boundaries of TCUs, neighbourhoods and wards are formally mapped. The basis for the technical mapping is a digital aerial picture in color. For the first 15 wards that have been mapped, the picture used was from 2002, with a ground resolution of 0.5 m (produced by Geospace International, Pretoria, South Africa). The relevant segments of the picture are color printed as a mosaic of A4 pages at a scale of 1:3,000. The prints are laminated in order to protect them

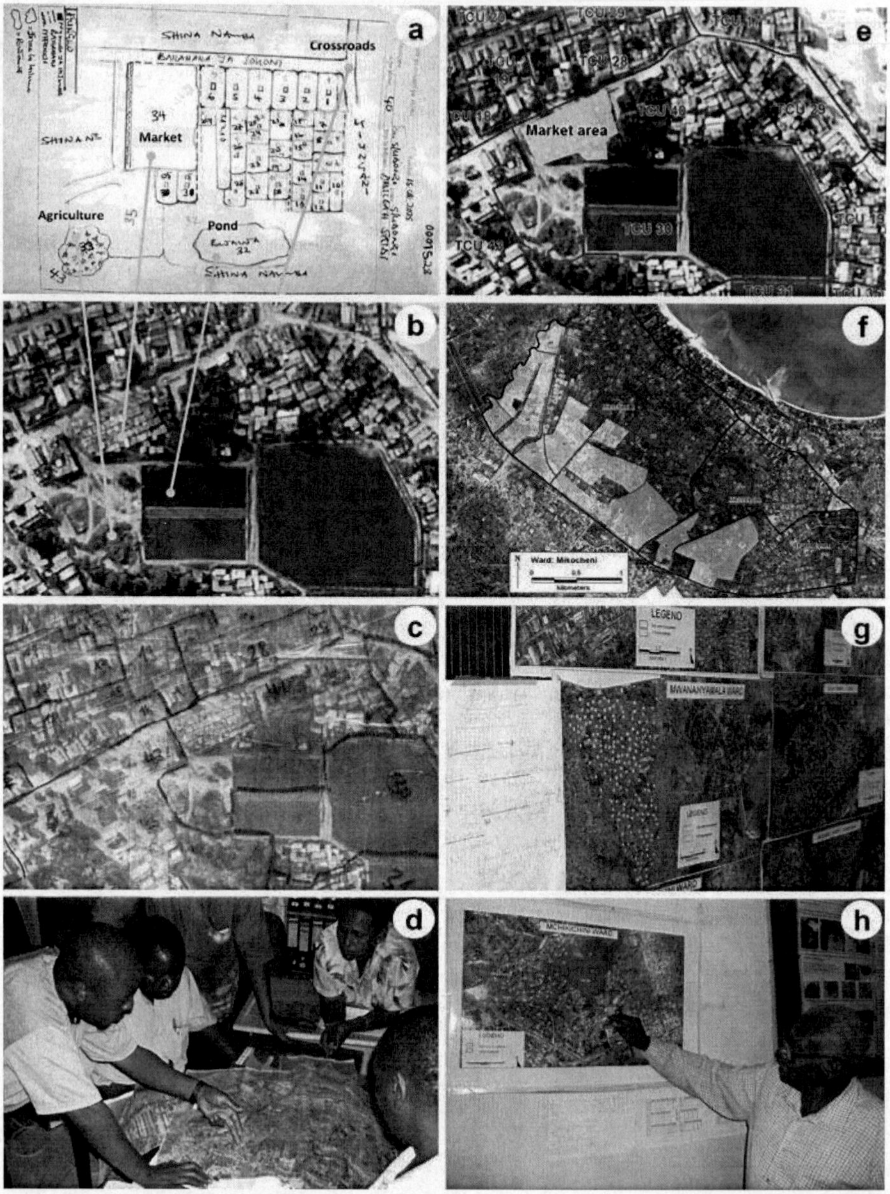

Fig. 16.3 **a** – Sketch map of a ten-cell-unit drawn by and used as a reference for community-based larval control staff. Features comprise individual plots, streets, drains, agricultural areas and ponds. **b** – The same area on an aerial picture. **c** – The same area on a laminated aerial photograph used for verification mapping in the field. The features to be mapped (boundaries of ten-cell-units and their numbers) were marked with non-permanent marker pens. **d** – Project management team discussing over result map, and deciding on necessary follow-up actions. **e** and **f** – Final maps after digitization in a GIS. The shaded regions indicate areas that were not part of the sketch map system yet, but could be included into the malaria control programme after their identification by the participatory mapping strategy. **g** – Digitized, printed and laminated maps of programme intervention areas in the UMCP management office at the city council. **h** – Supervisory programme staff explaining the laminated ward map displayed in the ward office

during intensive use in the field, and to allow drawing on the transparent surface with non-permanent marker pens that can easily be erased again for corrections (Fig. 16.3c). Finished parts of the map are covered with transparent sticky tape for protection of the drawings. At initial meetings with all stakeholders at the local government offices, the technical team shows a sample map so that everybody can understand how the technical map should appear in the end (Fig. 16.3d). The technical team and the responsible field workers then go to all TCUs he or she is working in, one after the other. After reaching a TCU boundary, the position is marked on the laminated photograph as the starting point. The team then walks along the boundary with the neighbouring TCU, while the boundary is continuously marked on the photograph. As soon as another border with a different adjoining TCU is reached, the team marks the three-way intersection of the TCU being mapped and the two adjacent TCUs. This procedure is continued until the starting point is reached again. If it is not possible to walk along the boundary due to construction or other obstacles, it is ensured that what is marked in the technical map represents the actual agreed border. With the same procedure, all existing TCUs within a ward are mapped. By doing so, previously unsurveyed areas are identified and included into the sketch maps (Fig. 16.3e, area shaded in grey).

16.3.3 Identification of Missing Areas and Correction of Sketch Maps

Some of the identified unsurveyed areas are relatively small and easy to integrate into the sketch maps, whereas others are quite large and require a more complex follow-up action by inclusion into newly created TCUs. Most problems can be solved directly on the spot, while more complex ones are discussed by the project management team. After the technical mapping of each single TCU, the team thoroughly checks for unsurveyed areas within that TCU. The sketch map has to cover exactly the same area as marked on the aerial photograph, and all areas within the TCU have to be assigned to specific plots. Omissions of certain areas from the sketch maps are immediately corrected by assigning a new plot number or by adding an area to an existing plot on the sketch map. Any unsurveyed areas included by the technical team are included in the sketch maps and description forms immediately. In the case of relatively large unsurveyed areas that do not belong to any TCU, new TCUs are created. Thus, the TCUs defined by the UMCP are not always identical to administrative TCUs in terms of their boundaries. Finally, the new sketch maps are formalized and corrected in exactly the same way as described above.

16.3.4 Digitization of Technical Maps and Provision for Operational Teams

As the last step, the technical maps based on the aerial imagery are digitized on screen. In our case, this is done by using the GIS software package MapInfo Professional® 7.0 (MapInfo Corporation, One Global View, Troy, New York 12180). Separate polygon layers are created for TCUs, neighbourhoods, wards, and unsurveyed areas, with attribute data such as TCU numbers, names of wards, neighbourhoods, and automatically calculated sizes of each polygon. After digitization, each ward and the mapped features are printed as color maps (Fig. 16.3f). One color map per ward is kept on file at the city office (Fig. 16.3g) together with copies of all corrected sketch maps and description forms. A large-scale color print of each ward map is laminated and returned to the respective local government offices (Fig. 16.3h), where the originals of the sketch maps and description forms are stored while not in use. During operations, the color maps are mostly used by supervisory staff for evaluation of the field workers' performance and assurance of complete larval control coverage.

16.4 Results

16.4.1 Phase 1 – Pilot Areas

The procedure was first tested in the year 2005 in three wards (Fig. 16.4), covering an area of 16.8 km^2, consisting of 12 neighbourhoods with a total of 128,000 inhabitants (National Bureau of Statistics, 2003). We identified 589 TCUs in that area. The total time needed for the actual work was six months, with a technical mapping team consisting of two persons. The costs for the participatory mapping of one TCU during this pilot phase proved to be relatively modest with about USD 25 per TCU. During this phase, it was found that before the technical mapping, only 83% of the study area had been included in TCUs, and only 68% of the study area had been surveyed for mosquito larval habitats by UMCP field staff. All shortcomings were solved by either adding areas to existing sketch maps, or by creating new TCUs and corresponding sketch maps where necessary.

In the course of the technical mapping, shortcomings of the TCU-based control system were identified and eliminated. All of them initially contributed to gaps in terms of areas where mosquito larval habitats were not treated with larvicide. Non-residential areas such as industrial areas, commercial areas and open spaces are not usually part of any TCU or residential lists. Therefore, they were often not included in preliminary sketch maps. Other initially unsurveyed areas resulted from misinterpretation of actual TCU boundaries by the UMCP field workers, for example where the boundaries between TCUs did not coincide with intuitive landmarks such as roads, but were located in less structured areas such as river valleys without residential areas. In such cases, all responsible staff members including those from adjacent TCUs and the technical team revisited the area. The borders between their respective areas could then be assessed properly with full participation by all responsible for and familiar with the area. In general, it was found that the collaboration with staff on all administrative levels stimulated creative, participatory and solution-oriented action.

The participatory mapping procedure led to a more accurate mapping of mosquito breeding sites and to a more equitable distribution and allocation of the work areas per field worker. Additional field workers were recruited and trained

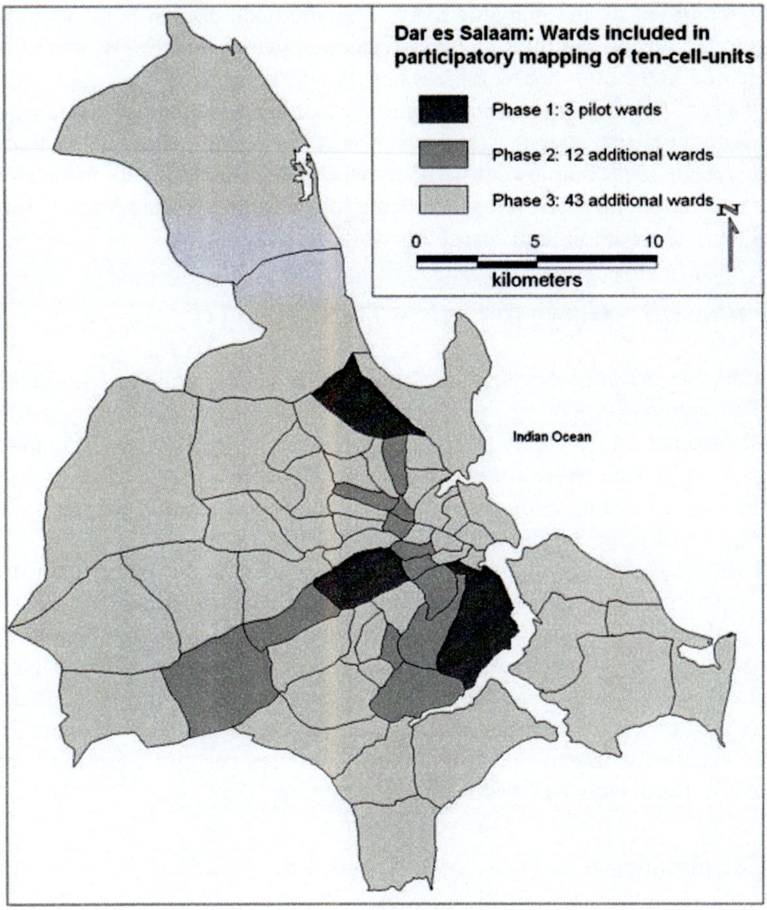

Fig. 16.4 Areas of Dar es Salaam included in participatory mapping procedure

where necessary, which very likely improved the overall quality of work. Before the sketch maps were corrected, some UMCP field workers had been assigned relatively small areas, whereas others were responsible for much larger areas.

16.4.2 Phase 2 – Whole UMCP Area

After the test phase involving three wards, the mapping approach was expanded and validated in twelve additional wards in 2006 and 2007 (Fig. 16.4). The maps are now used by project staff in the local government offices as well as the programme management at the city council. The total fifteen mapped wards represent the whole intervention area of the UMCP, covering an area of 56 km^2, 67 neighbourhoods with more than 610,000 inhabitants (National Bureau of Statistics, 2003) and about 3200

TCUs as identified by the mapping team. The approach used was identical to the one developed during the pilot phase described above. Interestingly, it was found that compared to the pilot phase, far less unsurveyed areas were identified. This was probably a result of the informal information dissemination that took place amongst UMCP staff members after the pilot phase of the mapping. It is likely that as a consequence, the responsible staff in the remaining areas proactively made an extra effort to ensure that all areas were included and covered by sketch maps before the technical teams arrived in their respective wards.

16.4.3 Phase 3 – Going to Scale – The City Level

Since 2008, the participatory mapping approach is being applied in 43 additional wards. This activity is ongoing and expected to be fully accomplished by mid 2010. The total mapped area including 58 wards will then cover almost 450 km^2 of the Dar es Salaam region. Based on the most recent census data available, 2.1 million people were residing in this area in 2002 (National Bureau of Statistics, 2003). A total of about 11,500 TCUs were identified by the mapping team.

With the start of this mapping phase, some adaptations were made to the approach. The reason for these adaptations was to not only provide the basis for scaling up the UMCP, but also to make the database more valuable for other programmes, sectors and applications. It was therefore decided to include the names of the respective TCU leaders and the number of households per plot in the data collection. The latter can serve as an indicator for the population density in each TCU. Both the TCU leader's names and household numbers per plot were retroactively included also for the areas mapped in phases 1 and 2.

16.5 Conclusions

16.5.1 Usefulness for Community-Based Malaria Vector Control

For the purpose of community-based malaria vector control, corrected sketch maps, description forms and formalized color maps based on aerial photographs are available for and used in the complete intervention area. This serves as the basis for achieving 100% spatial coverage of community-based mosquito larval habitat control.

From the point of view of the UMCP field workers, the sketch maps and associated detailed plot descriptions are indispensable guidance tools. The sketch map system accommodates the different cognitive abilities of the field workers, as the map style can be adapted according to their personal preferences in order to achieve optimal orientation. However, only few field workers are generally comfortable to use an aerial photograph as a basis for their work, which rules out the option of replacing all sketch maps with formalized maps.

From a programme management perspective, the sketch maps are an ideal method to assign a unique number to each plot, whereas the technical mapping approach with aerial imagery is essential for the verification and correction of the sketch maps. Moreover, the georeferenced color maps that show the demarcations and locations of TCUs enable management staff to assess and analyze the data collected by the UMCP field workers, and to conduct targeted spot checks. The use of GIS software in the mapping approach is extremely helpful for programme management and supervision of field activities, although only basic functions are utilized. The approach does not require any electronic devices such as GPS receivers in the field, as aerial imagery is sufficient for orientation in urban areas. In addition, if digital aerial imagery is available, costly equipment like digitizing tablets or large format scanners are not needed.

The use of GIS has also proven to be useful in other malaria control programmes, although on lower spatial resolutions. For example, a Malaria Information System (MIS) has been established for parts of South Africa (Martin et al., 2002), where a GIS-supported management system has been used to monitor insecticide consumption and spraying coverage (Booman et al., 2003). In the context of public health challenges in general, another GIS feature considered to be valuable is the spatial modelling capacity for understanding spatial variations of diseases, and their relationship to environmental factors and the health care system (Tanser and Le Sueur, 2002).

16.5.2 Potential and Restrictions for Other Applications

The entire GIS database as well as all subsequent updates thereof has been made available to the Dar es Salaam City Council and the Ifakara Health Institute, one of the leading institutions in East Africa in applied health research. It can be shared with other sectors and organizations and thus be used as a basis for a variety of applications, including health research and interventions, waste management programs, and urban planning, to name a few. In general, in can be used for spatially-explicit analyses for research and evaluation purposes with an unprecedented level of detail, as it allows aggregating datasets on the level of TCUs.

The participatory mapping procedure has been developed specifically for the purpose of optimizing community-based malaria control, notably funded to large parts by international donors in the health sector. One priority of the mapping was to unambiguously assign all areas of the city to certain TCUs. In some cases, for project operational purposes, this implied assigning uninhabited areas such as river valleys, swamps or industrial areas to specific TCUs, although these are not included on residential lists. Using the data for other purposes than operational larviciding of mosquitoes is therefore possible, but limited to certain purposes. For example, the data is ideal for scientific use related to spatial analysis, which has already been done in relation to malaria control (Castro et al., 2009; Geissbühler et al., 2009). Other potential scientific applications would be to georeference the homes of health facility attendants suffering from specific diseases. This could be done by including the TCU number and TCU leader's name in the respective study questionnaires. Figure 16.5 illustrates the high spatial resolution that can be achieved merely by

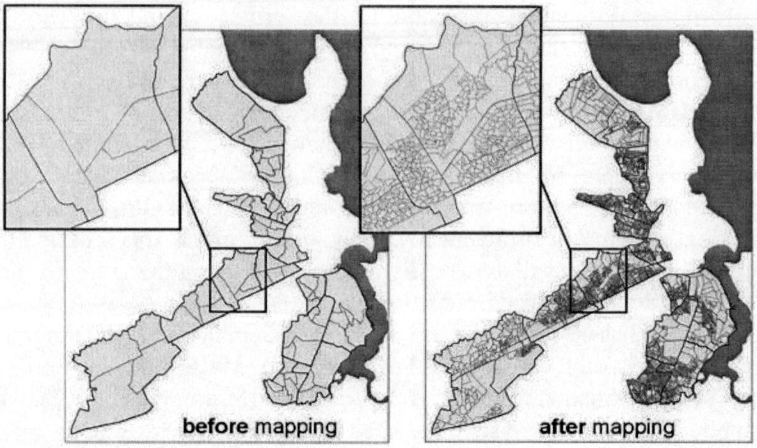

Fig. 16.5 Spatial resolution of administrative boundaries in Dar es Salaam available before and after participatory mapping. While the highest available resolution before the participatory mapping was wards and neighbourhoods (*left*), now the boundaries of ten-cell-units are available as maps and data layers, with an average of almost 50 ten-cell-units per neighbourhood (*right*). The figure also illustrates differences between planned and unplanned settlements: where the latter are found, the boundaries of the ten-cell-units appear much more dense and irregular on the map (*right*)

this system, without having to make any additional GPS measures. Given that a neighbourhood was found to consist of an average of almost 50 TCUs, the available spatial resolution is now about 50 times higher than before. However, governmental use for example related to censuses will probably be restricted unless it will be validated by the relevant authorities.

The mapping approach adheres to the existing administrative boundary system in Tanzania, referring to the ten-cell-units. In a dynamic environment such as the rapidly growing City of Dar es Salaam, this allows optimal orientation for community-based programme staff in the field, without having to create entirely new sets of artificial boundaries. It is argued that this approach has practical programmatic advantages over imposed raster grid systems, because it considers user-definable boundaries that can be agreed in a participatory manner on the ground and that can be readily recognized by community-based staff without access to, or the necessary education to use, GIS technology. In this way, GIS can be participatory, with the potential to enhance community involvement. In the operational context of malaria control Dar es Salaam, this rather basic but straightforward way of applying GIS is advantageous, as resources in terms of available data and expert personnel are limited.

The system of TCUs in Dar es Salaam is probably slightly different to the administrative systems in countries other than Tanzania. Therefore, applying this mapping approach to other regions of Africa and beyond will require the adaptation to the particular systems of each country. In such cases, the smallest existing administrative units in those countries can be used as adequate substitutes for ten-cell-units.

The TCU system has been established by the ruling party in Tanzania, which has been governing the country since its independence in the 1960s. Therefore, the TCUs are not officially recognized as administrative units. Nevertheless, the TCUs do exist and have identifiable boundaries. Basically everyone knows and uses the system, which makes it an ideal reference system for a wide range of purposes, provided that potential changes are regularly updated in the maps.

Regular updating of the maps is crucial in order to account for dynamic changes. As an example for this, since the time the mapping procedure has started in Dar es Salaam, some informal settlements have been demolished and turned into industrial areas. In addition, TCU leader's names change regularly where new leaders are being elected. In those parts of the city where the UMCP programme is operational, such changes are constantly updated. The field staff informs the project management where changes took place, and this information is integrated in the database. However, for the newly mapped expansion areas in phase 3, a different routine updating procedure will be needed. UMCP mapping staff will regularly consult the respective local government officials on ward level. After integrating any changes in the GIS database, the respective wards will be provided with new printouts of the ward maps if necessary.

The community-based participatory mapping represents a very useful tool for urban mosquito larval control, and has become an integral part of the Dar es Salaam Urban Malaria Control Programme. The resulting map data is most valuable also for a variety of other applications. After completion of the participatory mapping process in the city of Dar es Salaam, digital GIS layers with a spatial resolution of administrative boundaries that is almost 50 times higher than that of previously available data is are now available. This level of detail can be very useful for a wide range of other purposes rather than merely malaria control, for example implementation of council programmes in a variety of sectors and spatially-explicit analyses for research and evaluation purposes.

Acknowledgments We would like to thank the community members of the study areas in Dar es Salaam and all members of the Urban Malaria Control Programme involved in the implementation of this study. We thank the National Census Bureau for providing us the digital aerial imagery which was the basis for this study. The underlying research was financially supported by the Bill and Melinda Gates Foundation; the United States Agency for International Development through its Environmental Health Project, its Dar es Salaam mission and the United States President's Malaria Initiative; Valent BioSciences Corporation; the Swiss Tropical and Public Health Institute; the Wellcome Trust; and the Swiss National Centre of Competence in Research (NCCR) North-South.

Author Biographies

Dr. Stefan Dongus holds a Ph.D. in Public Health and Epidemiology from the Swiss Tropical and Public Institute, University of Basel (Switzerland), and a MSc degree in Geography and Mathematics from the University of Freiburg (Germany). In 2010, he joined the Liverpool School of Tropical Medicine (UK), where he took on a postdoctoral position as Senior Research Assistant. Currently, Stefan Dongus is seconded full-time to the Ifakara Health Institute in Dar es Salaam, Tanzania, where he is working on a large project examining new methods for the control of disease-transmitting mosquitoes.

Being a health geographer with a research focus in developing countries, Stefan Dongus' main interest is the participatory use of geographical tools such as basic geographical information system (GIS) platforms for research and disease control interventions. More specifically, he is trying to adapt such GIS tools in a way that they are simple and flexible enough to be useful for implementers with limited geographic background, in operational settings such as national malaria control programmes. Other research activities include the impact of environmental factors on malaria transmission risk, spatial aspects of disease control, urban food security and dynamics of urban land uses.

Victoria Mwakalinga is an urban planner and resources expert. Since 2008, she is employed as a research scientist and health geographer in the Biomedical and Environmental Thematic Group at the Ifakara Health Institute (IHI) in Dar es Salaam. She obtained a MSc degree in Urban Planning and Management in 2007 at Ardhi University in Dar es Salaam. Victoria Mwakalinga's main interest is to employ basic and advanced geographical tools such as GIS, participatory GIS, remote sensing, web-based GIS and spatial modelling in assisting planning interventions towards better urban and rural population health. Furthermore, she is interested in spatial and environmental determinants for health, and integrated disease surveillance systems such as mobile phone and web-based surveillance systems.

Khadija Kannady is the City Malaria Control Officer for the Dar es Salaam Urban Malaria Control Program, coordinating all program activities. She has worked as the Ilala District Research and health management information systems coordinator in Dar es Salaam from 1998 to 2004. Khadija Kannady holds a Master's degree in International Health from the Institute of Public Health in Copenhagen, Denmark. She has been involved in the former Urban Malaria Control Project in Dar es Salaam as in-charge of the epidemiological unit for 5 years. Her main interest is to practice and be involved in community-based approaches for malaria control, especially enabling local communities to implement and own malaria control activities.

Marcel Tanner obtained a Ph.D. on medical biology from the University of Basel and a MPH from the University of London. He is Director of the Swiss Tropical Institute and Professor of Epidemiology and Medical Parasitology at the University of Basel and at the Federal Institute of Technology. Since 1977, his research ranges from basic research on the cell biology and immunology on malaria, schistosomiasis, trypanosomiasis and filariasis to epidemiological and public health research on risk assessment, vulnerability, health impact and district health planning. His research, teaching and health planning expertise are based on substantial long term experience from working in rural and urban areas in Africa (mainly Tanzania, Chad, Burkina Faso and Côte d'Ivoire) and Asia (China, Thailand, Laos). He was co-investigator and coordinator of the first African malaria vaccine trial in 1992 and participated as co-principal investigator in several major intervention trials on malaria (iron supplementation, intermittent preventive treatment) and schistosomiasis. Besides research the capacity building and North–South partnership was a main interest as reflected in the development of the Ifakara Health Institute in Tanzania. He has published extensively in the many fields (>500 original papers). He also acts as advisor on communicable diseases research and control, health systems strengthening and capacity building in various national and international agencies/bodies and in boards/committees such as e.g. WHO/STAC-TDR, Wellcome Trust, DNDi, INCLEN-Trust.

Dr. Gerry Killeen began his career in science as a biochemist, pursuing a doctoral thesis describing the physiological influences of secondary metabolites in plant extracts used as supplements in animal feedstuffs. He graduated with a Ph.D. from University College Galway in the Republic of Ireland in 1996 and moved to Tulane University in New Orleans, USA where he investigated phage-display selection techniques for the identification of antibodies which kill malaria vector mosquitoes. During his 3 years at Tulane, he also began developing mathematical models of malaria transmission and this line of research led him to the field. In 2000, he moved to the shores of Lake Victoria in Kenya, working as a visiting scientist at the Mbita Point Research and Training Centre of the International Centre of Insect Physiology and Ecology. Here he developed his interests in field-based ecology, epidemiology and operational aspects of malaria control. In 2002, he moved to the Swiss Tropical Institute in Basel, Switzerland and was seconded to the Ifakara Health Institute (IHI) in the United Republic of Tanzania. Dr. Killeen has since promoted the development of entomology and environmental sciences at IHI. Dr. Killeen joined Durham University as a Wellcome Trust International Research Career Development Fellow in 2005 and transferred this fellowship to the Liverpool School of Tropical Medicine in January 2009. He remains based at IHI, focusing on integrated malaria control strategies in urban African settings and supporting the Dar es Salaam Urban Malaria Control Programme and the National Malaria Control Programme of Tanzania. Dr. Killeen has acted as a visiting lecturer at University of Limerick and the Royal College of Surgeons in Ireland and remains a Honorary Senior Research Fellow at University College Cork in the Republic of Ireland and a Research Associate Professor at Tulane University in the USA.

References

Booman M, Sharp BL, Martin CL, Manjate B, La Grange JJ, Durrheim DN (2003) Enhancing malaria control using a computerised management system in southern Africa. Malar J 2:13

Castro MC, Tsuruta A, Kanamori S, Kannady K, Mkude S (2009) Community-based environmental management for malaria control: evidence from a small-scale intervention in Dar es Salaam, Tanzania. Malar J 8:57

Castro MC, Yamagata Y, Mtasiwa D, Tanner M, Utzinger J, Keiser J, Singer BH (2004) Integrated urban malaria control: a case study in Dar es Salaam, Tanzania. Am J Trop Med Hyg 71 (2 Suppl):103–117

Chaki PP, Govella NJ, Shoo B, Hemed A, Tanner M, Fillinger U, Killeen GF (2009) Achieving high coverage of larval-stage mosquito surveillance: challenges for a community-based mosquito control programme in urban Dar es Salaam, Tanzania. Malar J 8(1):311

Dongus S, Nyika D, Kannady K, Mtasiwa D, Mshinda H, Fillinger U, Drescher AW, Tanner M, Castro MC, Killeen GF (2007) Participatory mapping of target areas to enable operational larval source management to suppress malaria vector mosquitoes in Dar es Salaam, Tanzania. Int J Health Geogr 6(1):37

Fillinger U, Kannady K, William G, Vanek MJ, Dongus S, Nyika D, Geissbühler Y, Chaki PP, Govella NJ, Mathenge EM, Singer BH, Mshinda H, Lindsay SW, Tanner M, Mtasiwa D, de Castro MC, Killeen GF (2008) A tool box for operational mosquito larval control: preliminary results and early lessons from the Urban Malaria Control Programme in Dar es Salaam, Tanzania. Malar J 7:20

Geissbühler Y, Kannady K, Chaki PP, Emidi B, Govella NJ, Mayagaya V, Kiama M, Mtasiwa D, Mshinda H, Lindsay SW, Tanner M, Fillinger U, de Castro MC, Killeen GF (2009) Microbial larvicide application by a large-scale, community-based program reduces malaria infection prevalence in urban Dar es Salaam, Tanzania. PLoS ONE 4(3):e5107

Gillies MT, De Meillon B (1968) The Anophelinae of Africa South of the Sahara (Ethiopian Zoogeographical Region). South African Institute for Medical Research, Johannesburg

Greenwood BM, Bojang K, Whitty CJ, Targett GA (2005) Malaria. Lancet 365(9469): 1487–1498

Martin C, Curtis B, Fraser C, Sharp B (2002) The use of a GIS-based malaria information system for malaria research and control in South Africa. Health Place 8(4):227–236

Mukabana WR, Kannady K, Kiama GM, Ijumba JN, Mathenge EM, Kiche I, Nkwengulila G, Mboera L, Mtasiwa D, Yamagata Y, van Schayk I, Knols BG, Lindsay SW, Caldas de Castro M, Mshinda H, Tanner M, Fillinger U, Killeen GF (2006) Ecologists can enable communities to implement malaria vector control in Africa. Malar J 5:9

National Bureau of Statistics (2003) The 2002 population and housing census general report. Dar es Salaam, Government of Tanzania

Roberts L, Enserink M (2007) Malaria. Did they really say eradication? Science 318(5856): 1544–1545

Sattler MA, Mtasiwa D, Kiama M, Premji Z, Tanner M, Killeen GF, Lengeler C (2005) Habitat characterization and spatial distribution of Anopheles sp. mosquito larvae in Dar es Salaam (Tanzania) during an extended dry period. Malar J 4:4

Tanner M, de Savigny D (2008) Malaria eradication back on the table. Bull World Health Organ 86(2):82

Tanser FC, Le Sueur D (2002) The application of geographical information systems to important public health problems in Africa. Int J Health Geogr 1(1):4

UN (2008) World Urbanization Prospects: The 2007 Revision Population Database. United Nations, Department of Economic and Social Affairs, Population Division, New York, NY

Warrell DA, Gilles HM (eds) (2002) Essential Malariology, London, Arnold

WHO (1982) Manual on environmental management for mosquito control with special emphasis on Malaria Vectors. WHO Offset Publication no. 66. Geneva, World Health Organization

WHO (2004) Global strategic framework for integrated vector management. Geneva, World Health Organization

WHO (2009) World Malaria Report 2009. Geneva, World Health Organization

Chapter 17
Revisiting Tobler's First Law of Geography: Spatial Regression Models for Assessing Environmental Justice and Health Risk Disparities

Jayajit Chakraborty

Abstract Multivariate regression has been used extensively to determine if race/ethnicity or socioeconomic status is related to presence of pollution sources, quantity of pollutants emitted, toxicity of emissions, and other indicators of environmental health risk. Most previous studies assume observations and error terms to be spatially independent, thus violating one of the standard regression assumptions and ignoring spatial effects that potentially lead to incorrect inferences regarding explanatory variables. This chapter focuses on the problem of spatial autocorrelation in geospatial analysis of environmental justice and explores the application of simultaneous autoregressive (SAR) models to control for spatial dependence in the data. A case study uses both traditional and SAR models to examine the distribution of cancer risk from exposure to vehicular emissions of hazardous air pollutants in the Tampa Bay MSA, Florida. Several approaches are explored to augment the standard regression equation, identify the neighborhood structure of each tract, and specify the spatial weights matrix that accounts for variations in cancer risk not predicted by explanatory variables. Results indicate that conventional regression analysis could lead to erroneous conclusions regarding the role of race/ethnicity if spatial autocorrelation is ignored, and demonstrate the potential of SAR models to improve geospatial analysis of environmental justice and health disparities.

Keywords Air pollution · Cancer · Environmental justice · Spatial regression · Spatial statistics

List of Abbreviations

AIC	Akaike Information Criterion
ASPEN	Assessment System for Population Exposure Nationwide
EPA	Environmental Protection Agency
HAPEM5	Hazardous Air Pollution Exposure Model 5
NEI	National Emissions Inventory

J. Chakraborty (✉)
Department of Geography, University of South Florida, Tampa, FL 33620, USA
e-mail: jchakrab@usf.edu

J.A. Maantay, S. McLafferty (eds.), *Geospatial Analysis of Environmental Health*, Geotechnologies and the Environment 4, DOI 10.1007/978-94-007-0329-2_17,

NATA	National-Scale Air Toxic Assessment
SAR	Spatial autoregressive model
TFL	Tobler's First Law of Geography

17.1 Introduction

During the last 2 decades, increasing attention has been paid to the fact that geospatial data analysis requires specialized approaches that are different from those used to analyze non-spatial data. A number of conventional techniques documented in statistics textbooks and taught in classrooms confront substantial difficulties when applied to the analysis of geographically referenced data. Traditional statistical tests of inference, for example, do not consider that locational proximity often results in value similarity for geospatial data. This fundamental concept was articulated in a simple statement by Waldo Tobler (1970, p. 236) as "everything is related to everything else, but near things are more related than distant things" and is known as Tobler's first law (TFL) of geography (Sui, 2004). The practical implication of TFL is that observations from nearby locations are often more similar than would be expected on a random basis. This phenomenon is also known as spatial dependence, and more formally as positive spatial autocorrelation. The presence of such autocorrelation can be problematic for standard statistical tests such as correlation and regression which assume independently distributed observations and errors. Regression analysis of spatially distributed variables can thus lead to incorrect statistical inference regarding model coefficients when spatial autocorrelation is present and when model specifications fail to include proper corrections for spatial dependence.

Empirical research on the disproportionate distribution of environmental pollution and concomitant health risks has traditionally relied on correlation or regression analysis to determine the existence of racial/ethnic and socioeconomic inequities. In conjunction with geographic information systems (GIS) software and US Census data, multivariate regression models have been used in numerous environmental justice studies to identify population characteristics that are spatially and statistically associated with the location of toxic emission sources, proximity to emission sources, quantity or toxicity of emissions, and other indicators of environmental exposure or health risk. Most prior studies, however, have assumed observations and error terms to be spatially independent, thus violating at least one of the classical regression assumptions and ignoring spatial effects that potentially lead to incorrect inferences about the significance of key explanatory variables such as the presence of racial/ethnic minorities or people in poverty.

This chapter examines the problem of spatial autocorrelation in geospatial analysis of environmental justice and health risk disparities, and explores the application of regression models that account for spatial dependence in the data. It discusses the nature of the problem, methods that are used to detect it, and statistical techniques appropriate for analyzing spatially aggregated data. The case study used to accomplish these objectives seeks to evaluate racial/ethnic and socioeconomic inequities in the distribution of cancer risk from inhalation exposure to vehicular emissions of hazardous air pollutants in the Tampa Bay Metropolitan Statistical Area (MSA), Florida. The goal is to systematically investigate the potential of spatial regression models to address spatial autocorrelation, as well as the implications of various analytical choices associated with the practical application of such models. The emphasis on cancer also addresses the urgent need for environmental justice analysis to make a systematic connection between disproportionate exposure to toxic pollution and its public health outcomes (Grineski, 2007; Maantay, 2007).

17.2 Spatial Autocorrelation in Geospatial Analysis of Environmental Justice

Formally defined as the correlation of a variable with itself in space (Anselin, 2009), spatial autocorrelation represents a basic concept in geospatial analysis that has been discussed for several decades. The earliest mention can be found in brief paper by census statistician who was concerned with the use of census data in social research and introduced the problem as: "data of geographic units are tied together, like bunches of grapes, not separate, like balls in an urn" (Stephan, 1934, p. 165). While TFL formalized the notion of spatial dependence, Cliff and Ord (1973, 1981) were the first to recognize that models which required traditional statistics for their evaluation were misspecified if spatial autocorrelation in the data was not considered, now referred to as the misspecification problem in geospatial analysis (Getis, 2007). In the presence of significant spatial autocorrelation, analytical units do not satisfy a formal statistical test of randomness and thus fail to meet a key assumption of classical statistics: independence among observations. With respect to statistical analyses that presume such independence (e.g., linear regression), positive autocorrelation implies that the spatially autocorrelated observations bring less information to the model estimation process than the same number of independent observations (Voss et al., 2006). A bigger problem relates to the inflation of type I errors, which means that confidence intervals are incorrectly estimated when observations are dependent and classical tests of significance for regression coefficients become biased and inconsistent (Legendre, 1993).

Since the 1980s, quantitative studies on environmental justice and risk disparities have used multivariate regression to analyze the presence of toxic emission sources (Burke, 1993; Fricker and Hengartner 2001; Pastor et al. 2004), proximity to toxic emissions (Pollock and Vittas 1995; Stretesky and Lynch 1999; Margai 2001), quantity of emitted pollutants (Bowen et al., 1995; Ringquist, 1997; Daniels and Friedman 1999), toxicity-weighted risk scores (Ash and Fetter 2004; Sicotte and Swanson 2007), and estimated health risks of exposure to air toxics (Morello-Frosch et al. 2001; Pastor et al., 2005; Gilbert and Chakraborty 2008). For these studies, independent variables describing the racial/ethnic and socioeconomic characteristics of the residential population have been derived from the US Census at the county, census tract or block group levels. Although environmental justice analysis is based on geospatial data and spatially distributed variables, only a few published studies have utilized regression techniques that explicitly account for spatial autocorrelation (Pastor et al., 2005; Grineski and Collins 2008; Chakraborty, 2009). Statistical associations between environmental health risk and race/ethnicity, for example, could be spurious if these variables individually tend to cluster, or are dependent on their values in neighboring spatial units. Regression techniques that control for potential spatial dependence or mitigate spatial autocorrelation are thus necessary to clarify whether racial/ethnic and socioeconomic variables are truly robust in the analysis of disproportionate exposure to environmental health risks.

The incorporation of spatial effects in a statistical model can overcome the complications of space and error dependence, improve the specification of models based on geospatial data, and provide parameter estimates that are less subject to statistical bias, inconsistency, or inefficiency. In addition, such approaches can contribute to theoretical notions regarding the role of space in social relationships and processes (Voss et al., 2006). Although several statistical techniques for incorporating spatial effects in regression analysis have been suggested since the late 1970s (Paelinck and Klaassen 1979; Cliff and Ord 1981; Anselin, 1988), their empirical application has been limited for a long time because these techniques are computationally intensive and mathematically complex (Kissling and Carl 2008). The use of spatial regression models, however, has increased since the advent of geographic information systems (GIS) and the recent availability of user-friendly spatial analysis software capable of implementing the spatial econometric techniques derived from earlier theoretical work. This chapter aims to systematically explore the methods and analytical choices associated with the measurement of spatial autocorrelation and the use of regression techniques which address this problem. Since the emphasis here is on practical application instead of the theoretical concepts related to spatial econometrics, the case study utilizes an open-source spatial analysis software program, GeoDa (Anselin, 2005), which is free to download and is thus easily accessible to researchers and practitioners.

17.3 Data and Variables

17.3.1 Study Area

The case study focuses on the Tampa-St. Petersburg-Clearwater Metropolitan Statistical Area (MSA), also known as the Tampa Bay MSA, which occupies approximately 6,616 km^2 on Florida's west central coast. As shown in Fig. 17.1, the MSA includes four counties, is bordered on the west by the Gulf of Mexico, and is intersected by three major interstate highways. With a total population of about 2.4 million inhabitants (2000), the Tampa Bay MSA is the second largest MSA in Florida and the 19th largest in the US Although its official census name is derived from the three major cities shown in Fig. 17.1, this MSA also encompasses five other unincorporated urban areas with a population between 50,000 and 100,000, and a variety of farmlands and preserves. Non-Hispanic Whites comprise about 75% of this MSA's population, with Hispanics (12%) and non-Hispanic Blacks (11%) representing the two largest minority groups.

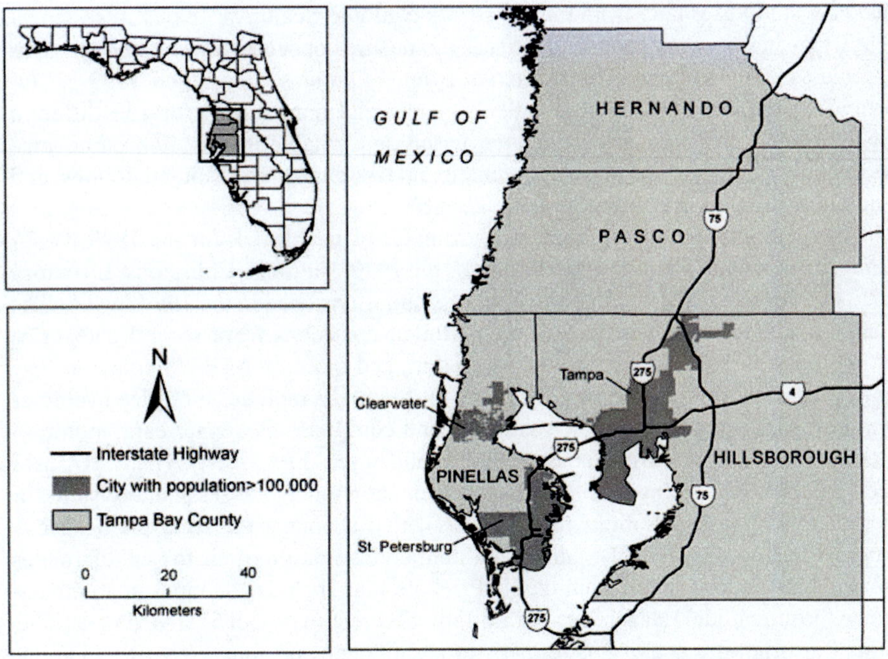

Fig. 17.1 Location of the Tampa Bay Metropolitan Statistical Area (MSA), Florida

Transportation-related air pollution, the focus of this study, is also a public health concern in this area. The Tampa Bay MSA is ranked fifth among all US metropolitan areas in terms of the increase (1990–2005) in total vehicle miles traveled (VMT) annually and eighth in the US for daily VMT per capita in 2005 (FHWA, 2008). While the population of this area has increased substantially since 1990, the annual VMT in Tampa Bay increased by about 83% between 1990 and 2005, potentially exposing a large segment of the population to various adverse health risks. In terms of hazardous air pollutants, the average individual's added cancer risk from mobile source emissions is at least three times higher than cancer risk from point source emissions in all four counties in the Tampa Bay MSA (Environmental Defense Fund 2008). There is thus an urgent need to investigate if minority and low-income communities in this metropolitan area are disproportionately exposed to adverse health risks associated with vehicular air pollution.

17.3.2 Cancer Risks from On-Road Emission Sources of Air Toxics

Hazardous air pollutants, also known as air toxics, include 188 specific substances identified by in the Clean Air Act Amendments of 1990 that are known to or suspected of causing cancer or other serious health problems, including respiratory, neurological, immune, or reproductive effects (EPA, 2008a). In order to measure cancer risks from ambient exposure to vehicular sources of hazardous air pollutants, this case study uses data from the US EPA's National-Scale Air Toxics Assessment (NATA) – an important tool for estimating exposure concentrations and health risks associated with inhalation of air toxics from different sources. The 1999 NATA provides exposure and risk assessment (cancer and non-cancer) for 133 different air toxics and diesel particulate matter, based on available information on adverse health effects, current EPA risk assessment and risk characterization guidelines, and estimated population exposures (EPA, 2008b).

The methodology used to generate estimates of health risk for the 1999 NATA comprises several steps (EPA, 2008c). The 1999 National Emissions Inventory (NEI) serves as the data source on air toxics emissions used in the 1999 NATA. The NEI provides estimates of annual air pollutant emissions from several categories of outdoor emission sources (point, non-point, and mobile) in all US counties. The on-road mobile source category, the focus of this study, includes motorized vehicles that normally operate on public roadways and comprises passenger cars, motorcycles, minivans, sport-utility vehicles, trucks, and buses (EPA, 2004). The NATA uses the 1999 NEI data as input to a Gaussian dispersion model [Assessment System for Population Exposure Nationwide (ASPEN)] that accounts for atmospheric decay to provide an estimate of the annual ambient concentration of air toxics. Estimates of ambient concentrations from the ASPEN are then included as input in an inhalation exposure model [Hazardous Air Pollution Exposure Model 5 (HAPEM5)]. This model incorporates activity patterns that may influence personal exposure to ambient pollutants. From these concentration estimates, the NATA estimates potential public health risks (e.g., cancer) from inhalation of air toxics following the EPA's risk characterization guidelines which assume a lifelong exposure to 1999 levels of outdoor air emissions. The census tract is the smallest spatial unit for which estimates of health risk are provided.

Cancer risk estimates are computed using inhalation unit risk (IUR) factors, which are a measure of carcinogenic potency for each pollutant (EPA, 2008c). For each census tract, the individual lifetime cancer risk associated with each toxic air pollutant is calculated by multiplying the concentration of the pollutant by its IUR estimate. Although the type of cancer (e.g., liver, blood, lung) and available evidence (known, suspected, or possible) varies by chemical, the cancer risks of different air pollutants are assumed to be additive and are summed to estimate an aggregate lifetime cancer risk for each tract, measured in persons per million. A lifetime cancer risk of N in a million, for example, implies that N out of one million equally exposed people would contract cancer if exposed continuously (24 h/day) to the specific concentration over a lifetime (70 years), in addition to those cancer cases that would normally occur in an unexposed population of one million people (EPA, 2008a).

For this study, estimates of lifetime cancer risk (persons per million) from inhalation exposure to on-road sources of air toxics were obtained from the 1999 NATA (EPA, 2008b) for all census tracts in the Tampa Bay MSA and used to represent the dependent variable. Descriptive statistics for this variable are provided in Table 17.1. The values of estimated cancer risk range from 2.7 to 33.2 per million and exceed the Clean Air Act goal (1990) of one in a million in all tracts in this MSA.

Table 17.1 Summary statistics for variables, Tampa Bay MSA, 2000 ($N = 547$)

	Min	Max	Mean	Std. Dev.
Lifetime cancer risk (persons/million)	2.683	33.243	10.388	4.173
Population per square mile	17	16,088	3,122	2,107
Proportion African-American	0.000	0.985	0.115	0.201
Proportion Hispanic	0.002	0.706	0.104	0.109
Proportion below poverty	0.000	0.768	0.120	0.095
Proportion owner-occupied homes	0.000	1.000	0.646	0.224

17.3.3 Explanatory Variables

Inequities in the distribution of estimated cancer risks are analyzed using a set of demographic and socioeconomic variables from US Census 2000 (Summary File 3) for the Tampa Bay MSA at the census tract level. Instead of including a wide and exhaustive range of variables, the focus here was on developing a regression model with census variables that are frequently used in studies on environmental justice and health disparities. To examine the effect of race/ethnicity, the analysis includes separate variables representing the two largest minority groups in the MSA: the proportion of persons identified as non-Hispanic Black and the proportion of persons identified to be of Hispanic or Latino origin. Socioeconomic variables include the proportion of the population with an annual income below the federal poverty level (poverty rate) and the proportion of occupied housing units that are owner-occupied (home ownership rate). Population density is a commonly used control variable in environmental justice analysis because densely populated areas are more likely to contain more pollution-generating activities. While population density is typically measured as the number of people per square mile, the natural logarithm of this value was taken in order to account for the diminishing effect of higher numbers, as suggested in previous studies (e.g., Pastor et al., 2005; Chakraborty, 2009). Descriptive statistics for the explanatory variables are provided in Table 17.1; all variables show substantial variability within the Tampa Bay MSA.

17.4 Detecting Spatial Autocorrelation

In any analysis of geospatial data, cartographical visualization is an important first step in investigating whether a given distribution suggests any patterns or relationships among mapped features. In this context, GIS software serves as a particularly useful tool for mapping the dependent and explanatory variables relevant to the analysis, and exploring how their values vary spatially within the study area. The geographic distribution of estimated cancer risk from on-road sources of air toxics, the dependent variable for this study, is depicted in Fig. 17.2. For this choropleth representation, census tracts are classified into four quartiles based on the values of lifetime cancer risk. Tracts facing the greatest cancer risk (top 25%) are located in the most densely populated urban areas of this region and near major highway

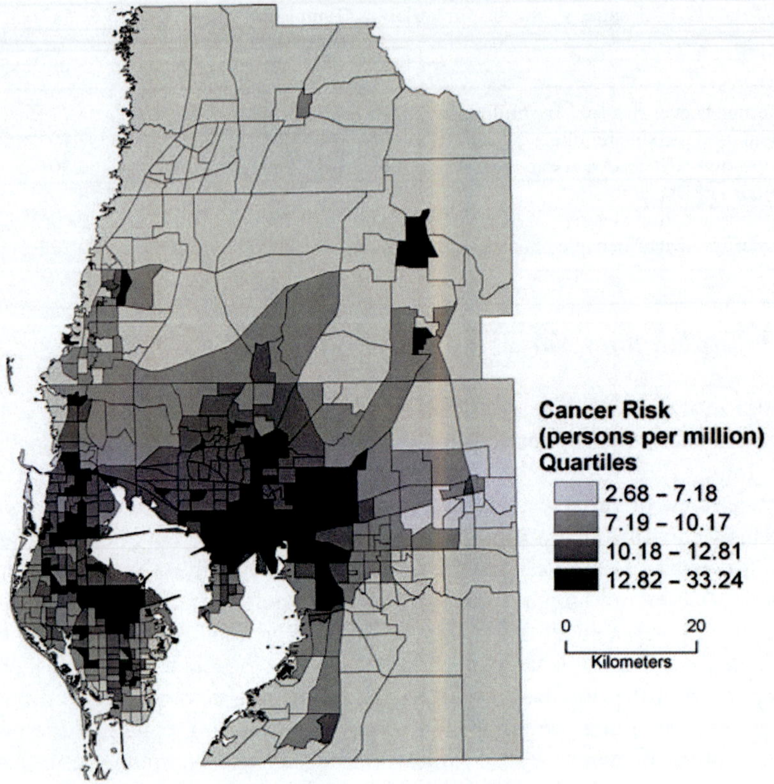

Fig. 17.2 Lifetime cancer risk from on-road sources of air toxics by census tract, Tampa Bay MSA, 1999

interchanges. A majority of tracts in highest quartile of cancer risk are immediately adjacent to tracts in the third quartile and most tracts in the lowest quartile are contiguous with tracts in the second quartile. Although the map suggests clustering of similar values across space, visual evidence can be tenuous and the presence of spatial dependence needs to be validated with formal quantitative evidence or statistical testing.

17.4.1 Spatial Definition of Neighbors

Before testing a variable for spatial autocorrelation, it is necessary to determine which units of observation are contiguous or sufficiently close together for spatial interaction to occur. For this purpose, a connectivity or spatial weights matrix is used to specify, for each spatial unit, which other units are "neighbors" and may influence its values (Cliff and Ord 1981; Pastor and Scoffins 2008). Since the structure of spatial dependence among analytical units in a study area is not known in advance, the specification of this weights matrix is a matter of considerable arbitrariness and a wide range of suggestions can be found in the literature (Anselin and Bera 1998). Software programs such as GeoDa, however, provide two basic approaches for defining the neighbors of a spatial unit: contiguity-based or distance-based. The contiguity-based approach to neighbor identification is usually based on either the rook or queen selection method. A queen weights matrix defines neighbors as spatial units that either share a common boundary or vertex, while a rook weights matrix only includes units that share boundaries. A second approach to neighbor identification is based on measuring the centroid-to-centroid distance between spatial units in a study area. Neighbors can be defined as units whose centroids fall within a user-specified radius (distance band) around the centroid of each unit, or by using a more complex technique that gives closer neighbors higher weights, such as scaling the weights according to the inverse of the distance between centroids (Pastor and Scoffins 2008). Regardless of the approach utilized to define neighbors, the formal result is a spatial weights matrix (W) that summarizes the nature and degree of interdependence among analytical units in a study area. While several methods are available for coding a spatial weights matrix, row standardization is commonly used to determine weights (Kissling and Carl 2008). In GeoDa, neighbors are defined on the basis of a binary (0, 1) row-standardized weights matrix. A weights matrix is row-standardized when the values of each of its rows sum to one. Each observation is represented by a row and a column in the matrix, with neighbors coded as 1 and all other units including the location itself coded as 0.

The application of both contiguity-based and distance-based approaches are explored to define the spatial weights matrix in this study. For the contiguity-based approach, the first-order queen method is used to define neighbors as adjacent census tracts that share a common boundary or vertex with a given tract. The difference between the rook and queen selection methods is expected to be minimal when units are irregular in size and shape, unlike squares on a grid (Anselin, 2009). For the distance-based approach, the extent or radius of the distance band for selecting neighbors is an important consideration. Since census tracts vary widely in size according to population density, a small radius will cause larger tracts in less urbanized areas to have no neighbors, while a large radius will cause smaller tracts in densely populated areas to have an excessively high number of neighbors. The literature suggests that the choice of this distance should reflect the theoretical understanding of the process or variable being mapped (Odland, 1988; Anselin, 1988). For ambient exposure to traffic-related air pollution, 2,000 m from a roadway has been assumed to represent the maximum distance for adverse health effects in previous studies (Venn et al., 2001; Wu and Batterman 2006). The distance-based spatial weights matrix selected for this study thus considers any two tracts to be neighbors if their centroids are within 2,000 m from each other. The results from the spatial regression models used in this study were also found to be robust to minor increases or decreases of this distance. An inverse distance weighting scheme was not utilized because this option is not available in GeoDa.

It is important to consider the spatial implications of the two different approaches for neighbor definition in the Tampa Bay MSA. Figure 17.3 provides connectivity histograms that depict the characteristics of the two spatial weights matrices.

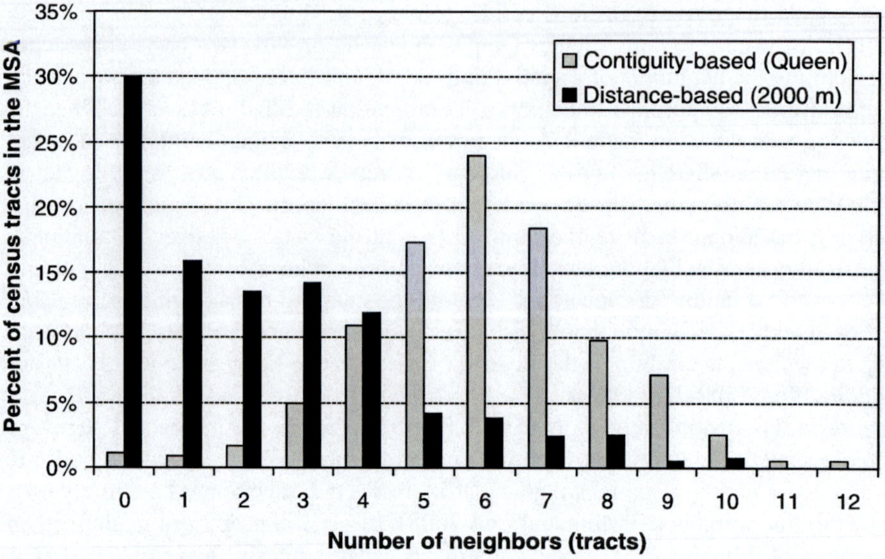

Fig. 17.3 Frequency distribution of neighbor relationships for contiguity-based and distance-based spatial weight matrices in the Tampa Bay MSA

When the contiguity-based (first-order queen) approach is used, the distribution of neighbors is reasonably symmetrical. With the exception of few island tracts near the western coastline, almost all tracts in the MSA have at least three neighbors. Nearly 75% of tracts in this area contain five to seven neighbors and about 20% of tracts encompass eight or more neighbors. In contrast, the frequency distribution of neighbors based on the distance-based (2,000 m) approach is highly skewed to the right. Almost 75% of tracts in the MSA contain less than four neighbors, and 30% of tracts do not have a tract centroid within 2,000 m. Only 4% of all tracts in the MSA, located predominantly in the densely populated areas, encompass eight or more neighbors for this distance band.

These differences can be explained, in part, by the fact that tracts vary substantially in shape and size across this metropolitan area, as shown in Fig. 17.2. Tracts that are larger in size and located in the less populated northern counties of this MSA do not contain neighbors when the 2,000 m distance band is utilized. However, the same set of tracts selects multiple contiguous tracts as neighbors when the queen method is applied. Contiguity-based methods for selecting neighbors are thus likely to overstate spatial effects in areas when spatial units are larger in size, compared to the use of fixed distance bands.

17.4.2 Measuring Spatial Autocorrelation

All variables listed in Table 17.1 were first tested for the presence of spatial autocorrelation. This was based on the global Moran's I statistic – a measure describing the general extent of spatial clustering of a variable in a given area, conditional on the specific neighborhood structure imbedded in the chosen weights matrix (Moran, 1950). This Moran's I is typically scaled to an interval ranging from –1.0 to +1.0, where a large positive value indicates value similarity among neighbors (clustering, or positive spatial autocorrelation), a large negative value indicates value dissimilarity (dispersion, or negative spatial autocorrelation), and a value near zero suggests the lack of spatial autocorrelation. In GeoDa, the pseudo statistical significance of the global Moran's I is estimated using a permutation approach that compares the observed value to a reference distribution of Moran's I values generated under conditions of spatial randomness (Anselin, 2009).

For both the contiguity-based and distance-based approaches to defining neighbors, Table 17.2 indicates that the global Moran's I for lifetime cancer risk from on-road air toxics is highly significant and greater than zero. This strong statistical evidence of positive spatial autocorrelation confirms the clustering of similar values initially suggested by the map (Fig. 17.2). The queen method produces greater spatial autocorrelation than the distance band, possibly because the use of contiguity exaggerates spatial effects for larger tracts in the MSA. While the most important autocorrelation test is for the dependent variable, the Moran's I for each independent variable is also provided in Table 17.2. The inclusion of variables on the right-hand side of the regression equation that are spatially dependent and correlated with the response variable can partially mitigate the extent of spatial autocorrelation in the resulting error term, making the regression coefficients and standard errors more reliable (Odland, 1988). Regardless of the approach used to define neighbors, all explanatory variables show significant and positive spatial autocorrelation within the Tampa Bay MSA, suggesting their potential effectiveness as spatial controls. While home ownership rate produces the lowest Moran's I, the highest amount of spatial clustering is indicated by the Black and Hispanic proportions.

17.4.3 Traditional Multiple Regression

The next step consisted of conducting a standard multiple regression analysis of lifetime cancer risk from on-road sources of air toxics, based on the six independent variables and the ordinary least squares (OLS) technique commonly utilized in prior

Table 17.2 Moran's I tests for spatial autocorrelation

	Contiguity-based (1st order queen)	Distance-based (2,000 m radius)
Lifetime cancer risk (persons/million)	0.728*	0.523*
Population per square mile	0.532*	0.260*
Proportion Black	0.682*	0.695*
Proportion Hispanic	0.596*	0.491*
Proportion below poverty	0.566*	0.579*
Proportion owner-occupied homes	0.390*	0.340*

$^*p < 0.001$.

Table 17.3 Conventional least squares regression of lifetime cancer risk (natural log)

Variable	Coefficient	*t*-value
Population per sq mile (natural log)	1.210	9.138*
Proportion black	3.909	3.656*
Proportion hispanic	9.470	6.803*
Proportion below poverty	3.218	1.261
Proportion owner-occupied homes	−2.388	−3.803*
Constant	0.858	0.656
N	–	547
F statistic	–	58.656*
Adjusted R-squared	–	0.352
Multicollinearity condition index	–	23.940
Residual Moran's I (1st order queen)	–	0.556**
Residual Moran's I (2,000 m radius)	–	0.450**

*$p < 0.01$; **$p < 0.001$.

studies on environmental justice and health disparities. This basic multivariate OLS model can be summarized as:

$$y = \alpha + \sum_k \beta_k x_k + e \tag{17.1}$$

where y equals the estimated tract level cancer risk, α equals the intercept (constant), x represents each independent variable, k equals the number of independent variables, and e represents the random error term (difference between the observed and predicted values of y).

The results of this conventional regression analysis are summarized in Table 17.3. The OLS regression model appears to perform reasonably well for this data. The ANOVA F-test indicates statistical significance for the overall model ($p < 0.001$) and the value of the adjusted R-squared (0.352) suggests a respectable goodness-of-fit. The multicollinearity condition index is also smaller than 30, indicating the absence of serious collinearity problems among the independent variables. With the exception of the proportion below poverty, parameter estimates for all explanatory variable are strongly significant with anticipated signs. Population density, proportion Black, and proportion Hispanic are significantly and positively associated with lifetime cancer risk, while the proportion of owner-occupied homes is negatively related to cancer risk. The traditional multivariate regression analysis provides strong evidence to suggest that cancer risk from on-road sources of air toxics is distributed inequitably with respect to race, ethnicity, and home ownership in the Tampa Bay MSA.

It is important to investigate if the error from this OLS model satisfies the standard linear regression assumption of independence. For this purpose, the Moran's I test for residual spatial autocorrelation is utilized (Anselin and Bera 1998). The regression residuals associated with both the contiguity-based and distance-based spatial weights matrices show a significantly positive Moran's I of 0.556 and 0.450, respectively. The spatial autocorrelation tests clearly indicate that the residuals are spatially dependent with respect to their values in neighboring tracts. Since this is a serious violation of an OLS assumption, inferences drawn from the traditional multiple regression model in Table 17.3 could be seriously flawed.

A comparison of the residual spatial autocorrelation ($I = 0.556/0.450$) in the OLS model with the spatial autocorrelation for the dependent variable ($I = 0.728/0.523$) suggests that spatial dependence in one or more independent variables potentially explains a portion of the spatial autocorrelation in the dependent variable. As mentioned previously, it is possible for explanatory variables in a regression model to completely account for the spatial autocorrelation in a dependent variable, thus removing a problematic spatially autocorrelated residual. Since the residual spatial autocorrelation in this OLS model remains statistically significant, the independent variables do not adequately account for spatial dependence in the data and a correction to the traditional OLS model is necessary to incorporate spatial externalities.

17.5 Addressing Spatial Autocorrelation with Spatial Regression Analysis

Spatial regression models, such as simultaneous autoregressive (SAR) models, are statistical models that consider spatial autocorrelation as an additional variable in the regression equation and estimate its effect simultaneously with effects of the other explanatory variables. This additional term is implemented with the spatial weights matrix which accounts for patterns in the dependent variable that are not predicted by explanatory variables, but are instead related to values in neighboring locations. Most software packages such as GeoDa provide two ways to incorporate spatial dependence in a regression model, depending on where spatial autocorrelation is assumed to occur (Anselin, 2005). The spatial error model assumes that only the error term is spatially autocorrelated, while the spatial lag model assumes that autocorrelation is an inherent characteristic of the dependent variable. The parameters for both types of SAR models are estimated using the maximum likelihood method.

For the spatial error model, the standard OLS regression equation is augmented by a term (λWe) which represents the spatial structure (λW) of the spatially dependent error term (e).

$$y = \alpha + \sum_{k} \beta_k x_k + \lambda We + u \tag{17.2}$$

where λ is the spatial autoregressive coefficient (error parameter); W is the spatial weights matrix; e is the random error term in the OLS model; and u is the spatially independent error term.

The spatial lag model includes a term (ρW) for the spatial autocorrelation in the dependent variable (y), in addition to the standard terms used in the OLS regression model.

$$y = \alpha + \sum_k \beta_k x_k + \rho W y + u \tag{17.3}$$

where ρ is the spatial autoregressive coefficient (lag parameter).

The choice of SAR model specification (lag or error) to account for spatial autocorrelation should be based on the theory of the process under investigation. In most practical situations, however, this decision is guided by the Lagrange Multiplier test statistics provided by GeoDa (Anselin, 2005). Since this case study focuses on exploring the implications of different analytical choices associated with spatial regression, both lag and error SAR models are utilized to analyze the distribution of cancer risk in the Tampa Bay MSA.

17.5.1 Comparison of Regression Model Performance

The results from the two SAR models, based on the two different spatial weights matrices, are summarized in Table 17.4. The first column provides the corresponding coefficients and diagnostics from the OLS regression model to allow a comparison of the statistical results. Following Kissling and Carl (2008), the comparison of SAR models is based on the following criteria: (a) minimum residual autocorrelation; (b) maximum model fit; and (c) the Akaike Information Criterion (AIC). The Moran's I test statistic for the regression residuals in all four SAR models is smaller than 0.10, pointing to a substantial decline in spatial autocorrelation with respect to the original OLS model. With the exception of the distance-based lag model, the residual Moran's I in the SAR models is not significantly different from zero ($p > 0.05$). This suggests that the inclusion of the spatial autoregressive term in these models has effectively eliminated spatial dependence of residuals. Compared to the conventional regression model which yielded an adjusted R-squared of 0.352, Table 17.4 shows that the R-squared for the SAR models range from 0.509 to 0.759, suggesting a considerable improvement in model fit. Since the SAR model parameters are estimated using the maximum likelihood method, the pseudo R-squared from a SAR model is not directly comparable to the adjusted R-squared from OLS regression. The AIC is considered to be a more appropriate metric for comparing the SAR model with the OLS model because it provides a compromise between model fit and the number of parameters (parsimony). While AIC is a relative measure and cannot be interpreted on its own, a lower AIC score between competing models is indicative of the more correct or valid model (Grove et al., 2006). Table 17.4 indicates that the AIC scores from all four SAR models are substantially lower than the AIC from the OLS model, suggesting a sizeable improvement in model performance. The contiguity-based SAR models show larger declines in AIC than the distance-based SAR models.

Table 17.4 also provides the results of the Lagrange Multiplier (LM) test statistics, which can be used to guide our choice of SAR model specification. Although both the robust LM error and LM lag statistics are significant for all models, the decision rule suggested by Anselin (2005) would lead to the selection of the distance-based spatial error model because it provides the largest

Table 17.4 Comparison of conventional and spatial regression results for lifetime cancer risk (natural log)

		Contiguity-based (1st order queen)		Distance-based (2,000 m radius)	
	OLS	Lag	Error	Lag	Error
Coefficients:					
Population per sq mile (natural log)	1.210**	0.420**	0.673**	0.218	1.136**
Proportion black	3.909**	0.361	0.886	2.065*	2.590*
Proportion hispanic	9.470**	1.628	0.475	5.564**	4.117**
Proportion below poverty	3.218	3.197*	5.178**	2.792	2.556
Proportion owner-occupied homes	2.388**	–1.216*	–0.513	–2.091**	–2.007**
Constant	0.858	–0.489	2.742*	6.246**	1.404
Spatial lag parameter (λ)	–	0.764**	–	0.328**	–
Spatial error parameter (ρ)	–	–	0.883**	–	0.617**
Model comparison:					
Residual Moran's I: (1st order queen)	0.556**	0.018	–	0.097*	–
Residual Moran's I: (2,000 m radius)	0.450**	–	–0.067	–	0.014
Robust lagrange mulitiplier (lag)	–	53.583**	–	22.039**	–
Robust lagrange mulitiplier (error)	–	–	53.242**	–	66.201**
Adjusted or Pseudo R-squared	0.352	0.746	0.759	0.509	0.597
Akaike information criterion (AIC)	2,889	2,456	2,466	2,756	2,700

*$p < 0.05$; **$p < 0.01$.

value of the LM statistic. This is also the SAR model which yields the smallest Moran's I (closest to zero), or is the most effective in eliminating residual spatial autocorrelation.

17.5.2 Comparison of Regression Model Coefficients

The most statistically significant coefficient ($p < 0.001$) in all four SAR models is the spatial lag or error parameter, confirming again the utility of including this additional term in the equation and the limitation of OLS regression to account for spatial autocorrelation in the data. While the inclusion of the spatial autoregressive coefficient minimizes residual spatial autocorrelation and improves overall model fit considerably, the estimated parameters for most explanatory variables in all SAR models show a decrease in absolute value with respect to those in the OLS model. The largest declines in SAR model coefficients are indicated by the racial/ethnic variables that were significant and positive in the OLS model. While the parameters for Black and Hispanic proportions remain large enough to be significant at the 5% level in the distance-based SAR models, these are not significantly different from zero ($p > 0.10$) in the contiguity-based models. In contrast, the proportion below poverty was non-significant in the OLS model, but its coefficient is large enough to be positive and significant in the contiguity-based SAR model.

While the significance and signs of model parameters in the distance-based SAR models are generally consistent with those in the original OLS model, contiguity-based SAR models yield very different results. An analysis of environmental justice or health risk disparity utilizing a contiguity-based SAR model would lead to the conclusion that poverty is a significant predictor of cancer risk from on-road air toxics in the Tampa Bay MSA, instead of race or ethnicity. It is important to consider, however, that contiguity-based spatial weights matrix is likely to overemphasize spatial effects when tracts are irregular in size across the area, or unusually large in certain areas. The overestimation of the effect of spatial neighbors in the regression equation, as represented by a large and highly significant spatial autoregressive coefficient, could lead to a potential underestimation of another model parameter (e.g., Black proportion) resulting in a non-significant coefficient. The extent of this underestimation appears to be smaller in the distance-based SAR model, based on the relative decline in estimated parameters associated with the explanatory variables.

17.6 Concluding Discussion

The concept of positive spatial autocorrelation is viewed as the operationalization of Tobler's first law (TFL) of geography which suggests that closer areas are more similar in value than distant ones. The presence of such autocorrelation violates the key assumption of independence for traditional statistical tests such as linear regression. When spatial dependence exists in the data but is ignored in regression analysis, standard tests of significance for model coefficients are likely to be biased. Although the concepts of TFL and spatial autocorrelation have been known for several decades, numerous empirical studies have used ordinary least squares (OLS) regression of spatially aggregated data to determine the existence of racial/ethnic and socioeconomic inequities in the distribution of environmental pollution and related health risks. To derive reliable results for geospatial analysis of environmental injustice and health risk disparities, there is growing need to utilize analytical techniques that acknowledge TFL and control for the effects of spatial autocorrelation.

This chapter has explored the implementation of regression models that can be used to minimize spatial autocorrelation and provide a more valid statistical basis for inferences regarding the significance of explanatory variables such as race or poverty, using a user-friendly spatial analysis software package. The results suggest that the application of conventional OLS regression to analyze cancer risk from on-sources of air toxics in the Tampa Bay MSA could lead to erroneous conclusions, because the variables and model residuals display significant and positive spatial autocorrelation. The findings of this study clearly demonstrate the potential of simultaneous autoregressive (SAR) models in reducing or removing spatial dependence of residuals and thus satisfying the assumption of independently distributed errors in regression analysis of geospatial data. The results also support previous findings that type I errors and parameters estimated from traditional regression analysis are strongly inflated when spatial autocorrelation is present (Kissling and Carl 2008). The statistical effect of race and ethnicity on cancer risk, for example, declines when a spatial autoregressive parameter is included in the model. This shift in parameter estimates between non-spatial and spatial regression can be explained, in part, by the fact that both Black and Hispanic proportions show the greatest spatial autocorrelation within the MSA compared to other independent variables.

The study indicates that SAR model performance and results depend on several analytical choices related to model specification (lag or error) and the approach used to select neighbors for defining the spatial weights matrix. Although the SAR models based on the contiguity-based approach show the largest improvement in model fit, diameters of census tracts vary considerably across this MSA, ranging from 500 to 10,000 m. Unless spatial units are represented by square grids of similar size, the use of rook or queen contiguity could lead to an inconsistent estimation of neighbor effects. This issue has also been mentioned in recent environmental justice studies to justify a distance-based weights matrix for spatial regression with census tract data (Pastor et al., 2005; Chakraborty, 2009). In this study, spatial autocorrelation in model residuals is found to be non-significant and smallest in the distance-based spatial error model, which is also the recommended model according to Anselin's (2005) decision rule. The spatial error specification operationalized by a distance-based (2,000 m) spatial weights matrix provides the most theoretically and empirically valid SAR model for this data and study area.

While this chapter emphasizes the implementation of spatial regression models, it also demonstrates the utility of GIS for geospatial analysis of disproportionate environmental health impacts. GIS software allows a researcher to: (a) integrate various data sets necessary for such analysis (e.g., pollution indicators and population characteristics); (b) represent such data in cartographic or map form; (c) apply spatial analytic techniques that can be used to identify neighbors and compute spatial weights; and (d) visually examine the spatial distribution of statistical analysis results (e.g., regression model performance and errors). Additionally, it is important to consider that spatial analysis software programs such as GeoDa are designed to complement GIS functionalities, and not serve as a substitute. Several fundamental GIS operations that are relevant to the construction of spatial weight matrices and regression analysis (e.g., map projections, merging, dissolving, or modifying data/shape files) are not available in GeoDa.

In conclusion, more systematic empirical research is clearly necessary to evaluate the usefulness and performance of spatial regression models for different data resolutions, neighbor definitions, spatial weights matrices, and geographic scales. Instead of choosing either the spatial lag or error specification, future work should also explore the application of models that includes both types of spatial autoregressive coefficients, as suggested in recent research (Kissling and Carl 2008). The problems and analytical techniques discussed in this chapter, however, are critically important for researchers and practitioners involved with geospatial analysis of environmental health data.

Author Biography

Jayajit Chakraborty is an Associate Professor of Geography at the University of South Florida, Tampa. He received his Ph.D. in Geography in 1999 from the University of Iowa, Iowa City. His research interests lie at the intersection of geographic information science, environmental geography, and urban geography, and include air pollution, environmental health, racial/ethnic disparities, and vulnerability to environmental hazards.

Dr. Chakraborty is particularly interested in applications of geographic information science and spatial statistical techniques. He has authored articles in prominent academic journals such as *The American Journal of Public Health, Annals of the Association of American Geographers, Environment and Planning A, Health and Place, Risk Analysis*, and *The Professional Geographer*. His research activities have been supported by grants from the US Environmental Protection Agency (EPA), National Science Foundation (NSF), and the Florida Department of Transportation. Dr. Chakraborty is also the editor of a book titled *Spatial and Environmental Injustice in an American Metropolis: A Study of Tampa Bay, Florida* (Cambria Press, 2010) and an editorial board member for the *International Journal of Population Research*. Additionally, he has served as a Chair of the Association of American Geographers Hazards Specialty Group (2005–2007) and was commissioned by the EPA to co-author and present a paper for their *Strengthening Environmental Justice Research and Decision Making Symposium* (2010).

References

Anselin L (1988) Spatial econometrics: Methods and models. Kluwer Academic, Dordrecht, The Netherlands

Anselin L (2005) Exploring spatial data with GeoDa: A workbook. Spatial Analysis Laboratory, Department of Geography, University of Illinois, Urbana-Champaign, IL

Anselin L (2009). Glossary of key terms. The GeoDa Center for Geospatial Analysis and Computation. Arizona State University. http://geodacenter.asu.edu/node/390. Accessed 13 May 2009

Anselin L, Bera A (1998) Spatial dependence in linear regression models with an introduction to spatial econometrics. In Ullah A, Giles D (eds) Handbook of applied economic statistics. Marcel Dekker, New York, pp. 237–289

Ash M, Fetter TR (2004) Who lives on the wrong side of the environmental tracks? Evidence from the EPA's Risk-Screening Environmental Indicators model. Soc Sci Q 78: 793–810

Bowen W, Salling M, Haynes K, Cyran E (1995) Toward environmental justice: Spatial equity in Ohio and Cleveland. Ann Assoc Am Geogr 85:641–663

Burke LM (1993) Race and environmental equity: A geographic analysis in Los Angeles. Geo Info Syst 3:44–50

Chakraborty J (2009) Automobiles, air toxics, and adverse health risks: Environmental inequities in Tampa Bay, Florida. Ann Assoc Am Geogr 99:674–697

Cliff AD, Ord JK (1973) Spatial autocorrelation. Pion Limited, London

Cliff AD, Ord JK (1981) Spatial processes: Models and applications. Pion Limited, London

Daniels G, Friedman S (1999) Spatial inequality and the distribution of industrial toxic releases: Evidence from the 1990 TRI. Soc Sci Q 80:244–262

Environmental Defense Fund (2008). The environmental defense scorecard. http://www.scorecard.org. Accessed 28 Jan 2008

EPA (Environmental Protection Agency) (2004) Air toxics from motor vehicles. EPA400-F-92-004. Office of Transportation and Air Quality, US Environmental Protection Agency, Washington, DC

EPA. (2008a). National-scale air toxics assessment – glossary of key terms. http://www.epa.gov/ttn/atw/nata1999/gloss1.html. Accessed 10 Jan 2008

EPA. (2008b). National-scale air toxics assessment – 1999 assessment results. http://www.epa.gov/ttn/atw/nata1999/nsata99.html. Accessed 10 Jan 2008

EPA. (2008c). National-scale air toxics assessment – The four steps of the assessment. http://www.epa.gov/ttn/atw/nata1999/4steps99.html. Accessed 10 Jan 2008

FHWA (Federal Highway Administration). (2008). State and metro-level vehicle miles traveled data. http://www.njfuture.org/Media/Docs/VMT Data for States and Cities.pdf. Accessed 10 Jan 2008

Fricker RD, Hengartner NW (2001) Environmental equity and the distribution of Toxic Release Inventory and other environmentally undesirable sites in Metropolitan NYC. Environ Ecol Stat 8:33–52

Getis A (2007) Reflections on spatial autocorrelation. Reg Sci Urban Econ 37:491–496

Gilbert A, Chakraborty J (2008) Health risks from hazardous air pollutants: Analyzing environmental injustice in Florida. Fla Geographer 39:71–94

Grineski S (2007) Incorporating health outcomes into environmental justice research: The case of children's asthma and air pollution in Phoenix, Arizona. Environ Hazards 7:360–371

Grineski S, Collins T (2008) Exploring environmental injustice in the Global South: Maquiladoras in Ciudad Juárez. Popul Environ 29:247–270

Grove JMAR, Troy JP, O'Neil-Dunne M, Burch WR Jr., Cadenasso ML, Pickett STA (2006) Characterization of households and its implications for the vegetation of urban ecosystems. Ecosystems 9:578–597

Kissling WD, Carl G (2008) Spatial autocorrelation and the selection of simultaneous autoregressive models. Glob Ecol Biogeogr 17:59–71

Legendre P (1993) Spatial autocorrelation: Trouble or new paradigm? Ecology 74:1659–1673

Maantay J (2007) Asthma and air pollution in the Bronx: Methodological and data considerations in using GIS for environmental justice and health research. Health Place 13:32–56

Margai FL (2001) Health risks and environmental inequity: a geographical analysis of accidental releases of hazardous materials. Prof Geogr 53(1):422–434

Moran PAP (1950) Notes on continuous stochastic phenomena. Biometrika 37:17–23
Morello-Frosch R, Pastor M, Sadd J (2001) Environmental justice and Southern California's "riskscape" – the distribution of air toxics exposures and health risks among diverse communities. Urban Aff Rev 36:551–578
Odland J (1988) Spatial autocorrelation. Sage, Newbury Park, CA
Paelinck JHP, Klaassen. LH (1979) Spatial econometrics. Saxon House, Farnborough, UK
Pastor M, Sadd JL, Morello-Frosch R (2004) Waiting to inhale: the demographics of toxic air releases in 21st century California. Soc Sci Q 85(2):420–440
Pastor M, Morello-Frosch R, Sadd J (2005) The air is always cleaner on the other side: Race, space, and ambient air toxics exposures in California. J Urban Aff 27:127–148
Pastor M, Scoffins J (2008) 'Space... the final frontier': Autocorrelation and small area income forecasting models. J Plan Educ Res 27:431–443
Pollock PH, Vittas ME (1995) Who bears the burden of environmental pollution? Race, ethnicity, and environmental equity in Florida. Soc Sci Q 76:294–310
Ringquist EJ (1997) Equity and the distribution of environmental risk: The case of TRI facilities. Soc Sci Q 78:811–829
Sicotte D, Swanson S (2007) Whose risk in Philadelphia? Proximity to unequally hazardous industrial facilities. Soc Sci Q 88:515–534
Stephan FF (1934) Sampling errors and interpretations of social data ordered in time and space. J Am Stat Assoc 29(185 Supplement):165–166
Stretesky P, Lynch MJ (1999) Environmental justice and the predictions of distance to accidental releases in Hillsborough County, Florida. Soc Sci Q 80:830–846
Sui DZ (2004) Tobler's first law of geography: A big idea for a small world? Ann Assoc Am Geogr 94:269–277
Tobler WR (1970) A computer movie simulating urban growth in the Detroit region. Econ Geogr 46:234–240
Venn AJ, Lewis SA, Cooper M, Hubbard R, Britton J (2001) Living near a main road and the risk of wheezing illness in children. Am J Respir Crit Care Med 164:2177–2180
Voss PR, White KJC, Hammer RB (2006) Explorations in spatial demography. In Kandel WA, Brown DL (eds) Population change and rural society. Springer, Dordrecht, pp. 407–429
Wu YC, Batterman S (2006) Proximity of schools in Detroit, Michigan to automobile and truck traffic. J Expo Sci Environ Epidemiol 16:457–470

Chapter 18
A Spatially Explicit Environmental Health Surveillance Framework for Tick-Borne Diseases

Aldo Aviña, Chetan Tiwari, Phillip Williamson, Joseph Oppong, and Sam Atkinson

Abstract We demonstrate how applying a spatially explicit context to an existing environmental health surveillance framework is vital for more complete surveillance of disease, and for disease prevention and intervention strategies. To illustrate this framework, we present a case study that involves estimating the risk of human exposure to Lyme disease. The spatially explicit framework divides the surveillance process into three components: hazard surveillance, exposure surveillance, and outcome surveillance. The components are used both collectively and individually, to assess risk of exposure to infected ticks. By utilizing all surveillance components, we identify different areas of risk which would not have been identified otherwise. Hazard surveillance uses maximum entropy modeling and Geographically Weighted Regression analysis to create spatial models that predict the geographic distribution of ticks in Texas. Exposure surveillance uses GIS methods to estimate the risk of human exposures to infected ticks, resulting in a map that predicts the likelihood of human-tick interactions across Texas, using LandScan 2008™ population data. Lastly, outcome surveillance uses kernel density estimation-based methods to describe and analyze the spatial patterns of tick-borne diseases, which results in a continuous map that reflects disease rates based on population location. Data for this study was obtained from the Texas Department of Health Services and the University of North Texas Health Science Center. The data includes disease data on Lyme disease from 2004 to 2008, and the tick distribution estimates are based on field collections across Texas from 2004 to 2008.

Keywords Environmental health surveillance · Tick-borne disease · Medical geography · GIS

A. Aviña (✉)
Department of Geography, University of North Texas, Denton, TX, USA
e-mail: aldo.avina@unt.edu

J.A. Maantay, S. McLafferty (eds.), *Geospatial Analysis of Environmental Health*, Geotechnologies and the Environment 4, DOI 10.1007/978-94-007-0329-2_18,

18.1 Introduction

18.1.1 Environmental Public Health Surveillance Framework

Several factors must be taken into consideration when performing environmental public health surveillance or examining how environmental factors influence public health. Public health surveillance is the "ongoing systematic collection, analysis, and interpretation of data on specific health events affecting a population, closely integrated with the timely dissemination of these data to those responsible for prevention and control (Thacker et al., 1996, p. 633)." According to Thacker et al. (1996), environmental public health surveillance can be divided into three interrelated components: hazard, exposure and outcome. Hazard surveillance tracks a disease agent in the environment; exposure surveillance identifies transmission/contact mechanisms between the agent and human populations; outcome surveillance tracks the observed number of cases that are clinically apparent as a result of such exposures. This framework provides an effective set of guidelines, rooted in public health and epidemiology, for performing environmental public health surveillance.

Although this model of the surveillance process provides an important framework, it does not explicitly address the geographical context of each component. Space and location are central to each component, but efforts towards developing such geographical dimensions have been relatively limited. Nuckols et al. (2004) provide a comprehensive review of exposure assessment studies that utilize a GIS-based approach to address at least one of the components stated in the environmental public health surveillance framework. Although these studies represent a diversity of GIS and spatial analytic approaches, most focus on exposure assessment as the end goal and do not attempt to combine the other components of Thacker's model within a spatial framework. This chapter illustrates such a comprehensive spatial framework in examining the incidence of Lyme disease in Texas. Specifically, we develop spatial analysis methods that use ecological variables to model the detailed geographic distribution of ticks in Texas and link this with information with exposures and outcomes. This research shows that adding a spatial component enhances the disease prediction and surveillance capabilities of the Thacker et al.'s framework, thereby improving its utility for environmental health surveillance.

18.1.2 Lyme Disease Within Texas

In the United States, Lyme disease is the most common vector-borne disease with over 20,000 confirmed new cases reported each year (Centers for Disease Control, 2007). The black-legged tick, *Ixodes scapularis*, is the primary vector of *Borrelia burgdorferi*, the bacteria that causes Lyme disease in humans (Dennis et al., 1998). In 2008 alone, New Hampshire and Delaware saw the highest rates of Lyme disease at 92.0 and 88.4 cases per 100,000 people, respectively, up from 17.4 and 40.8 per 100,000 in 2004. The national rate was 9.4 per 100,000 in 2008 (DVBID, 2010). The presence of Lyme disease in a population presents both an epidemiological and economic burden due to the high cost of diagnosis and treatment. In a study

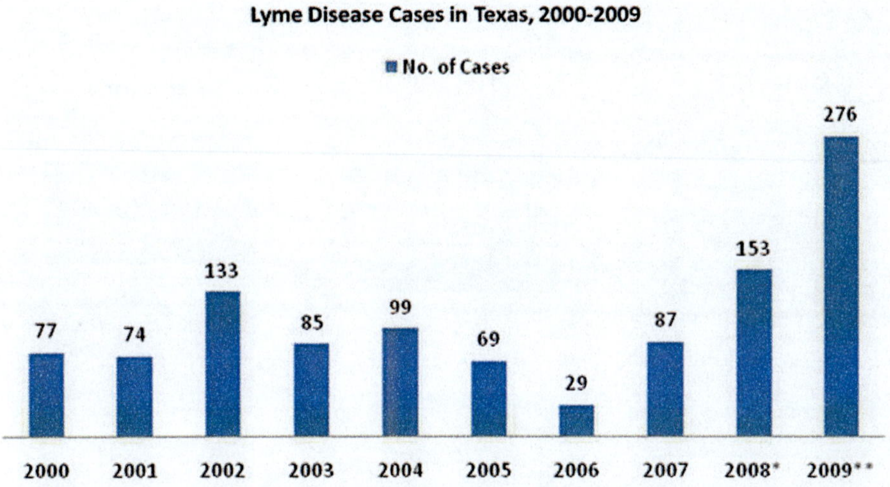

Fig. 18.1 Trend of Lyme disease in Texas, 2000–2009. *Totals 2008 and after include number of probable cases and number of confirmed cases. **Case definition change

performed in eastern Maryland, the average expected cost related to Lyme disease was $1,965 per patient, in year 2000 dollars (Zhang et al., 2006).

Figure 18.1 shows the trend of Lyme disease in Texas from 2000 to 2009. It should be noted that beginning in 2008, both probable and confirmed cases are used to calculate the prevalence of the disease, whereas earlier only confirmed cases were included. Additionally, a change in the case definition for Lyme disease in 2009 resulted in a significant increase in the case count from 2008 to 2009, although the trend in the latter part of the decade indicates cases may be on the rise.

The uncertainties in understanding the burdens of Lyme disease due to under-reporting and changes in case definitions are further compounded by the fact that the spatial distribution of ticks that are vectors for Lyme disease has not been investigated at high geographic resolution in Texas. Knowing the detailed spatial distribution of *I. scapularis* in Texas may help identify areas of increased disease risk. Geographically-driven prevention and control measures can mitigate the risk of Lyme disease and other tick-borne diseases in Texas.

18.2 Data and Methods

A spatially-explicit framework involves understanding where a disease agent exists in the environment, where exposure between the agent and the population occurs, and where the burden of the agent on the population is most significant. There are several considerations to be made when fully developing such a framework – the questions asked, the data used, and the statistics applied all play a role in how well the study addresses an environmental health problem (Elliot and Wartenberg, 2004). Following the public health surveillance framework discussed earlier, we provide a spatial context to the surveillance of Lyme disease in Texas. A spatially explicit predictive model is used to determine the geographic distribution of infected

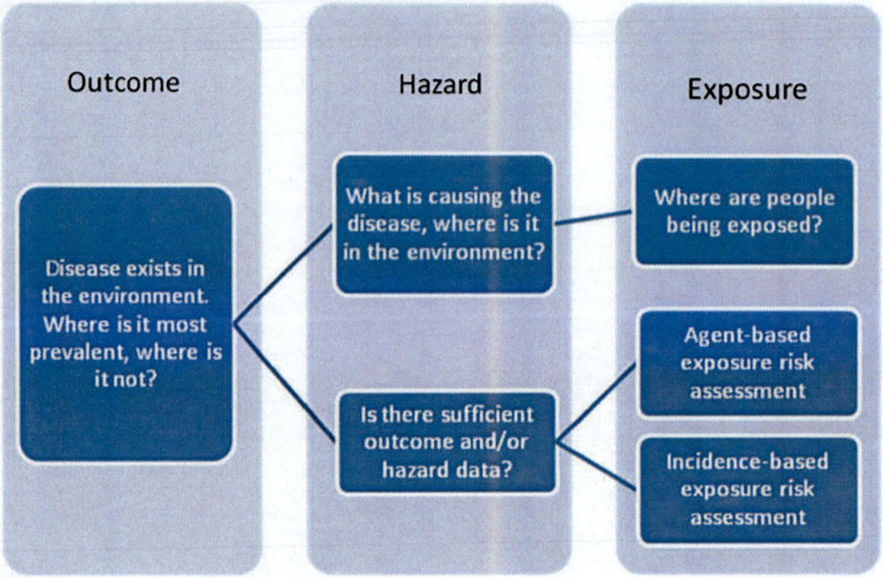

Fig. 18.2 A flowchart showing a possible decision-making process when performing spatially explicit environmental health surveillance

I. scapularis ticks across Texas. Estimates of human exposure to infected ticks are determined using a GIS-based approach. Finally, disease maps representing the incidence of Lyme disease across Texas are produced using kernel density estimation methods. From a public health perspective, the results of each surveillance component can be compared to develop relevant intervention and prevention strategies. This approach, shown in Fig. 18.2, may be applied by public health officials for more effective surveillance, using a geographic point of view.

18.2.1 Hazard Surveillance

It has been shown that the presence of *I. scapularis* increases the risk of Lyme disease (Eisen et al., 2006; Glass et al., 1995; Kitron and Kazmierczak, 1997). However, public health efforts to determine the spatial distributions of tick populations in Texas and in other parts of the country remain inadequate due to insufficient data collection (Dennis et al., 1998). Ticks have certain ecological and climatic requirements that create a suitable habitat for existence (Glass et al., 1995; Brownstein et al., 2003). In short, to predict the distribution of tick populations, we have to relate their occurrence in the environment, through observation, to their environmental requirements (Guisan and Zimmermann 2000). Previous literature suggests that the following environmental variables are most indicative of *I. scapularis* presence: Maximum, minimum, and mean temperature and precipitation, vapor pressure, soil types, and forest cover.

In a study in Baltimore, the environmental make-up surrounding the residences of Lyme disease cases was analyzed to determine the environmental variables associated with Lyme disease exposure (Glass et al., 1995). The study showed that residential areas that are near forested areas and sit on soils with medium- to high-water capacity correlate with areas of high Lyme prevalence. Since ticks are the primary vector for Lyme disease, these variables are also indicative of environments that are suitable for tick populations. In another study, Brownstein et al. (2003) predict the distribution of *I. scapularis* using maximum, minimum, and mean temperature, and vapor pressure. These studies show that specific environmental factors are indicative of tick presence.

This study utilizes the environmental determinants mentioned above and tick occurrence data in conjunction with a maximum-entropy modeling method, Maxent, to predict the distribution of *I. scapularis* in Texas. Maxent works well with limited, presence-only datasets, and quickly processes data to create probability distributions based on environmental parameters and species occurrence (Phillips et al., 2006). The model computes an area under the curve (AUC) value, which is a calculation of how well the model predicts the observed data on a scale from 0 to 1. This AUC value is a measure of model performance based on observed data, but because Maxent is so novel, we also test the data using other proven methods. To increase our confidence and to test the accuracy of the results, we also use Geographically Weighted Regression (GWR), which analyzes the spatial relationship between the dependent variable and the independent variables, and how the relationship differs across geographic space (Fotheringham et al., 2002). GWR yields a map of the predicted presence of *I. scapularis* at the zip-code level, which is compared to the Maxent results. A chi-squared goodness of fit test is then used to determine how well the GWR predicts actual tick data.

The tick occurrence data was generated between 2004 and 2008 from submissions to the Texas Department of State Health Services (TX DSHS). The occurrence data was aggregated to the zip-code level. Molecular screening for tick-borne pathogens, including *Borrelia burgdorferi*, was performed by the University of North Texas Health Science Center (Williamson et al., 2010). Maxent requires point data, while the existing data is aggregated to the zip code. The data is scaled down to the point level by randomly distributing tick occurrences within the corresponding zip code. This allows us to distribute the tick data across the entire zip code, rather than just at its centroid, which can cause autocorrelation problems. The points are randomly distributed and run through Maxent 30 times, and the results are averaged. Scaling down the tick data does present problems, which are discussed later, but is done due to the lack of point-level data. Performing a GWR, however, allows us to compare the results of scaling down the data.

The environmental parameters input into Maxent are shown in Table 18.1 (National Elevation Dataset, 2010; WorldClim.org, 2005; PRISM Climate Group, 2009; National Land Cover Dataset, 2009; Soil Survey Geographic Database, 2009). These parameters were chosen based on the studies discussed earlier, which demonstrated key ecological variables associated with *I. scapularis* presence.

Table 18.1 Data, sources, and purpose for usage

Data	Source	Purpose
Elevation	National elevation dataset	Determine if elevation is a factor of tick presence in Texas
Temperatures	WorldClim.org	Ticks sensitive to temperature extremes
Precipitation	PRISM climate group	Tick presence is dependent on moisture availability
Land use/land cover	National land cover dataset	Forest cover relates to tick suitability
Soil texture	Soil survey geographic database	Soil texture influences moisture content

By mapping the distribution of *I. scapularis*, something which has not been done at this scale before, we can perform an exposure risk assessment based on the location of these ticks. Populations living in areas that have increased probability of tick suitability will have a higher risk of tick exposure.

18.2.2 Outcome Surveillance

To better understand the characteristics of Lyme disease in Texas, we must map the disease incidence. Understanding the spatial patterns of disease allows researchers to make more informed observations and decisions regarding the prevention and control of the disease. Our Lyme disease case data is aggregated to the zip code level, and is susceptible to instable incidence rates due to the small population denominators in zip codes with low populations – the small numbers problem. In order to reduce the influence of the small numbers problem, it is necessary to smooth the disease count data using GIS and kernel density estimation methods. Using kernel density estimation allows us to normalize incidence rates that would otherwise be over or underestimated, creating a more accurate representation of disease incidence (Tiwari and Rushton, 2005). For example, a zip code with one Lyme disease case with a population of 75 results in a rate of 1,333 cases per 100,000 people – an extraordinarily high incidence rate. To address this problem, instead of only counting the population of that one zip code, we also count the populations of the surrounding zip codes, up to a certain threshold, and produce a more representative rate.

To implement this procedure, we lay a 10-mile grid across the study area. Each grid point has a rate calculated based on the number of cases and the population that falls within its spatial filter (kernel). In this case, a population threshold of 10,000 is chosen. The spatial filter increases in size outward from the grid point until it contains a minimum of 10,000 people counted from the surrounding zip code centroids. Next, the total numbers of disease cases in the zip codes whose centroids fall within the filter are used to calculate the incidence rate at that respective grid point. Each grid point in the study area now has a calculated incidence rate, and these rates are interpolated using the Inverse Distance Weighted method, creating a smoothed map. For a more complete description of this method, including diagrams, see Tiwari and Rushton (2005).

The size of the kernels or spatial filters depends on the density of the underlying population. Note that larger filters are used in sparsely populated areas and smaller filters are used in densely populated areas. Larger filters provide more smoothing, but also result in loss of geographic detail whereas smaller filters provide greater geographic detail. Because the adaptive filter method is based on varying filter sizes that adapt to population density, we are able to produce maps that provide high geographic resolution in those areas where such detail is expected while maintaining rate stability in rural areas with sparse populations.

Web-based GIS software (Tiwari, online) is used to apply the spatially adaptive filters method to Lyme disease cases in Texas, and create a smoothed representation of Lyme disease burden in Texas. Such maps can be used to identify areas of high or low disease burdens. The case data, which are reported to the CDC, were obtained from the Texas DSHS at the zip-code level for the study period 2004–2008. For national reporting, the CDC classifies a confirmed case of Lyme disease as a case with either typical skin lesions (erythema migrans) or a case with late manifestations (non-specific symptoms) confirmed by laboratory testing.

18.2.3 Exposure Surveillance

In tick-borne disease studies, exposure risk can be estimated by performing an agent-based study, incidence-based study, or both (LoGiudice et al., 2003; Glass et al., 1995; Eisen et al., 2006). An agent-based approach assesses exposure risk based on the presence and location of a disease-causing agent; an incidence-based approach assesses exposure risk based on the occurrence of disease. This case study uses the former method to estimate the risk of exposure to *I. scapularis* infected with *B. burgdorferi* in Texas. Eisen et al. (2006) previously used such an approach to identify the spatial patterns of Lyme disease risk in California by identifying the relationship between locations of disease cases and locations where infected ticks are likely to be present. It should be noted that the presence of measurement errors, mobility in human populations, and uncertainties in the geocoding process make it difficult to accurately estimate human exposures to environmental risks (Eisen et al., 2006; Glass et al., 1995; Kitron and Kazmierczak, 1997). The availability of high-resolution data on environmental hazards and human populations reduces the uncertainty in such agent-based risk assessment approaches (Elliot and Wartenberg, 2004).

The goal of our exposure surveillance is to produce an exposure risk map that allows us to identify areas of greater population exposure to infected *I. scapularis* ticks. Populations in those areas can be warned about their increased risk of coming into contact with infected ticks. Exposure risk is calculated by multiplying for each pixel (1000 m^2 small area) the infection rate of *I. scapularis* with *B. burgdorferi* in Texas by the estimated number of *I. scapularis* in the pixel,

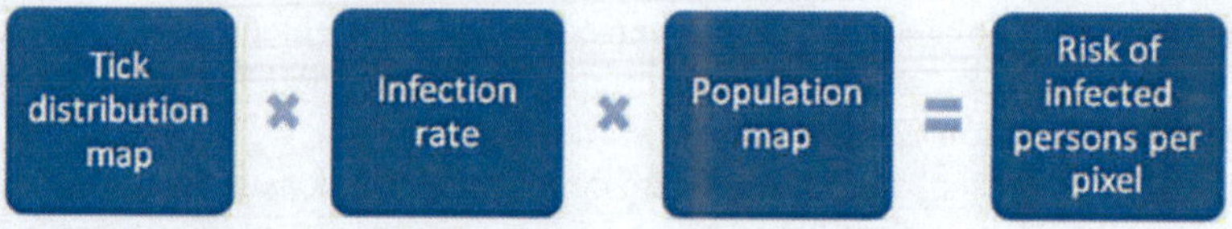

Fig. 18.3 Agent-based exposure risk assessment of Lyme disease

generated by Maxent. This results in a detailed map that portrays the estimated distribution of infected *I. scapularis* ticks. To estimate disease risk, the distribution of infected ticks obtained from the previous step is multiplied by population of the pixel (Fig. 18.3). This results in a map the shows the potential number of infections per pixel. Population data were obtained from Oak Ridge National Laboratory's LandScan data set. Note that the geographic resolutions of the two rasters (infected tick distribution and human population) are consistent at approximately 1,000 m^2 (LandScan 2008^{TM}).

18.3 Results

18.3.1 Hazard Surveillance

Figure 18.4 shows the Maxent output of the estimated distribution of *Ixodes scapularis*, based on environmental suitability. The areas of highest suitability occur in eastern and northeastern Texas.

The reliability of the Maxent output is assessed using an Average Area Under the Curve (AUC) number. This represents probability that a randomly selected pixel would be correctly classified by the model as being positive or negative for occurrence of ticks (Phillips et al., 2004). The AUC number of the map in Fig. 18.4 is 0.92. Note that a high AUC does not necessarily mean that the output is valid, rather that the model predicted the tick distribution well, according to the available data.

To check the robustness of these results, we compare the predicted tick distribution based on a Geographically Weighted Regression (GWR) model with the results of the maximum entropy model. Figure 18.5 shows the results of running GWR on the zip code level tick data; the map on the left in Fig. 18.5 shows the distribution of predicted tick locations, classified by standard deviation. Areas with higher values are expected to have higher numbers of ticks. The GWR model shows a similar pattern to that of our Maxent results, with higher tick suitability in eastern Texas. Results from the chi-square goodness-of-fit test, which measures the difference between what is predicted by the GWR and what is observed, is shown on the right in Fig. 18.5. The chi-square test outlines areas of over-prediction, especially a small cluster in east Texas.

Because of the favorable results from the chi-square test and the strong correspondence between the Maxent and GWR results, we continue to use the maximum entropy results to estimate the presence of tick hazard.

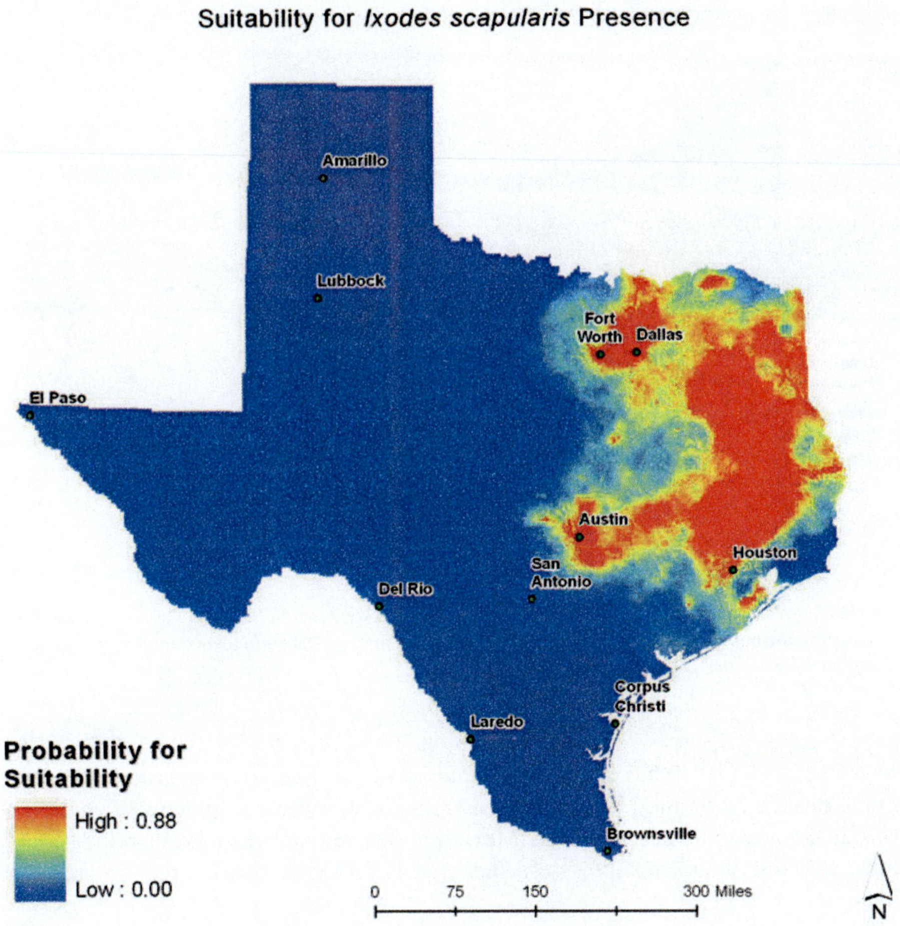

Fig. 18.4 Estimated distribution of *Ixodes scapularis*, based on environmental suitability

18.3.2 Outcome Surveillance

Figure 18.6 presents the smoothed map of Lyme disease incidence rates in Texas. It indicates that the highest burdens occur in central and southern Texas, outside of the most populous cities and away from areas of high tick suitability, according to our Maxent results. At this point it is unknown why some areas experience higher incidence rates than others. Some can be due to populations' accessibility to hike and bike trails, greenbelts, and campgrounds – areas where people are exposed to ticks. Another reason is that areas that would normally not be suitable for *I. scapularis* become suitable when agricultural and irrigation processes change the landscape in a way that creates a suitable and sustainable habitat for the ticks or its reservoirs. These agricultural processes were not incorporated in the tick hazard model. Also,

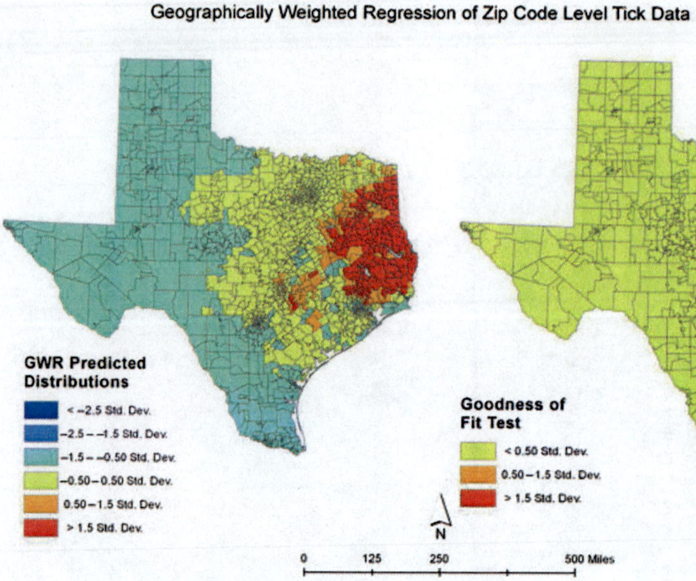

Fig. 18.5 Results from running Geographically Weighted Regression (GWR). The map on the *left* shows the distribution of predicted tick occurrences. The map on the *right* tests model performance, based on a chi-square goodness of fit test

Martinez et. al. (1999) has discovered *B. burgdorferi* in deer in northeastern Mexico. Lyme disease cases may be transported from (or to) Mexico, affecting Lyme disease incidence estimates in the US-Mexico border region. More localized research must be done to accurately assess the origins of Lyme disease risk in specific regions.

18.3.3 Exposure Surveillance

Exposure risk is defined as the risk associated with living in an area with high infected tick suitability. Figure 18.7 shows the geographic variation in risk of Lyme disease, expressed as a rate for every 100,000 people. The output of the model is initially the raw number of people that could be potentially exposed to *B. burgdorferi*, but changed to a rate to reflect whole numbers rather than decimal numbers. For example, a pixel with a value of 110 would indicate that for every 100,000 people living in the pixel, an estimated 110 would be exposed to an infected *I. scapularis* tick. These values are based on the assumption that every person living in the pixel has equal chance of coming into contact with an infected tick – an unrealistic assumption. However, the values provide a measure of relative population exposure in different geographic areas that may be useful in planning Lyme disease education and prevention programs.

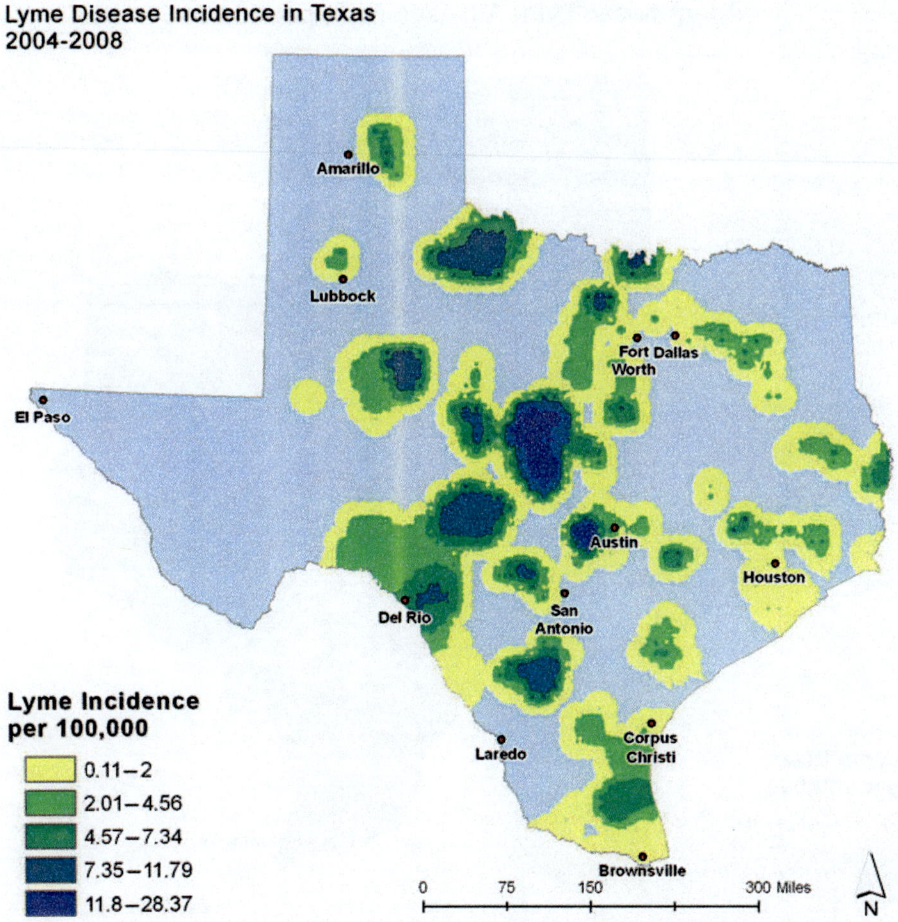

Fig. 18.6 Rates of Lyme disease based on cumulative cases from 2004 to 2008, standardized by the 2007 population

18.4 Discussion

Using Lyme disease as a case-study, we have shown that implementing a spatially explicit context to an existing environmental health surveillance framework allows for a more well-rounded surveillance study. Many studies assess exposure risk based on one surveillance component or another, while this study suggests that considering all the surveillance components is necessary to appropriately understand the patterns of disease, and the potential risk associated with it.

Agent-based risk maps are useful in identifying populations at greatest risk of exposure to the Lyme disease vector. Because the exposure risk is based on tick presence, it naturally follows the estimated tick distribution. This agent-based risk map serves as tool for guiding public health intervention strategies. The agent-based

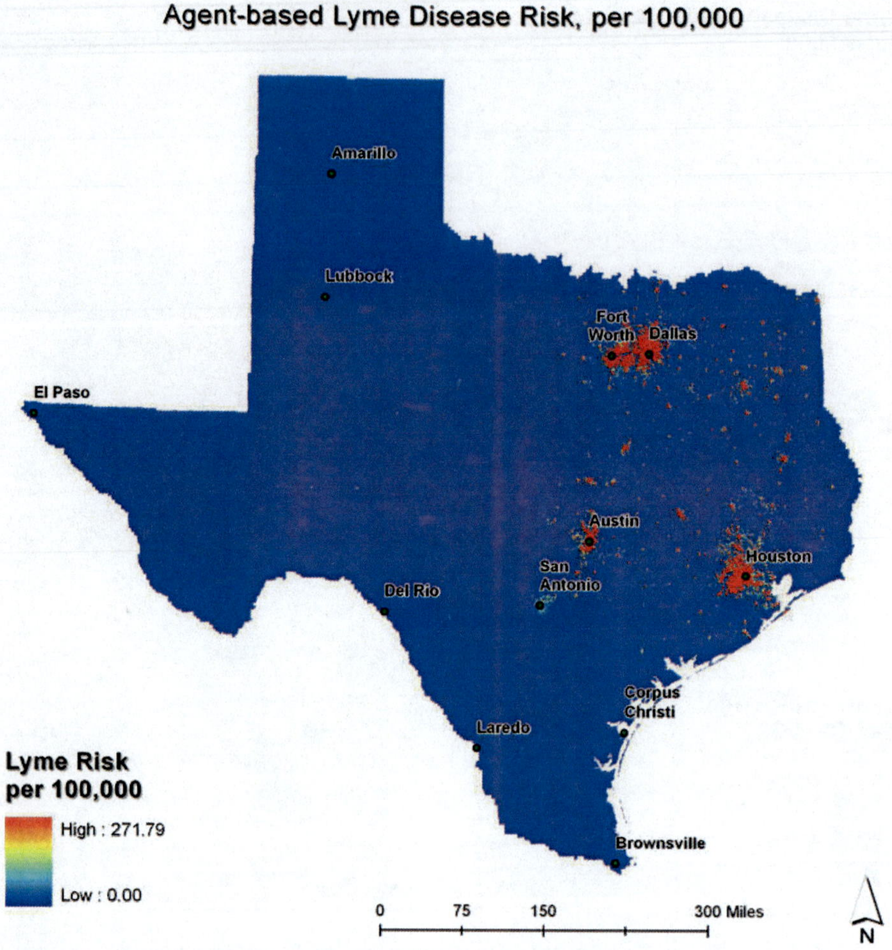

Fig. 18.7 Agent-based risk map of Lyme disease, multiplied by a population of 100,000. Based on the cumulative number of ticks collected from 2004 to 2008 and the population per pixel. Source: LandScan, 2008™, ORNL, UT-Battelle, LLC

map shows that areas of increased risk are located in the eastern portions of Texas, while the pattern of actual Lyme disease incidence is more prevalent in central to western Texas. Reasons for this difference need to be further investigated. As of now, we can only conjecture that it may be due mainly to insufficient tick sampling, people traveling to and becoming exposed in high-risk areas or the small possibility that *B. burgdorferi* is being spread by another species of tick. Areas that are highly suitable for tick presence that overlap areas with high Lyme disease prevalence could be targeted for intervention and prevention strategies.

Adding more data, both on tick locations and disease cases, will help in understanding why there is a geographic dissimilarity between *I. scapularis* distribution and Lyme disease incidence. Yet, these results indicate that by utilizing several components of surveillance, many more observations can be made about Lyme disease in Texas than would be possible with a single surveillance approach. Some observations include the geography of the overlap of Lyme disease incidence and high agent-based risk, and the fact that most areas of high Lyme disease incidence and tick suitability fall in different portions of Texas. Also, applying a spatial context to the environmental health surveillance framework, allows for more selectively-targeted and specific disease intervention and prevention strategies.

Although the Geographically Weighted Regression (GWR) results in Fig. 18.5 showed a promising degree of confidence in the results from the maximum entropy model, tick field surveys should be the next step, not only to confirm the results of the tick distribution map, but also to improve them. Scaling zip code data down to point data, as done in this study, is not realistic or practical, and creates errors in modeling tick distributions. With finer scale GPS data, one can better model the environmental determinants of tick distributions, leading to better statistical validation.

18.4.1 Conclusion

By implementing a spatially explicit context to an existing environmental health surveillance framework, and utilizing hazard, exposure, and outcome surveillance as a whole, rather than parts of a whole, it has been possible to examine the burdens of Lyme disease in Texas. Even though Texas is not heavily affected by the disease, the disease represents a potential threat that needs to be understood. Lyme disease in Texas has not been investigated in great detail, nor has the distribution of *I. scapularis* in the state been estimated and mapped. Although the results cannot be held with a high degree of confidence because of insufficient data, the concept of a spatially explicit environmental health surveillance framework is realistic, practical, and practicable. The distribution of the agent in question was predicted, the current distribution of Lyme disease incidence was mapped, and an exposure assessment was performed, uniting the three components of environmental health surveillance for understanding Lyme disease in Texas.

This study also reinforces the importance of using geographic information science in environmental health studies, as GIS was a critical component of the analysis and framework. Such a spatially explicit framework can be applied to many environmental health studies and to many kinds of environmental health concerns. Adding the dimension of space to environmental health analyses allows investigators to discover more meaningful results that can result in more effective public health intervention and prevention strategies.

Author Biographies

A first-generation American, **Aldo Aviña** currently attends the University of North Texas in Denton, Texas, completing a Master's of Science degree in Applied Geography, with a concentration in Applied GIS. In the fall of 2010, he began his doctoral studies in the

Environmental Health Science program in the School of Public Health at the University of North Texas Health Science Center in Fort Worth, Texas. Aldo also attended UNT as an undergraduate, performing research as a Ronald E. McNair Scholar. Over the past few years, Aldo has worked as a graduate assistant at UNT for the Department of Geography and the Institute of Applied Science, working on projects such as a campus tree inventory, and the characteristics of tuberculosis in Tarrant County, Texas. Also, Aldo is a research assistant for the UNT Health Science Center, studying the characteristics of several tick-borne diseases in Texas, and their respective vectors. While employed by Frito-Lay's Supply Chain & Logistics, he mapped route and delivery data.

Aldo became interested in geographic information systems as a McNair Scholar. While attempting to track and analyze the habitat requirements of a woodpecker in a wildlife management area, Aldo was introduced to the utility of a GIS, and was immediately convinced in the adaptability of a GIS to fit one's research needs. Much of the cutting-edge research done by many would be difficult, or nearly impossible, without the value, and usefulness of a GIS.

Dr. Chetan Tiwari is Assistant Professor of Geography at the University of North Texas. His research interests are in disease mapping, geographic information science, and computational geography. Specifically, he is interested in developing methods for mapping, analyzing, and visualizing the spatial patterns of disease burdens using spatial analysis methods and geographic information systems. Dr. Tiwari is also affiliated with the Computational Epidemiology Research Laboratory (CERL) at the University of North Texas.

Phillip Williamson, Ph.D., is an Assistant Professor in the Department of Forensic and Investigative Genetics at the University of North Texas Health Science Center. He is the Director for the Center for Biosafety and Biosecurity which includes the UNTHSC Tick-borne Disease Research Laboratory. His primary research focus is the development of methods and tools for rapid assessment of disease outbreaks (whether naturally occurring or intentionally released) and the study of efficient mechanisms to research epidemiology, genetics and associated clinical manifestations of emerging infectious disease.

Joseph Oppong is a Professor in the Department of Geography at University of North Texas, Denton, TX.

Dr. Sam Atkinson, Regents Professor at the University of North Texas, has a background in biology and environmental science/engineering. His research interests revolve around understanding the capabilities and limitations of current and future satellite remote sensing systems to examine ecosystem scale environmental questions. He has authored or co authored more than one-hundred scientific papers, technical reports, and books chapters, including two books on the effects of human activities on the environment. He was named Decker Scholar in 1997 for outstanding research in science and technology and Toulouse Scholar in 2008 for outstanding teaching and scholarly/creative achievements of the faculty of the graduate school.

References

Brownstein JS, Holford TR, Fish D (2003) A climate-based model predicts the spatial distribution of the Lyme disease vector Ixodes scapularis in the United States. Environ Health Perspect 111(9):1152–1157

Centers for Disease Control & Prevention (2007). Lyme disease-United States, 2003–2005. Morbidity and mortality weekly report [Online] 15 Jun, 56 (23):573–576. Available at: http://www.cdc.gov/mmwr/preview/mmwrhtml/mm5623a1.htm. Accessed 14 Oct 2009

Dennis DT et al (1998) Reported distribution of *Ixodes scapularis* and *Ixodes pacificus* (Acari: Ixodidae) in the United States. J Med Entomol 35(5):629–638

Division of Vector-Borne Infectious Diseases – Centers for Disease Control and Surveillance. Lyme Disease Statistics. [Online] (Updated 29 Jan 2010) Available at: http://www.cdc.gov/ncidod/dvbid/lyme/ld_statistics.htm. Accessed Feb 2010

Eisen RJ, Lane RS, Fritz CL, Eisen L (2006) Spatial patterns of Lyme disease risk in California based on disease incidence data and modeling of vector-tick exposure. Am J Trop Med Hyg 75(4):669–676

Elliot P, Wartenberg D (2004) Spatial epidemiology: current approaches and future challenges. Environ Health Perspect 112(9):998–1006

Fotheringham AS, Brunsdon C, Charlton ME (2002) Geographically weighted regression: the analysis of spatially varying relationships. Wiley, Chichester

Glass GE et al (1995) Environmental risk factors for Lyme disease identified with geographic information systems. Am J Public Health 85(7):944–948

Guisan A, Zimmermann NE (2000) Predictive habitat distribution models in ecology. Ecol Modell 135(2000):147–186

Kitron U, Kazmierczak JJ (1997) Spatial analysis of the distribution of Lyme disease in Wisconsin. Am J Epidemiol 145(6):558–566

LandScan 2008™ (2008) LandScan 2008™ high resolution global population dataset. Oak Ridge National Laboratory, UT-Battelle, LLC, Oak Ridge, TN

LoGiudice K, Ostfeld RS, Schmidt KA, Keesing F (2003) The ecology of infectious disease: Effects of host diversity and community composition on Lyme disease risk. Proc Natl Acad Sci USA 100(2):567–571

Martinez A, Salinas A, Martinez F, Cantu A, Miller DK (1999) Serosurvey for selected disease agents in white-tailed deer from Mexico. J Wildl Dis 35(4):799–803

National Elevation Dataset, US Geological Survey. The National Map Seamless Server. [Online] (Updated 01 Feb 2010) Available at: http://seamless.usgs.gov. Accessed Oct 2009

National Land Cover Dataset, US Geological Survey. Multi-Resolution Land Characteristics Consortium. [Online] (Updated 26 Jan 2010) Available at: http://www.mrlc.gov. Accessed Oct 2009

Nuckols JR, Ward MH, Jarup L (2004) Using geographic information systems for exposure assessment in environmental epidemiology studies. Environ Health Perspect 112(9):1007–1015

PRISM Climate Group, Oregon State University (2010). Precipitation 2004–2008. [Online] (Updated Apr 2010) Available at: http://www.prism.oregonstate.edu/index.phtml. Accessed 10 June 2009

Phillips SJ, Anderson RP, Schapire RE (2006) Maximum entropy modeling of species geographic distributions. Ecol Modell 190(2006):231–259

Phillips SJ, Dudík M, Schapire RE (2004). A maximum entropy approach to species distribution modeling. In Proceedings of the Twenty-First International Conference on Machine Learning. Banff, Alberta, Canada 04–08 July 2004. ACM: New York

Soil Survey Geographic (SSURGO) Database, US Department of Agriculture. [Online] (Updated May 2010) Available at: http://soils.usda.gov/survey/geography/ssurgo. Accessed 15 Oct 2009

Thacker SB, Stroup DF, Parrish RG, Anderson HA (1996) Surveillance in environmental public health: issues, systems, and sources. Am J Public Health 86(5):633–638

Tiwari C. Web-based disease mapping and analysis program. [Online] (Updated May 2010) Available at: http://www.webdmap.com. Accessed 1 Sep 2009

Tiwari C, Rushton G (2005) Using spatially adaptive filters to map late stage colorectal cancer incidence in Iowa. In Fisher P (ed) Developments in spatial data handling. Springer, London, pp. 665–676

Williamson PC et al (2010) Borrelia, Ehrlichia, and Rickettsia spp. in ticks removed from persons, Texas, USA. Emerging infectious diseases, [Online]. 16(3):441–446. Available at: http://www.cdc.gov/EID/content/16/3/441.htm. Accessed 25 May 2010

WorldClim.org. Global climate data. [Online] (Updated Unkown) Available at: http://www.worldclim.org. Accessed 15 Feb 2010

Zhang X et al (2006) Economic impact of Lyme disease. Emerging infectious diseases, [Online]. 12(4):653–660. Available at: http://www.cdc.gov/ncidod/EID/vol12no04/05-0602.htm. Accessed 15 April 2010

Chapter 19
Using Distance Decay Techniques and Household-Level Data to Explore Regional Variation in Environmental Inequality

Liam Downey and Kyle Crowder

Abstract This chapter links individual- and household-level data from the nationally representative Panel Study of Income Dynamics (PSID) with neighborhood-level environmental hazard data derived from the Environmental Protection Agency's (EPA) Toxics Release Inventory (TRI) in order to determine whether regional differences in environmental inequality exist at the household level. The data cover nearly every metropolitan area in the contiguous US from 1990 to 2005, we divide the contiguous US into nine regions, and we use Geographic Information System (GIS) software to weight the potential impact of each TRI facility inversely according to geographic distance. Results indicate that the existence and magnitude of environmental racial inequality, as well as the role that race, income and other household characteristics play in shaping this inequality, vary in important ways across the nine regions of the country. This has important implications for environmental inequality and public health research.

Keywords Environmental inequality · Health · Geographic information systems · Distance decay indicators

19.1 Introduction

The goal of this chapter is to determine whether environmental racial inequality levels vary across regions of the United States, using facility-based environmental hazard data drawn from the Environmental Protection Agency's (EPA) Toxics Release Inventory (TRI), individual- and household-level demographic data drawn from the nationally representative Panel Study of Income Dynamics (PSID), and a Geographic Information System (GIS) technique that weights the potential impact of each TRI facility inversely according to geographic distance. This distance

---

L. Downey (✉)
Department of Sociology and IBS Population Center, University of Colorado at Boulder, Boulder, CO 80309, USA
e-mail: Liam.Downey@colorado.edu

J.A. Maantay, S. McLafferty (eds.), *Geospatial Analysis of Environmental Health*, Geotechnologies and the Environment 4, DOI 10.1007/978-94-007-0329-2_19,

decay technique generates more precise hazard proximity estimates than those used in most prior environmental inequality research, thereby providing important advantages for researches interested in studying environmental inequality and the relationship between environmental inequality and public health.

We begin by discussing the background and rationale for the study. We then describe the theoretical models and GIS technique we use in the chapter and present our empirical analyses. The results of these analyses point to profound racial and ethnic differences in proximity to neighborhood pollution but suggest that the magnitude and sources of this environmental inequality vary sharply across regions of the country.

19.2 Background and Rationale for Study

A relatively new approach to explaining persistent racial and ethnic health disparities focuses on the role that residential segregation and environmental inequality play in placing members of different racial and ethnic groups in neighborhoods with varying environmental health risks (Evans and Kantrowitz, 2002; Gee and Payne-Sturgess, 2004; Lopez, 2002; Morello-Frosch and Jesdale, 2006; Morello-Frosch and Lopez, 2006; Northridge et al., 2003). This new approach draws upon a large body of environmental inequality research that shows that neighborhood environmental quality tends to be negatively associated with neighborhood income levels (Ash and Fetter, 2004; Derezinski et al., 2003; Downey, 2005; Morello-Frosch et al., 2001) and positively associated with neighborhood percent minority (Mohai and Saha, 2007; Morello-Frosch et al., 2001; Stretesky and Lynch, 2002).

However, while providing basic evidence of environmental inequality along both race and class lines, past research on the topic suffers from a number of important shortcomings. First, prior research relies almost exclusively on neighborhood-level demographic data, allowing researchers to assess the correspondence between neighborhood sociodemographic composition and neighborhood hazard levels, but making it impossible for them to (a) link specific individuals to specific neighborhood environmental conditions or (b) determine the role that individual and household characteristics play in shaping race/ethnic differences in exposure to neighborhood environmental health risks. Thus, while prior environmental inequality research provides an important foundation for the argument that unequal exposure to neighborhood environmental health risks likely shapes racial and ethnic health disparities, it does not utilize the individual and household-level data needed to substantiate this claim.

Using neighborhood-level demographic data to conduct environmental inequality research is also problematic because conclusions based on such data may be subject to ecological fallacy and because several of the most widely employed theoretical models in the literature make predictions about individual- and household-level outcomes and behaviors (Crowder and Downey, 2010). As a result, using neighborhood-level demographic data to conduct environmental inequality research is likely to produce erroneous conclusions and prevent researchers from (a) testing extant theory and (b) examining the individual- and household-level mechanisms that this theory suggests most strongly shape environmental inequality.

A second shortcoming in the literature is that researchers tend to focus on either the nation as a whole (Hunter et al., 2003; Perlin et al., 1995), the nation's urban areas as a whole (Ash and Fetter, 2004; Oakes et al., 1996;) or a single or limited number of metropolitan areas at a time (Brown et al., 1997; Downey, 2005; Pastor et al., 2002). This is potentially problematic because prior research suggests that patterns of environmental inequality may vary according to region of the country (Downey, 2006a). Thus, research that examines one or a few metropolitan areas at a time is not broadly generalizable while research that examines the nation as a whole may gloss over regional differences that have important implications for public health outcomes.

Finally, environmental inequality researchers have yet to fully explore the many ways in which Geographic Information System (GIS) software can be used to estimate local environmental health risks. Important exceptions do, of course, exist. For example, Mennis (2002, p. 281) uses GIS to create "statistical surface representations of both socioeconomic character and environmental risk," Mohai and Saha (2006) develop an aerial apportionment method to measure neighborhood proximity to hazardous waste sites, and Maantay (2007) uses proximity buffers to examine the association between residential proximity to hazards and asthma. But as Downey (2006b) and Mohai and Saha (2006) point out, most environmental inequality researchers simply sum up the number of environmental hazards or pounds of pollutants in each of their units of analysis and assume that anyone living in an analysis unit with a hazard is residentially proximate to that hazard, while anyone living outside that unit is not residentially proximate to the hazard, even if they live across the street from it. As a result, most environmental inequality research relies on measures that only weakly capture residential proximity to environmental hazards.

The study presented in this chapter addresses these shortcomings in the literature by linking individual- and household-level data from the nationally representative Panel Study of Income Dynamics with neighborhood-level environmental hazard data derived from the Environmental Protection Agency's Toxics Release Inventory. These data cover nearly every metropolitan area in the contiguous US from 1990 to 2005, and allow us to examine regional differences in environmental racial inequality while simultaneously controlling for individual and household-level characteristics that likely shape *household* proximity to environmental hazards. Thus, in this chapter, we examine differences in neighborhood hazard levels between Hispanic, non-Hispanic black, and non-Hispanic white households both within and across 9 regions of the contiguous US, and present results both before and after controlling for theoretically relevant individual and household-level characteristics.

Finally, in order to better estimate neighborhood environmental hazard levels, we utilize a GIS technique described in Downey (2006b) to weight the potential impact of each TRI facility inversely according to geographic distance. As previously noted, this technique not only allows us to measure hazard proximity more precisely than has been possible in most prior research, it also allows us to demonstrate the advantages of incorporating GIS into environmental inequality and public health research.

19.3 Theoretical Explanations for Environmental Inequality

Key theoretical models of environmental inequality tend to focus on hazardous facility siting decisions and household-level residential attainment processes. Unfortunately, however, the TRI, like other environmental hazard datasets employed in the literature, does not provide researchers with facility siting data. Thus, we restrict our theoretical discussion to two models that make predictions about the household-level residential attainment processes that likely concentrate socioeconomically disadvantaged groups in areas with high levels of pollution. The first of these models, the *racial income inequality thesis* (Downey, 2005), holds that property values and rents tend to be relatively low in environmentally hazardous neighborhoods, making such neighborhoods more attractive to lower-income families, among which non-white families are overrepresented, and less attractive to higher income families, among which white families are overrepresented. This argument is consistent with the more general *spatial assimilation model* found in residential attainment research that holds that residential differentiation by social class emerges as persons match their own socioeconomic status with that of their neighborhood. Thus, these models suggest that racial differences in residential proximity to environmental hazards are due largely to group differences in socioeconomic resources such as income and education.

In contrast, the *residential discrimination thesis* (Bullard, 1993) holds that racial differences in mobility into and out of environmentally hazardous neighborhoods result from housing market discrimination. Consistent with the broader *place stratification perspective* that informs research on residential attainment and mobility (Alba et al., 1999; Crowder and South, 2005), the residential discrimination thesis assumes that discriminatory actions by real estate agents (Yinger, 1995), local governments (Shlay and Rossi, 1981), and mortgage lenders (Ross and Yinger, 2002) create barriers to residential attainment for minority homeseekers (Massey and Denton, 1993). These barriers are assumed to reduce the ability of minority families to move out of, or avoid moving into, hazardous neighborhoods, thereby creating or maintaining environmental racial inequality. Thus, in direct contrast to the *racial income inequality thesis*, the *residential discrimination thesis* suggests that racial differences in residential proximity to pollution should persist even after controlling for household socioeconomic status.

Finally, prior research indicates that residential attainment is also shaped by the age and sex of the household head (Crowder and South, 2005; South and Crowder, 1997). These characteristics, as well as respondent income and education and the average neighborhood pollution proximity value in each respondent's metropolitan area, will be controlled for in the regression models presented below.

19.4 Data and Methods

19.4.1 Environmental Hazard Data

This study relies on data from the Panel Study of Income Dynamics linked to neighborhood-level environmental hazard data derived from the EPA's Toxics Release Inventory. The TRI, which is the most comprehensive and detailed, publicly available national record of industrial facility activity available to researchers, records the number of pounds of specified toxic chemicals released into the environment each year by industrial facilities that fall into one of seven industrial categories (manufacturing, metal mining, coal mining, electric generating facilities that combust coal or oil, chemical wholesale distributors, petroleum terminals, and bulk storage), employ the equivalent of ten or more full-time workers, and manufacture, process, or otherwise use the specified chemicals in specified quantities.

TRI data were first collected in 1987, but because there are some questions about the accuracy of the first few years of the data, our study utilizes only 1990–2005 TRI data. In addition, in order to improve the accuracy of our hazard estimates, we include only those TRI facilities that the EPA estimates were located within 200 m of the latitude and longitude coordinates provided in the TRI dataset. Thus, our data incorporate information from a total of 38,212 facilities in the contiguous United States between 1990 and 2005, with facility counts ranging from 18,758 to 21,462 per year.

We use the TRI to create a continuous, tract-level measure of neighborhood proximity to TRI facility air, water, and land pollution that weights the potential effect of each TRI facility inversely according to geographic distance from the facility, thereby incorporating both the level of toxic emissions produced by each TRI facility and facility proximity to the center of each census tract. The variable is calculated as follows. First, for each year of TRI data, we locate each TRI facility on a census tract map of the contiguous US, using latitude-longitude coordinates provided by the EPA to locate each facility. This map is then overlaid with a rectangular grid made up of 400-foot square grid cells. For each grid cell we then calculate a distance-weighted sum of the pounds of air, water, and land pollutants emitted that year by all TRI facilities located within 1.5 miles of that grid cell (we combine air, water, and land pollution in a single measure, rather than examining them separately, in order to be consistent with most prior research). For example, if two TRI facilities, emitting 10,000 and 2,000 pounds of pollutants per year respectively, are located within 1.5 miles of grid cell A, and the distance-based weights for these facilities are 1.0 and 0.62 respectively, then grid cell A receives a proximate industrial pollution value of $(1.0\times10{,}000) + (0.62\times2000)$, or 11,240.

Figure 19.1 illustrates this GIS estimation process for a small sample of California census tracts. The first map in Fig. 19.1 shows a set of census tract boundaries, two TRI facilities (facilities A and B), and a circle with a 1.5 mile radius that is centered over facility A. Map 2 places a set of 400-ft^2 grid cells over Map 1. One of these grid cells (grid cell A) is centered around facility A. Due to software limitations, not every grid cell in Map 2 can be individually represented. Nevertheless,

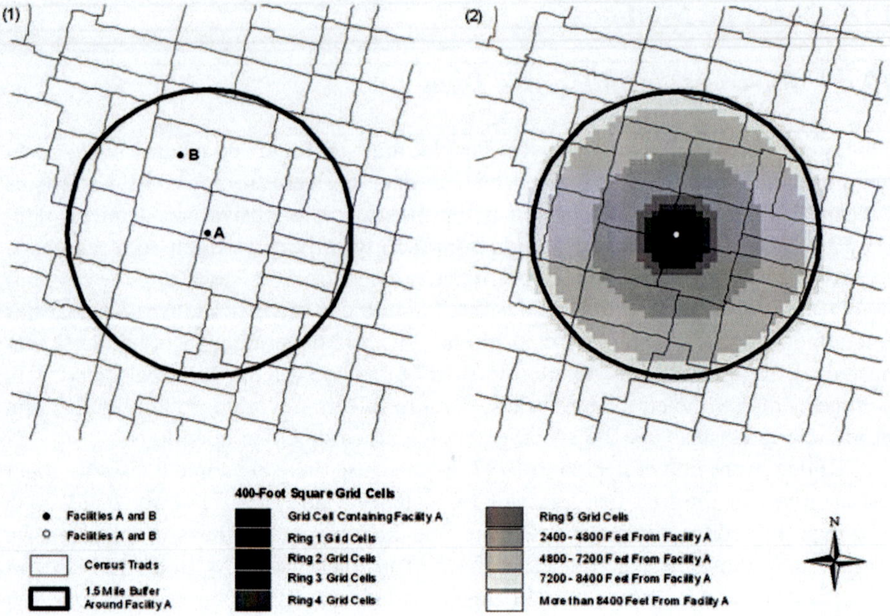

Fig. 19.1 Illustration of the distance decay method

it is possible to identify grid cell A (it is the cell hidden by facility A), the set of grid cells immediately surrounding grid cell A, the set of cells immediately surrounding these cells, and so on.

In Fig. 19.1, facility A has a weight of 1.0 for grid cell A because it is located in grid cell A, while facility B, which is approximately 4,600 ft from grid cell A, has a weight of 0.62 for grid cell A. Conversely, the grid cell containing facility B (grid cell B, which is hidden by facility B) is assigned a weight of 1.0 for facility B and a weight of 0.62 for facility A. Thus, as previously calculated, grid cell A receives a pollution value of 11,240 if facilities A and B emit 10,000 and 2,000 pounds of pollutants per year respectively, while grid cell B receives a value of $(0.62\times10{,}000)$ + $(1.0\times2{,}000)$, or 8,200.

The weights used in this study and in the previous set of calculations were derived from a 1.5 mile curvilinear distance decay function that is graphed in Fig. 19.2 (see the curve that is furthest to the right in the figure). Figure 19.2 demonstrates that various distance-decay functions can be used to calculate distance decay hazard proximity indicators and shows that the curvilinear function we use in this study declines from one to zero as distance from the grid cell increases until distance reaches 1.5 miles (after which the weight remains constant at zero).[1]

---

[1]In this study, the average number of grid cells per tract is 54 with a minimum of 1 and a maximum of 1,176. Since researchers have not developed a commonly accepted distance decay weighting scheme, we have experimented with alternative distance decay functions to estimate proximity

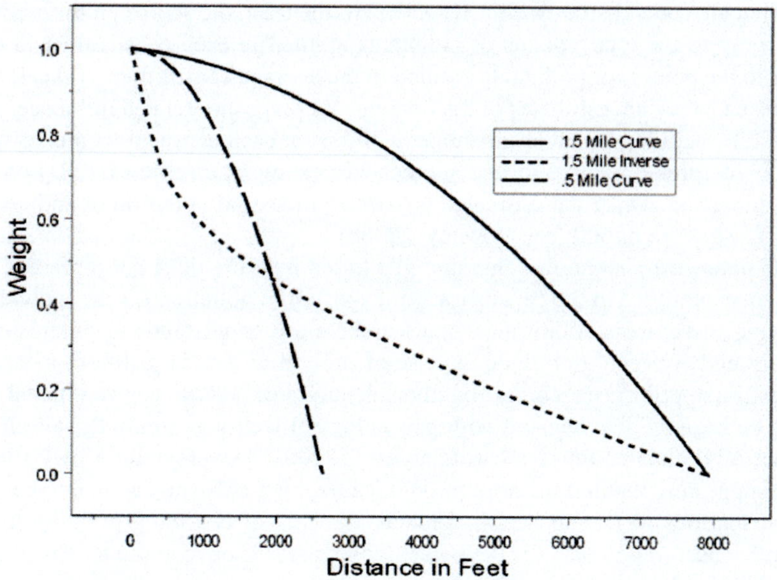

Fig. 19.2 Three distance decay curves

After assigning pollution values to each grid cell in the contiguous US, we calculated an average grid cell value for each *census tract* in the country. We did this separately for each census tract by summing up the grid cell values in each tract and then dividing this tract-specific sum by the number grid cells in that tract, assigning to each tract only those grid cells whose center points were located inside the tract (this included cells that crossed tract boundaries).

Finally, because these grid cell and census tract values can be very large, and because emissions from specific facilities sometimes fluctuate considerably from year to year, we (a) calculated the average level of industrial pollution in and around each census tract across the 3 years bounding the observation year (for example, for 2000, we calculated this average using pollution values from 1999 to 2001) and (b) divided these 3 year averages by 1000. The resulting tract-level, hazard-proximity score provides a more precise estimate of the level of proximate industrial pollution in and around US census tracts than has been utilized in most prior research.

Before proceeding, it is important to note that our proximate industrial pollution measure cannot be interpreted in absolute terms. Because the measure incorporates distance-weighted information about pollution from TRI facilities located not only

---

to industrial hazards. In doing this, we have altered not only the equations, but also the size of the grid-cells and the distance at which industrial sites are no longer considered influential (the threshold distance at which the distance decay weights reach zero). However, in prior research, altering the distance decay model in these ways has had only minor substantive impact on our results (see Crowder and Downey, 2010).

inside, but also outside (but within 1.5 miles) of the tract, the scores on this variable do not refer to the total pounds of pollutants emitted in each census tract in each year or to the pounds of pollutants emitted in the average census tract grid cell each year. Instead, they are estimates of the relative, non-exposure related influence of all nearby TRI facilities on environmental conditions in each census tract and as such, must be interpreted relative to one another. For example, a score of 1,000 on this variable indicates twice the estimated proximate industrial pollution in and around the tract as a score of 500 (see Downey, 2006b).

Also noteworthy is the fact that our TRI-based measure does not represent pollution concentrations in the census tract or exposure to pollution for the individuals occupying these areas. While health researchers are understandably interested in exposure and concentration data, our use of proximity data is justified on several grounds. First, prior research has found residential proximity to environmental hazards to be negatively associated with psychological well-being in both Detroit and a subset of Illinois counties (Boardman et al., 2009; Downey and Van Willigen, 2005), suggesting that hazard proximity may affect not only mental health, but also racial disparities in mental health. Second, the cost of obtaining pollution exposure and concentration data for large geographic areas is beyond the means of most researchers, and those datasets that do provide concentration data for the nation as a whole are either available for only a small number of years (the EPA's Cumulative Exposure Project and National Air Toxics Assessment datasets) or are available in a form that is not useful to most researchers (the EPA's Risk Screening Environmental Indicators dataset, or RSEI). This, of course, severely limits the ability of researchers to examine the effect of environmental inequality on health outcomes over large geographic areas, making it critical that researchers who examine these kinds of effects include in their methodological toolboxes GIS techniques such as the one employed in this chapter.

19.4.2 Individual and Household Data

We merge our environmental hazard data with individual- and household-level sociodemographic data by attaching our tract-level measure of local industrial pollution to the individual records of the Panel Study of Income Dynamics (PSID). The PSID is a well-known longitudinal survey of US residents and their families begun in 1968 with approximately 5,000 families (about 18,000 individuals). Members of panel families were interviewed annually between 1968 and 1997 and every 2 years thereafter. New families have been added to the panel as children and other members of original panel families form their own households.

PSID data are collected for a diverse national sample and contain rich information on a variety of individual- and household-level characteristics that are central to the study of residential attainment. In addition, the PSID's supplemental Geocode Match Files make it possible to link the addresses of individual respondents at each interview to their corresponding census tract identifiers, allowing us to merge respondent information with our tract-level pollution proximity scores. As a result, we know the proximate industrial pollution value associated with each respondent's tract of residence at each interview from 1990 to 2005.

Our effective sample consists of 15,653 PSID household heads, including 3,747 Hispanics, 5,048 non-Hispanic blacks, and 7,358 non-Hispanic whites who were interviewed between 1990 and 2005 and resided in a census-defined Metropolitan Statistical Area (MSA) at the time of the interview. One key drawback of the PSID for this study is that Hispanics are substantially underrepresented in the sample since the original panel was selected in 1968, just prior to the rapid increase of this population. The representation of Hispanics in the sample was increased between 1990 and 1995 with the incorporation of members from the Latino National Political Survey (LNPS) and in 1997 and 1999 with the addition of 511 panel families headed by post-1968 immigrants or their adult children. Thus, our sample includes Hispanics who were incorporated as part of these sample additions as well as those who were part of, or married into, original PSID panel families. Nevertheless, given the small size and questionable representativeness of the Hispanic sample, inferences drawn about the residential experiences of this group should be made with caution.

As previously noted, we focus on racial and ethnic differences in proximal pollution at the household level between 1990, the first year in which reliable TRI data are available, and 2005, the latest year for which PSID data are available. In order to minimize the effect of group differences in household size on calculated differences in average neighborhood pollution, we include in our sample only those individuals who were household heads at the time of the PSID interview. In addition, we focus exclusively on metropolitan residents in order to enhance comparability with past environmental inequality research, much of which focuses on aggregate population patterns within metropolitan areas. Focusing on metropolitan residents also allows us to calculate more precise pollution proximity estimates because metropolitan area census tracts tend to be smaller than non-metropolitan area census tracts. With these sample restrictions our analyses include data on households distributed across 314 of the 329 metropolitan areas in the contiguous US.

Finally, assessing racial disparities in residential proximity to neighborhood hazards is complicated by the fact that respondents' location near industrial pollution can change over time as a result of residential mobility and fluctuations in the level of pollution in and around the neighborhood of residence. Thus we take full advantage of the longitudinal nature of the PSID and pollution data by structuring the data as a series of person-periods with each observation describing the characteristics of an individual panel member and their neighborhood at the time of a particular PSID interview. This eliminates the need to focus on a single, arbitrary point in time, allowing us to examine differences in average levels of neighborhood pollution for members of different racial and ethnic groups across the entire span of the available data. The 15,653 individual householders in the sample each contribute, on average, just over 4.6 person-period observations for a total sample size of 72,938 observations.

19.4.3 Analytic Strategy

In order to examine regional differences in environmental racial inequality, we divide the contiguous US into nine regions based on the US Census Bureau's "census division" classification system. The nine regions are the Pacific, Mountain, West North Central, West South Central, East North Central, East South Central, New England, Mid-Atlantic, and South Atlantic divisions (see Fig. 19.3). It is important to note that some of the race/ethnic groups included in this study are poorly represented in some of these regions. In particular, black respondents are poorly represented in New England and the Mountain region and Hispanic respondents are poorly represented in the East South Central, West North Central, and New England divisions. Thus, results for these groups in these regions should be interpreted with caution.

We begin by comparing the proximate industrial pollution values of the average non-Hispanic black (hereafter "black"), non-Hispanic white (hereafter "white"), and Hispanic respondent (household head) in each region of the contiguous US. This initial analysis does not include any control variables, which (a) allows us to easily compare gross race/ethnic differences in pollution proximity both within and across the nine regions and (b) provides us with a baseline description of environmental racial inequality in each region of the US.

We then test potential explanations for these region-specific racial disparities by controlling for theoretically-relevant socioeconomic, life-cycle, and demographic characteristics shown in past research to affect residential attainment. Our primary indicators of socioeconomic status are education, measured by years of schooling completed by the household head, and total family (head and spouse) taxable

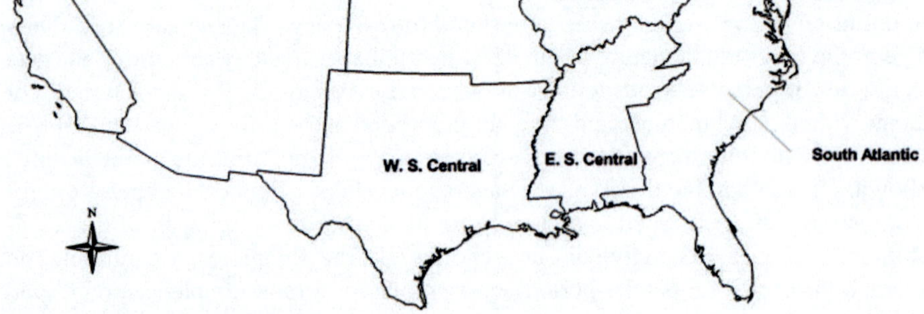

Fig. 19.3 US census divisions

income, measured in thousands of constant 2,000 dollars. Inserting these variables into the regression models allows us to test the *racial income inequality thesis* prediction that race/ethnic differences in pollution proximity will be attenuated with controls for socioeconomic status.

Key demographic and life-cycle predictors of residential attainment include age, age-squared, and the sex of the household head, which is captured by a dummy variable scored 1 for females and 0 for males.[2] In all models we also control for the year of observation to account for both temporal trends in pollution levels and the uneven distribution of observations for Hispanic householders across the years of the PSID data. We also control for the pollution level of the average tract in each respondent's metropolitan area. We include this control because overall pollution levels vary widely across metropolitan areas, and blacks, Hispanics, and whites are not distributed uniformly across these areas.

Thus, the regression analyses presented below allow us to assess differences in the level of neighborhood industrial pollution experienced by black, Hispanic, and white householders with similar sociodemographic characteristics and living in metropolitan areas with similar levels of pollution.

19.5 Results

19.5.1 Do Environmental Inequality Outcomes Vary Across Regions of the United States?

We begin our analysis with an assessment of gross racial differences in proximity to industrial pollution at the household level in each of the nine regions of the US (see Fig. 19.4 and Table 19.1). For example, the first row of data on the left-hand side of Table 19.1 shows that during the average observation period in New England, the average black, average Hispanic, and average white respondent lived in a neighborhood with a pollution proximity value of 33,058, 29,093, and 16,214 respectively.[3]

---

[2]Residential mobility researchers also routinely employ variables such as marital status, the presence of children in the family, home ownership, and household crowding. However, we do not include these controls in our models because they are predictors of the decision to move rather than of residential location.

[3]A small, but statistically influential, number of Hispanic respondents in the East North Central region lived in a single tract with extremely high pollution proximity values (in our sample, very few census tracts have multiple respondents during any single year). Thus, to avoid biasing the results for this region, we restricted the region's observations to respondents living in neighborhoods with pollution proximity scores of less than 1,000. Restricting the data in this way greatly reduces the gross Hispanic pollution proximity value for the region, while reducing the proximity values for blacks and whites in the region only slightly. As a result, these findings and the regression results reported below may underestimate the Hispanic/black and Hispanic/white pollution gap in the East North Central region.

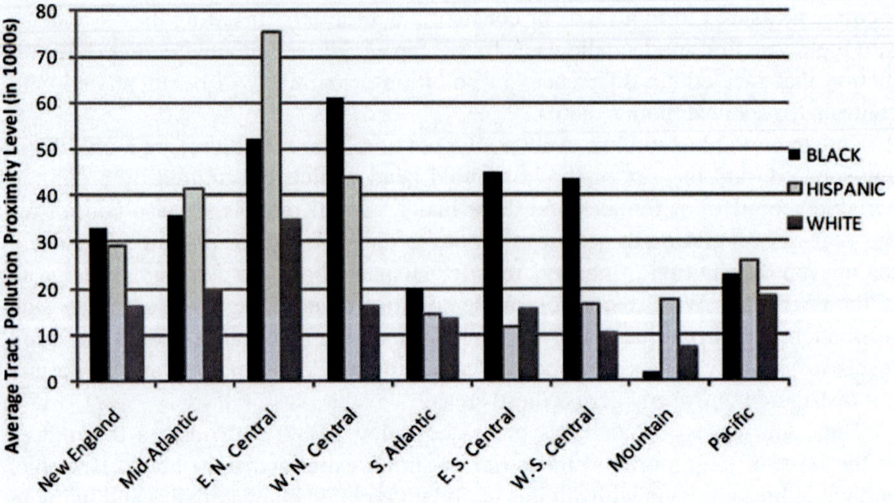

Fig. 19.4 Gross racial differences in pollution levels by region

The findings in Fig. 19.4 and Table 19.1 are notable in at least two respects. First, there are substantial inter-regional differences in the levels of proximate industrial pollution experienced by households of the same race/ethnic group. For example, the level of neighborhood pollution experienced by the average black respondent in the West North Central region (61,227) is over three times as high as the level of pollution experienced by the average black respondent in the South Atlantic region (20,200). Pollution values for a pooled sample of all respondents also vary across regions, as indicated by the fourth data column in Table 19.1.

Second, and more important for our purposes, environmental racial inequality levels (as represented by race/ethnic differences in pollution proximity) vary considerably across regions. The last three columns of Table 19.1 provide the clearest picture of these variations by presenting the ratios of black to white, Hispanic to white, and black to Hispanic pollution proximity and by indicating whether inter-group differences within each region are statistically significant as determined by Scheffe's test for Analysis of Variance results (significant results indicate significantly different means rather than ratios that are significantly different from one). For example, the second row of data on the right-hand side of Table 19.1 shows that in the Mid-Atlantic region, the average black respondent lived in a neighborhood with 1.83 times the pollution proximity value of the average white respondent and that this difference in pollution proximity is statistically significant at the 0.001 alpha level.

Overall, Fig. 19.4 and Table 19.1 show that black respondents lived in neighborhoods with higher pollution proximity scores than did white respondents in eight of the nine regions, with the environmental inequality ratio in these eight regions ranging from 1.28 in the Pacific region to 4.26 in the West South Central region. Conversely, in the Mountain region, where black and white pollution proximity scores are fairly low, white respondents lived in neighborhoods that had proximity scores that were, on average, 4.17 times greater (1/0.24 = 4.17) than the score for

Table 19.1 Environmental racial inequality within regions

Region	Average pollution proximity value (in 1000s)				Pollution proximity ratio		
	Black	Hispanic	White	Pooled[a]	Black/White	Hispanic/White	Black/Hispanic
New England	33.058	29.093	16.214	17.576	2.04**	1.79*	1.14
Mid-Atlantic	35.956	41.536	19.675	26.069	1.83***	2.11***	0.87
E. N. Central	52.294	75.430	34.539	42.094	1.51***	2.18***	0.69***
W. N. Central	61.227	43.788	16.037	23.390	3.82***	2.73***	1.40*
S. Atlantic	20.200	14.292	13.251	17.095	1.52***	1.08	1.41*
E. S. Central	45.194	11.513	15.275	30.406	2.96***	0.75	3.93
W. S. Central	43.757	16.305	10.283	23.739	4.26***	1.59	2.68***
Mountain	1.752	17.377	7.220	9.643	0.24	2.41***	0.10***
Pacific	23.270	26.028	18.196	21.503	1.28†	1.43***	0.89

† $p < 0.1$, * $p < 0.05$, ** $p < 0.01$, *** $p < 0.001$.

[a] These results are for a pooled sample of all respondents within each region.

the average black respondents' neighborhood. However, given that there were only 167 black respondents in the Mountain region, this difference is not statistically significant and should be interpreted with caution.

Turning our attention to the Hispanic/white pollution proximity ratios, we see that the average white respondent lived in a more polluted neighborhood than the average Hispanic respondent in the East South Central region, but again, this difference is not statistically significant. Hispanic respondents, on the other hand, lived in neighborhoods with significantly larger pollution proximity scores than did their white counterparts in six of the nine regions, with Hispanic to white ratios in these regions ranging from 1.43 in the Pacific region to 2.73 in the West North Central region.

Finally, the last column in Table 19.1 shows that in two regions – the East North Central and Mountain regions – the average Hispanic householder lived in a neighborhood with a significantly higher pollution proximity score than did the average black householder. However, the level of pollution experienced by the average black householder was significantly higher than that of the average Hispanic householder in the West North Central, West South Central, and South Atlantic regions.

The data clearly show, then, that environmental racial inequality is pervasive across the nine regions of the country, with significant group differences in neighborhood pollution proximity existing for at least one group pair in each of the nine regions. However, there is also tremendous variation in pollution proximity and environmental inequality across the regions, with (a) the magnitude of inequality, the overall and race-specific levels of pollution proximity, and the hierarchy of environmental burden among the race/ethnic groups differing greatly across the regions and (b) all three group contrasts statistically significant in only two of the nine regions (the East and West North Central Regions).

Thus, any effect that environmental inequality and pollution proximity might have on racial and ethnic health disparities is likely to vary according to region of the country, with the effect negligible for some groups in some regions and possibly highly significant for some groups in some regions.

19.5.2 Do Differences in Household Characteristics Explain Environmental Inequality at the Regional Level?

The results presented in the previous section clearly demonstrate that environmental inequality outcomes vary greatly across the nine regions of the country, suggesting that theoretical explanations of environmental racial inequality may not work equally well in each of these regions. In this section we ask whether this is the case by presenting regression analyses for each region of the country that test predictions drawn from the *racial income inequality* and *racial discrimination theses*.

As noted earlier, the *racial income inequality thesis* suggests that environmental racial inequality exists because minorities tend to possess lower levels of socioeconomic resources than do whites. Prior research also suggests that group differences in life-cycle and demographic characteristics may play a role in creating and maintaining environmental racial inequality. Alternatively, the *racial discrimination thesis* suggests that even when compared to whites with similar characteristics, minority households will still be restricted to neighborhoods with relatively high levels of pollution. Thus, controlling for household characteristics and examining the net effect of race on proximity to neighborhood pollution provides an opportunity to assess the relative veracity of these perspectives within different regions of the country.

To this end, Table 19.2 presents a series of regression models in which proximate industrial pollution (in 1000 s) is regressed on a set of dummy variables indicating the race/ethnic status of the respondent (with white respondents as the omitted category), controlling (sequentially) for the year of observation, individual and household characteristics, and the average pollution level in each respondent's metropolitan area. We present separate models for each of the nine regions, all the

Table 19.2 Explaining race/ethnic differences in pollution proximity within regions

Region	Independent variables	Model 1[a]	Model 2	Model 3	Model 4
New England	Hispanic	8.15	3.36	1.95	2.60
	Black	18.46*	15.25†	14.89†	11.63
	Constant	25.70***	43.35***	52.72**	46.96**
Mid-Atlantic	Hispanic	18.22***	14.75**	13.02**	13.05**
	Black	16.92*	14.69†	12.11†	14.40*
	Constant	26.75***	42.21***	56.40**	11.25
E. N. Central	Hispanic	37.18***	32.09***	30.74***	32.68***
	Black	17.58***	12.96**	11.27*	13.82*
	Constant	54.33***	84.12***	102.46***	106.73***
W. N. Central	Hispanic	24.44**	25.41**	23.72**	32.47**
	Black	44.27***	44.42***	42.02***	37.02***
	Constant	24.89***	13.21	21.64	51.06†
South Atlantic	Hispanic	−1.27	−1.92	−1.53	3.94
	Black	6.83*	5.83†	4.24	5.48
	Constant	17.92***	18.58***	24.49*	12.67
E. S. Central	Hispanic	−2.96	−2.26	−5.09	−1.85
	Black	30.78***	29.64***	25.71**	28.12*
	Constant	39.92***	31.58*	39.21	12.99
W. S. Central	Hispanic	0.14	−3.86	−4.43	−6.52
	Black	32.85**	29.63**	28.71***	10.98†
	Constant	21.66***	45.39**	39.50*	44.13†
Mountain	Hispanic	8.77**	7.67**	7.21**	10.17**
	Black	−5.37***	−5.94***	−7.41***	−4.77*
	Constant	10.11***	16.35***	23.51**	14.03†
Pacific	Hispanic	5.04†	4.47	3.70	3.40
	Black	4.33	4.37	4.57	4.07
	Constant	28.46***	32.31***	56.63***	22.73†

†$p < 0.1$, *$p < 0.05$, **$p < 0.01$, ***$p < 0.001$.
[a]Control variables: Model 1 (Year), Model 2 (Year, Income, Education), Model 3 (Year, Income, Education, Age, Age squared, Female), Model 4 (Year, Income, Education, Age, Age Squared, Female, Average pollution value in metropolitan area).

variables in these models are measured at the time of the PSID interview, and we use Stata's "cluster" procedure to account for the non-independence of observations related to the same individual householder (as previously noted, we use individual- and household-level data, rather than census tract demographic data, because conclusions drawn from tract-level demographic data are subject to ecological fallacy and because the theoretical perspectives tested in this chapter make predictions about individual- and household-level outcomes).

Due to space limitations, we only present coefficients representing the association between the race/ethnicity dummy variables and the dependent variable, with the constant term representing the proximate industrial pollution value for whites in the region. Model 1 of Table 19.2 shows that controlling for the year of observation has only a small effect on the overall pattern of results found in Table 19.1. For example, in every region except the Mountain region, black householders experience higher pollution levels than do white householders, and these differences are statistically significant in all of these regions except the Pacific even after controlling for temporal trends in the data. Interestingly, the relative pollution *advantage* for black householders observed in the Mountain region in Table 19.1 becomes statistically significant with the control for year of observation.

In contrast, once the year of observation is controlled for, the Hispanic/white pollution proximity gap becomes statistically non-significant in New England and marginally significant in the Pacific region; and there is now virtually no Hispanic/white pollution gap in the West South Central region. But otherwise the basic pattern of results for Hispanics and whites differs little from that found in Table 19.1.

Model 2, which controls for the income level of the family and the level of education of the individual householder, as well as for the year of observation, provides some support for the *racial income inequality* thesis in some regions of the country but not others. Unreported coefficients show, for example, that individual-level income and/or education are significantly and negatively associated with pollution proximity in four regions – the New England, Mid-Atlantic, East North Central, and South Atlantic regions. Moreover, the second column of Table 19.2 shows that in three of these regions (New England, the Mid-Atlantic, and the South Atlantic), the coefficients indicating the black/white pollution gap are reduced and become only marginally significant when group differences in socioeconomic characteristics are controlled.

But otherwise, inserting householder education and family income into the regression equations has virtually no effect on the pattern of significant results found in Model 1. The Hispanic coefficients and coefficient significance levels are virtually unchanged from Model 1 to Model 2 in two of the four regions where the significance of the black coefficient declined (the Mid-Atlantic and East North Central regions); in the other two regions where the significance of the black coefficient declined (New England and the South Atlantic), the Hispanic coefficient is non-significant in both Model 1 and Model 2; in the Pacific region, the only region in which the Hispanic coefficient declined to non-significance, the p-value in Model 1 already equaled.092; and in the remaining regions, the Hispanic coefficient changes little from Model 1 to Model 2. Moreover, unreported regression results indicate that inserting householder education and family income into the model has virtually no effect on the Hispanic/black pollution gap.

Thus, model 2 suggests that the *racial income inequality* thesis is partially supported for blacks and whites in the New England, Mid-Atlantic, and South Atlantic regions, unsupported for these groups in the remaining regions, and unsupported for Hispanics and whites in all nine regions. In addition, the unreported results indicate that the *racial income inequality* thesis is not supported for blacks and Hispanics in the five regions where the Hispanic/black pollution gap was statistically significant in model 1 (the East North Central, South Atlantic, East and West South Central, and Mountain regions).

Prior research suggests, of course, that race/ethnic differences in pollution proximity may be due to other respondent and family characteristics, such as the age and sex of the household head. However, as shown in Model 3 of Table 19.2, controlling for these characteristics has very little effect on most of the black and Hispanic coefficients, though it does tend to reduce their magnitude, such that in the South Atlantic region the black/white pollution gap is no longer statistically significant. The only other notable effect of inserting householder age and sex into the model is that the constant term for the East South Central Region is no longer statistically significant, suggesting that in this region, white pollution proximity is explained primarily by household characteristics that distribute whites differentially across neighborhoods. Finally, adding these controls has virtually no effect on the unreported regression model results comparing black and Hispanic pollution outcomes.

Of course, the unequal environmental distribution of whites, Hispanics, and blacks within regions may arise because these groups are not distributed evenly across metropolitan areas which, in turn, vary greatly in terms of the amount of pollution their industries produce (Downey, 2007). Model 4 examines this possibility by inserting a variable for the average tract-level proximate industrial pollution value in each respondent's metropolitan area.

After including this control, the black coefficients in New England and the West South Central region become non-significant and marginally significant respectively, suggesting that even after controlling for age, sex, education, and income, blacks in these regions still live in more polluted metropolitan areas than do their white counterparts. In contrast, controlling for metropolitan area pollution levels increases the statistical significance of the positive black coefficient in the Mid-Atlantic region, reduces the magnitude of the significant *negative* black coefficient in the Mountain region by approximately one-third, and increases the size of the significant positive Hispanic coefficients in the West North Central and Mountain regions by more than one-third. Moreover, controlling for metropolitan area pollution levels has little effect on the size and significance of the Hispanic coefficient in the Mid-Atlantic region or the black coefficients in the West North Central and East South Central regions.

This suggests that after controlling for age, sex, education, and income, blacks in the Mid-Atlantic and Mountain regions, and Hispanics in the West North Central and Mountain regions, live in less, not more, polluted metropolitan areas than do their white counterparts. It also suggests that net of these controls, Hispanics in the Mid-Atlantic region and blacks in the West North Central and East South Central regions live in similarly polluted metropolitan areas as their white counterparts.

The effect of metropolitan area pollution levels on the black/Hispanic pollution gap is also region-specific, with unreported results suggesting that net of individual- and family-level characteristics, Hispanics live in less polluted metropolitan areas than do blacks in the South Atlantic, East South Central, and West South Central regions, and in similarly polluted metropolitan areas as blacks in the remaining regions. As a result, after controlling for metropolitan area pollution levels, the black pollution disadvantage (relative to Hispanics) decreases in the South Atlantic and West South Central regions and becomes non-significant in the South Atlantic and East South Central regions.

Thus, like the preceding Models, Model 4 suggests that not only do environmental inequality outcomes vary significantly across regions of the country, so too do the factors that explain these outcomes.

But perhaps the most notable finding from Table 19.2 is that even after controlling for group differences in sociodemographic characteristics and the uneven distribution of pollution across metropolitan areas, there still remain significant racial and ethnic differences in proximate industrial pollution at the household level. Specifically, in four regions – the Mid-Atlantic, East North Central, West North Central, and Mountain – Hispanic householders still face significantly higher levels of neighborhood pollution than do their white counterparts even after controlling for these factors. Thus, these differences cannot be attributed to group differences in income, education, or other factors typically associated with residential attainment. Similarly, while black householders in the Mountain region (an admittedly small population) continue to face a slight but statistically significant pollution-proximity *advantage* relative to whites after controlling for these factors, black householders in five of the nine regions (the Mid-Atlantic, East North Central, West North Central, East South Central, and West South Central) face higher levels of pollution proximity relative to similarly positioned whites.

Thus, consistent with the *racial discrimination thesis*, race remains an important determinant of access to low-pollution neighborhoods in several regions of the United States even after controlling for other theoretically relevant factors. Nevertheless, the magnitude and nature of environmental racial stratification, as well as the role that race, income and other household and metropolitan characteristics play in shaping this stratification, varies in important ways across the nine regions of the country.

19.6 Conclusion

The findings reported in this chapter differ from those reported in prior environmental inequality research because this is the first study ever to compare environmental inequality outcomes across different regions of the country, and one of only two environmental inequality studies to employ PSID data and distance decay proximity indicators (the other study is Crowder and Downey, 2010). Thus, while several studies have found persistent associations between levels of pollution and concentrations of racial and ethnic minority populations at the neighborhood level, this study highlights the existence of racial/ethnic disparities in pollution proximity at the household level that persist after controlling for family income, householder education, and other factors affecting residential attainment. Moreover, this is the first study to show that these micro-level patterns of environmental inequality and the factors shaping them vary across regions of the country.

With the exception of Crowder and Downey (forthcoming in, 2010), this is also the only environmental inequality study that we are aware of that is able to directly test the individual- and household-level predictions made by the *racial discrimination* and *racial income inequality* theses. As a result, this is the first study ever to demonstrate that these theoretical models receive support in some regions of the country but not others.

Because this study is so unique, it has several potentially important (and fairly specific) implications for environmental inequality and health disparities research. First, the regional variation in overall and race-specific pollution proximity levels highlighted in Table 19.1 suggests that the public health implications of pollution proximity will be more serious in some regions of the country than others. Second, regional differences in the existence and magnitude of environmental racial inequality, as well as in the hierarchy of environmental burden among race/ethnic groups, suggest that the role that environmental inequality plays in shaping racial health disparities is also likely to vary greatly across regions, with the effect negligible for some groups in some regions and possibly highly significant for some groups in some regions. Third, regional variation in the ability of theoretical models and theoretically relevant predictors to explain racial differences in pollution proximity suggests that solutions to pollution-based racial health disparities will likewise vary across regions of the country.

Of course, further research is necessary to confirm these results. For example, our findings might have differed if we had examined a different type of environmental hazard, employed pollution concentration or exposure data, used a different set of distance decay equations to estimate pollution proximity, or defined the regions of the US according to a different classification scheme than that provided by the US Census Bureau.

Nevertheless, this study strongly suggests that environmental inequality and public health researchers need to take regional variation into account when developing and testing theoretical models of the relationship between environmental inequality and racial health disparities.

This study also demonstrates some of the important advantages of using GIS to estimate pollution proximity. Specifically, by employing a relatively low cost GIS technique (in terms of both time and money) that provides more accurate pollution proximity estimates than those typically found in the literature, this study illustrates how researchers can use GIS to estimate neighborhood environmental quality over large geographic areas for an extended period of years. This is important because as previously noted, there are only a limited number of national pollution concentration or exposure datasets, such datasets are very expensive and difficult to create, and those that exist tend to be available for either a limited number of years or in a form that is difficult for researchers to use. Moreover, it is very difficult to establish an empirical association between health outcomes and pollution proximity or exposure based on only 1 or 2 years of data.

However, using the kinds of GIS techniques employed in this study, researchers can create national, longitudinal datasets that merge precise neighborhood-level pollution proximity estimates with individual- and household-level sociodemographic and health data (such as that available in the PSID). Such data can be used to (a) track the movement, over time, of specific individuals and households into and out of environmentally hazardous neighborhoods; (b) estimate the length of time these individuals and households spent living in neighborhoods of varying environmental quality; and (c) link respondent's health data to their pollution proximity biographies. And this information, in turn, can be used to examine both the association between environmental inequality and racial health disparities *and* the residential mobility processes that likely play an important role in shaping these two racially inequitable outcomes.

Thus, while this is only one among several GIS techniques that researchers can use to estimate pollution proximity, and while researchers should put pressure on the EPA to release more usable national and longitudinal pollution concentration datasets, it should be apparent that GIS provides researchers with several important tools for conducting environmental inequality and racial health disparities research.

Acknowledgments Direct all correspondence to Liam Downey, Department of Sociology, University of Colorado, 219 Ketchum Hall, UCB 327, Boulder, CO 80309. 303-492-8626 (Liam.Downey@colorado.edu). This research was supported by a development grant to the first author from the NICHD funded University of Colorado Population Center, and grants to the second author (R21 HD049610) and to both authors (R21 HD058708) from the National Institute of Child Health and Human Development.

Author Biographies

Liam Downey is Associate Professor of Sociology and faculty research associate in the Institute of Behavioral Science at the University of Colorado at Boulder. His research examines environmental inequality in metropolitan America and the structural determinants of environmental degradation.

Kyle Crowder is Professor of Sociology and a fellow in the Carolina Population Center at the University of North Carolina, Chapel Hill. His research focuses on processes of residential stratification and the effects of physical and social contexts on individual behavior.

References

Alba RD, Logan JR, Stults BJ, Marzan G, Zhang. W (1999) Immigrant groups in the suburbs: a reexamination of suburbanization and spatial assimilation. Am Sociol Rev 64:446–460

Ash M, Fetter. TR (2004) Who lives on the wrong side of the environmental tracks? evidence from the EPA's risk-screening Environmental Indicators Model. Soc Sci Q 85:441–462

Boardman J, Downey L, Jackson JS, Merrill JB, Saint Onge JM, Williams DR (2009) Proximate industrial activity and psychological distress. Popul Environ 30:3–25
Brown P, Ciambrone D, Hunter. L (1997) Does "Green" mask grey? environmental equity issues at the metropolitan level. Int J Contemp Sociol 34:141–158
Bullard R (1993) Confronting environmental racism. South End Press. Boston, MA
Crowder, K, Downey L (2010) Inter-neighborhood migration, race, and environmental hazards: modeling microlevel processes of environmental inequality. Am J Sociol 115(4):1110–1149
Crowder KD, South. SJ (2005) Race, class, and changing patterns of migration between poor and nonpoor neighborhoods. Am J Sociol 110:1715–1763
Derezinski DD, Lacy MG, Stretesky. PB (2003) Chemical accidents in the United States, 1990–1996. Soc Sci Q 84:122–143
Downey L (2005) The unintended significance of race: environmental racial inequality in detroit. Soc Forces 83:305–341
Downey L (2006a) Environmental inequality in metropolitan America in 2000. Sociol Spectr 26:21–41
Downey L (2006b) Environmental racial inequality in Detroit. Soc Forces 85:771–796
Downey L (2007) US metropolitan-area variation in environmental inequality outcomes. Urban Stud 44:953–977
Downey L, Van Willigen. M (2005) Environmental stressors: the mental health impacts of living near industrial activity. J Health Soc Behav 46:289–305
Evans GW, Kantrowitz E (2002) Socioeconomic status and health: the potential role of environmental risk exposure. Annu Rev Public Health 23:303–331
Gee GC, Payne-Sturgess. DC (2004) Environmental health disparities: A framework integrating psychosocial and environmental concepts. Environ Health Perspect 112: 1645–1653
Hunter LM, White MJ, Little JS, Sutton. J (2003) Environmental hazards, migration, and race. Popul Environ 25:23–39
Lopez R (2002) Segregation and black/white differences in exposure to air toxics in 1990. Environ Health Perspect 110(Suppl 2):289–295
Maantay J (2007) Asthma and air pollution in the Bronx: methodological and data considerations in using GIS for environmental justice and health research. Health Place 13:32–56
Massey DS, Denton. NA (1993) American apartheid: segregation and the making of the underclass. Harvard University Press. Cambridge, MA
Mennis J (2002) Using geographic information systems to create and analyze statistical surfaces of population and risk for environmental justice analysis. Soc Sci Q 83:281–297
Mohai P, Saha. R (2006) Reassessing racial and socioeconomic disparities in environmental justice research. Demography 43:383–399
Mohai P, Saha. R (2007) Racial inequality in the distribution of hazardous waste: a national-level reassessment. Soc Probl 54:343–370
Morello-Frosch R, Lopez R (2006) The Riskscape and the color line: examining the role of segregation in environmental health disparities. Environ Res 102:181–196
Morello-Frosch R, Pastor M, Sadd J (2001) Environmental justice and southern California's "Riskscape": the distribution of air toxics exposures and health risks among diverse communities. Urban Aff Rev 36:551–578
Morrello-Frosch R, Jesdale. BM (2006) Separate and unequal: residential segregation and estimated cancer risks associated with ambient air toxics in US metropolitan areas. Environ Health Perspect 114:1–8
Northridge ME, Stover gn, Rosenthal JE, Sherard. D (2003) Environmental equity and health: understanding complexity and moving forward. Am J Public Health 93:209–214
Oakes JM, Anderton, DL, Anderson AB (1996) A longitudinal analysis of environmental equity in communities with hazardous waste facilities. Soc Sci Res 25:125–148
Pastor M, Sadd JL, Morello-Frosch R (2002) Who's minding the kids? pollution, public schools, and environmental justice in los angeles. Soc Sci Q 83:263–280

Perlin SA, Setzer RW, Creason J, Sexton K (1995) Distribution of industrial air emissions by income and race in the United States: an approach using the toxic release inventory. Environ Sci Technol 29:69–80

Ross SL, Yinger J (2002) The color of credit: mortgage discrimination, research methodology, and fair-lending enforcement. The MIT Press. Cambridge

Shlay AB, Rossi. PH (1981) Keeping up the neighborhood: estimating net effects of zoning. Am Sociol Rev 46:703–719

South SJ, Crowder. KD (1997) Escaping distressed neighborhoods: individual, community, and metropolitan influences. Am J Sociol 102:1040–1084

Stretesky P, Lynch. M (2002) Environmental hazards and school segregation in Hillsborough County, Florida, 1987–1999. Sociol Q 43:553–573

Yinger J (1995) Closed doors, opportunities lost: the continuing costs of housing discrimination. Russell Sage Foundation. New York, NY

Chapter 20
Merging Satellite Measurement with Ground-Based Air Quality Monitoring Data to Assess Health Effects of Fine Particulate Matter Pollution

Zhiyong Hu, Johan Liebens, and K. Ranga Rao

Abstract Geospatial technologies have been widely used in environmental health research, including air pollution and human health. This chapter demonstrates the potential of integrating satellite air quality measurement with ground-based $PM_{2.5}$ data to explore health effects of fine particulate air pollution. This study assesses the association of estimated $PM_{2.5}$ concentration with chronic coronary heart disease (CCHD) mortality. Years 2003 and 2004 daily MODIS (Moderate Resolution Imaging Spectrometer) Level 2 AOD images were collated with US EPA $PM_{2.5}$ data covering the conterminous USA. Pearson's correlation analysis and geographically weighted regression (GWR) found that the relationship between $PM_{2.5}$ and AOD is not spatially consistent across the conterminous states. GWR predicts well in the east and poorly in the west. The GWR model was used to derive a $PM_{2.5}$ grid surface for the eastern US (RMSE = 1.67 $\mu g/m^3$). A Bayesian hierarchical model found that areas with higher values of $PM_{2.5}$ show high rates of CCHD mortality: $\beta_{PM_{2.5}} = 0.802$, posterior 95% Bayesian credible interval (CI) = (0.386, 1.225). Aerosol remote sensing and GIS spatial analyses and modelling could help fill pervasive data gaps in ground-based air quality monitoring that impede efforts to study air pollution and protect public health.

Keywords Remote sensing · GIS · Aerosol optical depth · MODIS · Particulate matter · Air pollution · Chronic coronary heart disease

List of Abbreviations

AIC	Akaike Information Criterion
AOD	Aerosol optical depth
AQS	Air Quality System
BUGS	Bayesian inference using Gibbs sampling
CAR	Conditional auto-regression
CCHD	Chronic coronary heart disease

Z. Hu (✉)
Department of Environmental Studies, University of West Florida, Pensacola, FL 32514, USA
e-mail: zhu@uwf.edu

J.A. Maantay, S. McLafferty (eds.), *Geospatial Analysis of Environmental Health*, Geotechnologies and the Environment 4, DOI 10.1007/978-94-007-0329-2_20,

EPA	Environmental Protection Agency
GOES	Geostationary Operational Environmental Satellite
GWR	Geographically weighted regression
ICD-10	International Classification of Disease, 10th Revision
LIDAR	Light detection and ranging
MCMC	Markov chain Monte Carlo
MODIS	Moderate Resolution Imaging Spectrometer
PM	Particulate matter
SMR	Standardized morbidity/mortality rate
USGS	US Geological Survey

20.1 Introduction

Geospatial technologies have been widely used in environmental health research, including air pollution and human health. This chapter demonstrates the potential of integrating satellite air quality measurement with ground-based $PM_{2.5}$ data to explore health effects of fine particulate air pollution. To this end, a suite of geospatial technologies has been used in a geographic information system (GIS) environment which provides a framework for integrating layers of spatially referenced air pollution and health data, analyzing the data to reveal spatial patterns, and modelling the spatial relationship between environment and health.

Air pollution epidemiological studies often rely on ambient observations from pollution monitoring sites to provide metrics of exposure. Methods of exposure assessment in those studies include averaging multiple monitors within each enumeration unit or study site (Klot et al., 2005; Murakami and Ono, 2006; Zanobetti and Schwartz, 2005), assigning the exposure value of the nearest monitor to each case/control (Mille et al., 2007; Peters et al., 2001) and spatial interpolation/modelling methods (Maheswaran et al., 2005). Such ground monitoring data lack spatially complete coverage. Ground monitors are rare in rural areas. Assessment of the exposure to air pollution using in situ observations is hampered by the sparse and unbalanced spatial distribution of the monitors.

The repetitive and broad-area coverage of satellites may allow atmospheric remote sensing to offer a unique opportunity to monitor air quality at continental, national and regional scales. Recent studies have established quantitative relationships between satellite derived aerosol optical depth (AOD), which describes the mass of aerosols in an atmospheric column, and fine particulate matter (particles smaller than 2.5 μm, $PM_{2.5}$) using linear regression models (Kumar et al., 2007; Liu et al., 2007; Schafer et al., 2008; van Donkelaar et al., 2006; Vidot et al., 2007). Except in long-range dust or pollution transport events, AOD is dominated by near-surface emissions sources (Seinfeld and Pandis, 1998). AOD retrieved at visible wavelengths is most sensitive to particles between 0.1 and 2 μm (Kahn et al., 1998). Several studies have merged AOD with ground $PM_{2.5}$ measurements to derive $PM_{2.5}$ surfaces (Apituley et al., 2008; Helix-Atlanta, 2009; Liu et al., 2009). A study in a region centered in Massachusetts (Liu et al., 2009) examined the benefits of using AOD retrieved by the Geostationary Operational Environmental Satellite (GOES) in conjunction with land use and meteorological information to estimate ground-level $PM_{2.5}$ concentrations. Another project (Helix-Atlanta, 2009) combined MODIS (Moderate Resolution Imaging Spectrometer) AOD data with US EPA $PM_{2.5}$ data to estimate a $PM_{2.5}$ surface in Atlanta, Georgia. Existing studies estimating $PM_{2.5}$ surfaces using AOD data use uniform linear relationships between AOD and $PM_{2.5}$. However, studies have found that the correlation between $PM_{2.5}$ and AOD is not spatially consistent (Engel-Cox et al., 2004) due to variation in terrain, land cover, selection of aerosol model in the AOD retrieval algorithm and meteorological factors such as mixing height.

Numerous epidemiological studies indicate that exposure to $PM_{2.5}$ is associated with asthma, respiratory infections, lung cancer, cardiovascular problems, and premature death (Anderson et al., 2003; Hu et al., 2008; Peters et al., 2001; Pope, 2000; Zanobetti and Schwartz, 2005). A few studies have examined coronary heart disease, finding evidence for acute effects on mortality and hospital admissions (Le Tertre et al., 2002; Poloniecki et al., 1997; Schwartz, 2000). Recently, attention has focused on whether there is an association between chronic exposure to air pollution and coronary heart disease (Maheswaran et al., 2005). An ecological study at the census enumeration district level found an association between nitrogen oxides, and to a lesser extent particulate matter (PM_{10}) and carbon monoxide, and coronary heart disease mortality in Sheffield, UK (Maheswaran et al., 2005).

This chapter assesses the association between $PM_{2.5}$ and chronic coronary heart disease (CCHD). The study quantitatively examines the relationship between $PM_{2.5}$ ground measurements and MODIS AOD data in the conterminous USA using Pearson's correlation analysis and geographically weighted regression (GWR). For the region with high correlations, the GWR model was used to calculate a $PM_{2.5}$ surface based on the AOD data and the spatially varying relationships between $PM_{2.5}$ and AOD. A Bayesian hierarchical model was used to link $PM_{2.5}$ with CCHD.

20.2 Methods

20.2.1 MODIS Data

The MODIS sensor flies on polar-orbiting and sun-synchronous Terra and Aqua satellites. MODIS performs measurements in the visible to thermal infrared spectrum region. The MODIS sensor was expected to be the key for monitoring global aerosol properties. Not only have MODIS aerosol products been used to answer scientific questions about radiation and climate, they are being used for applications not previously intended, including monitoring surface air quality for health (Al-Saadi et al., 2005; Chu et al., 2003; Gupta et al., 2006; Hutchison, 2003; Hutchison et al., 2005).

One of the fundamental aerosol products from MODIS is spectral AOD. MODIS files are produced every day at a spatial resolution of 10×10 km (at nadir). To retrieve aerosol information, the aerosol retrieval algorithm makes use of seven wavelength bands (channels 1–7) and a number of other bands to help with cloud rejection and other screening procedures. Different dynamic aerosol models (biomass burning, dust aerosol, and aerosol from industrial/urban origin) are used to determine the aerosol optical properties used in the algorithm for different areas of the US. For the eastern US, the urban/industrial aerosol model is used while the biomass burning/dust model is used in the west. The splitting line is at approximately –100° longitude.

Daily level 2 MODIS data (2003–2004) were obtained from the NASA Level 1 and Atmosphere Archive and Distribution System (LAADS Web). Using a map algebra function in a GIS environmental, a 2-year average AOD raster data layer (10 km by 10 km grid) was calculated. Data from both Terra and Aqua satellites were used. MODIS AOD data are not available every day due to cloud cover. Data for cold seasons (October to March) were not used since cloud cover, snow reflectivity, and diminished vertical mixing all reduce the accuracy of ground-level pollutant levels measured in winter. During warm seasons, vertical columns in the atmosphere are more integrated. AOD measures correlate best with ground-based monitoring in warm months, likely because of stronger boundary layer mixing during the warmer months (Tinkle et al., 2007).

20.2.2 Associations Between AOD and Ground $PM_{2.5}$

To determine if AOD measurements accurately reflect ground $PM_{2.5}$ levels, we examined correlations between AOD levels and corresponding ground PM measurements. $PM_{2.5}$ ground data was obtained from US EPA Air Quality System (AQS) online Data Mart (US EPA). There were 877 monitoring sites for the conterminous US. $PM_{2.5}$ data was collated with AOD both temporally and spatially. For each MODIS AOD image scene and each monitor within the scene, a $PM_{2.5}$ measurement within 1 h of the imaging time was assigned to the pixel containing the monitor using a GIS spatial join function. Pearson's correlation coefficient was calculated for each monitoring site. The Pearson's correlation analysis examined the temporal relationship between $PM_{2.5}$ and AOD for each site. The relationship shows how AOD changes as $PM_{2.5}$ changes over time at a sampling site.

A geographically weighted regression (GWR) model was also fitted to examine the relationship between $PM_{2.5}$ (dependent variable) and AOD (independent variable) using the 2-year average $PM_{2.5}$ and AOD values. GWR is a local form of regression modelling used to model spatially varying relationships among variables (Fotheringham et al., 2002). GWR is one of several spatial regression techniques, increasingly used in geography and other disciplines. GWR involves fitting a regression equation to every feature in the dataset based on data within a local "window" of the feature. Values of the dependent and explanatory variables of features falling within the kernel window are used in estimating the local regression equation. GWR accounts for the effect of local spatial autocorrelation which occurs when the values for a particular explanatory variable cluster spatially. GWR creates coefficient raster surfaces for the model intercept and coefficients for each explanatory variable.

In this study, GWR analyzed the spatially varying relationship between 2-year mean AOD and mean $PM_{2.5}$. GWR revealed how the relationship between AOD and $PM_{2.5}$ changes across the space. In this study, a fixed kernel was used to fit the regression. The bandwidth (kernel radius) value was chosen by using the corrected Akaike Information Criterion (AICc). To account for varying number of air quality monitors in different areas, spatial weights were applied to individual monitoring sites so that places with more samples are weighted higher.

The GWR model was used to estimate $PM_{2.5}$ values across the study area based on AOD measurements. Using map algebraic function in a GIS environment, Local AOD values were substituted into the local regression equation to compute the predicted $PM_{2.5}$ level yielding a surface of $PM_{2.5}$ values. The accuracy of the estimated $PM_{2.5}$ surface was assessed using a "bootstrapping" (or "leave-one-out") procedure. The procedure was developed using ArcGIS 9.1 ArcObjects and Arc Macro Language (AML). The procedure omitted one observation, fit GWR and calculated a 2-year average $PM_{2.5}$ surface using N-1 observations, then compared the value of the surface at the location of the omitted observation with the observed 2-year average $PM_{2.5}$ value. A root mean square error (RMSE) was calculated after the process was repeated for all observations.

20.2.3 Health Outcomes Data

Health outcomes (CCHD) data at the county level were extracted for the period from 2003 to 2004 from the National Center for Health Statistics Compressed Mortality File 1999–2005 in the CDC WONDER online database (US CDC, 2008). CCHD was defined by the International Classification of Disease version 10 codes: I25.0–I25.6, I25.8, and I25.9. CCHD count and population at risk were retrieved by county, race (White, Black or African-American, Other race) and age. CCHD data has age groups at 5 year intervals from 1–4 to 85+. Aggregated CCHD mortality count and population at risk were also retrieved by race and age groups for the whole focused study area to be used in calculating mortality rates by age and race.

Race and age adjusted rates were calculated using indirect standardization (Mausner and Kramer, 1985) for each county. Rate adjustment is a technique for removing the effects of race and age from crude rates, so as to allow meaningful comparisons across populations with different underlying race and age structures. An internal standard population (a super-population containing the counties in the eastern US) was used to standardize rates. The indirect standardization first calculated expected number of CCHD for each county. The calculated count is the number of cases that would be expected in the county if people in the county died from CCHD at the same rate as people in the standard population. Standardized mortality rates (SMRs) were calculated by dividing the observed count by the expected value.

According to Curtin and Klein (1995), one of the problems with rate adjustment is that rates based on small numbers of deaths will exhibit a large amount of random variation. In the aggregate CCHD mortality count and population data set, observed counts of twenty or less were flagged as statistically "unreliable". CCHD counts were "suppressed" when the data meets the criteria for confidentiality constraints. CCHD counts for counties with census year populations of less than 100,000 were replaced with "suppressed" if the number of cases is five or less and the count is based on only 1 or 2 years of data. All unreliable and suppressed data were not used in calculating standardized rates and spatial analysis and modelling thereafter.

20.2.4 Calculating County Average $PM_{2.5}$ for Ecological Modelling of Health Effects

To link health outcomes with air pollution, the 2003–2004 mean $PM_{2.5}$ raster grid surface calculated using the GWR model was first resampled so that each 10 km by 10 km grid cell was subdivided into 10 by 10 smaller cells retaining the original $PM_{2.5}$ values. The purpose of the resampling procedure was to divide 10 km cells located on county boundaries into separate parts for neighbouring counties to achieve higher accuracy in calculating county average $PM_{2.5}$. The resampled $PM_{2.5}$ grid was then overlaid with the health outcome maps. A GIS zonal statistical function was used to calculate the mean $PM_{2.5}$ value for each county. The mean $PM_{2.5}$ value was calculated by averaging $PM_{2.5}$ values of all cells whose centroids are within the county.

20.2.5 Bayesian Hierarchical Modelling of CCHD and $PM_{2.5}$

A Bayesian hierarchical model was used to explore the association between CCHD mortality and $PM_{2.5}$. Simulation-based algorithms for Bayesian inference allow us to fit very complicated hierarchical models, including those with spatially correlated random effects. The following model was fitted allowing a convolution prior for the random effects:

$$O_i \sim Poisson(\mu_i) \tag{20.1}$$

$$\log \mu_i = \log E_i + \beta_0 + \beta_1 PM_{2.5} + b_i + h_i \tag{20.2}$$

where i is the index for a county, O is observed CCHD death count, E is expected death count reflecting race-age-standardized values. For model specification, an improper (flat) prior for the intercept parameter β_0 and a uniform prior distribution for the fixed-effect parameters (β_1) were assumed. By fixed effect we mean it applies equally to all the counties. Two sets of county-specific random effects were included in the model. The first set b_i is a spatially structured random effect assigned an intrinsic Gaussian conditional auto-regression (CAR) prior distribution (Besag et al., 1991). The second set of random effects h_i is assigned an exchangeable (non-spatial) normal prior. The random effect for each county is thus the sum of a spatially structured component b_i and an unstructured component h_i. This is termed a convolution prior (Besag et al., 1991; Mollie, 1986). The model is more flexible than assuming only CAR random effects, since it allows the data to decide how much of the residual disease risk is due to spatially structured variation, and how much is unstructured over-dispersion. To complete the model specification, conjugate inverse-gamma prior distributions were assigned to the variance parameters associated with the exchangeable and/or CAR priors. The Markov chain Monte Carlo (MCMC) simulation computation technique and Gibbs sampling algorithm were used to fit the Bayesian model. Summaries of the post-convergence MCMC samples provide posterior inference for model parameters. The result of such analysis is the posterior distribution of an intensity function with covariate effects.

The model was fitted using the *WinBUGS* software – an interactive Windows version of the *BUGS* (Bayesian inference Using Gibbs Sampling) program for Bayesian analysis of complex statistical models using MCMC techniques (Lunn et al., 2000). A queen's case spatial adjacency matrix ($w_{ij} = 1$ when county i and j share a boundary or a vertice, $w_{ij} = 0$ otherwise) that is required as input for the conditional autoregressive distribution was created using the Adjacency for WinBUGS Tool developed by the Upper Midwest Environmental Sciences Center of the US Geological Survey (USGS).

20.3 Results

20.3.1 Pearson's Correlation and GWR

The average correlation between $PM_{2.5}$ measured at fly-over time and AOD is 0.67 to east of the –100° longitude line and 0.22 to the west. Of the 20 monitoring sites with top significant correlations ($r > 0.8$, $p < 0.05$), 18 are located east of the –100° longitude line.

Results from the GWR using the 2-year average $PM_{2.5}$ and AOD values show that all the monitoring sites have a condition number less than 30. The condition number evaluates local collinearity. In the presence of strong local collinearity, results become unstable. Results associated with condition numbers larger than 30 may be unreliable. Figure 20.1 shows a map of local R square. R^2 values range between 0.0 and 1.0 and indicate how well the local regression model fits observed $PM_{2.5}$ values. Very low values indicate the local model is performing poorly. It can be seen that GWR generally predicts well in the eastern USA and poorly in the west. Figure 20.2 shows the coefficient raster surface for the explanatory variable AOD. The map exhibits regional variation in the explanatory variable. The relationship between $PM_{2.5}$ and AOD is not spatially consistent (stationary) across the conterminous states. Like Pearson's correlation, AOD coefficient values are higher in the eastern USA, while values in the west are generally lower. Even negative values are found in part of the western region. The Pearson's correlation and GWR analyses revealed that it is appropriate to estimate a $PM_{2.5}$ surface using AOD for

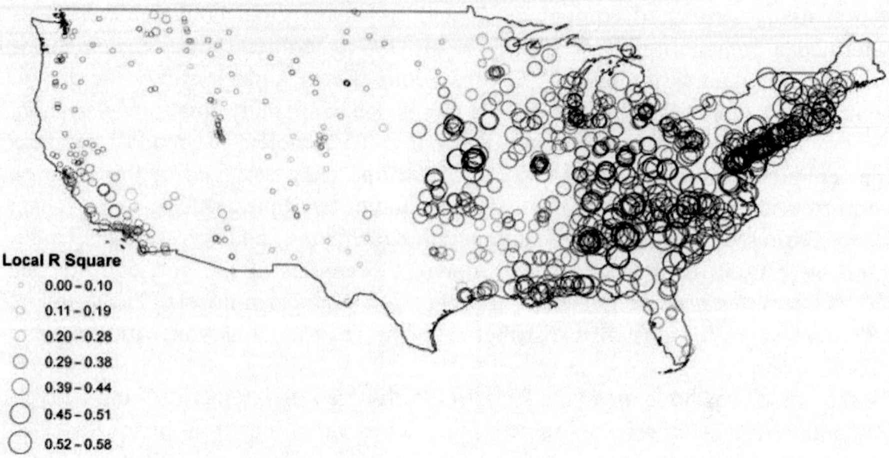

Fig. 20.1 Local R square of geographically weighted regression

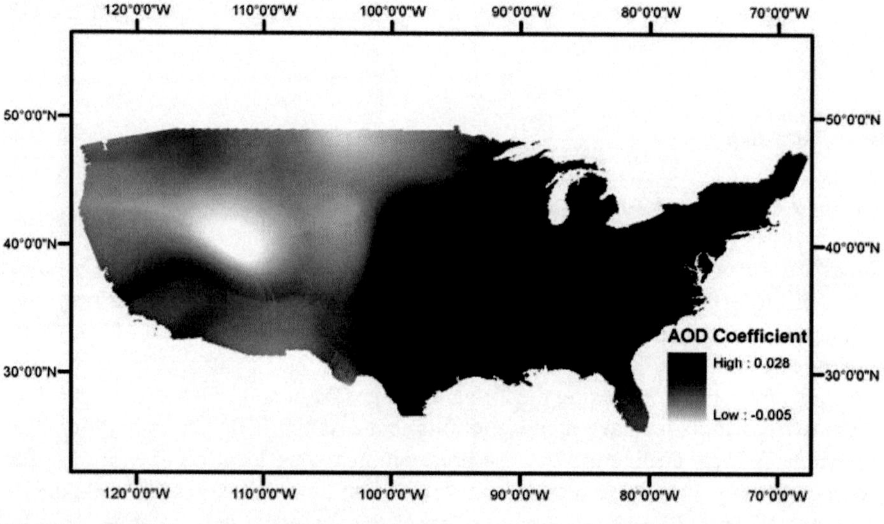

Fig. 20.2 Coefficient raster surface for AOD from geographically weighted regression of $PM_{2.5}$ against AOD

disease modelling for the region to the east of –100° longitude. Application of the GWR model to the 2-year mean AOD raster resulted in a continuous $PM_{2.5}$ surface (Fig. 20.3) with a RMSE of 1.67 $\mu g/m^3$. Compared to the average $PM_{2.5}$ value of 10.18 $\mu g/m^3$ for all the monitors in the east ($PM_{2.5\ min}$= 3.71 $\mu g/m^3$, $PM_{2.5\ max}$ = 26.77 $\mu g/m^3$, standard deviation = 3.25 $\mu g/m^3$), the $PM_{2.5}$ estimation achieved an accuracy of 84%.

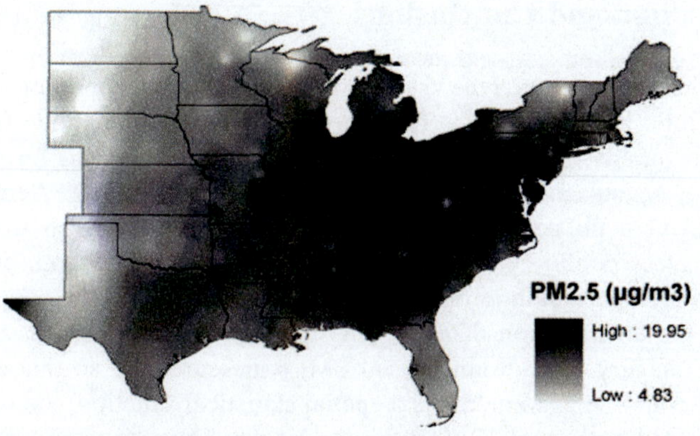

Fig. 20.3 $PM_{2.5}$ surface calculated by merging MODIS AOD and EPA $PM_{2.5}$ ground measurements

20.3.2 Bayesian Model of CCHD SMR and $PM_{2.5}$

Modelling of health outcomes and $PM_{2.5}$ was focused on the east region where AOD and ground $PM_{2.5}$ measurements were most strongly correlated. The number of counties to the east of the –100° longitude is 2,506. With 200 counties which have suppressed or unreliable data omitted from the analysis, the number of data points (counties) in the Bayesian modelling is 2,306. Table 20.1 provides the estimated posterior mean, median, and associated 95% credible set for each of the fixed effects. A 95% credible set defines an interval having a 0.95 posterior probability of containing the parameter of interest. Standard deviations and Monte Carlo (MC) errors were calculated to assess the accuracy of the simulation. As a rule of thumb, the simulation should be run until the Monte Carlo error for each parameter of interest is less than about 5% of the sample standard deviation. The MC errors calculated from iterations 60,001 to 80,000 for both parameters are less than 5% of the corresponding standard deviations, suggesting an accurate estimate for each parameter. The result in Table 20.1 shows an association between $PM_{2.5}$ pollution and CCHD. The 95% credible set covers positive values: $\beta_{PM_{2.5}} = 0.802$, CI = (0.386, 1.225). The positive boundary values of the CI indicate that areas with higher values of $PM_{2.5}$ also show high rates of CCHD mortality.

Table 20.1 Results of Bayesian hierarchical modelling[a]

Fixed effects	Posterior mean	Posterior median	Standard deviation	MC error	95% credible set
β_0	0.264	0.273	0.064	0.003	(–0.366, –0.117)
β_1	0.802	0.812	0.223	0.010	(0.386, 1.225)

[a] Posterior means, medians, and 95% credible sets are based on 20,000 post-convergence iterations (from 60,001 to 80,000). Fixed effects are: β_0 – intercept, β_1 – effect of $PM_{2.5}$.

20.4 Discussion and Conclusions

This chapter has demonstrated the use of a suite of geospatial technologies in air pollution and public health research. The geospatial technologies range from remote sensing of air quality, integration of data of varying sources and formats, spatial data handling, spatial analyses, and spatial statistical modelling. Aerosol remote sensing could help fill pervasive data gaps that impede efforts to study air pollution and protect public health. The GIS's ability to manage both attribute and spatial data was important in joining the health outcome data to the county map and linking it to the air pollution data. A GIS spatial join operation was used to collate satellite imagery with ground-based $PM_{2.5}$ measurement so that AOD-$PM_{2.5}$ relationships could be examined. As a spatial statistical function, the GWR model is tightly coupled with ArcGIS, and the model directly uses integrated GIS data. GIS made it possible to define the kernel and spatial weights for running the GWR model. Map algebra functions in a GIS environmental were adopted to calculate the 2-year average AOD raster data layer and the $PM_{2.5}$ surface using the AOD data and the GWR model. The GIS zonal statistical function was used to calculate the mean $PM_{2.5}$ value for each county. The bootstrapping procedure developed using ArcGIS 9.1 ArcObjects and Arc Macro Language (AML) demonstrates how repetitive tasks can be easily implemented by using GIS' customized programming capability.

The Pearson's correlation and GWR analyses found that the relationship between $PM_{2.5}$ and MODIS AOD is not spatially consistent across the conterminous states. In the west AOD poorly correlates with $PM_{2.5}$. The east region exhibits high positive correlations between $PM_{2.5}$ and AOD, and a $PM_{2.5}$ surface can be estimated using AOD data for assessment of $PM_{2.5}$'s effect on disease. The difference in correlation is likely due to differences in terrain, the AOD retrieval algorithm (Engel-Cox et al., 2004) and meteorological conditions. The algorithm is based on dark surface pixels and contains assumptions about the types of pollutants and the terrain (Remer et al., 2006). The lower correlations in the more arid parts of the western US reflect the higher surface reflectance which reduces contrasts and the fact that the model assumes more dust and smoke than typically exists in the west. The MODIS aerosol retrieval algorithm uses the urban/industrial aerosol model for eastern USA and other densely populated regions, such as Western Europe and eastern China (Remer et al., 2005). Urban – industrial aerosols are mainly from fossil fuel combustion in populated industrial regions and are dominated by fine particles. The positive relationship in the east is likely to exist in other regions of the world where the aerosol retrieval algorithm uses the urban/industrial aerosol model. The inherent differences in the datasets may also explain some of the geographic variation in association (Engel-Cox et al., 2004). MODIS AOD measures aerosol scattering in a total column from ground to satellite while EPA PM monitors measure a size-based subset of PM in the boundary layer in a specific location.

The Bayesian model found a spatial association between age-race-standardized mortality rate of chronic coronary heart disease and $PM_{2.5}$. There is an excess risk of

CCHD mortality in areas with high $PM_{2.5}$ levels. Fine aerosol particles tend to penetrate into the gas-exchange regions of the lung, and ultrafine particles are able to pass through the lungs to enter the blood circulation and affect other organs (Nemmar et al., 2002). Studies indicate that $PM_{2.5}$ leads to high plaque deposits in arteries, causing vascular inflammation and atherosclerosis – a hardening of the arteries that reduces elasticity, which can lead to heart attacks and other cardiovascular problems (Pope et al., 2002; Suwa et al., 2002). The associations between CCHD mortality and $PM_{2.5}$ revealed by the ecological models can be taken as indicative (though not necessarily causative) of a potential air pollution effect. The associations call for further toxicological studies to investigate the biological mechanisms by which fine particulate air pollution adversely affects CCHD. The evidence of elevated incidence of CCHD mortality risks in high pollution areas would support targeting of policy interventions in such areas to reduce pollution levels.

The study calculated a $PM_{2.5}$ surface by merging satellite derived AOD with ground measurements using GWR. To our knowledge, no applications of satellite remote sensing data in spatial modeling of regional $PM_{2.5}$ concentrations have considered spatially varying relationships. Existing studies often use a uniform linear relationship to estimate $PM_{2.5}$ using AOD (Helix-Atlanta, 2009; Kumar et al., 2007; Liu et al., 2007). While the uniform relationship could apply to a smaller region such as a metropolitan area, it might not work for a larger region such as the eastern US. The MODIS aerosol data lends itself to population-based exposure assessment and the ecological approach. Moderate resolution satellite image pixels and disease data enumeration districts are both area features and thus can easily be compared. Derivation of a continuous $PM_{2.5}$ surface using AOD alleviates the challenges of comparing discrete point monitoring data with area-aggregated disease data. A continuous $PM_{2.5}$ surface accounting for spatially-varying correlations between AOD and ground measurements better represents exposure than one based on a single global correlation.

There are also several limitations to this study. First, the use of aggregated data means that inferences cannot be directly transferred to the individual level. An inherent limitation of an ecological study is that it uses aggregate data and does not incorporate individual information. This ecological study also suffers the modifiable areal unit problem. Second, the correlation and GWR analyses did not account for meteorological parameters (such as mixing height, relative humidity, air temperature, and wind speed), aerosol vertical distribution, and aerosol properties. Those parameters could also modify the relationship between AOD and $PM_{2.5}$ concentrations, as shown by existing research (van Donkelaar et al., 2006; Liu et al., 2009). Third, the study estimated an ambient $PM_{2.5}$ surface to assess the association of fine particulate air pollution with county-level rates for CCHD. However, ambient concentrations do not necessarily represent actual individual exposures, which can be influenced by the infiltration of ambient pollution into indoor facilities (such as automobiles, homes, schools, and work places) and the activity of individuals (such as outdoor exercise, walking, commuting, etc.). This population-based ecological analysis also did not consider population dynamics, for example, population migration.

Furthermore, the use of county-scale health indicators is problematic in that county populations vary significantly in socio-economic characteristics and environmental exposures. Other social-economic variables known to influence CCHD risk have not been controlled in this study. Last, it must be noted that there are limitations of AOD data. AOD is a quantitative measure of total column aerosol, which is the mass of aerosols within a measured column extending from Earth to the satellite sensor. AOD does not correlate well with $PM_{2.5}$ during cold seasons. This study used warm season AOD data but the disease data covered all seasons. In addition, AOD does not differentiate between fine and course particles. Ideally, fine-mode optical depth should be used for the health effect assessment because fine mode particles which dominate urban/industrial pollution are thought to cause the most severe health problems. However, land-based measures of fine-mode fraction of AOD can only be used as a qualitative indicator of whether AOD values are dominated by natural or anthropogenic emissions (Remer et al., 2006). Additionally, AOD does not specify the location of aerosols within a column and the AOD values do not necessarily represent ground conditions. Long-range dust or pollution transport events can contribute PM_{25} in the column well above ground level. This lack of vertical information emphasizes the importance of combining the satellite image with vertical profiles. Emerging Light Detection and Ranging (LIDAR) systems could provide vertical resolution for AOD and quantify pollutant levels on the ground.

Acknowledgments The studies reported in this chapter have been partially supported by US EPA Cooperative Agreement Award X-9745002 to the University of West Florida. The content of this report are solely the responsibility of the authors and do not necessarily represent the official views of the US EPA.

Author Biographies

Zhiyong Hu (Ph.D. in Geography, 2004, University of Georgia, USA; M.S. 1995 and B.S. 1992 in Geology, Northeast University, China) is Assistant Professor in the Department of Environmental Studies at the University of West Florida. Dr. Hu specializes in geospatial science and technology and spatial statistics. His research interests are GIS and remote sensing theories, methods, and applications in solving environmental issues. His work includes land use/cover classification from remote sensing imagery, GIS-based modelling of spatio-temporal dynamics and human dimensions in land use/cover changes, and spatial analysis of environmental health. Recent research projects have focused on remote sensing and health effect assessment of air pollution. He has published in *Computer Engineering and Application, Computers Environment and Urban Systems, Environmental Health, GIScience and Remote Sensing*, and in *International Journal of Health Geographics.* Dr. Hu is a member of the Association of American Geographers, and American Society of Photogrammetry and Remote Sensing.

Johan Liebens (Ph.D., 1996, Michigan State University; M.S., 1983, B.S., 1981, Free University Brussels, Belgium) is Professor in the Department of Environmental Studies at the University of West Florida. Dr. Liebens is a broad-based physical geographer with research interests in geography of soils, heavy metal pollution of sediments, and environmental soil science. His work includes both basic and applied research. GIS-based spatial analyses are part of most of his research. He has extensive experience with field mapping, surveying methods and physical and chemical analyses of soils and sediments. Recent research projects have focused on organic carbon pools in soils, and pollution of surface soils, sediments in storm water retention ponds, and water and sediments in urban bayous along the Gulf Coast. His research has been funded, among others, by NSF, GSA, USDA, EPA, and Florida Center for Solid and Hazardous Waste Management. He has published in *Annals of the AAG, Physical Geography, Geomorphology, Environmental Geology, Communications in Soil Science and Plant Analysis, Historical Archaeology, Shore and Beach, American Midlands Naturalist, Marine Pollution Bulletin, Water Air and Soil Pollution, and in Soil Use and Management*. Dr. Liebens is a member of the Association of American Geographers, the Soil Science Society of America, and the Association for Environmental Health and Sciences.

K. Ranga Rao (Ph.D., 1967; M.Sc., 1961; B.Sc., 1958: Andhra University, Visakhapatnam, India) is a Distinguished University Research Professor in the Center for Environmental Diagnostics and Bioremediation at the University of West Florida. He is a broad-based biologist, with research interests in comparative physiology, biochemistry, and endocrinology; environmental toxicology and public health. Recent projects have focused on assessments of pollution in various media and potential impacts on environmental and public health. Dr. Rao's research over the years has been supported by various federal agencies such as NSF, EPA, and CDC. He published an edited volume. "Pentachlorophenol: Chemistry, Pharmacology, and Environmental Toxicology," (Plenum Press, 1978), and published more than 150 articles in various journals, including *Marine Biology, Marine Environmental Research, Environmental Research, Environmental Pollution, Biological Bulletin, Comparative Physiology and Biochemistry, General and Comparative Endocrinology, Proceedings of the National Academy of Sciences, USA*. He is a Fellow of the American Association for Advancement of Science, and a member of Crustacean Society, Society of Integrative and Comparative Biology, and Society of Neuroscience.

References

Al-Saadi J, Szykman J, Pierce RB, Kittaka C, Neil D, Chu DA, Remer. L, Gumley L, Prins E, Weinstock L, MacDonald C, Wayland R, Dimmick F, Fishman J (2005) Improving national air quality forecasts with satellite aerosol observations. Bull Am Meteorol Soc 86(9):1249–1261

Anderson HR, Atkinson RW, Bremner SA, Marston L (2003) Particulate air pollution and hospital admissions for cardiorespiratory diseases: are the elderly at greater risk? Eur Respir J 21(Suppl):39–46

Apituley MS, Koelemeijer R, Timmermans R, Schoemaker R, de Leeuw G (2008). Construction of satellite derived $PM_{2.5}$ maps using the relationship between AOD and $PM_{2.5}$ at the Cabauw experimental site for atmospheric research (CESAR) – the Netherlands. In: IEEE international geoscience & remote sensing symposium, Boston, MA, 6–11 July 2008

Besag J, York Y, Mollie A (1991) Bayesian image restoration with two applications in spatial statistics (with discussion). Ann Inst Stat Math 43(1):1–59

Chu DA, Kaufman YJ, Zibordi G, Chern JD, Mao J, Li C, Holben BN (2003) Global monitoring of air pollution over land from EOS-Terra MODIS. J Geophys Res 108(D21):4661

Curtin LR, Klein RJ (1995) Direct standardization (age-adjusted death rates). statistical notes, No. 6. National Center for Health Statistics, Hyattsville, Maryland

Engel-Cox JA, Holloman CH, Coutant BW, Hoff RM (2004) Qualitative and quantitative evaluation of MODIS satellite sensor data for regional and urban scale air quality. Atmos Environ 38:2495–2509

Fotheringham SA, Chris B, Martin C (2002) Geographically weighted regression: the analysis of spatially varying relationships. New York, NY, Wiley

Gupta P, Christopher SA, Wang J, Gehrig R, Lee Y, Kumar N (2006) Satellite remote sensing of particulate matter and air quality assessment over global cities. Atmos Environ 40:5880–5892

HELIX-Atlanta (Health & Environment Linked for information exchange, Atlanta). http://www.cdc.gov/nceh/tracking/helix.htm. Accessed 21 March 2009

Hu Z, Liebens J, Rao KR (2008) Linking stroke mortality with air pollution, income, and greenness in northwest Florida: an ecological geographical study. Int J Health Geogr 7:20

Hutchison KD (2003) Applications of MODIS satellite data and products for monitoring air quality in the state of Texas. Atmos Environ 37:2403–2412

Hutchison KD, Smith S, Faruqui SJ (2005) Correlating MODIS aerosol optical thickness data with ground-based $PM_{2.5}$ observations across Texas for use in a real-time air quality prediction system. Atmos Environ 39:7190–7203

Kahn R, Banerjee P, McDonald D, Diner D (1998) Sensitivity of multiangle imaging to aerosol optical depth and to pure-particle size distribution and composition over ocean. J Geophys Res 103(D24):32195–32213

Klot S, Peters A, Aalto P, Bellander T, Berglind N, D'Ippoliti D, Elosua R, Hörmann A, Kulmala M, Lanki T, Löwel H, Pekkanen J, Picciotto S, Sunyer J, Forastiere F (2005) Ambient air pollution is associated with increased risk of hospital cardiac readmissions of myocardial infarction survivors in five European cities. Circulation 112:3073–3079

Kumar N, Chu A, Foster A (2007) An empirical relationship between $PM_{2.5}$ and aerosol optical depth in Delhi Metropolitan. Atmos Environ 41:4492–4503

Le Tertre A, Medina S, Samoli E, Forsberg B, Michelozzi P, Boumghar A, Vonk JM, Bellini A, Atkinson R, Ayres JG, Sunyer J, Schwartz J, Katsouyanni K (2002) Short-term effects of particulate air pollution on cardiovascular diseases in eight European cities. J Epidemiol Community Health 56:773–779

Liu Y, Franklin M, Kahn R, Koutrakis P (2007) Using aerosol optical thickness to predict ground-level $PM_{2.5}$ concentrations in the St. Louis area: a comparison between MISR and MODIS. Remote Sens Environ 107(1–2):33–44

Liu Y, Paciorek CJ, Koutrakis P (2009) Estimating regional spatial and temporal variability of $PM_{2.5}$ concentrations using satellite data, meteorology, and land use information. Environ Health Perspect. doi:10.1289/ehp.0800123

Lunn DJ, Thomas A, Best N, Spiegelhalter D (2000) WinBUGS – a Bayesian modelling framework: concepts, structure, and extensibility. Stat Comput 10:325–337

Maheswaran R, Haining RP, Brindley P, Law J, Pearson T, Fryers PR, Wise S, Campbell MJ (2005) Outdoor air pollution, mortality, and hospital admissions from coronary heart disease in Sheffield, UK: a small-area level ecological study. Eur Heart J 26:2543–2549

Mausner RB, Kramer S (1985) Epidemiology: an introductory text. W. B. Saunders, Philadelphia, PA

Mille KA, Siscovick DS, Sheppard L, Shepher K, Sullivan JH, Anderson GL, Kaufman JD (2007) Long-term exposure to air pollution and incidence of cardiovascular events in women. N Eng J Med 356:447–458

Mollie A (1986) Bayesian mapping of disease. In Gilks WR, Richardson S, Spiegelhalter DJ (eds) Markov chain monte carlo in practice. Chapman and Hall, New York, NY, pp. 359–379

Murakami Y, Ono M (2006) Myocardial infarction deaths after high level exposure to particulate matter. J Epidemiol Community Health 60:262–266

NASA Level 1 and Atmosphere Archive and Distribution System (LAADS Web). http://ladsweb.nascom.nasa.gov/. Accessed 20 July 2008

Nemmar A, Hoet PH, Vanquickenborne B, Dinsdale D, Thomeer M, Hoylaerts MF, Vanbilloen H, Mortelmans L, Nemery B (2002) Passage of inhaled particles into the blood circulation in humans. Circulation 105:411–414

Peters A, Dockery DW, Muller JE, Mittleman MA (2001) Increased particulate air pollution and the triggering of myocardial infarction. Circulation 103:2810–2815

Poloniecki JD, Atkinson RW, de Leon AP, Anderson HR (1997) Daily time series for cardiovascular hospital admissions and previous day's air pollution in London, UK. Occup Environ Med 54:535–540

Pope CA III (2000) Epidemiology of fine particulate air pollution and human health: biologic mechanisms and who's at risk? Environ Health Perspect 108(Supl 4):713–723

Pope CAIII, Burnett RT, Thun MJ, Calle EE, Krewski D, Ito K, Thurston GD (2002) Lung cancer, cardiopulmonary mortality, and long-term exposure to fine particulate air pollution. J Am Med Assoc 287:1132–1141

Remer LA, Kaufman YJ, Tanre D, Mattoo S, Chu DA, Martins JV, Li RR, Ichoku C, Levy RC, Kleidman RG, Eck TF, Vermote E, Holben BN (2005) The MODIS aerosol algorithm, products, and validation. J Atmos Sci 62:947–972

Remer LA, Tanré D, Kaufman YJ (2006). Algorithm for remote sensing of tropospheric aerosol from MODIS: Collection 005. http://modis-atmos.gsfc.nasa.gov/_docs/MOD04:MYD04_ATBD_C005_rev1.pdf. Accessed 12 Oct 2008

Schafer K, Harbusch A, Emeis S, Koepke P, Wiegner M (2008) Correlation of aerosol mass near the ground with aerosol optical depth during two seasons in Munich. Atmos Environ 42:4036–4046

Schwartz J (2000) Harvesting and long-term exposure effects in the relationship between air pollution and mortality. Am J Epidemiol 151:440–448

Seinfeld JH, Pandis SN (1998) Atmospheric chemistry and physics: from air pollution to global change. Wiley, New York, NY

Suwa T, Hogg JC, Quinlan KB, Ohgami A, Voncent R, van Eeden SF (2002) Particulate air pollution induces progression of atherosclerosis. J Am Coll Cardiol 39(6):935–942

Tinkle S, Gant M, Humble M, Gary Foley G, Garcia V, Bond A (2007). Integrated earth observations: Application to air quality and human health. http://www.epa.gov/geoss/pdf/ieo-workshop.pdf. Accessed 3 Nov 2008

U.S. CDC 2008. Centers for Disease Control and Prevention, National Center for Health Statistics. compressed mortality file 1999–2005. CDC WONDER On-line Database, compiled from Compressed Mortality File 1999–2005 Series 20 No. 2 K. http://wonder.cdc.gov/cmf-icd10.html. Accessed 10 Oct 2008

US EPA. Air Quality System (AQS) online Data Mart. http://www.epa.gov/ttn/airs/aqsdatamart/index.htm. Accessed 10 Oct 2008

USGS. Adjacency for WinBUGS Tool. http://www.umesc.usgs.gov/management/dss/adjacency_tool.html. Accessed 8 Feb 2009

van Donkelaar A, Martin RV, Park RJ (2006) Estimating ground-level $PM_{2.5}$ using aerosol optical depth determined from satellite remote sensing. J Geophys Res 111(D21). doi:10.1029/2005JD006996

Vidot J, Santer R, Ramon D (2007) Atmospheric particulate matter (PM) estimation from SeaWiFS imagery. Remote Sens Environ 111:1–10

Zanobetti A, Schwartz J (2005) The effect of particulate air pollution on emergency admissions for myocardial infarction: a multicity case-crossover analysis. Environ Health Persp 113(8): 978–982

Chapter 21
Poverty Determinants of Acute Respiratory Infections in the Mapuche Population of Ninth Region of Araucanía, Chile (2000–2005): A Bayesian Approach with Time-Space Modeling

Flavio Rojas

Abstract This chapter highlights the relationship between poverty and disease among *Mapuche* indigenous peoples *vis-à-vis* the local population and ultimately, tests comparative differentials in mortality rates. First of all, we offer an overview of the destitute poverty in which *Mapuche* live and the consistency among all measurements including Census data, educational achievement scores and vulnerability index from school children. Although aggregate information gives a valuable and fair description of the problem, additional tests and GIS-based maps highlight the internal structures of inequalities among neighborhoods and the sharp territorial contrasts between *Mapuche* and non-*Mapuche* living conditions. GIS-based poverty maps display the territorial distributions of deprivation, whereas specific clusters of diseases are tested to verify whether such configurations are random or spatially dependent. Tobbler's "first law of geography" is discussed and eventually tested in this section. Since the study data is longitudinal, test for autocorrelation is introduced with Bayesian time-space modelling. Conclusively, *Mapuche* people die at higher rates than non-*Mapuches* as well as show significantly higher rates of disease. Consistently, this ethnic group also represents the most impoverished and marginalized one of Chilean society, both at regional and national levels.

Keywords Mapuche-Chile · Poverty-based disease · GIS-maps · Tobbler's first law of geography · Indigenous peoples · Deprivation and epidemiological maps · Bayesian time-space models

List of Abbreviation

Bayesian CI	Bayesian Credible Interval
CAR	Conditional Auto-Regression
CAS-2	Caracterización Social, Versión 2 (Social Characterization)

F. Rojas (✉)
University of North Carolina at Chapel Hill, Odum Institute for Research in Social Science
Manning Hall, CB #3355, Chapel Hill, NC 27599, USA
e-mail: frojas@email.unc.edu

J.A. Maantay, S. McLafferty (eds.), *Geospatial Analysis of Environmental Health*, Geotechnologies and the Environment 4, DOI 10.1007/978-94-007-0329-2_21,

CASEN	Caracterización Socio-económica Nacional (National Socio-Economic Characterization)
CELADE	Centro Latinoamericano de Demografía (United Nations Center for Latin American Demography)
C.I.	Coefficient of Interval
FPS	Ficha de Protección Social (Social Protection Records)
GLMM	General Linear Mixed Models
ICD-10	International Code for Diseases, Version 10
ILPES	Instituto Latinoamericano y del Caribe de Planificación Economica y Social (Latin American and Caribbean Institute of Economic and Social Planning)
INE	Instituto Nacional de Estadísticas (National Institute of Statistics)
JUNAEB	Junta Nacional de Auxilio Escolar y Becas (National School Assistance and Scholarship. Board)
MAUP	Modifiable Area Unit Problem
MIDEPLAN	Ministerio de Planificación (Ministry of Planning)
MINEDUC	Ministerio de Educación (Ministry of Education)
MINSAL	Ministerio de Salud (Ministry of Health)
PROC MEANS	Procedure Means
REDATAM	Recuperación de Datos de Areas Pequeñas por Microcomputador (Microcomputer Based System for Small Area Data Retrieval)
RR	Relative Risk
RUT	Rol Unico Tributario (Unique Tax Roster)
SIMCE	Sistema Nacional de Medición de la Calidad de la Educación (National System for Measurment the Quality of Education)
SMR	Standardized Morbidity Rates or Standardized Mortality Rates.

21.1 Introduction

This chapter concerns health inequalities in rates of respiratory infectious diseases affecting the *Mapuche* population of *Araucanía* Region, Chile. Only very recently has it been possible to obtain a count of Chile's largest ethnic population, and only very recently are there medical records to establish the ethnicity of patient populations accessing the hospital system. In this way, the present study has brought out from the closet an important, yet historically neglected segment of the Chilean population: the *Mapuche* People.

Differing perspectives have been taken to study this ethnic group. For many years anthropologists and historians alone offered exclusive accounts of the long and fierce struggles of these people to preserve their identity. Never conquered by the Spanish *Conquistador* and never subjugated by the Inca Empire, the *Mapuche* People remained free and autonomous until 1875, when the Chilean army invaded their territories in the so called "Pacification of *Araucanía*". Their bravery however, proved no match when ultimately fighting against a well organized army which only a few years later would defeat Peru and Bolivia in a bloody war (1879–1884).

The Chilean nationally awarded (1992) historian, Sergio Villalobos, proposed a controversial thesis of the "assimilation" of *Mapuches* to the predominant Chilean society. Assimilation implied willingness from within *Mapuches* as well as the results of frontier contacts, bribes from the Spanish *Conquistador*, trading (looting) and "sexual contact" on both sides leading to racially mixed offspring –the *mestizo* (Villalobos, 1995). From this viewpoint, there would be "no ancestral rights" (Villalobos, 2000) and no true *Mapuche* identities today. *Mapuches* or "*Araucanos*" as he calls them "are not indigenous peoples but racially mixed *mestizos* just like anybody else in Chile". Therefore there would be no "historical debts" of Chilean society with them (Villalobos, 2009). At a time when Mapuche leaders are presenting a lawsuit for defamation before the Chilean Justice system against Villalobos, the chapter questions Villalobos' assimilation thesis by means of its empirical research concerning environmental poverty-related issues. First of all, this study operationally distinguishes *Mapuche* people as an ethnically distinct group by using the Population Census' "self-definitions of ethnic belonging" and by the use of individual (persons) ancestral surnames from Hospital Discharge Records. Secondly this study applies a statistical framework with the goal of testing poverty attributes across individuals' above-defined ethnicity and via aggregates, specifically geographic clusters of poverty. Finally, this chapter analyzes whether such poverty aggregates are linked to disproportionally higher rates of respiratory infections and mortality burden among *Mapuche* vis-à-vis the rest of the population. Such empirical findings and the statistical testing of ethnic inequalities in morbidity and mortality rates may provide initial evidence to think of *Mapuches* as an indigenous population unable to reproduce itself and doomed to perish rather than a population in a continuous process of "assimilation" as proposed by Villalobos.

Intrinsically joined to this population analysis is the concept of randomness. If one ethnic group displays comparatively higher risks of disease and mortality than the other, it is necessary to test statistically whether these events are random or actually reveal historical and geographic (or both) patterns across-time. It is reasonable not only to test for a covariate, (in this case, poverty) to evaluate possible statistical associations between respiratory infections and material deprivation, but such a probe should also rule out (adjust) for other confounding factors as geographical (spatial) proximity and time.

Finally, Bayesian methods for disease mapping are incorporated as a way to address all the above together and simultaneously while keeping rate estimates and relative risk stable across time. See further elaboration of these terms in the Methods Section 21.2.2. Multilevel methods to fit clustered data visualized by GIS are employed here to provide estimates of *Comunas*, or counties, displaying race-age-specific respiratory infections (spatial heterogeneity). Time-frames for these events are analyzed using a nested model where hierarchical specification is applied to each year of data separately (Leyland and Goldstein, 2001). This Bayesian methodological strategy constitutes an alternative to that performed in a previous investigation by Rojas (2007) based on a frequentist General Linear Mixed Model

(GLMM) approach for cross-sectional data. Bayesian methods provide greater flexibility and precision for simultaneously analyzing longitudinal and cross-sectional spatial data. Finally, dynamic models in a Bayesian framework are used to produce smoothness in time trends, so that estimates for any particular time could "borrow strength" from data at adjacent times (Knorr-Held and Besag, 1998).

21.1.1 The Study Area

There is little doubt that the subject of environmental justice has provided in the last 10 years a solid framework for understanding how certain groups of society and subpopulations can experience disproportionate burdens of pollution, dangerous waste, habitat deterioration and poverty. Hazardous components such as lead in water pipes and the presence of asbestos in house dwellings clearly link living conditions to unhealthy environments. When analyzed for racial and ethnicity attributes, there is evidence for spatial segregation which influences "the unequal distribution of environmental risk" (Brulle and Pellow, 2006). Behind this lens, social context and spatial segregation become intimately linked to environmental risks. Although both are associated, the direction of causality is elusive. Does poverty lead to environmental risk or is it the latter which results in human deprivation? The analysis of time-sensitive data stemming from individuals and contextual factors as they affect health outcomes over a time period allows an identification of directionality in the associations. To overcome this complex mix, one would require extensive statistics or costly surveys centered on individual risk behaviors.

In sum, traditionally, individual attributes have been hard to aggregate and validly establish inferences when data has been collected at an individual level with conclusions made at an aggregate level. These problems may introduce ecological bias and originate from a *modifiable areal unit problem* (MAUP) where the scale of data aggregated to a particular set of districts may change if one aggregates the same underlying data to a different set of districts (Waller and Gotway, 2004). With the recent development of hierarchical models in biostatistics, it is now possible to integrate in a simultaneous manner links between individuals and their social contexts while adjusting for proximity and variations over time (Wakefield and Shaddick, 2006). This chapter considers poverty (material deprivation) as a contextual factor defined by a set of hierarchies ranging from individuals, to neighborhoods, to *Comunas* (counties) of the Ninth Region of *Araucanía*, Chile. See Fig. 21.1.

Multilevel modeling allows for the exploration of individual and contextual aspects of variation in exposure Soobaden et al., (2006; Goldstein, 2003; Leyland and Goldstein, 2001; Subramanian and Kawachi, 2006). The Ninth Region of *Araucanía* is one of 15 administrative regions of Chile comprising an area of 31.842,3 km^2 (12,294.264 mi^2) and with a population of 869,535 according to the latest 2002 Population Census Instituto Nacional de Estadísticas, INE, (2002); of that total, 204,195 or 23.4% self-declared to be of *Mapuche* ethnicity. In turn, the *Araucanía* is administratively divided into 31 *Comunas* Instituto Nacional de Estadísticas, INE, (2002a). After the Census, one *Comuna* (*Nueva Imperial*) was in

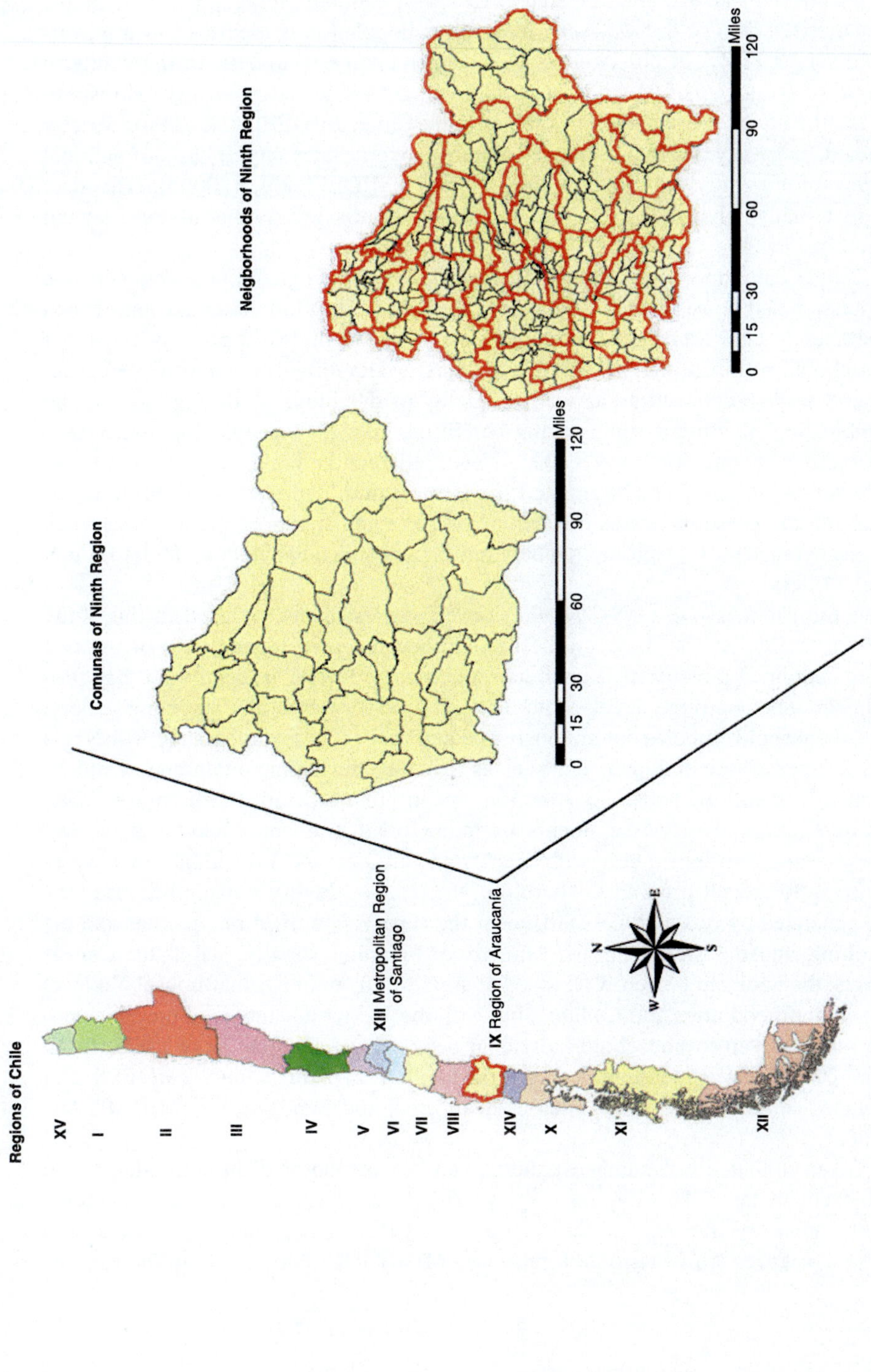

Fig. 21.1 Chile and geographical location of Araucania region, map of comunas and neighborhoods within comunas Chile and regions

turn subdivided and named *Chol-Chol*. The *Comuna* of Temuco, in which the capital of the Region is located, bears the largest population with 245,347 inhabitants. In terms of poverty rates, this Region ranks second highest in the country with a 20.1% poverty rate overall compared to 13.7 % of national average; however, almost a quarter (24.7%) of the *Mapuche* indigenous population is identified as impoverished versus 18.2% of non-Mapuches. This percentage is highest for the youngest *Mapuche* age-group of under 30 years old, with 57.7% living in poverty (Ministerio de Planificación y Cooperación, Gobierno de Chile, MIDEPLAN, 2006). Strictly defined, every 2 years MIDEPLAN compiles poverty rates on the basis of national surveys following "the poverty line method" MIDEPLAN, (2006b). The data present many advantages, but do not allow for disaggregating figures beyond municipal level.

Census data, on the other hand, are useful for providing complete counts of the population and housing and also are helpful in describing individual attributes such as ethnicity, age, literacy, economically active or unemployed persons and other household-level information such as crowding, availability of piped water, electricity and sewage connection, as well as quality of dwellings. A disaggregation and combination of attributes measuring conditions in which people live establishes the Index of Deprivation and a Basic Need Satisfaction Scale ranging from 0 (no deprivation, all basic needs satisfied) to 1 (maximum deprivation, all basic needs unsatisfied). These measures of deprivation have geo-spatial referents which may be disaggregated to permit rapid tabulation for any area down to city block Conning et al. (1989).

A third tool, *Ficha CAS-2/Familia* poverty assessments are used in this study area. In many aspects these are best suited because they incude both of the two aforementioned poverty measurement tools, but go further in-depth: (1) they provide continuous poverty levels from individual persons and families living under a same household, but on a continuously updated basis; (2) they integrate CASEN and Census measures and dimensions such as income, educational attainment, employment and quality of house construction, but in greater detail. Furthermore, *Ficha CAS-2/Familia* poverty assessments are municipal instruments filled out by social-workers and verified in-situ with data from families and individuals in poverty applying for social programs. Therefore, the data are highly reliable because they are generated by professional staff under the supervision of Municipalities and the resulting statistics stem from the Ministry of Planning. Finally, just as the Census Tracts, this tool can be geo-referenced at any level from individual-household level to neighborhood units and *Comuna*, but with the greater advantage of possible geo-referencing even to household/individual person level. A note: as of 2006, *Ficha CAS-2/Familia* was modified; it is now called *Ficha de Proteccion Social* (FPS) and includes additional variables, such as health risk and family needs (MIDEPLAN, 2006c).

Other similar disadvantaged patterns can be corroborated through educational achievement tests. The Region displays significantly lower educational achievement rates than the rest of the country in math, science, language, social and natural sciences (Ministerio de Educación, MINEDUC, 2007). Within the Region,

Mapuche students have significantly lower educational achievement scores than non-*Mapuches* as measured by SIMCE standardized achievement tests. In terms of the Index of Student Vulnerability (IVE) generated by the *Junta Nacional de Auxilio Escolar y Becas* (JUNAEB, 2006), *Mapuche* students also have higher vulnerability scores on average in every level of poverty compared to the same poverty gradients of the non-*Mapuche* student population. IVE evaluates biometric information such as number of dental cavities, weight/hight ratio, quality of eye vision and socio-economic indicators such as parental education attained, family income and employment and housing conditions JUNAEB, (2010).

Given these overall aggregate contextual attributes, one important aspect of the study consists in employing GIS tools to analyze spatial segregation, testing at *Comuna* levels whether cluster configurations of poverty pockets are linked to *Mapuche* Indigenous Populations. First, there is an important element of visualization, for which GIS-based maps are helpful particularly in their use of color contrasts which identify cluster-like poverty configurations. Second, GIS programs incorporate formal statistical testing methods for spatial configurations or "hot-spots"(i.e. Global Morans and Gettis-Ord). Third, in order to combine spatial modeling with time effects GIS offers mapping capabilities that may be combined with other Bayesian software (WinBUGS-GeoBUGS) to produce space-time estimates and interactions which ultimately may be exported back to GIS to take advantage of its accurate mapping capacities.

21.1.2 Use of Small Area Statistics

Employed here are small area statistics in order to identify subsets of the territory with the goal of scrutinizing distributions and event relationships which otherwise would be hidden from analysis. From a practical point of view, policy makers need a way to assess targeting, they need effective monitoring, and finally, timely follow-up based on information as disaggregated as possible geographically according to number of incidents and on an individual level. Practical disaggregation of data from larger geographic units to smaller units leads to some challenging methodological issues of event data variation. Variations of observed and expected rates of diseases in large areas with small populations and small areas with large populations may generate unstable rates over time, so that although the standardization in some sense yields comparability of means, the unequal base from region to region results in unequal variances (Cressie, 1992). Thus, variances larger than the mean in count data open up the existence of unknown, unmeasured, underlying residual variation of risk. Count data frequently display overdispersion. Hierarchical Poisson models have been found to be effective in capturing overdispersion (SAS Institute, 2010). In this study, questions could be raised with respect to nested hierarchies and the extent to which nearby observations are in fact, correlated whereas more distant events are less related. Tobler's "Fist law of geography" states that "everything is related to everything else, but near things are more related than distant things" (Tobler, 2004, 1970). This first law of geography provides a fundamental tool for testing spatially correlated data and clustered structures and the presence of over-dispersion which otherwise would remain hidden in the territorial aggregates. Over-dispersed data can have serious effects on testing whether regression coefficients are zero or not (Haining et al., 2009).

21.1.3 Spatial-Temporal Analysis

Another source of spatial autocorrelation comes not from geographical distributions of observations and the clusters which may be detected when moving from one to another level of aggregation or across areas, but is produced by dynamic variations over time. Unlike the simple spatial analysis described above and at one point in time, analysis of spatial-temporal data must take into account both spatial and temporal correlations (Suchindran and Rojas, 2009). Usually, when spatial clustering is assessed by the spatial distribution of incidence within a defined time period, or over other time frames, some evidence for the variation in spatial distribution will be lost (Lawson, 2006). In the present case, repeated measures of poverty over the years were modeled so that more robust estimates of the relative risks might be obtained Lawson (2006). Although the more conventional application of logistic models employs repeated measures of the same individual over time, here the repeated measures are applied to groups of individuals who share the same environment (poverty exposure) over time. The focus here primarily lies on the spatial-temporal variations of relative risk. The most common format for observations is a count of cases of disease within small areas that are then available for a sequence of T time periods (Lawson et al., 2003).

21.2 Data and Methods

The sections outlined before presented several ways to measure living conditions of a population. These conceptual tools were introduced as relevant for understanding the contextual milieu of the study area. Census data provide the most comprehensive and complete coverage of the population every 10 years. Operational aspects of Census such as hierarchies and the proper combination of variables (social attributes and housing quality conditions) are also important. A resultant Deprivation Index and geographical references have been created under the supervision of the *Instituto Latinoamericano de Planificación Económica y Social* (ILPES, 1995) and the United Nations, *Centro Latinoamericano de Demografía* (CELADE) which developed software for accessing Small-Area Census data as territorial referents (REDATAM). Other socio-economic deprivation indices constructed from the Census and applied for small areas using a GIS-based method can be found in (Bell et al., 2007). Of greatest methodological importance here is the linking of the Census Deprivation Index to the actual data provided by CAS-2/*Familia* poverty records in order to compare territorially how equivalent or divergent both measures may be. Census data have another important methodological contribution which is the possibility of generating standardized morbidity and mortality rates (SMR's) by means of direct and indirect methods, then using SMR's to compare different study groups and the relative risk between the standard population and the population under study (Pan American Health Organization, PAHO, 2002). Population Census serves in this study as the denominator, providing the total number of people by ethnicity, age-groups and gender at *Comuna* level.

Hospital Discharge Records (*Egreso Hospitalario*) are used to generate the numerator of crude (unstandardized) rates of disease (ICD-10 codes). A hospital discharge is defined by the formal release of a patient who has occupied a hospital bed and clinical services from a medical facility where he/she stayed with the purpose of observation, care, diagnostic or treatment. A discharge is produced by death or a cure to return home (Ministerio de Salud, MINSAL, 2004). It does not include clinical visits, ambulatory treatments or primary care visits. Emergency Room visits are also considered if the patient is later transferred to a bed or dies. Ethnicity is classified if either surname of a patient is *Mapuche* or non-*Mapuche* (Hispanic or European), by age (falling into one of the 8 age group categories) and gender. This information is collected at the time of discharge. In the *Araucanía* Region there are 28 hospitals (6 private, and 22 public) with differing levels of complexity. Administratively, these hospitals depend on two health services, "Araucanía Norte" and "Araucanía Sur". The compilation of data includes all Hospital Discharge Records (*Egreso Hospitalario*) from 2000 to 2005 as normed by the Chilean Ministry of Health (*Ministerio de* Salud, MINSAL, 2007) corresponding to diseases classified as "respiratory infections" (codes, J00-J06, J10-J18, J20-J22, H65-H66). A total of 14,202 geo-referenced records of individual patients treated for respiratory infections were compiled. A patient may have more than one record. Individual poverty records were obtained from *Ficha* CAS-2/*Famili*a for the 31 *Comunas* of *Araucanía* Region. These records have detailed socio-economic information and are filled individually by persons who apply for social programs, government subsidies or unemployment benefits. A total of 527,539 individuals and 94,131 families filed for social benefits in that Region since 1980. One important aspect allowing the integration of health records with *Ficha* CAS-2/*Familia* is that each individual person is identified by a unique ID labeled "RUT" (*Rol Unico Tributario*, Unique Tax Roster), equivalent to the Social Security Number of the United States. Geographical areas of *Ficha* CAS-2/*Familia* are classified according to "neighborhood units". Several neighborhood units aggregated constitute a *Comuna* level. Both, *Ficha* CAS-2/*Familia* and Hospital Discharge Records, are also longitudinally organized. A summary of the data structure generated and used in this chapter by levels and sources is presented on Fig. 21.2.

Figure 21.3 compares poverty scores as spatially distributed by *Ficha* CAS-2/*Familia* records in quintiles with the Index of Deprivation provided by Census Districts. Both are highly consistent and identify the same areas of poverty and deprivation. The Maps displayed align *Ficha* CAS-2/*Familia*'s 439 Neighborhood Units and their respective poverty gradients with Census 297 Disricts and corresponding deprivation values. While the Census Index combines several indicators of the quality of dwelling with social attributes such as age, literacy and economic

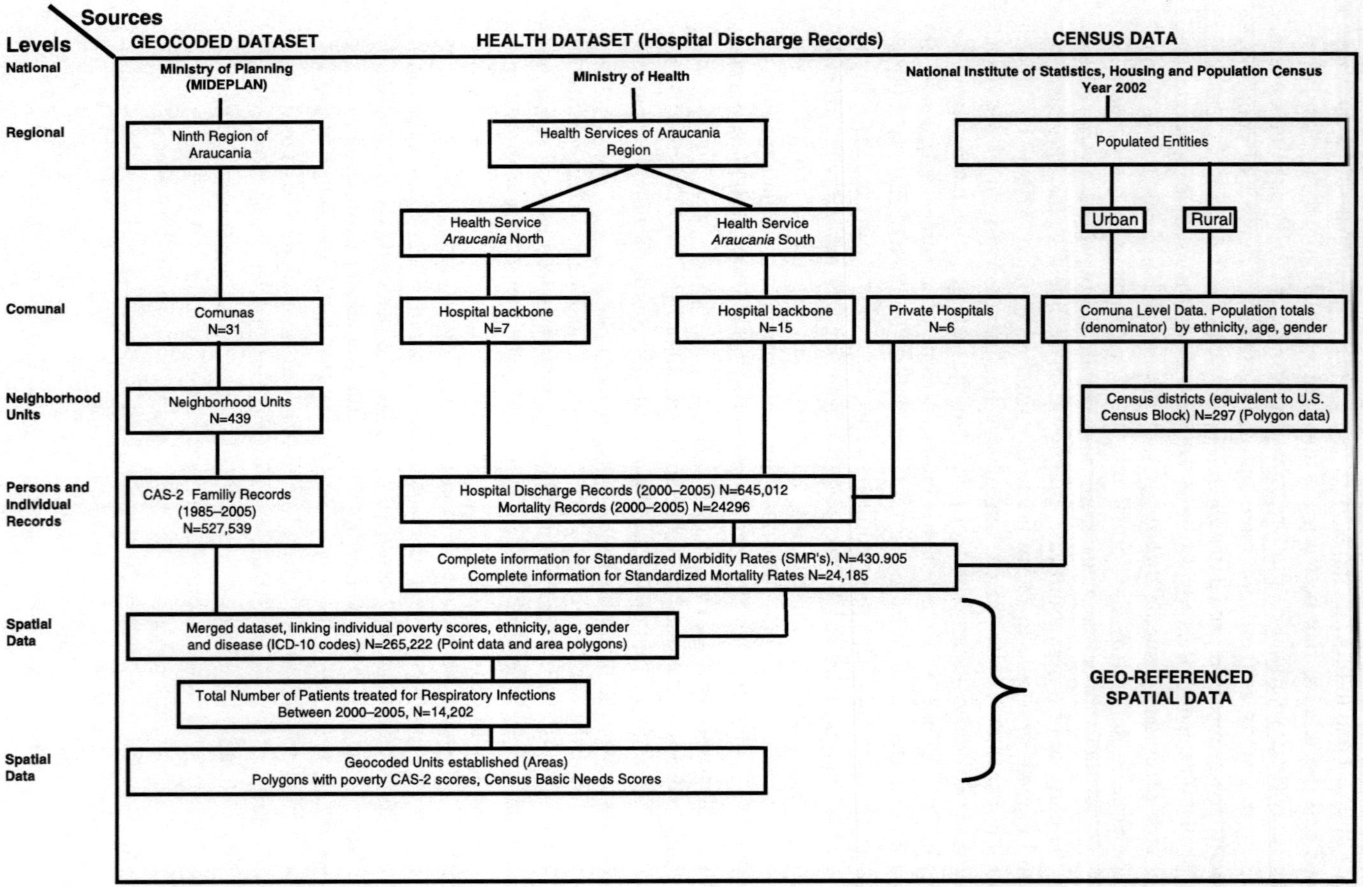

Fig. 21.2 Data structure by level and sources

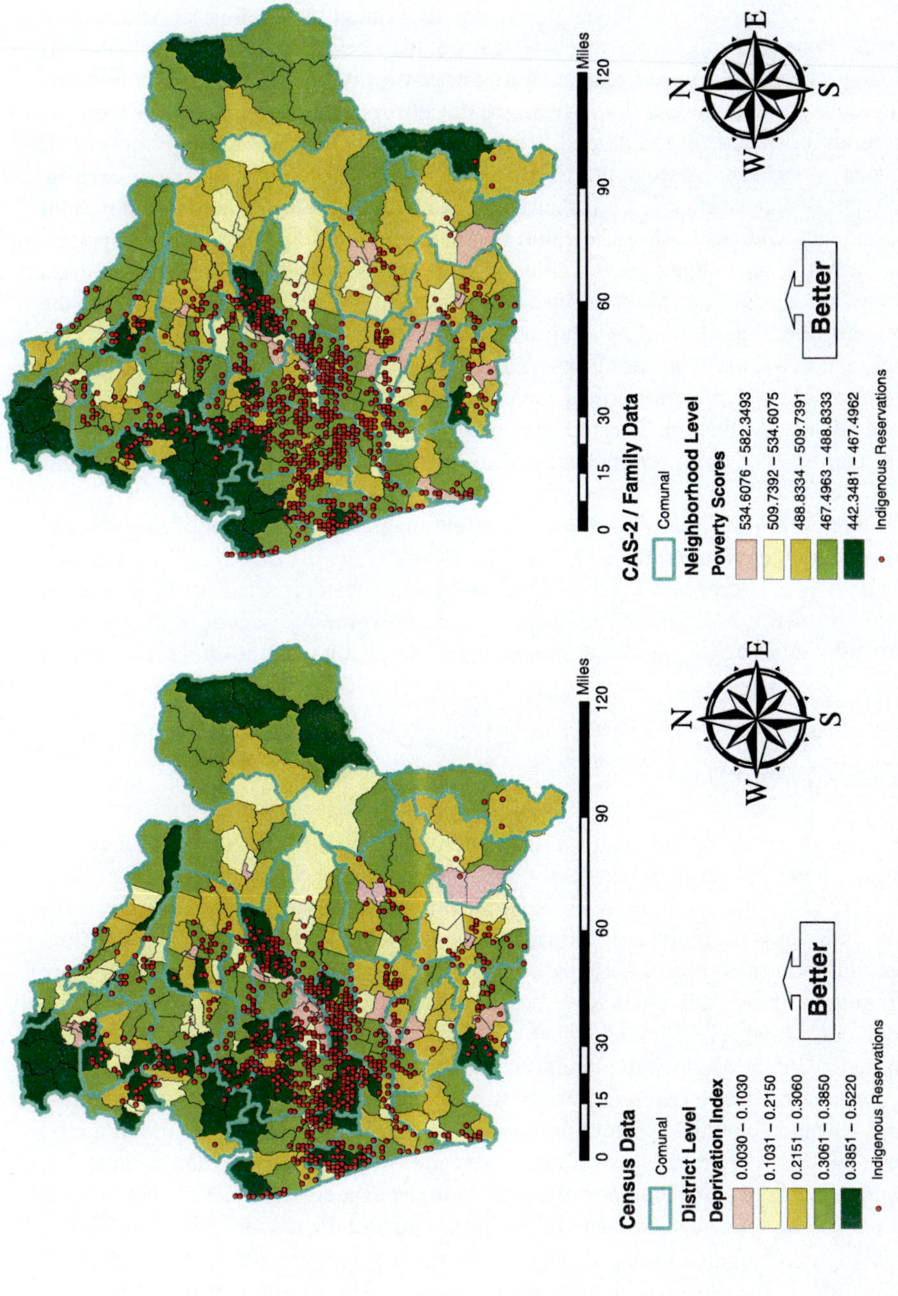

Fig. 21.3 Deprivation index scores from census and Ficha CAS 2/families scores compared

dependency of individual households, CAS-2 directly points to income, education, employment and housing quality as fundamental dimensions of poverty. The weights and variables are also different. The quality of a dwelling is estimated by *Ficha* CAS-2 according to fewer items directly related to building standards such as wall, floor and roof, water access, sewage and shower, and crowding, whereas the Census assesses more items such as ownership, types of dwelling, power line installation, telephone and TV, which are not necessarily related to precarious living conditions. Education attained is assessed by *Ficha* CAS-2 through number of years of schooling, whereas the Census focuses on assessing reading and writing and types of school attended. Ethnicity in Census data is self-declared by household respondents whereas CAS-2 the same idea is operationalized by whether a person lives or not in an Indigenous Reservation. Despite the aforementioned differences, Census and CAS-2 are highly consistent and identify the same areas towards the sea-coast (left side of the map) and towards the Andes Mountains (right side of the map). The few pink areas are urban ones, that is, cities which display better living conditions (low deprivation/lower poverty scores). These are islands of prosperity in the sea of deprivation. The red dots display locations of *Mapuche* reservations. Note that most of them are located within or overlap the most impoverished areas of the Region.

While descriptive maps of poverty distribution are useful for identifying contextual areas in which individuals reside, this chapter concerns the link of individuals' poverty scores aggregated in those areas as we move from individuals to *Comunas*, as well as their corresponding incidence rates of respiratory infections as we move from individuals aggregated to *Comunas*, while adjusting for time effects and spatial proximity.

21.2.1 Census Data

If Census data and the Deprivation Index are consistent with CAS-2/Family scores, then the latter scores may be validly used to test the association between poverty and health outcomes at individual patient levels. Because of privacy norms, Census data cannot provide individual information at household levels to link that information with individual health outcomes. *Ficha* CAS-2 does allow for this. Even more importantly, these individual level scores may be used to test the extent to with which contextual levels across-*Comunas* are associated with higher relative risks of respiratory infections for the population in them.

Data for the health outcome variable are generated by the application of indirect standardization methods to obtain standardized rates of respiratory infections by each *Comuna* over time, adjusted by age, sex and ethnicity. This method compares at *Comuna* level the actual number of events with the expected number of events when factor-specific rates, age, sex and ethnicity are taken into account from a reference population and applied to the local population (Roalfe et al., 2008). The Reference Population is the regional population of *Araucanía* according to the Population Census of 2002 by Comuna, age-groups, gender and ethnicity. It is essential that rates for different years be adjusted to the same standard population before making comparisons between Mapuches versus non-Mapuche population. Conceptually, any population distribution can be used as the standard, but that choice must not have a great effect on the relative levels of the age-sex-ethnic adjusted rates that are being compared. One author suggests that with indirect standardization, estimates have lower variances. This internal standard is especially important for small areas such as counties and Census tracts, and constitutes a method of choice for maps with estimates of multiple areas showing geographical variation (Kulldorff, 2003). In particular, if we wish to compare indirectly standardized rates between counties within the same state, we may wish to use data from the entire state as the standard population to obtain rates for each county (Waller and Gotway, 2004).

21.2.2 Methods

Figure 21.2 provides an organizational glimpse of levels, sources and data structure for this research. The methods used should account for a nested, yet hierarchical order of levels and how across-level data are related to health outcomes. Specifically, individuals are nested (part of) household units, household units are subsets of neighborhoods, and neighborhoods are components of Comunas, with 31 Comunas conforming the Region of study.

This element of nesting reveals that observations are not independent. Therefore, multilevel techniques explicitly are used to model correlated data where the assumption of independence between observations is violated and conventional ordinary least square techniques are not appropriate (Soobaden et al., 2006). Recently, many applied Bayesian methods using hierarchical models have appeared. For example, an important application of hierarchical models is "small area estimation" in which estimates of population characteristics for local areas are improved by combining the data from each area with information from neighboring areas (Gelman et al., 2003). These methods, "borrow strength" from adjacent areas, thus making local estimators more stable. A Conditionally Autoregressive model (CAR) is used here to provide spatial smoothing to improve local area estimates. CAR models are a more common way of dealing with the spatial correlation between neighboring areas, since we cannot fit spatially-correlated effects at differing levels of geography (Lawson et al., 2003). Generalized linear mixed models (GLMM) have been applied by both Classical Frequentist statistics and Bayesian methods to handle correlated data. Bayesian methods here incorporate CAR to smooth unstable disease rates. When handling spatial lattices, CAR is commonly used and accounts for random effects included for each subregion. Here, the CAR model accounts for real risk heterogeneity among subregions/Comunas, resulting from undermeasured risk factors, following (Escaramìs et al. 2008). Finally, this hierarchical order is subject to change over time and subsequently this research involves variabilities simultaneously occurring in a span of 6 years.

A combination of descriptive methods and Bayesian time-space nested models were used to respond to the questions of poverty, location of impoverished areas and severity of poverty gradients in the population of the *Araucanía* Region and the outcome variable: SMR's and the expected Relative Risk. The SMRs are the maximum likelihood estimates of the relative risk under a Poisson model for the observed number of deaths/disease (Heistercamp et al., 1993). Indirect standarized methods were used here. These methods compare the actual number of events in a local area (i.e. *Comunas*) with the number expected when factor-specific rates (age, sex and ethnicity) in a reference population (the population of *Araucania* Region) are applied to the local population.

Directly standardized rates were also obtained for aggregate comparisons of the incidence of morbidity and mortalities by years, age-groups and ethnicity. Since these are not samples, strictly-speaking, confidence intervals are not really necessary; nevertheless, the calculation of limits based on gamma distribution was accomplished using the Anderson-Rosemberg method recommended by the NCHS (1998), DUG2 for the directly standardized rates. I am thankful to Professor Emeritus, Dana Quade for his SAS macro programming and advice.

The scores from *Ficha* CAS-2/*Familia* records are routinely calculated by municipalities under the technical supervision of MIDEPLAN of Chile. These scores are applied to individuals and to each member of the family living in the same household. Scores are obtained at neighborhood-level units (*Unidades Vecinales*, UV) and *Comunas* are averaged using PROC MEANS statement SAS Version (9.2) by Year and *Comuna*. Whereas borders and jurisdictions provide an identification of the geographical distributions, poverty scores and the application of quintiles allow for assessment of the intensity of poverty and the establishment of orderly gradients at lowest levels of aggregation or *Unidades Vecinales,* as well as at *Comuna* levels.

These unique individual poverty records were later seamlessly merged with patients' Hospital Discharge Records as compiled by every hospital of the *Araucanía* Region, including information on patients' sex, age, ethnicity, and disease, home address, *Comuna* of residence and unique ID identity card number (RUT).

A second methodological step consisted in linking individual patients' attributes (observed count data) to standardized morbidity rates with geographical areas. We assume here that counts of observed respiratory infections have a Poisson distribution with expected value $E_{it}\theta_{it}$ where variations of disease occur by Comuna but also over time. Thus, $i = 1, \ldots \ldots 31$ *Comunas* and $t = 1, \ldots \ldots 6$ years. For each *Comuna* and for each year, the standardized morbidity rate defined as the observed divided by the expected number of cases – was calculated. Next, following Waller et al. (1997) and Lawson et al. (2003) nd, a nested model was used in which the hierarchical specification of Besag et al. (1991) was applied to each year-point separately. From a Bayesian approach perspective, the posterior mean of the relative risk (RR) in the overall map for the *i*th area is the weighted average of the SMR for the *i*th area. This weight is inversely related to the variance of the SMR (Lawson et al., 2003). As this variance is large for rare diseases and for small areas, this weight is small and the posterior mean tends towards a global mean thereby producing a smoothed map (Gilks et al., 1998). Previous work has incorporated observed health outcomes as well as social indicators of deprivation into a more comprehensive model, but for psychiatric morbidity, and also has allowed the data to determine the appropriate level of spatial smoothing (Congdon, 2009). An equivalent Bayesian approach establishing the association between socioeconomic and racial disparities as related to stroke mortality using spatial smoothing for unstable local rates is found in Tassone et al. (2009). Spatial smoothing techniques specified under CAR model also were used to link asthma incidence by time and positive effects for the percent Non-white variable in North Carolina (Suchindran and Rojas, 2009). Further contributions for the Bayesian analytical framework and methods are found in (Dunson, 2001). In this research, the log of the relative risk for the space-time model is parameterized as:

$$\text{Log}\,\theta_{it} = \alpha + \mu_i^{(t)} + \nu_i^{(t)} + \Sigma\beta_1 x_{1i}^{(t)} \tag{1}$$

where α is an overall level of the relative risk ${\mu_i}^{(t)}$ and ${\nu_i}^{(t)}$ are the correlated and uncorrelated heterogeneity terms that can vary in time (random effects for the *Comuna* i th in time *(t)*) and $\beta_1\ x_{1i}^{(t)}$ the poverty score of *Comuna* i th in time *(t)*. α and β1 are fixed effects. The software WinBUGS was used to fit the model of Spiegelhalter et al. (2003). With respect to number of iterations after convergence, the more samples that are saved, the more accurate the posterior means estimates (Spiegelhalter et al., 2003). Another additional 50,000 iterations were run, 501 was used as the starting point for the second run of final iterations, instead of 1 as used in the previous burn-in as the starting point. The WinBUGS code and corresponding convergence diagnostic for the overall level of relative risk α and poverty score parameter β are presented in Appendix 1. The historical time-series plot of Alpha chains 1:3 seems to converge very well since all 3 chains reveal to be overlapping one another. Note that the center of the parameter of Interest β_1 (poverty) appears to be around –0.03 with very small fluctuations. This means that the Gibbs Sampler has eventually reached a stationary condition. The aspects of stationarity that are most recognizable from a trace plot are of relatively constant mean and variance (SAS Institute, 2010); it is the absence of any systematic change in the mean or variance (King et al., 2009). Operationally, effective convergence of Markov chain simulation has been reached when inferences for quantities of interest do not depend on the starting point of the simulations (Brooks and Gelman, 1998). The data of relative risks (thetas) was generated and GeoBUGS was used to visualize and interpret the outcomes. Finally, the resulting information was exported to ArcView to generate the final maps.

21.3 Results

Differential rates of disease are exhibited by the *Mapuche* versus non-*Mapuche* population, both by age-groups and for the 6 consecutive years of respiratory infectious diseases investigated.

In Table 21.1, Relative Risk (RR) represents the relative risk of Mapuche to non-Mapuche to contract respiratory infections. The evidence indicates that *Mapuche* children younger than 5 years old experience recurrently higher rates of respiratory infections over time with a peak in the year 2001. *Mapuche* children also exhibit

Table 21.1 Respiratory infections-related, morbidity rates compared by year age group and ethnicity. Directly standardized rates per 1,000. Ninth Region of Araucanía, Chile 2000–2005

			Mapuche C.I. 95%			Non-Mapuche C.I. 95%			
			Mapuche	Lower	Upper	Non-Mapuche	Lower	Upper	RR
$N = 7753$	2000	< 5	61.4	57.5	65.5	54.7	52.6	56.8	1.12
		5–14	5.9	5.2	6.7	6.2	5.7	6.7	0.96
		15–29	1.0	0.7	1.3	1.1	0.9	1.3	0.90
		30–44	1.8	1.4	2.3	1.3	1.2	1.6	1.33
		45–59	3.0	2.4	3.8	2.1	1.8	2.5	1.42
		60–69	8.0	6.6	9.7	5.8	5.0	6.7	1.38
		70–79	21.1	17.6	25.0	16.3	14.7	18.1	1.29
		80 >	38.8	33.2	45.2	41.7	38.0	45.6	0.93
$N = 8735$	2001	< 5	81.0	97.8	85.8	62.2	58.6	64.5	1.30
		5–14	5.6	6.7	6.4	5.4	4.8	5.9	1.04
		15–29	1.6	1.6	2.0	1.2	1.0	1.4	1.31
		30–44	1.9	2.0	2.4	1.4	1.2	1.6	1.31
		45–59	4.1	5.6	4.9	3.0	2.3	3.4	1.38
		60–69	11.0	12.8	13.0	7.4	5.6	8.4	1.49
		70–79	19.4	23.0	23.4	16.9	14.1	18.7	1.15
		80 >	52.8	61.9	61.4	43.9	37.8	47.9	1.20
$N = 7097$	2002	< 5	64.3	60.3	68.5	50.0	48.0	52.15	1.28
		5–14	5.3	4.5	6.1	5.2	4.8	5.62	1.02
		15–29	1.0	0.7	1.4	1.0	0.9	1.20	0.96
		30–44	1.6	1.3	2.0	1.2	1.0	1.35	1.40
		45–59	2.7	2.2	3.4	2.0	1.7	2.35	1.35
		60–69	7.7	6.4	9.3	4.6	3.9	5.42	1.66
		70–79	17.2	14.2	20.6	13.9	12.4	15.51	1.24
		80 >	40.6	34.5	47.6	40.9	37.3	44.85	0.99

Table 21.1 (continued)

			Mapuche C.I. 95%			Non-Mapuche C.I. 95%			
			Mapuche	Lower	Upper	Non-Mapuche	Lower	Upper	RR
$N = 6284$	2003	< 5	77.0	72.2	82.1	27.4	25.9	28.8	2.82
		5–14	6.5	5.6	7.6	3.3	3.0	3.6	1.99
		15–29	1.2	0.9	1.6	0.6	0.5	0.7	2.13
		30–44	2.1	1.5	2.7	0.7	0.6	0.9	2.89
		45–59	4.3	3.3	5.6	1.6	1.3	1.8	2.76
		60–69	16.9	13.3	21.1	4.5	3.9	5.2	3.74
		70–79	36.8	30.1	44.6	13.6	12.3	15.1	2.70
		80 >	60.7	50.9	71.9	28.8	26.0	31.8	2.11
$N = 6410$	2004	< 5	76.8	71.4	82.4	28.4	26.9	29.9	2.71
		5–14	6.6	5.7	7.6	2.7	2.4	3.0	2.43
		15–29	1.4	1.1	1.8	0.7	0.6	0.8	2.02
		30–44	1.4	1.1	1.8	0.7	0.5	0.8	2.10
		45–59	4.8	3.8	6.1	1.6	1.3	1.8	3.07
		60–69	10.4	8.4	12.6	4.4	3.8	5.1	2.35
		70–79	42.2	34.7	50.8	13.4	12.1	14.9	3.14
		80 >	65.4	55.5	76.4	33.0	30.0	36.1	1.98
$N = 6281$	2005	< 5	76.4	71.0	82.0	28.4	27.0	29.9	2.69
		5–14	7.0	5.9	8.2	3.9	3.6	4.2	1.80
		15–29	1.4	1.0	2.0	0.6	0.4	0.7	2.59
		30–44	1.5	1.1	2.0	0.5	0.4	0.7	2.87
		45–59	4.6	3.2	6.4	1.5	1.2	1.7	3.14
		60–69	12.3	9.2	16.3	4.2	3.6	4.9	2.91
		70–79	40.8	32.8	50.3	11.9	10.7	13.2	3.44
		80 >	73.1	61.3	86.4	28.5	25.8	31.5	2.56

Source: Own, based on Health Services Araucania North and South Hospital discharge records.
Respiratory infections, ICD-10 Codes, J00-J06, J10-J18, J20-J22, H65-H66.

higher relative risks than the non-*Mapuche* for the same age cohort during all 6 consecutive years. Overall, Relative Risk shows a significantly larger value for the *Mapuches* (i.e. C.I.'s for the RR > 1, not shown here). Another important finding is that even for middle-age groups, which according to life-course perspectives are the least vulnerable groups, respiratory infections among *Mapuches* are also higher than that of non-*Mapuches* in all 6 years from 2000 to 2005, with peaks in years 2003–2005 or twice the historical trend of years 2000–2002. Conclusively, relative risks for respiratory infections across 2000–2005 are predominantly higher for all *Mapuche* age-groups, but worsen in the years 2003–2005.

An area of even greater concern is whether such diseases turn out more deadly when comparing groups. Evidence of ethnically-based mortality rates would provide an important argument against the "assimilation thesis" of Villalobos. In this case, directly standardized mortality rates were calculated using the same method and program as used for calculating morbidity rates. Table 21.2 indicates that mortality rates are also highest among Mapuche children, with the exception of year 2004. Years 2005 and 2002 display twice and thrice the relative risks as compared to the same non-Mapuche group. Middle age deaths are also higher for 30–44 and 45–59 age-group categories for the Mapuche group, when compared to the non-Mapuche. In 2001, 2002 and 2005, the relative risk of dying from respiratory infections for Mapuche middle-age group 30–44 peaks with values of 7.12, 4.79, and 8.70 respectively. Such high peaks of mortality lead the investigator to ponder time effects of higher relative risk of disease and death of Mapuches over the years 2000–2005. One final empirical probe remains: to establish the association between poverty and disease over time, and the extent to which higher relative risks of respiratory infections are most likely found among the poor.

The following model estimated the relative risks of respiratory infectious diseases (posterior expected risk) between 2000 and 2005 for the population of *Araucanía* Region. The practical understanding of disease as a preceding cause of death led us to estimate the association between poverty and respiratory infections over time.

Table 21.3 shows posterior values for the parameters of the model after 55,000 iterations for the alpha (overall level of the relative risk) and β_1 nodes. For a technical discussion of the required iterations of Gibbs sampler (See, Raftery and Lewis, 1991). Both fall between Bayesian Credible Intervals (CI). The latter displays a negative sign, indicating that the lower the scores (the poorer an area) the higher the relative risk of respiratory infections for those living in that area. Note that credible intervals are quite narrow, indicating precise estimates, that is, influence of the covariate is significant. A Credible Interval in the Bayesian approach is equivalent to the "Confident Interval" created through point estimation in Frequentist Statistics, as it constitutes a posterior probability interval used for interval estimation.

Table 21.4 presents the posterior statistics and 95% credible intervals of the variance components for the clustering effects model τu over time *(t)*. The values reveal that between 2000 and 2002 a variability in the relative risk is observed, suggesting that there is a temporal trend for the first three years. This increase is explained more by the clustering effects over time than spatially uncorrelated extra-variation. The

Table 21.2 Respiratory infections-related, mortality rates compared by year age group and ethnicity. Directly standardized rates per 1,000. Ninth Region of Araucanía, Chile 2000–2005

			Mapuche C.I. 95%			Non-Mapuche C.I. 95%			
			Mapuche	Lower	Upper	Non-Mapuche	Lower	Upper	RR
$N = 199$	2000	< 5	0.49	0.21	0.99	0.29	0.15	0.51	1.69
		5–14	0.00	0.00	0.19	0.01	0.00	0.05	0.00
		15–29	0.00	0.00	0.16	0.03	0.01	0.08	0.00
		30–44	0.03	0.00	0.18	0.05	0.02	0.11	0.72
		45–59	0.07	0.01	0.22	0.03	0.01	0.09	2.34
		60–69	0.13	0.02	0.38	0.23	0.11	0.43	0.55
		70–79	1.59	0.57	3.51	0.83	0.49	1.30	1.92
		80 >	5.09	2.71	8.72	6.15	4.85	7.70	0.83
$N = 251$	2001	< 5	0.98	0.55	1.61	0.62	0.41	0.90	1.58
		5–14	0.00	0.00	0.19	0.00	0.00	0.06	0.00
		15–29	0.02	0.00	0.14	0.02	0.00	0.07	1.35
		30–44	0.14	0.04	0.36	0.02	0.00	0.06	7.12
		45–59	0.10	0.03	0.25	0.11	0.05	0.20	0.90
		60–69	0.33	0.13	0.70	0.21	0.10	0.39	1.56
		70–79	0.95	0.40	1.90	1.23	0.84	1.74	0.77
		80 >	3.90	2.27	6.24	6.68	5.25	8.39	0.58
$N = 172$	2002	< 5	0.52	0.15	1.30	0.15	0.06	0.29	3.50
		5–14	0.01	0.00	0.06	0.00	0.00	0.06	0.00
		15–29	0.02	0.00	0.10	0.02	0.00	0.05	0.98
		30–44	0.03	0.00	0.17	0.01	0.00	0.04	4.79
		45–59	0.12	0.01	0.50	0.04	0.01	0.11	2.92
		60–69	0.14	0.01	0.57	0.16	0.06	0.33	0.84
		70–79	1.82	0.93	3.19	0.71	0.41	1.13	2.57
		80 >	6.64	3.97	10.42	5.31	4.05	6.84	1.25

Table 21.2 (continued)

			Mapuche C.I. 95%			Non-Mapuche C.I. 95%			
			Mapuche	Lower	Upper	Non-Mapuche	Lower	Upper	RR
$N = 186$	2003	< 5	0.32	0.03	1.19	0.24	0.12	0.43	1.30
		5–14	0.00	0.00	0.19	0.00	0.00	0.06	0.00
		15–29	0.01	0.00	0.04	0.01	0.00	0.03	1.24
		30–44	0.03	0.00	0.16	0.03	0.01	0.07	1.10
		45–59	0.07	0.01	0.20	0.04	0.01	0.11	1.82
		60–69	0.26	0.04	0.86	0.26	0.13	0.46	1.03
		70–79	0.31	0.09	0.75	0.77	0.47	1.18	0.40
		80 >	7.42	4.82	10.93	5.71	4.38	7.31	1.30
$N = 184$	2004	< 5	0.12	0.01	0.50	0.23	0.11	0.43	0.50
		5–14	0.00	0.00	0.19	0.00	0.00	0.06	0.00
		15–29	0.02	0.00	0.09	0.02	0.00	0.05	0.91
		30–44	0.04	0.00	0.15	0.03	0.01	0.08	1.34
		45–59	0.00	0.00	0.29	0.04	0.01	0.10	0.00
		60–69	0.13	0.02	0.40	0.20	0.08	0.41	0.65
		70–79	1.20	0.52	2.36	0.56	0.31	0.94	2.13
		80 >	7.57	4.94	11.10	6.05	4.71	7.66	1.25
$N = 162$	2005	< 5	0.53	0.17	1.23	0.23	0.11	0.44	2.28
		5–14	0.00	0.00	0.19	0.00	0.00	0.06	0.00
		15–29	0.01	0.00	0.08	0.02	0.00	0.05	0.85
		30–44	0.06	0.01	0.21	0.01	0.00	0.04	8.70
		45–59	0.02	0.00	0.11	0.03	0.01	0.09	0.65
		60–69	0.00	0.00	0.00	0.20	0.08	0.42	0.00
		70–79	0.72	0.26	1.59	0.31	0.14	0.59	2.31
		80 >	6.46	4.05	9.76	6.25	4.88	7.89	1.03

Source: Own, based on Health Services Araucania North and South mortality records.
Respiratory infections, ICD-10 Codes, J00-J06, J10-J18, J20-J22, H65-H66.

Table 21.3 Posterior statistics for the estimates of the relative risks time-space model

Node	Mean	SD	2.50%	97.50%
alpha	0.01014	0.003746	0.002834	0.01756
beta1	–0.02665	0.003692	–0.03391	–0.Ta01944

Table 21.4 Posterior statistics of the variance components for the clustering effects over time *T*

Node	Mean	SD	2.50%	97.50%
tau.u[2000]	2,491	1,895	422.0	7,481
tau.u[2001]	2,536	1,918	418.7	7,588
tau.u[2002]	2,046	1,737	281.1	6,689
tau.u[2003]	2,129	1,720	364.8	6,769
tau.u[2004]	2,311	1,820	382.6	7,173
tau.u[2005]	2,580	1,926	451.6	7,648

Table 21.5 Posterior mean of the un-correlated heterogeneity effect

Node	Mean	SD	2.50%	97.50%
tau.v[2000]	2,188	1,528	569.8	6,293
tau.v[2001]	2,467	1,698	617.9	7,016
tau.v[2002]	1,898	1,427	463.8	5,794
tau.v[2003]	3,197	2,065	772.7	8,569
tau.v[2004]	2,882	1,923	679.6	7,902
tau.v[2005]	3,691	2,204	987.6	9,341

posterior mean of the uncorrelated heterogeneity τv does not seem to provide additional extra variation in the data beyond that correlated with neighboring *Comunas* (Table 21.5).

Maps for the posterior expected relative risk are presented next.

The relative risk estimates presented include the locations of indigenous reservations. If contextual poverty overlaps the areas where indigenous reservations are located, it is important to visualize whether relative risk rates also overlap across the same areas. As displayed by Fig. 21.3, the highest relative risks are observed towards areas of the right and left side of these maps, thus overlapping with the areas of poverty identified before with Fig. 21.2. The Figure also illustrates red dots representing indigenous reservations which are mostly located in the areas where highest risk rates are observed. It seems that clustering effects over time explain the variability of the relative risk in the first 3 years 2000–2002, as only 11 or 12 *Comunas* exhibit expected relative risks over 1; however, after 2002 these variations are explained more by uncorrelated heterogeneity than by spatially structured cluster over-time. After 2002, 16–20 *Comunas* exhibit expected relative risks over 1. We may conclude that once respiratory infections gain ground over time, there are spurt-like clustered configurations that may reverberate and expand the disease to other areas (Fig. 21.4).

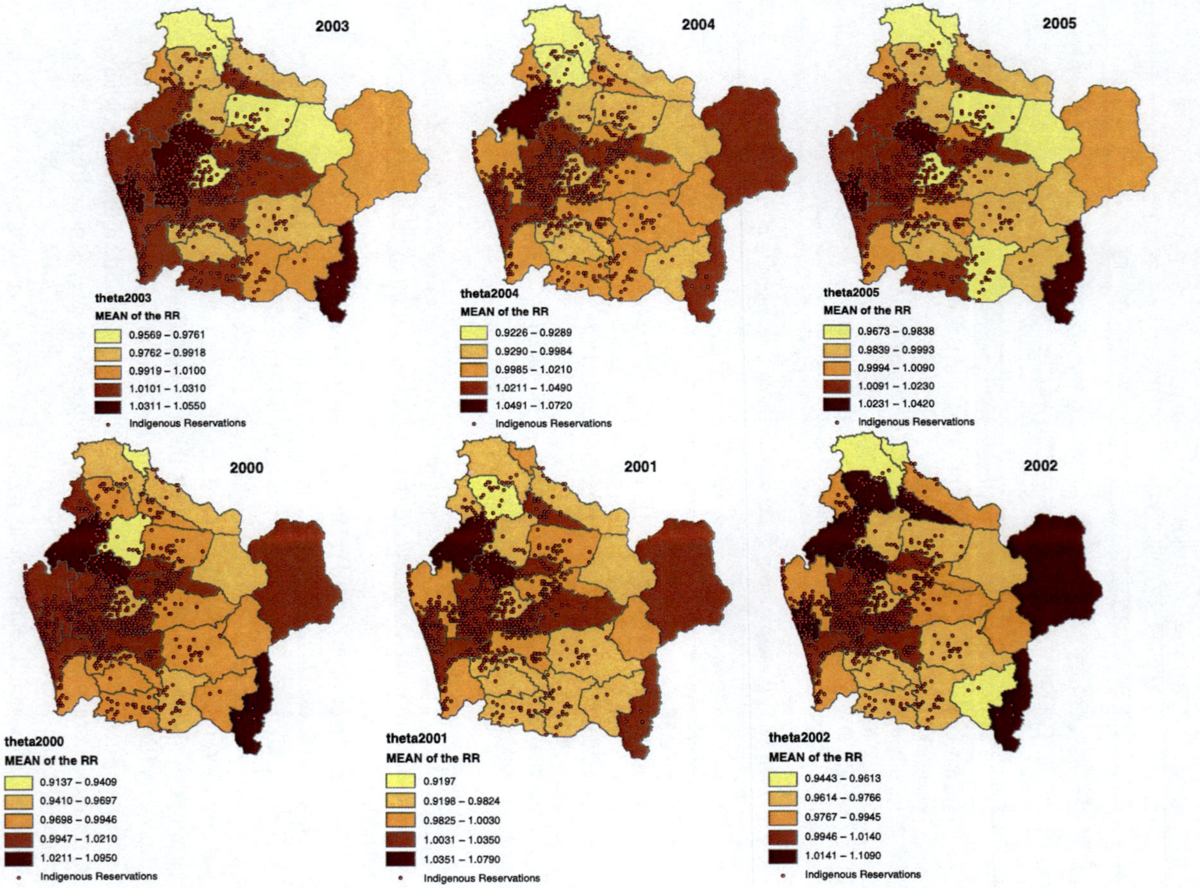

Fig. 21.4 Relative risk (RR) posterior expected relative risk for respiratory infections 2000, 2001, 2002, 2003, 2004, 2005 time-space model with poverty as predictor

21.4 Discussion

For many years, the ethnic components of the Chilean population remained distant, if not hidden from research scrutiny. Most approaches to the indigenous people came from anthropology with field work and case studies very difficult to compare and generalize. With the Population Census of 2002 and the establishment of questions concerning respondents' ethnic self-identification, including several categories of response, the result was a numeric count of several ancestral groups and cultures as well as their geographic locations. Similarly, after a long and tedious process of compilation of Hospital Discharge Records, it is now possible to incorporate real ethnic data into analyses of health inequalities. One caveat should be mentioned with respect to ethnicity, this is, that the definition of ethnicity by the Office of the Census differs in some respect from ethnicity as categorized by the health authorities. In the former, it is a self-definition from the respondent according to several alternatives of ethnicity provided; in the latter, ethnicity is determined by surnames of patients. Concordance statistics between Census and Hospital Records were not tested and this may be a limitation of the study; although Gini Coefficients here (Elliot et al., 2009) may have been used, the test would have required a hefty work of spatial concordance of Census districts with *Ficha* CAS-2/*Familia* UVs (Neighborhood Units). Given that *Mapuche* People now in fact, can be studied quantitatively, the important possibility exists of linking this group to other statistics and datasets provided by medical records, as well as to other social gradients, such as educational achievement tests. In many ways, this study confirms similar findings on ethnic minorities, such as research concerning the living conditions of ancestral vis-à-vis European, white populations in Australia and New Zealand or research examining non-white ethnicity and asthma incidences in North Carolina (Suchindran and Rojas, 2009). More specifically, in a previous study Rojas (2007) used General Linear Mixed Models (GLMM) to provide evidence that *Mapuche* people experience higher levels of poverty than the general population; in that case, spatial segregation measures indicated higher rates of respiratory infections as well as sadly, higher mortality rates produced by them. This chapter has taken a different approach and strategy, yet there are convergences in the outcomes: ethnicity and poverty are powerful predictors of health inequalities of disease.

Perhaps the most pervasive finding worth discussion in the present study is the vulnerability and actual mortality of *Mapuche* children < 5 years old vis-à-vis the non-*Mapuches*. In the beginning of this chapter, environmental risk was described as a disproportionate burden that some groups experience from unhealthy environments. In this investigation, we have also found that very early in their lives, *Mapuche* children disproportionately experience the burden of poverty by residing in highly precarious areas with higher material deprivation and which eventually may trigger respiratory infections leading to higher mortality rates. Other than higher material deprivation, there are other covariates and factors which are known to be associated with respiratory infections, for instance, winter-time. Respiratory infections peak during winter-time (June-August in Chile) when cold weather and temperatures are lowest. The space-time model used here is set-up on a yearly basis (2000–2005), meaning that seasonal factors are averaged out and not entered in the model. Cold weather in that Region facilitates indoor permanency around the fireplace, thus facilitating lengthy exposure to polluting agents as presented in a study of Temuco's daycare children (Rivas et al., 2008). Distance from hospitals, isolation of rural communities, poor and muddy roads in winter-time and cultural handling of diseases by the Mapuche healers known as "Machi" may also delay prompt treatment of patients and cause early deaths, particularly among children and the elderly. In fact, in 2005 *Mapuche* children displayed twice the risk of dying than non-*Mapuche* infants for the same age-group; in 2002, the risk was three times the risk for the same age-group. Moreover, if harsh conditions were found to be present at the earliest and most tender age in life, it is not surprising that as shown by data presented here, this vulnerability among the Mapuche people is carried over into middle-agers when the human body should be strongest, more immune to health hazards, and human life reaches its peak of productivity.

Precarious living and material poverty may be statistically linked to disease, but other covariates such as nutrition and soil productivity should also be explored. Proper representation of soil fertility and seasonal variations may also be integrated to the GIS and remote sensing techniques employed to assess the links between small-size property where Mapuche live, the nutritious value of their food intake and subsequent depletion of vegetal layer of their land. Disease may be linked to material deprivation and an ethnically-based disproportionate share of the relative risks, but also to other covariates that trigger disease events.

These grim statistical findings in ethnicity should alert policy experts as well as International Law practitioners concerned with human rights violations. Although individual human rights violations of political nature are easier to scrutinize, these collective violations of human rights which result from systematic neglect in health conditions are crucial and difficult, but not impossible to establish. We hope the avenues opened by this research may be widened by additional medical-spatial evidence.

The route from disease to the end of life caused by a preventable disease is a complex one. Here we have singled out one aspect, that is, higher prevalence of poverty rates among *Mapuches*. There may be other predictors, such as biomass fuels habitually used by *Mapuches* to cook inside their huts in wintertime; an interesting finding has been that respiratory infections peak precisely in winter time (data not included in this chapter). Indoor biomass fuel use may be another covariate that perhaps moderates the effects of poverty and disease on differential mortality rates found here. Other covariates may be related to cultural perceptions of disease and Mapuche initial use of ethnic medicine, causing delays in reaching the established "western" medical treatment provided by the hospital system in the *Araucania* Region.

With respect to the "assimilation" thesis by Villalobos with which we opened this chapter, it was possible beyond doubt to statistically identify a *Mapuche* population and establish their relative vulnerabilities (material and health), including a death toll significantly higher than the rest of the population. This evidence redefines the suggestion that this ancestral subpopulation has been absorbed by another dominant one, ceasing to exist as separate entity. Consequently, we must confront the possible fate of a minority ethnically-based materially deprived population which dies at higher rates of respiratory infections than that of the rest of the population. Furthermore, Jordan et al. (2006) and Prescott et al. (2003) have directly tested poverty in diverse populations with mortality as outcome variable, resulting in a statically significant fatal relationship. Additional research should directly link Mapuche individual mortality to the poverty gradients of that population.

Unless ethnically sensitive and targeted health interventions are employed in Chile, the Mapuche will likely fail to survive. If Mapuche birth rates are comparable to those of the general population, they are failing to reproduce themselves. Not the sword, but germs, poverty and inequalities are key factors resulting in the decimation of ancestral populations according to works by Diamond (2005, 1997) and Farmer (2005). The Mapuche People, besieged by poverty which then detonates into infectious communicable diseases, stands at risk of succumbing to the same fate of elimination.

Acknowledgments I am very grateful to Ingrid Q. Rojas for her editing assistance. I want to express my gratitude to Dr. Jaime Neira, Dr. Milton Moya, and Dr. Miguel Angel Solar, former Directors of Hospital Systems *Servicio de Salud Araucania Sur,* without whose help this research would not have been possible.

Author Biography

Flavio Rojas was born in Santiago, Chile in 1952. After completing high-school, he studied Sociology, receiving a Degree of *Licenciado* in Sociology from the *Instituto de Sociología, Pontificia Universidad Católica* of Chile in 1981. He has also Magister studies in Urban and Regional Planning from Centro Interdisciplinario de DesarrolloUrbano-Instituto de Planificación Urbana (CIDU-IPU), Universidad Católica of Chile. In 1982 he received the Regular Training Scholarship Award from the Organization of American States to conduct graduate studies in sociology at Duke University. His Masters of Art was awarded in 1985 and Ph.D. in 1991. He has lectured at University of Chile, Instituto de Estudios Internacionales, and Odum Insitute "GIS short course series." In 2002, he was admitted as a Post-doctoral Research Associate in the Department of Biostatistics, Gillings School of Global Public Health, University of North Carolina at Chapel Hill, where he conducted original research linking statistically and geographically hospital discharge records with poverty records of the ethnic population of Ninth Region of *Araucaria.* This research provided the basis for testing the inequalities in the relative risks of morbidity and inequalities in mortality rates between *Mapuche* indigenous peoples and the rest of the population. He has disseminated these findings in several peer-reviewed Conferences in Europe (Oslo and Lisbon) as well as other professional organizations, such as the Latin American Studies Association, LASA (San Juan and Rio de Janeiro). After completing a Post-doc in Biostatistics he joined The Odum Institute for Research in Social Science as a Statistical Consultant and Staff Researcher.

Dr. Rojas had the unique experience of living in one of Chile's most impoverished regions (1996–2002) and experiencing first-hand how material deprivation has shaped the lives of the *Mapuche* indigenous population and how Chilean society has pushed them to reduced and scattered territorial areas. Throughout this research it has been his goal to uncover and establish the relative risks and prevalence of preventable disease among this ethnicity and thereby question historical perspectives which propose that *Mapuche* have simply assimilated to Chilean society, mixing with their culture. It is his hope that with these findings and further research, he can shed light that may orient and effectively redefine efforts at enhancing their health and living conditions and ultimately their survival as an ancestral culture.

Flavio Rojas is a dedicated conservationist, wild-life/environment supporter and vowed to dedicate his life in the defense and protection of indigenous peoples. He is a member of the International Sociological Association, the American Sociological Association and the Latin American Studies Association, subgroup Ethnicity, Race and Indigenous Populations (ERIP). He lives in Chapel Hill, North Carolina, is married to Ingrid Q. Rojas and has 6 children.

Appendix 1 The Model Used, WinBUGS Code

```
model
{
for (t in 1:T)
{
for (i in 1:m)
{
        # Poisson likelihood for observed counts
        y[i,t]~dpois(mu[i,t])
        log(mu[i,t])<-log(e[i,t])+alpha+beta1*scoresPov[i,t]+u[i,t]+v[i,t]
        # Relative Risk
        theta[t,i]<-exp(u[i,t]+v[i,t])
}
}
# CAR prior distribution for spatial correlated heterogeneity
u1[1:m]~car.normal(adj[],weights[],num[],tau.u[1])
u2[1:m]~car.normal(adj[],weights[],num[],tau.u[2])
u3[1:m]~car.normal(adj[],weights[],num[],tau.u[3])
u4[1:m]~car.normal(adj[],weights[],num[],tau.u[4])
u5[1:m]~car.normal(adj[],weights[],num[],tau.u[5])
u6[1:m]~car.normal(adj[],weights[],num[],tau.u[6])
for(i in 1:m)
{
u[i,1]<-u1[i]
u[i,2]<-u2[i]
u[i,3]<-u3[i]
u[i,4]<-u4[i]
u[i,5]<-u5[i]
u[i,6]<-u6[i]
}
# Prior distributions for the Uncorrelated Heterogeneity
for(i in 1:m)
{
v1[i]~dnorm(0,tau.v[1])
v2[i]~dnorm(0,tau.v[2])
```

```
v3[i]~dnorm(0,tau.v[3])
v4[i]~dnorm(0,tau.v[4])
v5[i]~dnorm(0,tau.v[5])
v6[i]~dnorm(0,tau.v[6])
}

for (i in 1:m)
{
v[i,1]<-v1[i]
v[i,2]<-v2[i]
v[i,3]<-v3[i]
v[i,4]<-v4[i]
v[i,5]<-v5[i]
v[i,6]<-v6[i]
}
# Weights
for(k in 1:sumNumNeig)
{
        weights[k]<-1
}
# Improper prior distribution for the mean relative risk in the study region
alpha~dflat()
mean<-exp(alpha)
beta1 ~ dnorm(0.015, 0.002)  # vague prior on covariate effect

# Hyperprior distributions on inverse variance parameter of random effects
for (i in 1:T)
{
        tau.v[i]~dgamma(0.5,0.0005)
        tau.u[i]~dgamma(0.5,0.0005)
}
}
```

FINAL 50000 ITERATIONS for alpha

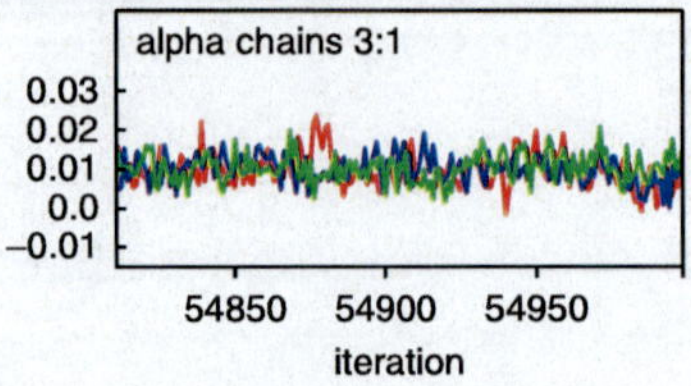

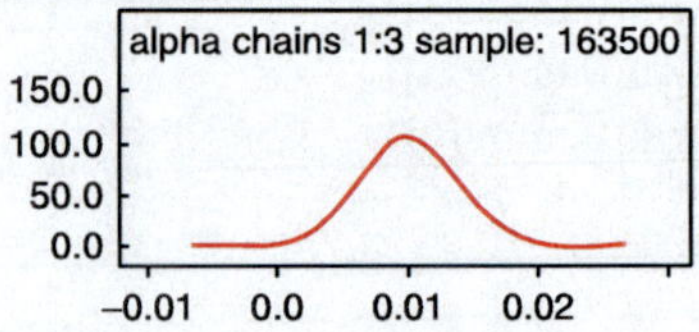

Historical time-series

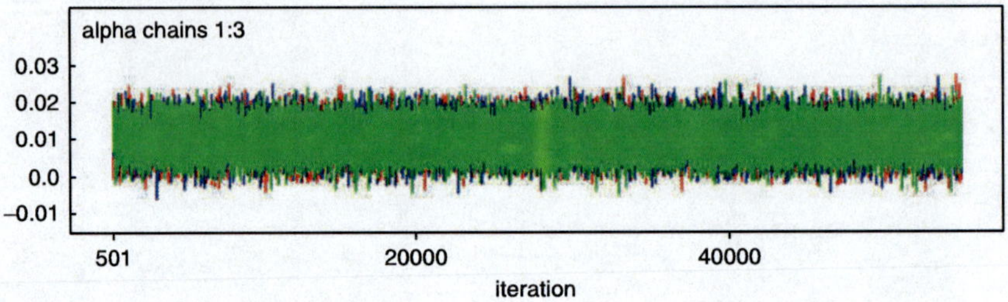

node	mean	sd	MC error	2.5%	median	97.5%	start	sample
alpha	0.01014	0.003746	1.791E-5	0.002834	0.01011	0.01756	501	163500

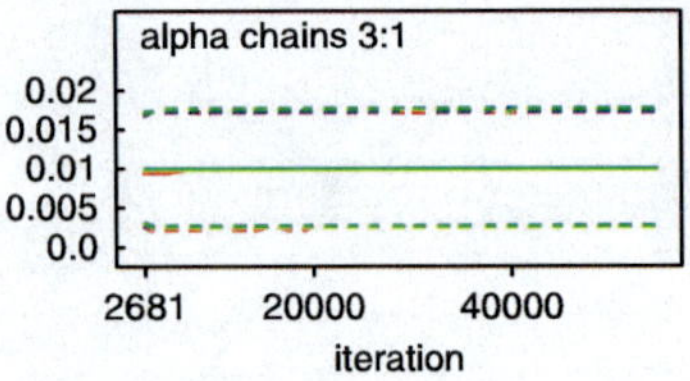

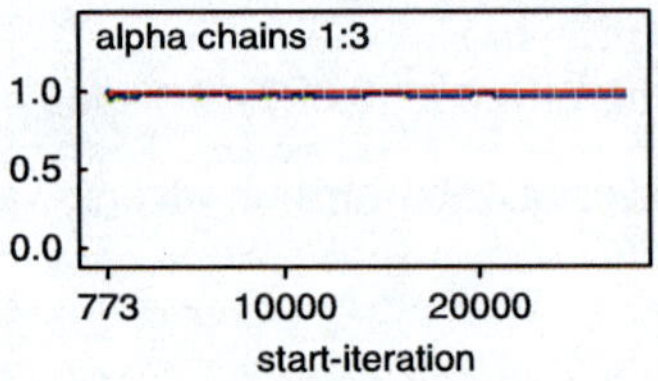

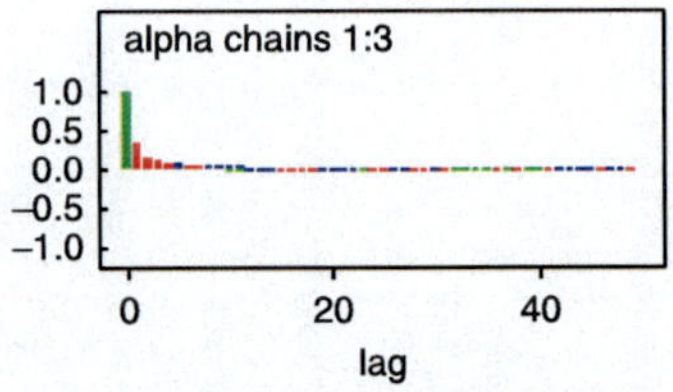

FINAL 50000 ITERATIONS for Beta1

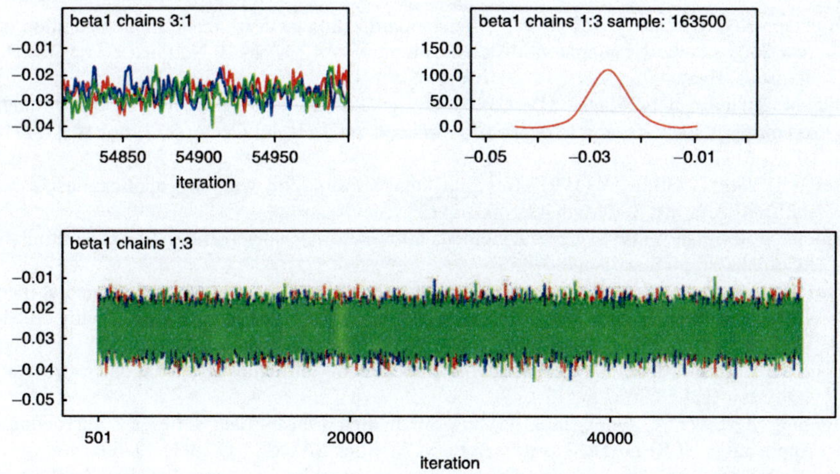

node	mean	sd	MC error	2.5%	median	97.5%	start	sample
beta1	–0.02665	0.003692	1.597E-5	–0.03391	–0.02664	–0.01944	501	163500

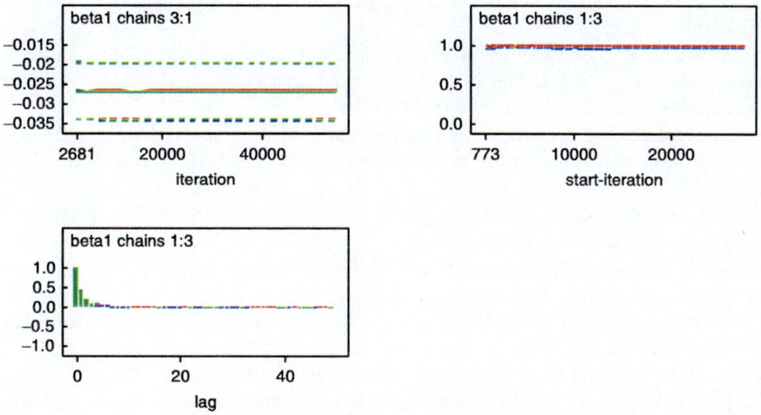

References

Anderson RN, Rosenberg HM (1998). Age standardization of death rates: implementation of the year 2000 standard, national vital statistics reports, Vol 37, No. 3. National Center for Health Statistics, Hyattsville, MD

Bell N, Schuurman N, Hayes MV (2007) Using GIS-based methods of multicriteria analysis to construct socio-economic deprivation indices. Int J Health Geogr 6:17. doi:10.1186/1476-072X–6–17

Besag J, York J, Mollie A (1991) Bayesian image restoration with two applications in spatial statistics. Ann Inst Stat Math 43:1403–1412

Brooks SP, Gelman A (1998) General methods for monitoring convergence of iterative simulations. J Comput Graph Stat 7(4):434–455

Brulle RJ, Pellow DN (2006) Environmental justice: a human health and environmental inequalities. Ann Rev Public Health 27:107–124. doi: 10.1146/annurev.publichealth.27021405. 102124

Congdon P (2009) Modelling the impact of socioeconomic structure on spatial outcomes. Comput Stat Data Anal 53:3047–3056

Conning A, Silva A, Finnegan L (1989) Small area Census data series by microcomputer: Applications of REDATAM system in Latin America. IASSIST Q 13(1):39–50

Cressie N (1992) Smoothing regional maps using empirical Bayes estimates. Geogr Anal 24:75–95

Diamond J (1997) Guns, germs and steel the Fate of human societies. W.W. Norton and Company, New York, NY

Diamond J (2005) Collapse, how societies choose to fail or succeed. Penguin Books, New York, NY

Dunson DB (2001) Commentary: practical advantages of bayesian analysis of epidemiologic data. Am J Epidemiol 153:12

Elliot MN, Morrison PA, Fremont A et al (2009) Using the Census Bureau's surname list to improve estimates of race/ethnicity and associated disparities. Health Ser Outcomes Res Methodol 9:69–83. doi 10.1007/s10742-009-0047–1

Escaramís G, Carrasco JL, Ascaso C (2008) Detection of significant disease risks using a spatial conditional autoregressive model. Biometrics 64(4):1043–1053. doi: 10.1111/j.1541-0420.2007.00981.x

Farmer P (2005) Pathologies of power. health, human rights and the new war on the poor. University of California Press, Berkeley, CA

Gelman A, Carlin JB, Stern JS, Robin DB (2003) Bayesian data analysis. *Texts in Statistical Science*. Chapman and Hall/CRC, Boca Raton, FL

Gilks WR, Richardson S, Spiegelhalter DJ (1998) Markov chain Monte Carlo in practice. Chapman and Hall/CRC, Boca Raton, FL

Goldstein H (2003) Multilevel statistical models. Wiley, New York, NY

Haining R, Law J, Griffith D (2009) Modelling small area counts in the presence of overdispersion and spatial autocorrelation. *Computational Statistics and Data Analysis.* Elsevier BV, Amsterdam. doi:10.1016/j.csda.2008.08.014

Heistercamp SH, Doornbos G, Gankema M (1993) Disease mapping using empirical Bayes and Bayes methods on mortality statistics in the Netherlands. Stat Med 12:1895–1913

Instituto Latinoamericano y del Caribe de Planificación Economica y Social, ILPES (1995). La Identificación de Proyectos y Bolsones de Pobreza a Nivel Local: Guía para Utilizar la Base de Datos del Censo en REDATAM+, LC/IP/L. 117, Distribución Limitada, Santiago de Chile.

Instituto Nacional de Estadísticas, INE (2002). XVII Censo de Población y VI de Vivienda. Población total por grupo étnico, según región, Santiago de Chile. http://www.ine.cl/cd2002/sintesiscensal.pdf (p.21)

Instituto Nacional de Estadísticas, INE (2002a). Chile, División Política Administrativa http://www.ine.cl/canales/chile_estadistico/territorio/division_politico_administrativa/pdf/DPA_COMPLETA.pdf

Junta Nacional de Auxilio Escolar y Becas, JUNAEB (2006) Sistema de Encuestas JUNAEB 2006. http://zeus.junaeb.cl/EncuestaIve/

Junta Nacional de Auxilio Escolar y Becas, JUNAEB (2010). Sistema de Encuestas JUNAEB 2010. http://zeus.junaeb.cl/EncuestaIve/

Jordan R, Verlander N, Olowokure B, Hawker JI (2006) Age, sex, material deprivation and respiratory mortality. Respir Med 100:1282–1285. doi:10.1016/j.rmed.2005.10.014

King MD, Gadigan DD, Clark CA (2009) A random effects modeling approach to the crossing-fibre problem in tractography. NeuroImage 44:753–768. doi:10.1016/j.neuroimage.2008.09.058

Knorr-Held L, Besag J (1998) Modelling risk from disease in time and space. Stat Med 17: 2045–2060

Kulldorff M (2003). Spatial statistics for cancer surveillance, See: http://www.satscan.org/presentation/

Lawson AB (2006) Statistical methods in spatial epidemiology. *John Wiley Series in Probability and Statistics*. Wiley, Sussex, UK

Lawson AB, Browne WJ, Vidal Rodeiro CL (2003) Disease mapping with WinBUGS and MLwiN. *Statistics in Practice*. Wiley, Sussex, UK

Leyland AH, Goldstein H (2001) Multilevel modeling of health statistics. *Wiley Series in Probability Statistics*. Wiley, Sussex, UK

Ministerio de Planificación y Cooperación, Gobierno de Chile, MIDEPLAN (2006). Encuesta CASEN. http://www.mideplan.cl/casen/publicaciones/2006/CASEN2006-LaAraucania.pdf

Ministerio de Planificación y Cooperación, Gobierno de Chile, MIDEPLAN (2006b). Serie Analisis de Resultados de la Encuesta de Caracterización Socioeconómica Nacional (CASEN 2006) No. 1 La Situacion de Pobreza en Chile 2006, Santiago de Chile. http://www.mideplan.cl/casen/publicaciones/2006/Pobreza.pdf

Ministerio de Planificación y Cooperación, Gobierno de Chile, MIDEPLAN (2006c). Apuntes de Protección Social No. 2, Santiago de Chile. http://www.fichaproteccionsocial.cl/upfile/documentos/26082008101743Apuntes_fps.pdf

Ministerio de Educación, MINEDUC (2007). Resultados Nacionales SIMCE 2007 8º Básico, Santiago Chile. http://www.simce.cl/index.php?id=262

Ministerio de Salud, MINSAL (2004). Indicadores Básicos de Salud, Notas Técnicas, Santiago de Chile. http://deis.minsal.cl/Indicadores/ind2004.pdf

Ministerio de Salud, MINSAL (2007). Norma General Técnica que Establece uso DEL FORMULARIO "INFORME ESTADÍSTICO DE EGRESO HOSPITALARIO" PARA LA PRODUCCIÓN DE INFORMACIÓN ESTADÍSTICA SOBRE CAUSAS DE EGRESO HOSPITALARIO Y VARIABLES ASOCIADAS", Santiago de Chile. http://juridico1.minsal.cl/DECRETO_127_07.doc

Pan American Health Organization, PAHO (2002) Standardization: A classical epidemiological method for the comparison of rates. Epidemiol Bull 23(3):9–12

Prescott E, Godtfredsen J, Vestbo J, Osler M (2003) Social position and mortality from respiratory diseases in males and females. Eur Respir J 21:821–826. doi: 10.1183/09031936.03.00047502

Raftery AE, Lewis S (1991) How many iterations in the Gibbs sampler? In: Bernardo JM, Berger JO, Dawid AP, Smith AFM (eds) Bayesian statistics, 4. Oxford University Press, Oxford, pp 763–773

Rivas ER, Barrios SC, Dorner AP, Osorio X (2008) Fuentes de contaminación intradomiciliaria y enfermedad respiratoria en jardines infantiles y salas cuna de Temuco y Padre Las Casas, Chile. Revista Médica de Chile 136(6):767–774. doi: 10.4067/S0034-98872008000600013

Roalfe AK, Holder RL, Wilson S (2008) Standardization of rates using logistic regression: a comparison with the direct method. BCM Health Serv Res 8:275. doc. 10.1186/1472–6963–8–275

Rojas F (2007) Poverty determinants of acute respiratory infections among Mapuche indigenous peoples in Chile's Ninth Region of Araucania, using GIS and spatial statistics to identify health disparities. Int J Health Geogr 6:26. doi: 10.1186/1476-072X-6-26. http://www.ij-healthgeographics.com/content/6/1/26

SAS Institute (2009). SAS/STAT(R) 9.2 User's Guide Second Edition, Carey North Carolina. Available at: http://support.sas.com/documentation/cdl/en/statug/63033/HTML/default/viewer.htm#/documentation/cdl/en/statug/63033/HTML/default/statug_introbayes_sect008.htm

SAS Institute (2010). Bayesian hierarchical Poisson regression model for overdispersed count data, Carey, North Carolina. Available at: http://support.sas.com/rnd/app/examples/stat/BayesSalm/bayessalm.pdf

Soobaden M, Cubbin C, Gee GC, Rosenbaum A, Laureson J (2006) Levels of analysis for the study of environmental health disparities. Environ Res 102(2):172–180. doi:10.1016/j.envres.2006.05.001

Spiegelhalter JJ, Thomas A, Best NG, Gilks WR (2003). Bayesian Inference using Gibbs sampling. MRC Biostatistics Unit, Cambridge England. Available at: http://www.mrc-bsu.cam.ac.uk/bugs/

Subramanian SV, Kawachi I (2006) Whose health is affected by income inequality? A multilevel interaction analysis of contemporaneous and lagged effects of state income inequality on individual self-rated health in the United States. Health Place 12:141–156. doi:10.1016/j.healthplace.2004.11.001

Suchindran CM, Rojas F (2009) Strategies for analyzing spatial-temporal emographic data. In: Singh KK, Yadava RC, Panday A (eds) Population, poverty and health. Hindustan Publishing, New Delhi

Tassone EC, Waller LA, Casper ML (2009) Small-area racial disparity in stroke mortality. An application of bayisian spatial hierarchical modeling. Epidemiology 20:234–241. doi: 10.1097/EDE.0b013e3181935aee

Tobler W (1970) A computer movie simulating urban growth in the detroit region. Econ Geogr 46:234–420

Tobler W (2004) On the fist law of geography: a reply. Ann Assoc Am Geogr 94(2):304–310. doi: 10.1111/j.1467–8306.2004.09402009.x

Villalobos S (1995) Vida Fronteriza en la Araucania, El Mito de la Guerra de Arauco. Editorial Andres Bello, Santiago de Chile

Villalobos S (2000). Araucanía: Errores Ancestrales, *El Mercurio*, May 14, 2000, page. A2 See: http://diario.elmercurio.cl/detalle/index.asp?id={f4e857f0-c2d6-45ea-9b1b-06927eee565e}

Villalobos S (2009). Falsedades sobre la Araucania, *El Mercurio,* August 30, 2009. See: http://diario.elmercurio.cl/detalle/index.asp?id={176103bb-ef71-4043-80d5-cc39a8bd4ba8}

Wakefield J, Shaddick G (2006) Health-exposure modeling and the ecological fallacy. Biostatistics 7(3):438–455

Waller LA, Carlin BP, Xia H, Gelfand AE (1997) Hierarchical spatio-temporal mapping of disease rates. J Am Stat Assoc 92:607–617

Waller LA, Gotway CA (2004) Applied spatial statistics for public health data. *Wiley Series in Probability and Statistics*. Wiley, Hoboken, NJ

Chapter 22
GIS and Atmospheric Diffusion Modeling for Assessment of Individual Exposure to Dioxins Emitted from a Municipal Solid Waste Incinerator

Jean-François Viel

Abstract The most potent dioxin congener (2,3,7,8-TCDD) is classified as a human carcinogen. Municipal solid waste incinerators (MSWI) are one of the major sources of dioxins and are therefore a cause of public concern. Blood dioxin levels are considered the best estimates of actual exposure, but they are costly and technically difficult to gather from individuals and to measure consistently. However, dioxins are good candidates for a combined GIS-modeling-based approach to simulate the ways in which they propagate in the environment, and the exposures that occur as a result. Dioxins are released into the air by a few known industrial point sources, and their environmental concentrations can therefore be estimated through plume modeling. Furthermore, they are known to be resistant to environmental and biological degradation, and accumulate in soils. We conducted a sequential epidemiologic investigation in the vicinity of a MSWI with high dioxin emission levels (Besançon, France). Contours of modeled ground-level air concentrations were used to assign a dioxin exposure category for any inhabitant of the town. Exposure accuracy was assessed through dioxin measurements from soil samples. In a mixed individual/ecological case-control study, a higher risk for non-Hodgkin lymphoma was found among individuals living in the area with the highest dioxin concentration (odds ratio 2.5, 95% confidence interval 1.4–4.5). The replication of these findings at the nationwide level added further evidence. GIS and exposure modeling can be considered innovative and appropriate for the assessment of dioxin exposure, moving from source identification to personal exposure estimates using environmental surrogates.

Keywords Atmospheric dispersion model · Dioxins · Environmental epidemiology · Exposure assessment · Municipal solid waste incinerator · Non-Hodgkin lymphoma

---

J.-F. Viel (✉)
CNRS n° 6249 "Chrono-Environment", Faculty of Medicine, Besançon, France
e-mail: jean-francois.viel@univ-fcomte.fr

J.A. Maantay, S. McLafferty (eds.), *Geospatial Analysis of Environmental Health*, Geotechnologies and the Environment 4, DOI 10.1007/978-94-007-0329-2_22,

List of Acronyms and Abbreviations

2,3,7,8-TCDD	2,3,7,8-tetrachlorodibenzo-*p*-dioxin
APC3	Air Pollution Control 3
CI	Confidence interval
DL-PCB	Dioxin-like polychlorinated biphenyls
EU	European Union
GIS	Geographic information system
IARC	International Agency for Research on Cancer
ICD-O	International Classification of Diseases for Oncology
I-TEQ	International toxic equivalency factor
MSWI	Municipal solid waste incinerator
NHL	Non Hodgkin lymphoma
OR	Odds ratio
PCB	Polychlorinated biphenyls
US-EPA	US Environmental Protection Agency

22.1 Introduction

Exposure in environmental epidemiology can be assessed in different ways. If available, personal measurements (personal exposure meter, biomarkers) are considered the best estimates of actual exposure, but they are costly and technically difficult to gather from individuals and to measure consistently. Environmental epidemiology has therefore generally relied on environmental measurements or models (Nieuwenhuijsen et al., 2006). Although these methods do not take into account differences in route of administration, activity, and physiology, a good correlation between environmental and personal exposure estimates can be achieved, provided that adequate data and models are used (Beyea and Hatch, 1999). According to the National Research Council (1994), such quantified area measurements rank second in the hierarchy of exposure measurements with respect to the true exposure.

Most pollution is derived from specific sources and tends to spread out with progressively lower concentrations, showing considerable systematic spatial variation. Different geographical approaches (greatly strengthened by the use of geographical information systems – GIS) can therefore be used to assess levels of exposure to environmental pollution. One important factor that distinguishes among these approaches is whether they are based on monitored pollution data, or whether they are estimated using data on source activities. In the first case, local pollution patterns are modeled on the basis of monitored data to fit a surface through the available monitored data, in order to predict pollutant concentrations at sites where measurements have not been taken (Nieuwenhuijsen et al., 2006). Geostatistical techniques are increasingly being used in this respect. However, these geographical methods require a wide range of retrospective data over large study areas. If available, these data, usually collected for other purposes (routine monitoring...), tend to be far from optimal (Colville et al., 2003).

Where there is little monitored data on pollution concentrations, source-receptor modeling becomes important. Simple proxies, such as distance from a source (factory, incinerator, road, or landfill) have frequently been used. However, distance and proximity measures imply that pollution spreads uniformly away from its source, whatever the pollutant or the exposure pathway. Inherently non-specific, these measures are therefore difficult to interpret (Briggs, 2003). When the distribution of pollutants is better understood, more sophisticated models can be used to simulate both the processes and the pathways of pollutant propagation in the environment, and to estimate historical levels of contaminants due to point sources (e.g. smokestack) or nonpoint sources (e.g. groundwater). Owing to the knowledge gained on the physics and environmental processes of air pollution transport in recent decades, atmospheric dispersion modeling is now more firmly established. Dispersion is often modeled as a plume that shows a normal distribution of concentration in the vertical and horizontal directions (although the vertical concentration distribution can be a skewed Gaussian in convective conditions). This so-called Gaussian model is appropriate when modeling in the near field (0–100 km), and provides a lot of information about the distribution of pollutant concentrations in the environment downwind of the source (Colville et al., 2003). GIS can be used in combination with dispersion models to simulate the ways in which pollutants propagate in the environment, and the exposures that occur as a result (Briggs, 2003). In a GIS environment, maps of iso-dose or iso-risk contours can be displayed. Resulting pollution surfaces can then be overlaid onto georeferenced data to assign exposure to individuals at their place of residence or work. Such exposures can then be used in other study designs, such as case-control (Vine et al., 1997).

The US Environmental Protection Agency and the International Agency for Research on Cancer have classified 2,3,7,8-TCDD (the most potent dioxin congener) as a human carcinogen (US-EPA, 1994; IARC, 1997). Dioxin emissions from municipal solid waste incinerators (MSWI) are one of the major sources of dioxins and are therefore an exposure source of public concern. Our team had previously examined the spatial distribution of non-Hodgkin lymphomas (NHL) around one of these polluting MSWIs (Besançon, eastern France). Using a spatial scan statistic, we found evidence for a cluster, composed of two electoral wards (one of them, Besançon city, containing the MSWI), with a standardized incidence ratio of 1.3 (95% confidence interval (CI) 1.1–1.4) (Viel et al., 2000). However, this cluster investigation relied on distance as a proxy for exposure. Our goal was therefore twofold: (1) to use GIS-based technology to refine exposure measurements, improving sensitivity and specificity beyond a simple proximity metric, and (2) to use these environmental exposure estimates as a tool to enhance epidemiologic case-control investigations.

Dioxins are indeed good candidates for a combined GIS modeling-based approach:

- while the majority of hazardous chemicals are emitted from countless, diffuse sources, dioxins are released into the air by a few known industrial point sources (residential burning of wood and road traffic constituting the background level),

and their environmental concentrations can therefore be estimated through plume modeling;
- they are known to be resistant to environmental and biological degradation, are chemically stable, poorly metabolized, and accumulate in soils; this matrix is therefore considered an adequate environmental monitor for assessing long-term exposure to dioxins.

22.2 Study Site

MSWIs are usually installed in broad industrial parks where other factories may also play a role in dioxin emission into the atmosphere, making the sources of the dioxins difficult to discern. The MSWI of Besançon is very unique in this respect. Located 4 km west of the city center, it has no adjacent industrial sources of exposure (no cement kilns, iron or steel works, or foundries). Polluting industries in this area were replaced two decades ago by small-scale advanced technologies.

Combustion chambers 1 and 2 (each with a capacity of 2.1 metric tons/h) were put into operation in 1971. In 1976, a third combustion chamber was opened (with a capacity of 3 metric tons/h). In 1998, approximately 67,000 metric tons of waste were processed. Some legal guidelines for incinerator emissions have not been followed at this location. For example, in 1997, dust and hydrogen emission levels were higher than prescribed and exhaust gases were not maintained at temperatures of more than 850°C for the legally prescribed time ($\geq$ 2 s), allowing dioxins to be emitted. The first time that the dioxin concentration of an exhaust gas was ever measured (in December 1997), it was found to be 16.3 ng international toxic equivalency factor (I-TEQ)/m^3, whereas the European guideline value is 0.1 ng I-TEQ/m^3. Combustion chamber 1 (the most polluting) was shut down on December 31, 1998. Combustion chamber 2 was replaced by a new one with up-to-date pollution controls (combustion chamber 4), which started operation in late 2003.

22.3 Dioxin Exposure Assessment Through Geographic Modeling

22.3.1 Plume Modeling

We took advantage of a first-generation Gaussian-type dispersion model created in 1999 with the Air Pollution Control 3 software (APC3 – Aria Technologies, Colombes, France). This subcontracting company intervened in the framework of an environmental impact statement (supervised by the district council) to predict the future impact of dioxin emissions both from the old (but renewed) combustion chamber 3 and from the new combustion chamber 4 with up-to-date pollution controls. Dispersion modeling being heavily influenced by factors that are stable over time (mean meteorological conditions, terrain elevations and stack height), we assumed that contour shapes (as a direct output from the prediction model), were reliable estimates of past dioxin deposition profiles, provided that relative figures rather than absolute figures were used.

The model took into account meteorological data (5 years of data for wind speed, wind direction, pressure, temperature, and Pasquill atmospheric stability classes), simplified surface topography, plume rise, stack characteristics, and the future dioxin emission rate from the MSWI to assess average concentrations in hundreds of meteorological conditions (one Gaussian plume for each particular meteorological condition). The respective contours of these modeled ground-level air concentrations, a priori determined by the subcontracting company (< 0.0001, 0.0001–0.0002, 0.0002–0.0004, 0.0004–0.0016 pg/m^3), were digitalized and transferred onto the surface of a map.

22.3.2 GIS-Based Exposure

Intending to use dioxin ground-level concentrations as relative figures rather than absolute figures to estimate past exposure, the exposure areas defined by contour lines were then classified as very low (modeled ground-level dioxin concentration < 0.0001 pg/m^3 zone), low (modeled ground-level dioxin concentration 0.0001–0.0002 pg/m^3 zone), intermediate (modeled ground-level dioxin concentration 0.0002–0.0004 pg/m^3 zone), and high (modeled ground-level dioxin concentration 0.0004–0.0016 pg/m^3 zone) exposure areas (Fig. 22.1).

The concentration map with a dragonfly wing shape clearly shows that:

- the concentration bands are not circular, but are stretched along the north-east and south-westerly direction due the foot-hills of the Jura mountains, which channel the wind preferentially in these two directions (Fig. 22.2);
- the maximum concentration is not immediately adjacent to the plant, but is rather about 2 km to the north-east and south-west, where the elevated plume touches down, because the pollution does not disperse vertically downwards from the elevated source;
- the concentration falls off rapidly with distance from the source.

Using GIS technology (Star GIS software, Star Informatic, Liege, Belgium), a dioxin concentration category was attributed to each of the 705 city blocks (the smallest level of geographic resolution in the French census database, typically a quadrangle bounded by four streets, and averaging 161 inhabitants), and 52 block groups (averaging 2,183 inhabitants) of the city of Besançon (provided that half or more of their area was within a given contour). Moreover, by matching a file containing addresses (street and number) against a street network file, we were able to pinpoint the location of any residence. We could therefore assign a dioxin exposure category and obtain a risk field classification for any inhabitant of Besançon, whatever the spatial resolution of the place of residence (exact coordinates, block, block group). The geographic coordinates were expressed in the Lambert two French plane coordinate system (conformal conic projection, Clarke 1880 spheroid, 1st standard parallel: 45°53′56.11″, 2nd standard parallel: 47°41′45.65″).

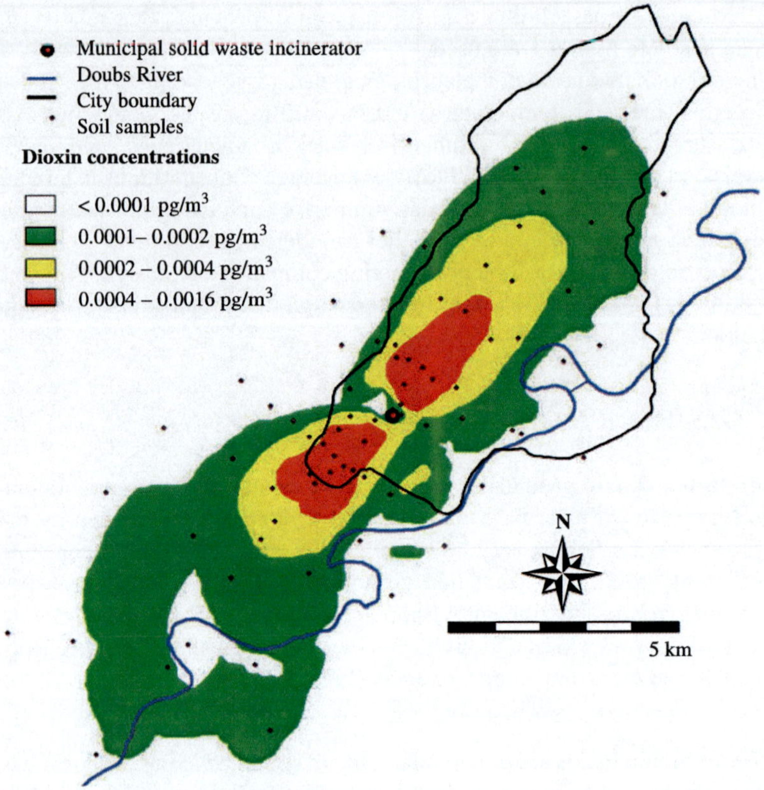

Fig. 22.1 Modeled ground-level dioxin concentrations and soil sample locations around the municipal solid waste incinerator in Besancon, France. Adapted from Floret et al. (2003) with permission from Lippincott Williams & Wilkins

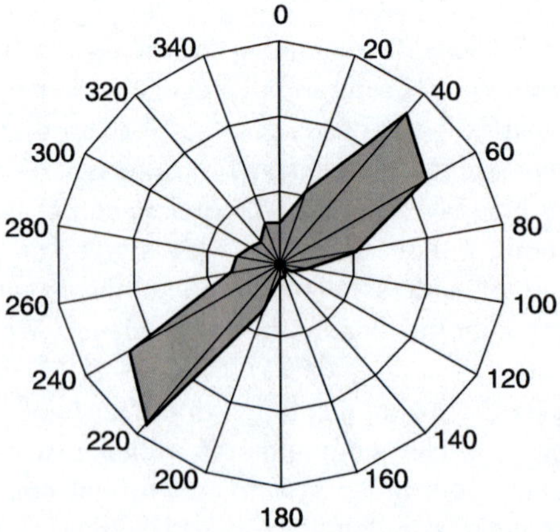

Fig. 22.2 Wind direction climatology (1993–1997) in Besançon, France, showing channeling of local wind by the foothills of the Jura mountains

These modeled ground-level concentrations represented the best available surrogates for past dioxin exposure measurements from the local MSWI, given that no earlier measurements had been taken (the first and only data available was one concentration in 1997).

22.4 Validation of GIS-Based Dioxin Exposure

Validation is often overlooked in the exposure assessment process. A fundamental rule in environmental modeling is to not transfer use of a model from one geographic region to another without validating it with measurement data from the new study area (Nuckols et al., 2004). The accuracy of this modeled exposure, therefore had to be assessed. We achieved this through dioxin measurements from soil samples, but unfortunately did so retrospectively (after having completed a NHL case-control study) due to fundraising difficulties.

Seventy-five sampling points were selected for their homogeneous geological and topographical conditions (Floret et al., 2006). To our knowledge, this is the only study ever performed with this large a number of soil samples, yielding fairly representative and precise estimates. Dioxin concentration, pH, organic carbon concentration, cation exchange capacity, and geomorphology and ecology features were assessed for each soil sample. The study design employed a stratified random selection process, involving the four quadrants around the incinerator, with an emphasis on sampling in the quadrants that were historically downwind from the incinerator: the northeast and the southwest quadrants (Fig. 22.1). The precise position of each sampling site was determined to maintain homogeneous geological and topographical conditions and vegetation across samples. The following conditions were sought during site selection: (1) level, undisturbed soil; (2) away from trees; (3) not adjacent to roads or railway lines, and (4) not known or suspected to have high dioxin concentrations for any other reason. In July 2002, 75 soil samples were collected in the vicinity of the facility, between 97 m and 12 km from the stack. Each sample site consisted of an area of 10×10 m. Aliquots were collected at each corner and at the center of the sites. Soil samples were taken from the upper 10 cm of soil, without vegetation. All the borehole aliquots collected in the same sampling site were mixed to obtain a composite sample of about 500 g.

Soil dioxin concentrations ranged from 0.25 to 28.06 pg I-TEQ/g dry matter. Dioxins measured in soil samples showed fairly similar exposure profiles to modeled air concentrations. In fact, an interaction between measured dioxin concentrations and topography complexity (simple terrain on the northeast side with gentle hills of moderate slope, and complex terrain on the southwest side with more pronounced hills and valleys) was found (Table 22.1). An upward trend was noticeable across modeled exposure categories, only in the northeast direction.

These results were confirmed by multivariate models, adjusting for organic carbon concentration and altitude after stepwise regression. In simple terrain, a significant association was observed between modeled ground-level air dioxin concentrations and log-transformed measured soil dioxin concentrations, with a strong

Table 22.1 Means (standard deviations) of dioxin soil concentrations (I-TEQ/g dry matter), by geographic-based exposure and topography complexity categories (Besançon, France)

Modeled dioxin exposure	Very low	Low	Intermediate	High
Complex topography	1.09 (1.76)	2.44 (3.53)	1.91 (1.12)	1.37(0.21)
Simple topography	1.81 (1.14)	1.99 (1.37)	3.53 (2.30)	11.25(12.39)

gradient across exposure categories. Conversely, in a complex topography situation, the model overestimated ground-level air concentrations, particularly in the high exposure zone.

Several limitations of APC3 software could explain these results. First, only a simplified topography was modeled, ignoring complex terrain with hills and channels. Second, the model assumed that turbulence generated in one place tends not to persist for any significant distance downwind (the so-called equilibrium of turbulence), not accounting for the turbulence boundary layer between surface and air. Third, surface roughness, which affects the vertical profiles of wind and temperature and the dispersion rates in the surface layer, was not accounted for. A short roughness length (0.2 m for open grassland and 2–3 m for arable crops, compared to 5–10 m typically in urbanized areas) leads to significant decreases in particle deposition velocity and, therefore, lower local deposition of dioxins. Fourth, the low height of the stack (40 m) represented an additional variable, making the fraction of dioxin emissions deposited locally very sensitive to the treatment of dispersion. We concluded that first-generation modeling provided a reliable proxy for dioxin exposure in simple terrain. However, a more advanced atmospheric diffusion model should have been used for a refined assessment in complex terrain.

A later study attempted to examine the nature of the dioxin soil contamination in the surroundings of the MSWI to characterize whether more than one potential emission source could explain the presence of the dioxins (Floret et al., 2007). Dioxin congener profiles were compared according to the most environmentally impacted zones and to various spatial contrasts. Two different clustering algorithms identified the same main cluster (consisting of 73 samples). The remaining two soil samples composed either one or two clusters. All clusters showed similar congener profiles. Moreover, no contrast was observed for congener distributions between complex and simple topographies, inside and outside the city boundary, or the two most and the two least exposed areas, reflecting a common fingerprint. Congener profiles therefore indicated that the area impacted by the MSWI was not subject to other point sources of dioxins.

Beside a potential aerial route, we assumed that exposure to airborne dioxins via the food chain could represent a significant pathway for people residing near the MSWI of Besançon. In this setting, the consumption of local foods, resulting from the practice of shopping in popular open markets where locally raised products are sold, may be elevated relative to more urban areas. To this aim, a pilot survey of dioxins and dioxin-like PCB (DL-PCB) congeners in eggs was conducted. The pathways of soil-related exposure of free-range chickens are mainly ingestion of

Table 22.2 Dioxin and dioxin-like PCB concentrations (pg I-TEQ/g fat) in 6-egg pooled samples (Besançon, France, 2006)

Sample	Modeled dioxin exposure	DL-PCBs[a]	Dioxins	Total
1	Intermediate	3.8	1.9	5.7
2	High	22.0	5.6	27.6
3	High	10.0	43.0	53.0

[a]Dioxin-like PCBs.

soil and ingestion of soil organisms (worms, insects). High concentrations in the soil lead to increased dioxin levels in the eggs of foraging chickens. The regular consumption of eggs from highly contaminated areas can result not only in increased intake, but also in significantly elevated concentrations in the bodies of exposed individuals.

We identified three private gardens located under the prevailing wind stream (northeast) where chickens were foraging: two in the high exposure area, and one in the intermediate exposure area. The chickens were allowed to forage, but were also fed table scraps and commercial grain. For each site, dioxins and DL-PCBs were measured in a 6-egg pooled sample. The two samples collected in the high exposure area contained dioxin and furan levels above the 3 pg I-TEQ/g fat maximum level applied in Europe, the second sample containing one of the highest dioxin levels ever reported in eggs laid by chickens reared in the vicinity of a MSWI (43 pg I-TEQ/g fat) (Table 22.2). Soil dioxin contamination profiles closely matched those in eggs for samples 1 and 3, but did so more loosely for sample 2. The DL-PCB levels exceeded the proposed EU limit in hen eggs (3 pg I-TEQ/g fat), and were the main contributors to the total I-TEQ values for samples 1 and 2.

When interpreting these results, one must bear in mind that the type of housing, grazing habits, contamination of feedstuffs, foraging behaviors..., can be significantly associated with pollutant concentrations in eggs. Moreover, the levels found in sample 3 could be partly explained by the fact that the site was formerly a local sawmill. Spills of chemicals or drippage from freshly treated timber could have resulted in localized contamination of the ground with both pentachlorophenol and its associated dioxin contaminants. Nevertheless, as a result of this pilot survey, chicken owners were advised by local authorities to no longer eat home-produced eggs.

At the end of this exposure assessment process, we concluded that modeled environmental concentrations of dioxins (originally expressed in pg/m^3) were good surrogates for exposure concentrations (i.e. presence of dioxins at the point of contact with receptors), expressed in a four-category ordinal scale.

22.5 Mixed Individual/Ecological Case-Control Study

To further explore the environmental route suggested by the initial cluster investigation, we carried out a population-based case-control study at the Besançon city scale (therefore excluding 29,000 inhabitants of the second electoral ward) since detailed census data (needed to sample population controls) were available only for this area (Floret et al., 2003). We capitalized on the existence of modeled data for the exposure measure, but used individual data to support the MSWI-dioxin-NHL hypothesis.

We compared 222 incident cases of NHL diagnosed between 1980 and 1995 and population controls. We obtained NHL incidence data from the Doubs cancer registry (International Classification of Disease for Oncology (ICD-O) morphology codes: 9590/3-9595/3, 9670/3-9723/3 and 9761/3). This registry was established in 1976 and is complete for NHL cases, as ascertained by the ratio of the number of deaths to the number of cases registered during 1983–1987, which, at 47% (for the Doubs region), is very similar to those reported in other Western countries. Virtually all cases were verified histologically (97% among men and 99% among women). We collected data concerning the patients' address at diagnosis, date of birth, gender, cancer diagnosis, and age at diagnosis from their medical records.

We selected controls from a reliable and accessible database, the 1990 population census. Because of French privacy laws and confidentiality requirements, the only individual data available to researchers were sex, age categories (0–19, 20–39, 40–59, 60–74 and 75+ years), and residence in a given block. We randomly selected population-based controls, according to a 10-to-1 matching procedure. Matching criteria were sex and age, yielding 10 strata. Risk factor data were limited to what was available through the census either on an individual level or on a block group level (educational, occupational, household-based indicators).

A dioxin exposure category was attributed to exact residential addresses (for cases), and blocks of residence (for controls) (see Section 22.3.2 GIS-based exposure). Multilevel models were run to explain the outcome (case/control status) defined at the individual level, while introducing risk factors at the individual level (dioxin exposure) and the block group level (socio-economic characteristics).

The risk of developing NHL was 2.3 times higher (95% CI 1.4–3.8) among individuals living in the area with the highest dioxin concentration than among those living in the area with the lowest dioxin concentration. No increased risk was observed for the intermediate dioxin exposure categories (Table 22.3). Adjustment

Table 22.3 Association of non-Hodgkin lymphoma with dioxin exposure categories (Besançon, France, 1980–1995)

Modeled dioxin exposure	Cases	Controls	OR (95% CI[a])
Very low	42	441	1.0
Low	91	952	1.0 (0.7–1.5)
Intermediate	58	681	0.9 (0.6–1.4)
High	31	146	2.3 (1.4–3.8)

[a]Confidence interval.

for a wide range of socioeconomic characteristics at the block group level did not alter the results.

We re-analyzed the data once the validation study was completed, to check whether the diffusion model's inaccuracy in complex terrain challenged the initial findings. We were fortunate that only 10.5% of cases and 9.3% of controls were concerned. When restricting the study to cases and controls residing on the north-east side, an increased odds ratio (OR) in the highest dioxin exposure area was still found (OR = 2.5, 95% CI 1.4–4.5).

These findings were in line with the results obtained by Bertazzi et al. (2001a) on the 20-year mortality of the Seveso population. People in the Seveso cohort had mean TCDD blood lipid concentrations of 136 ng TCDD/kg, which falls between the typical occupational dioxin levels (> 1,000 ng TCDD/kg) and background levels (2–3 ng TCDD/kg). Allowing for a latency time window of 15–20 years, results for NHL clearly did stand out, according to Bertazzi et al. (2001b), with a relative risk of 2.8 (95% CI 1.1–7.0).

We concluded that our results lend support to the hypothesis that environmental dioxins increase the risk of NHL among populations living in the vicinity of a MSWI.

22.6 External Consistency

External consistency is a key issue in environmental studies. We attempted therefore to replicate our findings at the nationwide level with a similar exposure assessment approach (Viel et al., 2008). The study area consisted of four French administrative departments, comprising a total of 2,270 block groups. NHL cases that were diagnosed during the period of 1990–1999, and were aged 15 years and over, were considered. Each case was assigned a block group by residential address geocoding. The Atmospheric Dispersion Model System version 3 (Cambridge Environmental Research Consultants, Cambridge, UK), a second generation Gaussian model, was used to estimate immissions in the surroundings of 13 incinerators that operated in the study area. Then, cumulative ground-level dioxin concentrations were calculated for each block group. Poisson multiple regression models, incorporating penalized regression splines to control for covariates and dealing with Poisson overdispersion, were used. Five confounding factors were considered: population density, urbanization, socio-economic level, airborne traffic pollution, and industrial pollution.

A total of 3,974 incident NHL cases were observed (2,147 among males, and 1,827 among females) during the 1990–1999 time period. A statistically significant relationship was found at the block group level between risk for NHL and dioxin exposure, with a relative risk (RR) of 1.1 (95% CI 1.0–1.3) for persons living in highly exposed census blocks compared to those living in slightly exposed block groups. Population density appeared to be positively linked to both risk for NHL and dioxin exposure. Subgroup multivariate analyses per gender yielded a statistically significant RR for females only (RR = 1.2, 95% CI 1.0–1.4).

This study, in line with previous results obtained in the vicinity of the incinerator located in Besançon, added further evidence to the link between NHL incidence and exposure to dioxins emitted by municipal solid waste incinerators.

22.7 Conclusion

Our goal was not to carry out a top-down human health risk assessment for various stakeholders involved. We rather used a bottom-up sequential epidemiologic approach, starting from crude investigations, gradually refining to specific, aimed studies to address the following issue: although intake from food is assumed to account for over 90% of the burden of dioxins in the *general* human population, could this assumption not hold for people living in the vicinity of a MSWI?

In no instance was actual individual exposure measured; rather, it was estimated using exposure zones. Residence location as a surrogate for exposure cannot distinguish between contributions from direct (vapor inhalation or dermal absorption) and indirect (ingestion of particulate emissions deposited onto soil and plants and subsequently eaten by grazing animals or foraging chickens) exposure pathways. At first glance, the so-called "geophysical plausibility" Nuckols et al., (2004) was challenged. Although the aerial route of transport for the contaminant between the source and the receptor was plausible, an unresolved issue was how well our individual exposure metric estimated personal dose (i.e. the amount of dioxins that actually entered the human body).

Since most of the meats and dairy products consumed are not produced locally, but have been transported over hundreds or thousands of kilometers, the majority of dioxin exposure does not come from local dioxin sources. However, local communities whose diets consist significantly of foodstuffs grown/reared in the vicinity of an incinerator may have significantly elevated serum dioxin levels. Goldman et al. (2000) showed that the consumption of both home-produced eggs and meat for 2–15 years was associated with a significant 2- to 6-fold increase in serum levels of dioxin-like chemicals. In another study, concentrations of dioxins in subjects living around two old incinerators in Belgium increased proportionally to their intake of local animal fat, with almost a doubling in subjects with a fat intake higher than 150 g of fat per week Fierens et al., (2003). Fruit and vegetables grown locally can also become contaminated by incinerator emissions. After analyzing fruit (apples) and vegetable (zucchini, lettuce, potatoes) samples, Lovett et al. (1997) concluded that consuming these foodstuffs would represent an additional 8% of the normal dietary intake for dioxins.

We lacked the necessary bioassay data (actual blood dioxin levels), which is expensive and time-consuming to collect, to validate the prediction of GIS-based exposure in terms of dose. We note, however, that the ecologic classification of exposure status based on soil levels around Seveso was not refuted by classification based on blood dioxin measurements available from a sub-sample (Bertazzi et al., 2001a).

Geographic modeling strives to create the equivalent of a hypothetical ideal monitoring system that would have measured the concentration of pollutants at all locations and times in the medium and domain under study (Beyea and Hatch, 1999). GIS and exposure modeling can be considered innovative and appropriate for the assessment of dioxin exposure, moving from source identification to personal exposure estimates using environmental estimates.

Acknowledgments This work would not have been possible without the assistance of numerous people who have contributed time and expertise in various fields. I owe particular thanks to Nathalie Floret for her commitment and enthusiasm during her PhD studies.

Author Biography

Jean-François Viel is a professor of Biostatistics and Epidemiology at the Faculty of Medicine, Besançon, France, and deputy director of the CNRS n° 6249 "Chrono-environment" team, Besançon, France. His research interests are in geographical epidemiology, exposure modeling, spatial statistics, and their interactions with the help of GIS technology. In his earlier work, he made contributions to the assessment of the impact of sources of pollution on the health status of communities.

References

Bertazzi PA, Consonni D, Bachetti S, Rubagotti M, Baccarelli A, Zocchetti C, Pesatori AC (2001a) Health effects of dioxin exposure: a 20-year mortality study. Am J Epidemiol 153(11):1031–1044

Bertazzi PA, Consonni D, Bachetti S, Rubagotti M, Baccarelli A, Zocchetti C, Pesatori AC (2001b) Bertazzi et al. respond to Smith and Lopipero. Am J Epidemiol 153(11):1048–1049

Beyea J, Hatch M (1999) Geographic exposure modelling: a valuable extension of geographic information systems for use in environmental epidemiology. Environ Health Perspect 107(S1):181–190

Briggs D (2003) Environmental measurement and modelling: geographical information systems. In: Nieuwenhuijsen MJ (ed) Exposure assessment in occupational and environmental epidemiology. Oxford University Press, Oxford, UK, pp. 55–70

Colville R, Briggs D, Nieuwenhuijsen MJ (2003) Environmental measurement and modelling: introduction and source dispersion modelling. In: Nieuwenhuijsen MJ (ed) Exposure assessment in occupational and environmental epidemiology. Oxford University Press, Oxford, UK, pp. 39–53

Fierens S, Mairesse H, Hermans C, Bernard A, Eppe G, Focant JF, De Pauw E (2003) Dioxin accumulation in residents around incinerators. J Toxicol Environ Health A 66(14):1287–1293

Floret N, Lucot E, Badot PM, Mauny F, Viel JF (2007) A municipal solid waste incinerator as the single dominant point source of PCDD/Fs in an area of increased non-Hodgkin's lymphoma incidence. Chemosphere 68(8):1419–1426

Floret N, Mauny F, Challier B, Arveux P, Cahn JY, Viel JF (2003) Dioxin emissions from a solid waste incinerator and risk of non-Hodgkin lymphoma. Epidemiology 14(4):392–398

Floret N, Viel JF, Lucot E, Dudermel PM, Cahn JY, Badot PM, Mauny F (2006) Dispersion modelling as a dioxin exposure indicator in the vicinity of a municipal solid waste incinerator: a validation study. Environ Sci Technol 40(7):2149–2155

Goldman LR, Harnly M, Flattery J, Patterson DG, Needham LL (2000) Serum polychlorinated dibenzo-*p*-dioxins and polychlorinated dibenzofurans among people eating contaminated home-produced eggs and beef. Environ Health Perspect 108(1):13–19

IARC (International Agency for Research on Cancer) (1997) Polychlorinated dibenzo-para-dioxins and polychlorinated dibenzo-furans. IARC Monographs on the evaluation of carcinogenic risks to humans, Vol 69. IARC, Lyon, France
Lovett AA, Foxall CD, Creaser CS, Chewe D (1997) PCB and PCDD/DF congeners in locally grown fruit and vegetable samples in Wales and England. Chemosphere 34(5–7):1421–1436
NRC (National Research Council) (1994) Science and judgment in risk assessment. National Academy Press, Washington, DC
Nieuwenhuijsen MJ, Paustenbach D, Duarte-Davidson R (2006) New developments in exposure assessment: The impact on the practice of health risk assessment and epidemiological studies. Environ Int 32(8):996–1009
Nuckols JR, Ward M, Jarup L (2004) Using geographic information systems for exposure assessment in environmental epidemiological studies. Environ Health Perspect 112(9):1007–1015
US-EPA (US Environmental Protection Agency) (1994) Health assessment document for 2, 3, 7, 8-tetrachlorodibenzo-p-dioxin (TCDD) and related compounds, Vol 1–3. US Environmental Protection Agency, Washington DC
Viel JF, Arveux P, Baverel J, Cahn JY (2000) Soft-tissue sarcoma and non-Hodgkin's lymphoma clusters around a municipal solid waste incinerator with high dioxin emission levels. Am J Epidemiol 152(1):13–19
Viel JF, Daniau C, Goria S, Fabre P, de Crouy-Chanel P, Sauleau EA, Empereur-Bissonnet P (2008) Risk for non Hodgkin's lymphoma in the vicinity of French municipal solid waste incinerators. Environ Health doi:10.1186/1476-069X–7–51
Vine MF, Degnan D, Hanchette C (1997) Geographic information systems: their use in environmental epidemiologic research. Environ Health Perspect 105(6):598–605

Chapter 23
Synthesizing Waterborne Infection Prevalence for Comparative Analysis of Cluster Detection Methods

Niko Yiannakoulias

Abstract When water is an important direct or indirect facilitator in the transmission of disease it is reasonable to expect that clusters of disease may occur near or along these water sources. As such, searching for water-related disease clusters can be an important part of spatial analysis process, particularly when there may be unknown spatial heterogeneities in the relationship between proximity to water and illness. We illustrate the value of using a new class of disease cluster detection methods in the spatial analysis of diseases suspected to emerge in unusual and irregular spatial patterns. Our experiment uses synthetic *Schistosoma mansoni* prevalence data created from information on environmental factors known to influence risk of infection. Our simulations suggest that cluster detection methods that assume circular cluster shapes are less precise in the delineation of cluster areas, even when the difference between cluster and non-cluster areas is large. We conclude that methods able to find irregularly shaped disease clusters are particularly well suited to applications in which features of the physical environment are suspected to influence risk of illness or infection.

Keywords Data synthesis · Cluster detection · Waterborne diseases

23.1 Background

23.1.1 Water-Related Disease and Distance to Water

For many diseases, nearness to surface water can be an important predictor of risk, particularly in parts of the world without indoor plumbing. Close proximity to contaminated surface water is important for describing the spatial variability in cholera prevalence in Rural Bangladesh (Ali et al., 2002). Living near some surface water sources is associated with greater infection by *Helicobacter pylori* in Ethiopia

---

N. Yiannakoulias (✉)
School of Geography and Earth Sciences, McMaster University, Hamilton, ON, Canada L8S4K1
e-mail: yiannan@mcmaster.ca

J.A. Maantay, S. McLafferty (eds.), *Geospatial Analysis of Environmental Health*, Geotechnologies and the Environment 4, DOI 10.1007/978-94-007-0329-2_23,

(Lindkvist et al., 1998). Living close to fresh or marine water sources contaminated with sewage can spread *Escherichia coli* and other pathogens through consumption of fish or seafood living near the shore (El-Shenaw and El-Shenaw, 2005). Even in regions of the world with indoor plumbing, breakdowns in water treatment infrastructure often result in geographic patterns that express a strong relationship between distance to water and infection.

Although drinking contaminated water is perhaps the most common route of exposure for infectious water-related diseases, other routes exist. Schistosomiasis, for example, is spread when persons come into physical contact with water populated by snails infected with the schistosome parasite. Proximity to surface water has shown a particularly strong relationship with infection by the schistosomiasis pathogen at local (Booth et al., 2004) and regional (Brooker and Clements, 2009) geographic scales. In several studies in different regions of Africa, persons living within roughly 4–5 km of permanent water sources have considerably higher levels of infection with schistosome parasites than persons living further away (Amazigo et al., 1997; Handzel et al., 2003; Booth et al., 2004; Clements et al., 2006).

For vectorborne diseases, proximity to water can be important for determining human exposure to a pathogen since disease-transmitting vectors often lay their eggs in water. Malaria prevalence is higher in densely populated areas near egg-laying and breeding sites (Kleinschmidt et al., 2001; Le Manach et al., 2005; Manga et al., 1993; Sabatinelli et al., 1986; Staedke et al., 2003). This relationship is most likely due to the travel behaviour of mosquitoes; greater distances between egg-laying sites and human blood meals decrease the efficiency of mosquitoes to infect humans (Le Manach et al., 2005). These observations may also be true for mosquitoes carrying dengue fever (Vanwambeke et al., 2006) and diseases spread by other arthropods. For example, while not a key factor in describing the geographic distribution of disease, distance to water may also influence the likelihood of exposure to ticks carrying Lyme disease (Bunnell et al., 2003) and even prevalence of the human plague (Eisen et al., 2007).

Distance may also be a consideration for health conditions associated with chemical contamination of water. Deep tube wells with lower arsenic contamination are more likely to be used by households within a reasonable walking distance (Hassan, 2005). Living in regions where groundwater is in contact with natural geological sources of radon may increase risks of cancer through inhalation of radon released into the air through domestic water use or the ingestion of dissolved radon (Bean et al., 1982; Hopke et al., 2000). Fish consumers living near water sources with high levels of mercury often have higher than average mercury exposure, and have been found to suffer disproportionately from the neurological effects of mercury poisoning (Boischio and Hensheld, 1996; Dolbec et al., 2000).

Health care planning and intervention often requires information about where risk of disease is particularly high; knowing that persons in a particular locale have disproportionately high risk allows for the allocation of resources to the people in need. When there is a strong relationship between distance to water and risk of disease, a map of communities within a particular distance of a water source may be the simplest method of communicating key risk information. However, if there are anomalies *within* the baseline association between risk of illness/infection and proximity to water – for example, because certain regions are better transmission sites than others – then these maps could be enhanced by spatial analysis methods that can identify such anomalies. This spatial exploration process is particularly important when the local factors that may interact with proximity to water are unknown, and cannot be explicitly modelled.

23.1.2 Finding Clusters of Water-Related Disease

Geographic information systems have a well established role in identifying the "where" of infectious disease, and in turn, supporting decision makers and the community in planning interventions Abdel-Rahman et al., (2001; Beck et al., 1994; Booman et al., 2000; Carter et al., 2000; Gamperli et al., 2006; Nihei et al., 2006). Within most geographic information systems are analytical tools that help explore and characterize spatial patterns in disease. Cluster detection methods are particularly useful for identifying spatial anomalies or "heterogeneities" in disease risk. Cluster detection methods are usually subdivided into several methodological groups based on the specific question being asked. Tests for *global* clustering (e.g., Tango, 1995; Whittemore et al., 1987) identify whether or not cases of disease tend to be close together in general, without specifically identifying cluster locations. *Focussed* cluster detection tests (e.g., Stone, 1988; Tango, 1995) identify whether or not cases of disease tend to occur around hypothesized sources of risk (such as contaminated wells, pulp mills or factories). *Local* cluster detection tests (e.g., Openshaw et al., 1988; Turnbull et al. 1990; Kulldorff, 1997) identify the location and statistical noteworthiness of local clusters or "hot-spots" of disease. Of these three general approaches, the local cluster detection tests are of the most use in exploring spatial heterogeneities of risk related to water since they can be used to help identify the locations where risk is higher than expected. When a local cluster is found, we are informed about the location where risk is anomalously high, and in turn, where the intervention may be most needed.

Over the past decade, the widespread use of the spatial scan cluster detection method (Kulldorff, 1997) illustrates a general interest in identifying local clusters in health research, surveillance, and a variety of other fields. In its most common form, the spatial scan involves searching for geographic clusters in a study area in order to determine which cluster is most likely to be a geographical anomaly. The original spatial scan uses a large number of circular "windows" of different sizes and at different locations in this search process; each window is evaluated as a potential cluster. The potential cluster most likely to reject a null hypothesis of constant risk is identified as a "primary" cluster. This primary cluster represents the geographic subset of observations that is most likely to be a geographic anomaly, and is tested for significance using Monte Carlo methods.

The circular search windows of the spatial scan make this approach less effective at identifying clusters that occur in non-circular shapes. This poses a particular problem when interested in exploring disease clusters associated with proximity to water, since the geometry of the real cluster is likely to at least partly reflect the geometry of the water source, which is often irregularly shaped. Fortunately, numerous recent methodological advancements have expanded the ability of the spatial scan to detect clusters of irregular shape (Duczmal and Assunção 2004; Tango and

Takahashi, 2005; Assunção et al. 2006; Yiannakoulias et al., 2007; Duczmal et al., 2008). Approaches outside the spatial scan paradigm have also been developed for detecting irregularly shaped clusters (Aldstadt and Getis, 2006; Wieland et al., 2007; Jacquez, 2009).

23.1.3 Geographic Information Systems as Data Synthesis Tools

While the theoretical benefits of irregularly-shaped cluster detection methods are clear, there is a dearth of research evaluating the ways in these methods may be useful for exploring patterns of disease that are likely to occur in the real world. Most comparative research to date has relied on a small number of arbitrary and unrealistic synthetic data sets as a baseline for comparison. In most of these applications, synthetic cluster areas are assigned a high risk level, all other areas a lower risk level, and methods are judged based on their ability to distinguish these areas from each other. This approach assumes that risk is spatially homogenous within these areas, which makes cluster detection relatively easy. The complex and highly localized spatial heterogeneities found in the real world differ greatly from most existing experiments using synthesized data, and in turn, most comparative research to date lacks the rigour to effectively compare different methods.

In this study, we use a geographic information system to build realistic synthetic data sets of *Schistosoma mansoni* infection prevalence. *S. mansoni* is a waterborne parasite responsible for schistosomiasis, an infection with serious and often long-term public health impacts, particularly in Africa. Our objective is to conduct a geo-computational experiment in which we evaluate how well two cluster detection methods are able to detect clusters of infection that are likely to follow the shape of permanent water sources. *S. mansoni* infection prevalence is strongly influenced by proximity to water, but also other features of the physical and social environment. A key component of the experiment is that these synthesized data are a realistic representation of infection prevalence, and display variations in *S. mansoni* infection prevalence that are likely to be seen in the real world. These data are created by combining environmental and socioeconomic information into a geographic information system to produce a surface of synthesized prevalence. From this surface we generate a large number of synthetic data sets which are then analyzed for clusters using the two different methods. We then use a geographic information system to display the results of the analysis, and compare the effectiveness of the different methods.

23.2 Methods

23.2.1 Data

This experiment is based on real data from two different regions of Kenya which have been used in previous empirical studies on prevalence and infection intensity

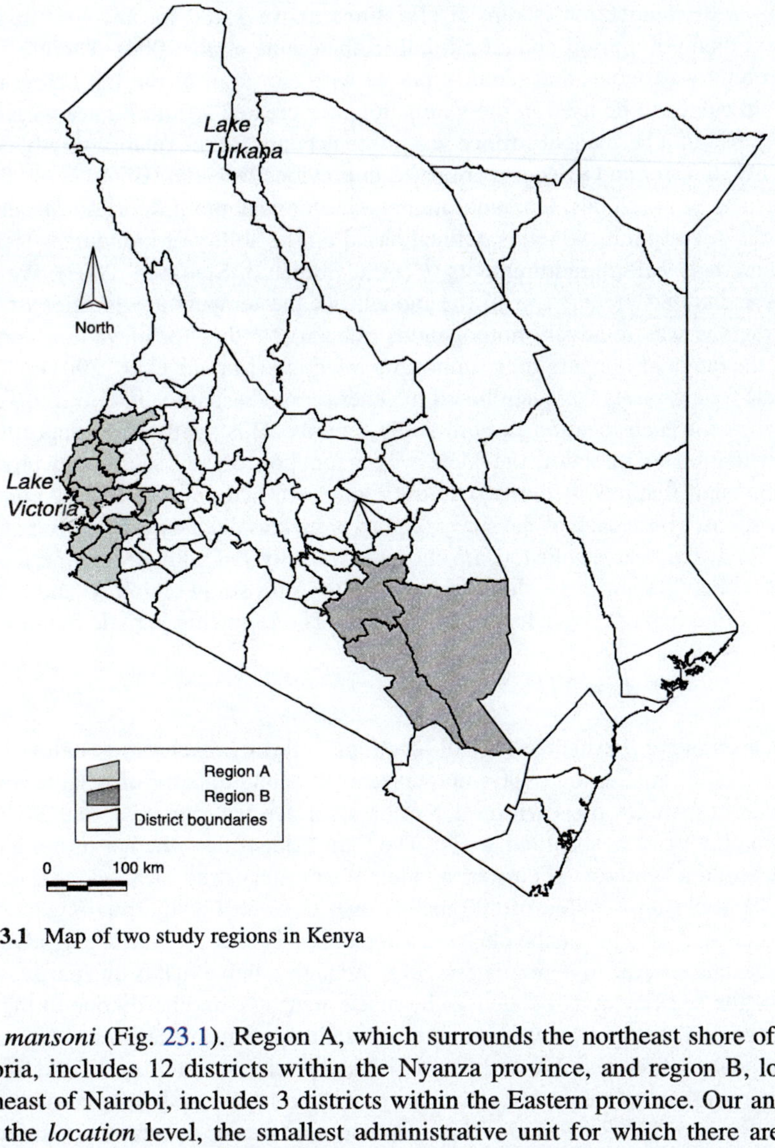

Fig. 23.1 Map of two study regions in Kenya

of *S. mansoni* (Fig. 23.1). Region A, which surrounds the northeast shore of Lake Victoria, includes 12 districts within the Nyanza province, and region B, located southeast of Nairobi, includes 3 districts within the Eastern province. Our analysis is at the *location* level, the smallest administrative unit for which there are data from the Kenyan Census. There are 343 locations in region A, and 168 locations in region B.

Population and socioeconomic data are from the 1999 Census, available for download from the World Resources Institute (http://www.wri.org/publication/content/9291). Elevation data are based on the USGS GTOPO30 global digital elevation model. Precipitation data were obtained from WorldClim (http://www.worldclim.org) (Hijmans et al., 2005). Although there is little population-based empirical research showing the general functional relationship between elevation, precipitation and *S. mansoni* infection prevalence, research in the region has shown that *S. mansoni* infection is rare at elevations above 1,400 m, and low in areas with less than 900 mm of annual rainfall (Kabatereine et al., 2004). Therefore, we created binary variables that identify places with elevation or rainfall below these threshold values to be used in the simulation. We created a third binary variable – within 5 Km of a permanent surface water – to account for the relationship between proximity to water and infection observed in previous research (Booth et al., 2004; Kabatereine et al., 2004). Our final model variable was the proportion of persons below the poverty line, which is defined based on the ability of individuals to meet minimum daily calorie requirements (Central Bureau of Statistics, 2003). We considered including temperature in the model, but the temperature profiles of both study regions was relatively homogenous after accounting for elevation, and fell within the range of temperatures supporting infection (Malone et al., 2001).

These four datasets were combined to generate synthesized *S. mansoni* infection prevalence for each location in both study regions. This synthesized data reflects the distribution of infection prevalence as a function of key social and physical environmental features in the two study regions. In addition, subsets of contiguous locations within each of the study regions were classified as synthesized cluster areas. These synthesized clusters are chosen arbitrarily, but both are near permanent water sources (the northeast shore of Lake Victoria for study region A, and a small network of permanent rivers in study region B). The following logistic function,

$$p_i = 1/\left(1 + e^{-\text{intercept} + E_i^{*}2.5 + SES_i^{*}1.5 + C_i^{*}2}\right), \tag{1}$$

provides synthetic prevalence p at all locations i. Term E is elevation below 1,400 m, R is rainfall greater than 900 mm per year, W is an indicator of whether or not the location includes or is within 5 Km of a surface water source, and SES is a measure of poverty as specified above. The term C identifies whether or not a location is within a synthesized cluster; a value of "1" means the location is inside the synthesized cluster, a value of "0" means that it is not. For the final synthesized surface, relative risk inside the cluster area in region A was 3.3, and the relative risk inside the cluster area in region B was 5.2. The high relative risk is due partly to the value of the coefficient associated with cluster areas (2), and partly due to the fact that the cluster areas are located in regions baseline prevalence is already high. The other model coefficients (2.5, 1, 2.5 and 1.5) and intercept were determined by trial and error so that the map of synthesized prevalence generated patterns of similar to findings in previous research. This map is not a true representation of infection prevalence, but we believe it represents a reasonable baseline for the experiment.

Based on these synthesized prevalence values, we randomly assign status as a case or non-case to persons within each location, where the probability of an individual being a case is a function of the value of p at the location i in which they reside. We used 0.1% samples of the population counts at the location level as our sampling population. We do this because comprehensive population-wide data of *S. mansoni* prevalence would be expensive and time consuming to collect, and therefore, any future surveillance would likely be based on population samples. A 0.1% sample results in populations of 4,107 in region A and 2,028 in region B. In total,

1,000 data sets are created for both study regions. Each of these synthesized data sets is then analyzed using the two different cluster detection methods. Results are mapped in order to identify the relative success of these methods at identifying the zones with highest risk.

23.2.2 Greedy Growth Scan

The traditional spatial scan and the greedy growth scan method are identical except in how they identify potential clusters. While the spatial scan is typically understood as a search using circular windows, it can also be understood as a search based on nearest neighbours. Starting at any zone (a "seed" zone), this zone is considered a potential cluster. Next, the nearest geographic neighbour (typically based on the distance between the centroids of the two zones) is added to form a new potential cluster that now includes two zones (the seed zone and a new zone). The next nearest neighbor to the seed zone is then added, forming yet another potential cluster now consisting of three zones. This process continues until some threshold size has been met (typically, 50% of the zone populations). This process is then repeated so that each zone in the study region is treated as a seed zone of this nearest-neighbour search procedure. This process creates a large set of "circular" potential clusters, from which the potential cluster with the largest log-likelihood ratio is identified as the most-likely to cause a rejection of a null hypothesis that there are no clusters.

The greedy growth scan also creates a large set of potential clusters, but adds neighbours that explicitly maximize the test statistic (e.g., the Poisson log-likelihood test statistic) rather than in order of closest neighbours. The method is greedy since it involves incrementally agglomerating zones into new potential clusters in a way that *immediately* maximizes the test statistic objective function. This is distinguishable from methods that make short-term sub-optimal potential clusters to find other potential clusters with more generally optimal characteristics (e.g., Duczmal and Assunção 2004). The only constraint to the search process is that all clusters must be topologically connected, which in most geographic applications, means geographically contiguous. To ensure that the algorithm searches an entire study area thoroughly, the search processes initiates from all zones within a study region. Once a search over a study area is completed, the potential cluster with the highest test statistic is tested for significance using Monte Carlo methods in a manner similar to the traditional spatial scan.

While all potential clusters found by the greedy growth scan are contiguous, topological connectivity can vary considerably between them. Topological connectivity can describe the shape and structure of a cluster based on the adjacency of zones rather than a particular geometric form. Some potential clusters consist of long chains in which each zone is neighbour to only one or two other zones, and other clusters are compact, with most zones adjacent to three, four or even more neighbouring zones. Cluster detection methods like the greedy growth scan sometimes find clusters that are highly-irregular in shape – often of large "octopus-like" forms. This is because there are no restrictions on what shape a cluster may take.

While these clusters may sometimes represent statistical anomalies, they are less conceptually meaningful, and difficult to interpret. To deal with this problem, we apply a non-connectivity penalty based on the *gamma index*, originally used in the structural analysis of transportation networks (Taaffe and Gauthier 1973). In this application, the gamma index is the ratio of the total observed neighbouring connections between zones in a potential cluster to the total possible number of neighbouring connections between zones in a potential cluster. We define a neighbourhing connection as a situation in which two locations are immediately adjacent to one another. Potential clusters with a high degree of topological connectivity (such as roughly circular-shaped clusters) have higher values of the gamma index than elongated or octopus-like clusters. One can apply the gamma index directly to the test statistic (the log-likelihood ratio) associated with potential clusters in order to offset the method's tendency to identify most-likely clusters of very irregular shape. This non-connectivity penalized likelihood ratio is

$$PLR(Z, \alpha) = LR(Z)^{K(Z)^{\alpha}} \tag{2}$$

where

$$K(Z) = \frac{e(Z)}{3(v(Z) - 2)}. \tag{3}$$

LR(Z) is the likelihood test statistic typically used in the spatial scan, and depends on the particular statistical model chosen (see Kulldorff, 1997). *K*(Z) is the ratio of the observed number of neighbours in a potential cluster (*e(Z)*) to the total possible number of neighbours in a potential cluster, which can be determined from the number of zones (*v(Z)*) in a potential cluster. The value of α determines the strength of the penalty; larger values indicate a stronger penalty, and a greater penalty against detecting highly irregular clusters. In this application, we apply a weak penalty ($\alpha = 0.25$), because the synthesized cluster areas are designed to be irregularly shaped; a stricter penalty would make it more difficult for the greedy growth scan to find clusters of irregular shape, resulting in clusters similar to those found by the circular spatial scan method.

23.3 Results

Figures 23.2 and 23.3 illustrate the spatial pattern of synthesized prevalence for regions A and B respectively. These prevalence levels are based on Eq. 1. Prevalence in region A is relatively geographically homogenous, and most locations have synthesized prevalence over 50%. Prevalence is lower in region B, and is more heterogeneous. In both regions, the synthesized cluster area is located where prevalence is already high.

Figures 23.4a, b and 23.5a, b display the sensitivity and specificity of the two cluster detection methods for each of the 1,000 simulated data sets for both study areas. If a method worked perfectly, it would identify all the locations in the cluster

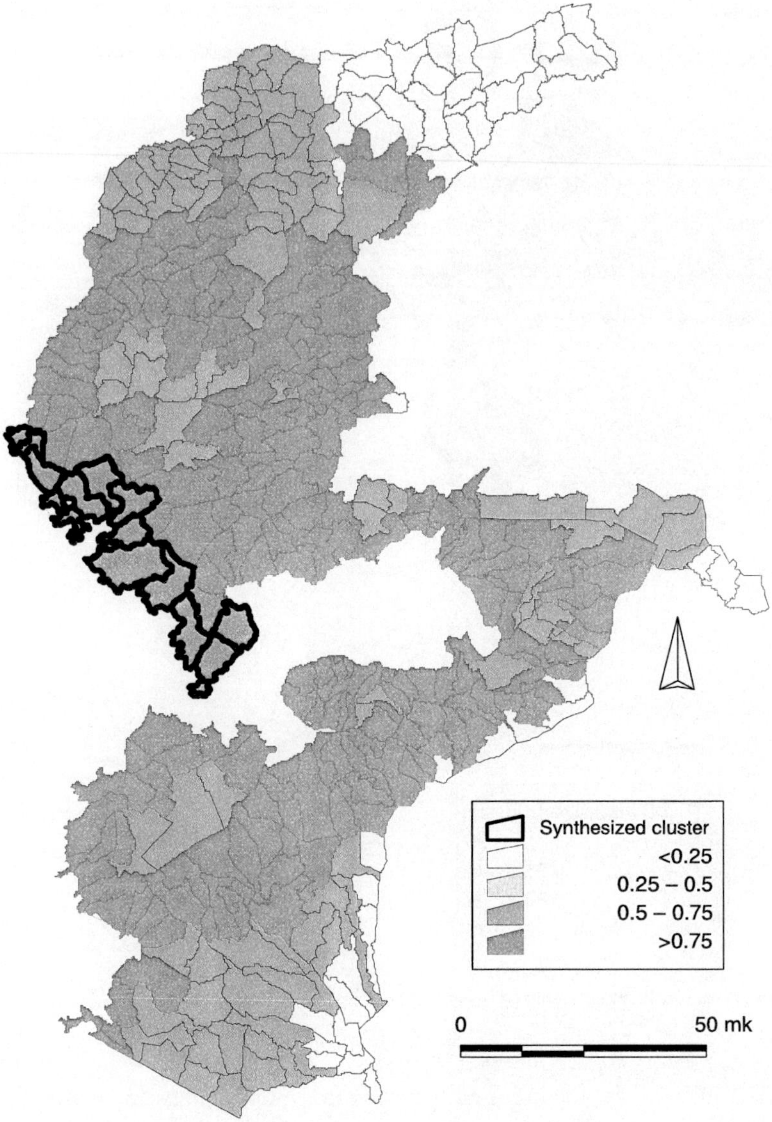

Fig. 23.2 Location-level synthesized prevalence for region A

area as part of the cluster area (perfect "sensitivity"), and all the locations outside the cluster area as not part of the cluster area (perfect "specificity") for each of the 1,000 data sets. This information is represented on the maps using shading: the higher the mapped values (and the darker the shading) in the synthesized cluster area, the greater the sensitivity of the method, and the higher the mapped values outside the synthesized cluster areas (the darker the shading), the greater the specificity of the

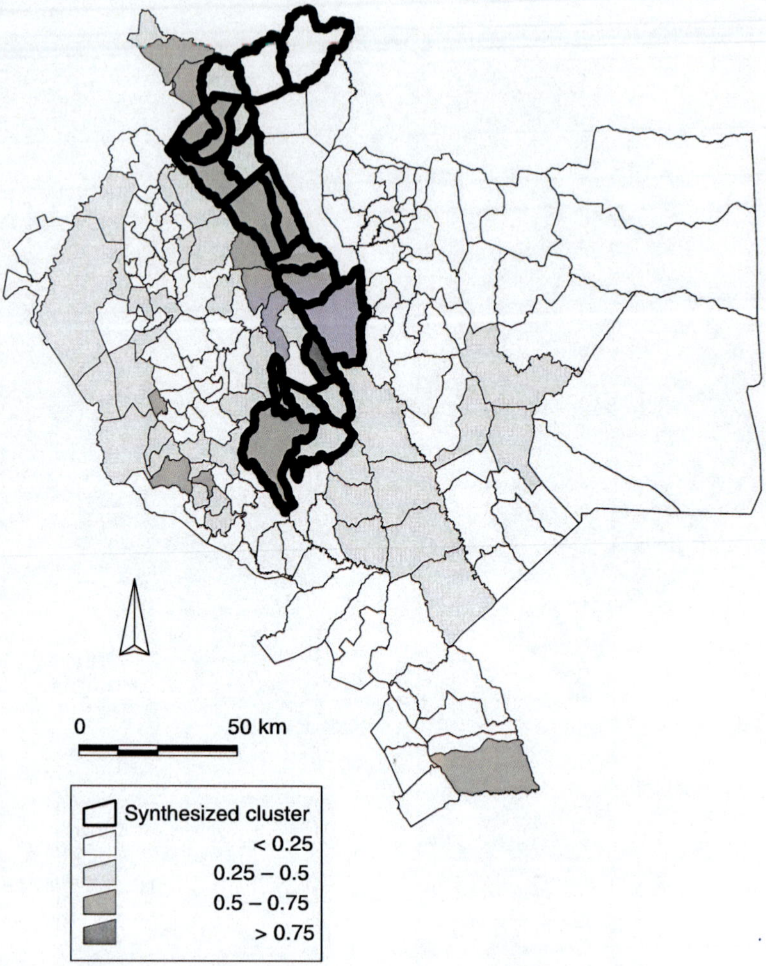

Fig. 23.3 Location-level synthesized prevalence for region B

method. A method that identifies all the locations in the synthesized cluster area as part of a cluster, and all other locations as not part of a cluster would produce a map with sensitivity and specificity equal to 1, and all locations would be shaded dark. A lightly shaded location inside the cluster area indicates low sensitivity for inclusion of that location in clusters found by the method. A lightly shaded location outside the cluster area indicates low specificity for exclusion in clusters found by the method.

For study region A, Both methods appear to have high sensitivity; most locations inside the synthesized cluster area have sensitivity, with most locations correctly included in found clusters over 90% of the time (Fig. 23.4a, b). Both methods appear able to successfully identify the locations inside the synthesized cluster areas. The

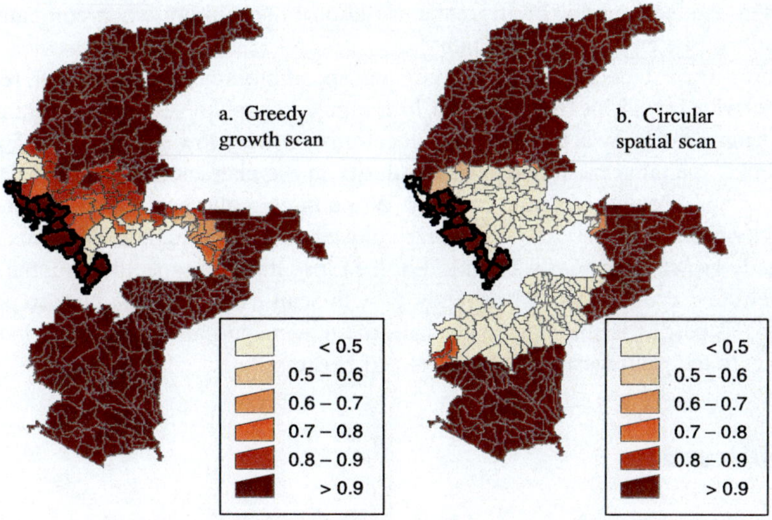

Fig. 23.4 Sensitivity and specificity of detection, region A

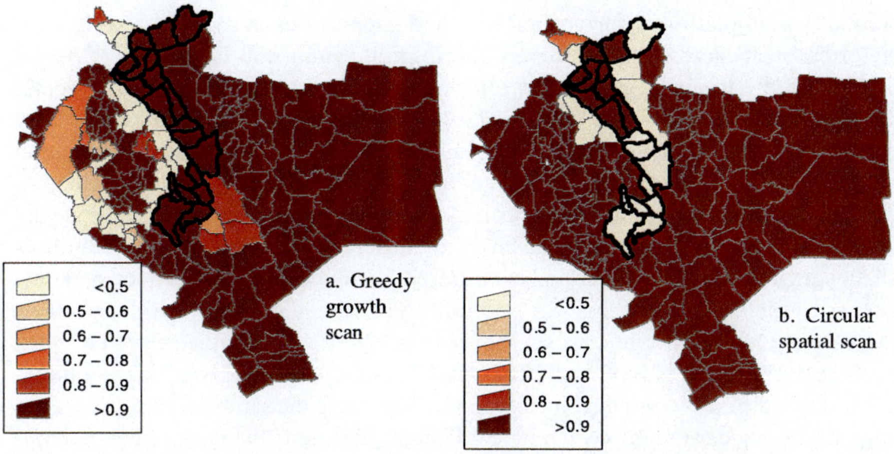

Fig. 23.5 Sensitivity and specificity of detection, region B

methods differ more with respect to specificity; the circular spatial scan has lower specificity in the regions along the shore of Lake Victoria, illustrated by a large lightly shaded circle of locations with low specificity. This suggests that the method probably found the same large cluster in most of the simulated data sets. The greedy growth scan has a less regular pattern of specificity. There are areas with moderate to low specificity, particularly on the north shore of Lake Victoria. However, the clusters found by the greedy growth scan were probably not the same across the simulated data sets, and exhibit greater variation in specificity when compared to the results of the circular spatial scan.

Figures 23.5a, b display the sensitivity and specificity for each location in region B. Sensitivity for all locations is high in the greedy growth scan. For the circular spatial scan, sensitivity is high for some locations, but very low for others. As above, the circular spatial scan shows more regularity in the characterization of locations as part of and not part of found clusters. While the circular spatial scan has lower sensitivity in some parts of the synthesized cluster area, it has higher specificity than the greedy growth scan. Only a handful of locations have low specificity in the case of the circular spatial scan. The greedy growth scan has high sensitivity to detect locations as part of found clusters, but has the tendency to also incorrectly include locations to the southwest as part of detected clusters.

23.4 Discussion

As spatial analytic methodology advances, and the number of methods available increases, comparing the effectiveness of different methods is becoming increasingly important. This is especially true for methods that are, or are likely to become, integrated into geographic information systems. Geographic information systems are most often platforms for general use, and support the analysis of a wide variety of problems across many different disciplines. Within sub-disciplines of health research likely to apply geographic information systems – such as environmental health and infectious disease epidemiology – specific analytical requirements may demand a particular subset of the tools available in standard geographic information systems. It is important that such methodological tools are evaluated and tested in ways that are meaningful to these specialised applications.

Formally evaluating the effectiveness of different spatial analytic methods is challenging, however. Comparing how different methodologies perform on real data does not provide enough range for generalization; a method may prove very effective in the analysis of one data set, but less effective with another. Further, when using real data, there is no formal benchmark for comparison; if two different methods provide different analytical information when using the same data set, there is no simple way of determining which method better represents the data, or best accomplishes the specific analytical goals. As a result, synthetic data are more frequently used for comparing different methodologies, though they are not without their own shortcomings. First, there are no widely accepted standards for creating synthetic data. Comparative studies using different synthetic data sets in their analysis are not often identical, so conflicting findings do not provide an obvious answer about which is correct. Second, there is often little assurance that synthetic data represent something that could occur in the real world. Experimenting with unrealistic synthetic data may create a false impression of how different methods would perform in real world applications.

Our goal was to generate synthesized data to provide a realistic framework for comparing how well two different disease cluster detection methods identify clusters that are likely to follow the shape of a permanent surface water source. We generated these synthesized data by combining four publicly available data sets of environmental and population data into a geographic information system. These data were then used to generate a simple model of prevalence of *S. mansoni* infection for each location in the two study areas. We then used this baseline map of prevalence to create 1,000 synthetic data sets. Each of these data sets was analyzed by the greedy growth scan and the circular spatial scan methods of cluster detection. Our results illustrate how these two methods perform on realistic synthetic data, and provide a credible comparison of how the methods may perform when used to analyze real data that exhibit similar patterns.

The results from our experiments provide evidence that when real clusters of infection prevalence are irregularly shaped, the greedy growth scan is a suitable cluster detection method. The circular spatial scan is restricted to a simpler geometry, and unable to identify the location of a synthesized cluster without falsely including a large number of non-cluster locations in identified clusters. However, the greedy growth scan may have lower specificity than the circular spatial scan. To some degree, the greedy growth scan is a victim of its own success; it is so good at identifying locations where risk is high that it will do so at the expense of finding clusters that have believable and environmentally meaningful shapes. While this can be controlled by imposing a stronger non-connectivity penalty, there are no general rules as to what these penalties should be a priori. This points to one of the great advantages of the circular spatial scan: it requires very little pre-specification, and is therefore less likely to introduce pre-specification biases.

Methods based on connectivity (rather than spatial proximity) have a capacity to explore patterns with more complex topological structures than traditional spatial modelling and cluster detection methods. Spatial connectivity is a particularly important concept for water-related diseases, since locations connected by water sources may exhibit greater similarity in risk than locations that are geographically proximate (Xu et al., 2006). Water flowing downstream from a site of contamination may transport risk downstream, thereby connecting distant places to a common source of hazard. Places along a shoreline or islands in a single body of water may be considered connected to a common hazard when water is a source of exposure, even if the places are geographically distant. The greedy growth scan is particularly suited to the analysis of clusters that may follow water sources because the search method, including penalization, is based on topological connectivity rather than distance. This flexibility is an advantage when the physical structure of the environment influencing risk of disease or infection may have distant, but important, connections.

23.5 Conclusion

Geographic information systems have many uses in applied environment and health. They can be used to combine different sources of information to build models of exposure, map variations in relative risk, and estimate relationships between environmental exposures and particular health outcomes. Here we used a geographic information system to generate synthetic data that allowed us to compare how well two different methods of cluster detection are able to find clusters of irregular shape. Given the many diseases for which water environments may be important in environmental exposure, it is important to know what methods may be best able to find spatial anomalies that could occur in the shape of surface water sources.

Exploratory methods of spatial analysis can be useful for detecting geographic patterns of disease related to features of the environment. These methods can be applied to routine surveillance activities, but also research in environment and health, particularly when disease aetiology is not well understood. The experiments presented here illustrate the value of using a geographic information system to generate synthetic spatial data for the comparison of different cluster detection methods. While cluster detection methods with a fixed geometry have a number of advantages (simplicity, speed, and greater power to detect a signal in sparse data) the information these methods offer is often geographically ambiguous. Relaxing the fixed geometry by basing searches on topological relationships allows cluster detection methods to find patterns or processes responsible for interesting geographic variations in disease.

Author Biography

Niko Yiannakoulias is an Assistant Professor in the School of Geography and Earth Sciences at McMaster University. His general research interests are in geo-computation, medical geography and environmental health, with a particular emphasis on public health surveillance, infectious disease processes, and cluster detection methods. He has applied interests, including injury epidemiology, urban health and crime, and waterborne illness.

References

Abdel-Rahman M, El-Bahy M, Malone J, Thompson R, El-Bahy N (2001) Geographic information systems as a tool for control program management for schistosomiasis in Egypt. Acta Tropical 79:49–57

Aldstadt J, Getis A (2006) Using AMOEBA to create a spatial weights matrix and identify spatial clusters. Geogr Anal 38:327–343

Ali M, Emch M, Donnay J, Yunus M, Sach R (2002) Identifying environmental risk factors for endemic cholera: a raster GIS approach. Health Place 8:201–210

Amazigo U, Anago-Amanze A, Okeibunor J (1997) Urinary schistosomiasis among school children in nigeria: cnsequences of ingigenous beliefs and water contact activities. J Biosoc Sci 29:9–18

Assunção R, Costa M, Tavares A, Ferreira S (2006) Fast detection of arbitrarily shaped disease clusters. Stat Med 25:723–742

Bean J, Isacson P, Hahne R, Kohler J (1982) Drinking water and cancer incidence in Iowa. Am J Epidemiol 116:924–932

Beck L, Rodriquez M, Dister S, Rodriguez A, Rejmankova E, Ulloa A, Meza R, Roberts D, Paris J, Spanner M, Washino R, Hacker C, Legters L (1994) Remote sensing as a landscape epidemiological tool to identify villages at high risk for malaria transmission. Am J Trop Med Hyg 51:271–280

Boischio A, Hensheld D (1996) Risk assessment of mercury exposure through fish consumption by the riverside people in the Madeira Basin, Amazon, 1991. Neurotoxicology 17:295

Booman M, Durrheim D, La Grange K, Martin C, Mabuza A, Zitha A, Mbokazi F, Fraser C, Sharp B (2000) Using a geographical information system to plan a malaria control programme in South Africa. Bull World Health Organ 78:1438–1444

Booth M, Vennervald B, Kenty L, Butterworth A, Kariuki H, Kadzo H, Ireri E, Amaganga C, Kimani G, Mwatha J, Otedo A, Ouma J, Muchiri E, Dunne D (2004) Micro-geographical variation in exposure to Schistosoma mansoni and malaria, and exacerbation of splenomegaly in Kenyan school-aged children. BMC Infect Dis 4:13

Brooker S, Clements A (2009) Spatial heterogeneity of parasite co-infection: Determinants and geostatistical prediction at regional scales. Int J Parasitol 39:591–597

Bunnell J, Price S, Shields T, Glass G (2003) Geographic information systems and spatial analysis of adult *Ixodes scapularis* (Acari: Ixodidae) in the middle Atlantic region of the USA. J Med Entomol 40:570–576

Carter R, Mendis K, Roberts D (2000) Spatial targeting of interventions against malaria. Bull World Health Organ 78:1401–1411

Central Bureau of Statistics (2003) Geographic dimensions of well-being in Kenya. The Regal Press Kenya Ltd, Nairobi, Kenya

Clements A, Moyeed R, Brooker S (2006) Bayesian geostatistical prediction of the intensity of infection with Schistosoma mansoni in East Africa. Parasitology 133:711–719

Dolbec J, Mergler D, Sousa Passos C, Sousa de Morais S, Lebel J (2000) Methylmercury exposure affects motor performance of a riverine population of the Tapajos river, Brazilian Amazon. Int Arch Occup Environ Health 73:195–203

Duczmal L, Assunção R (2004) A simulated annealing strategy for the detection of arbitrarily shaped spatial clusters. Comput Stat Data Anal 45:269–286

Duczmal L, Cançado A, Takahashi R (2008) Delinieation of irregularly shaped disease clusters through multiobjective optimization. J Comput Graph Stat 17:243–262

Eisen R, Reynolds P, Ettestad P, Brown T, Enscore R, Biggerstaff B, Cheek J, Bueno R, Targhetta J, Montenieri J, Gage K (2007) Residence-linked human plague in New Mexico: a habitat-suitability model. Am J Trop Med Hyg 77:121–125

El-Shenawy M, El-Shenawy M (2005) Enterohaemorrhagic Escherichia coli O157 in coastal environment of Alexandria, Egypt. Microb Ecol Health Dis 17:103–106

Gemperli A, Sogoba N, Fondjo E, Mabaso M, Bagayoko M, Briet O, Anderegg D, Liebe J, Smith T, Vounatsou P (2006) Mapping malaria transmission in West and Central Africa. Trop Med Int Health 11:1032–1046

Handzel T, Karanja D, Addis D, Hightower A, Rosen D, Colley D, Andove J, Slutsker L, Secor W (2003) Geographic distribution of schistosomiasis and soil-transmitted helminths in Western Kenya: implications for anthelminthic mass treatment. Am J Trop Med Hyg 69: 318–323

Hassan M (2005) Arsenic poisoning in Bangladesh: spatial mitigation planning with GIS and public participation. Health Policy 74:247–260

Hijmans R, Cameron S, Parra J, Jones P, Jarvis A (2005) Very high resolution interpolated climate surfaces for global land areas. Int J Climatol 25:1965–1978

Hopke P, Borak T, Doull J, Cleaver J, Eckerman K, Gundersen L, Harley N, Hess C, Kinner N, Kopecky K, McKone T, Sextro R, Simon S (2000) Health reisks due to radon in drinking water. Environ Sci Tech 34:921–926

Jacquez G (2009) Cluster morphology analysis. Spat Spattemporal Epidemiol 1:19–29

Kabatereine N, Brooker S, Tukahebwa E, Kazibwe F, Onapa A (2004) Epidemiology and geography of Schistosoma mansoni in Uganda: implications for planning control. Trop Med Int Health 9:372–380

Kleinschmidt I, Sharp B, Clarke G, Curtis B, Fraser C (2001) Use of generalized linear mixed models in the spatial analysis of small-area malaria incidence rates in KwaZulu Natal, South Africa. Am J Epidemiol 153:1213–1221

Kulldorff M (1997) A spatial scan statistic. Commun Stat Theory Methods 26:1481–1496

Le Manach A, McKenzie F, Flahault A, Smith D (2005) The unexpected importance of mosquito oviposition behaviour for malaria: non-productive larval habitats can be sources for malaria transmission. Malar J 4:23

Lindkvist P, Enquselassie F, Asrat D, Muhe L, Nilsson I, Giesecke J (1998) Risk factors for infection with Heliobacter pyloru – a study of children in rural Ethiopia. Scand J Infect Dis 30:371–376

Malone J, Yilma J, McCarroll J, Erko B, Mukaratirwa S, Zhou X (2001) Satellite climatology and the environmental risk of Schistosoma mansoni in Ethiopia and east Africa. Acta Tropica 79:59–72

Manga L, Fondjo E, Carnevale P, Robert V (1993) Importance of low dispersion of Anopheles gambiae (Diptera: Culicidae) onmalaria transmission in hilly towns in south Cameroon. J Med Entomol 30:936–938

Nihei N, Kajihara N, Kirinoki M, Chigusa Y, Matsuda H, Saitoh Y, Shimamura R, Kaneta H, Nakamura S (2006) Establishment of a GIS monitoring system for schistosomiasis japonica in Kofu, Japan. Ann Trop Med Parasitol 100:143–153

Openshaw S, Craft A, Charlton M, Birch J (1988) Investigation of leukemia clusters by use of a geographical analysis machine. Lancet 331:272–273

Sabatinelli G, Rossi P, Belli A (1986) Etude sur la dispersion d'Anopheles gambiae s.l. dans une zone urbaine à Ouagadougou, Burkina Faso. Parasitologia 28:33–39

Staedke A, Nottingham E, Cox J, Kamya M, Rosenthal P, Dorsey G (2003) Short report: proximity to mosquito breeding sites as a risk factor for clinical malaria episodes in an urban cohort of Ugandan children. Am J Trop Med Hyg 69:244–246

Stone RA (1988) Investigations of excess environmental risks around putative sources: statistical problems and a proposed test. Stat Med 7:649–660

Taaffe EJ, Gauthier HL (1973) Geography of transportation. Prentice Hall, Englewood Cliffs, NJ

Tango T (1995) A class of tests for detecting "general" and "focused" clustering of rare diseases. Stat Med 14:2323–2334

Tango T, Takahashi K (2005) A flexibly shaped spatial scan statistic for detecting clusters. Int J Health Geogr 4:11

Turnbull B, Iwano E, Burnett W, Howe H, Clark L (1990) Monitoring for clusters of disease: application to leukemia incidence in upstate New York. Am J Epidemiol 132:S136–S143

Vanwambeke A, van Benthem B, Khantikul N, Burghoorn-Maas C, Panart K, Oskam L, Lambin E, Somboon P (2006) Multi-level analyses of spatial and temporal determinants for dengue infection. Int J Health Geogr 5:5

Whittemore A, Friend N, Brown B, Holly E (1987) A test to detect clusters of disease. Biometrika 74:631–635

Wieland S, Brownsten J, Berger B, Mandl K (2007) Density-equalizing Euclidean minimum spanning trees for the detection of all disease cluster shapes. Proc Natl Acad Sci USA 104:9404–9509

Xu B, Gong P, Seto E, Liang S, Yang C, Wen S, Qiu D, Gu X, Spear R (2006) A spatial-temporal model for assessing the effects of intervillage connectivity in schistosomiasis transmission. Ann Assoc Am Geogr 96:31–46

Yiannakoulias N, Rosychuk R, Hodgson M (2007) Adaptations for finding irregularly shaped clusters. Int J Health Geogr 6:28

Chapter 24
Spatiotemporal Analysis of $PM_{2.5}$ Exposure in Taipei (Taiwan) by Integrating PM_{10} and TSP Observations

Hwa-Lung Yu, Chih-Hsin Wang, George Christakos, and Yu-Zhang Wu

Abstract Many studies have shown a significant association between human exposure to Particulate Matter (PM) and population health effects (premature mortality, respiratory and cardiovascular diseases, emergency room visits, and systemic inflammation). Fine PM particles ($PM_{2.5}$) are believed to be more dangerous than PM_{10} because fine particles are easier inhaled and accumulated deeply into human lungs. Taipei is the largest city in Taiwan, where a variety of industrial and traffic emissions are continuously generated across space and time. Thus, it is crucial for health agencies to improve their understanding of spatiotemporal $PM_{2.5}$ exposure of people living in Taipei city. The Bayesian Maximum Entropy (BME) theory of spatiotemporal statistics and science-based mapping provides valuable information about population exposure to $PM_{2.5}$ pollution in Taipei. PM-related data (PM_{10}, $PM_{2.5}$ and Total Suspended Particles, TSP) are collected by different central and local government institutes. BME analysis introduces concepts and techniques that have a number of important features (theoretical and computational): several kinds of site-specific data and core knowledge bases are integrated that are associated with different physical scales; a variety of uncertainty sources are taken into account; non-linear, in general, PM estimators are used at unobserved locations; non-Gaussian laws are automatically incorporated; and a complete characterization of the dynamic PM distribution is obtained in terms of the probability density function at each space-time point rather than a single PM value. These BME advantages have considerable consequences as far as health risk analysis is concerned. Detailed space-time $PM_{2.5}$ maps account for (*i*) composite space-time dependence structure of PM values, (*ii*) hard and soft datasets available about $PM_{2.5}$, PM_{10} and TSP, and (*iii*) empirical evidence about the $\frac{PM_{2.5}}{PM_{10}}$ and $\frac{PM_{10}}{TSP}$ ratios. PM measures are investigated, including the fraction of fine particles that vary considerably across space-time. BME analysis properly identifies and quantifies factors that influence the spatiotemporal patterns of these measures, such as weather conditions and land use (e.g., the PM distributions in highly-developed commercial or industrial areas

H.-L. Yu (✉)
Department of Bioenvironmental Systems Engineering, National Taiwan University,
Taipei 10617, Taiwan
e-mail: hlyu@ntu.edu.tw

J.A. Maantay, S. McLafferty (eds.), *Geospatial Analysis of Environmental Health*,
Geotechnologies and the Environment 4, DOI 10.1007/978-94-007-0329-2_24,

generally have higher fine particle fractions). Information generated by rigorous BME analysis and mapping across space-time constitutes valuable input to health management and decision-making under conditions of uncertainty.

Keywords Spatiotemporal analysis · BME · $PM_{2.5}$ · Exposure · $PM_{2.5}/PM_{10}$ ratio

24.1 Introduction

Many studies have shown a significant association between human exposure to Particulate Matter (PM) and population health effects (premature mortality, respiratory and cardiovascular diseases, emergency room visits, and systemic inflammation). Particulate matter (PM) refers to a suspension of solid, liquid or a combination of solid and liquid particles in the air Wilson et al., (2005). Fine PM particles ($PM_{2.5}$, particulate matter particles with aerodynamic diameter $\leq$ 2.5 μm) are believed to be more dangerous than PM_{10} because fine particles are easier inhaled and accumulated deeply into human lungs (Dockery et al., 1993; Pope 2000a, b; Pope et al., 2004). Taiwan Environmental Protection Agency (TWEPA) has set National Ambient Air Quality Standards (NAAQS) in 1992 for six criteria pollutants, including PM, ozone, nitrogen dioxide, sulfate dioxide, carbon monoxide, and lead. These pollutants are considered potentially the most harmful to human health and the environment. Among them, two PM indicators, total suspended particle (TSP) and PM_{10} (PM particles with aerodynamic diameter $\leq$ 10 μm), are used by TWEPA to assess the exposure level. TWEPA has regularly recorded the ambient pollutants and meteorological covariates throughout the island since September, 1993. However, the TWEPA $PM_{2.5}$ monitoring network did not begin to operate regularly until 2004.

From the analysis of air quality data during 1994–2003, Chang and Lee (2008) identified the three main contributors to air pollution in the Taipei area, which are traffic emissions, photochemical pollution, e.g. ozone, and transboundary pollution, e.g. sand storm. In addition, the meteorological condition (wind speed, wind direction, precipitation and temperature) are important contributors to the general air quality trend (e.g., PM concentration), as well. The bad air quality during wintertime can often be characterized by high atmospheric pressure, low wind speed, less precipitation. Chen et al., (1999; Tsai et al., 2007; Yang, 2002). Recent studies have indicated that the assumption of homogeneous intra-urban concentrations can lead to errors in long-term exposure assessment (Wilson and Zawar-Reza, 2006). To account for the spatial heterogeneity of air pollution, studies applying GIS and spatial analysis to exposure assessment are rapidly growing due to its capability to associate spatial information, e.g. land use, with health data and account for the uncertainty of the dataset collected in complex space-time environment Jerrett et al., (2005b; Liao et al., 2005). The application of GIS to environmental health studies can possibly investigate the health risks at specific space/time locations Jerrett et al., (2005a). Given that traffic emissions dominate air quality in Taipei, understanding the spatiotemporal pattern of air quality at the scale of the Taipei city is an important prerequisite for rigorous risk assessment of human health and ecological systems. Chen and Mao (1998) showed that significant fluctuations of PM_{10} concentrations exist among the eighty-four PM_{10} samples collected at different locations and times within Taipei city. At the same spatial location $\boldsymbol{s}$, e.g., at time t the PM concentration value can be 4–8 times the PM value at a different time $t' \neq t$. On the other hand, at the same time t, the concentration value at location $\boldsymbol{s}$ can be 9–10 times the value at a different location $s' \neq s$ (note that the monitoring sites along the roadside have the greatest level of PM concentration, which is due mainly to traffic emissions (Chen et al., 1999)).

In addition to the spatiotemporal variation of PM measurements, the PM particle size distribution can also vary across space and time. In general, fine particles are anthropogenic (e.g., industrial combustion and traffic emission). Coarse particles are mainly formed by mechanical processes, e.g., wind erosion and mineral dust, and significant portions of these particles originate in natural systems. The $\frac{PM_{2.5}}{PM_{10}}$ ratio is usually used to characterize the PM particle size distribution. The average ratio of $\frac{PM_{2.5}}{PM_{10}}$ in northern Taiwan is about 54–59% (Chen et al., 1999). However, the ratio values observed within Taipei change significantly from one location to another; e.g., the ratio is about 0.82 around the incinerator, 0.68 in high traffic areas, and 0.57 in downtown areas (Li and Lin, 2002). For comparison purpose, the average $\frac{PM_{2.5}}{PM_{10}}$ ratio in Taipei, Taiwan, is at a similar level as that of Los Angeles, USA; and it is higher than the corresponding ratio in Phoenix, USA (Li and Lin, 2002).

In order to obtain a more comprehensive and informative understanding of the spatiotemporal $PM_{2.5}$ distribution, the present study investigates the improvement of spatiotemporal $PM_{2.5}$ prediction by incorporating relevant PM information about PM_{10} and TSP. The spatiotemporal ratios of $\frac{PM_{2.5}}{PM_{10}}$ and $\frac{PM_{2.5}}{TSP}$ are considered as the indicators of the spatial emission patterns varying across time, which are closely associated with the local land use and meteorological conditions. We integrated PM_{10} and TSP data with $PM_{2.5}$ measurements by using the ratios to estimate the monthly $PM_{2.5}$ concentrations over Taipei city during 2004–2007. The spatiotemporal mapping of $PM_{2.5}$ and ratios were performed in Quantum GIS with the QtBME toolbox (Ku, 2010), in which Bayesian maximum entropy (BME) method (Christakos, 1990, 2000) is implemented to enhance the spatiotemporal functionality of the Quantum GIS software. The cross-validation was implemented to compare the $PM_{2.5}$ prediction performance between (*a*) the dataset of $PM_{2.5}$-only data and (*b*) the datasets of $PM_{2.5}$, PM_{10} and TSP observations.

24.2 Materials

24.2.1 Study Area

Taipei, located in northern Taiwan, is the largest metropolitan area in Taiwan, including Taipei city and Taipei county, with the vehicle density as high as over 6,000 vehicles per km^2. Except for traffic emissions, the three incineration plants are the major stationary sources in the area (Chang and Lee, 2007). Taipei is the second largest basin in Taiwan which is bounded by mountains, i.e. Yangming mountains to the north, Linkou mesa to the west, and ridge of Snow mountains to the southeast (Fig. 24.1).

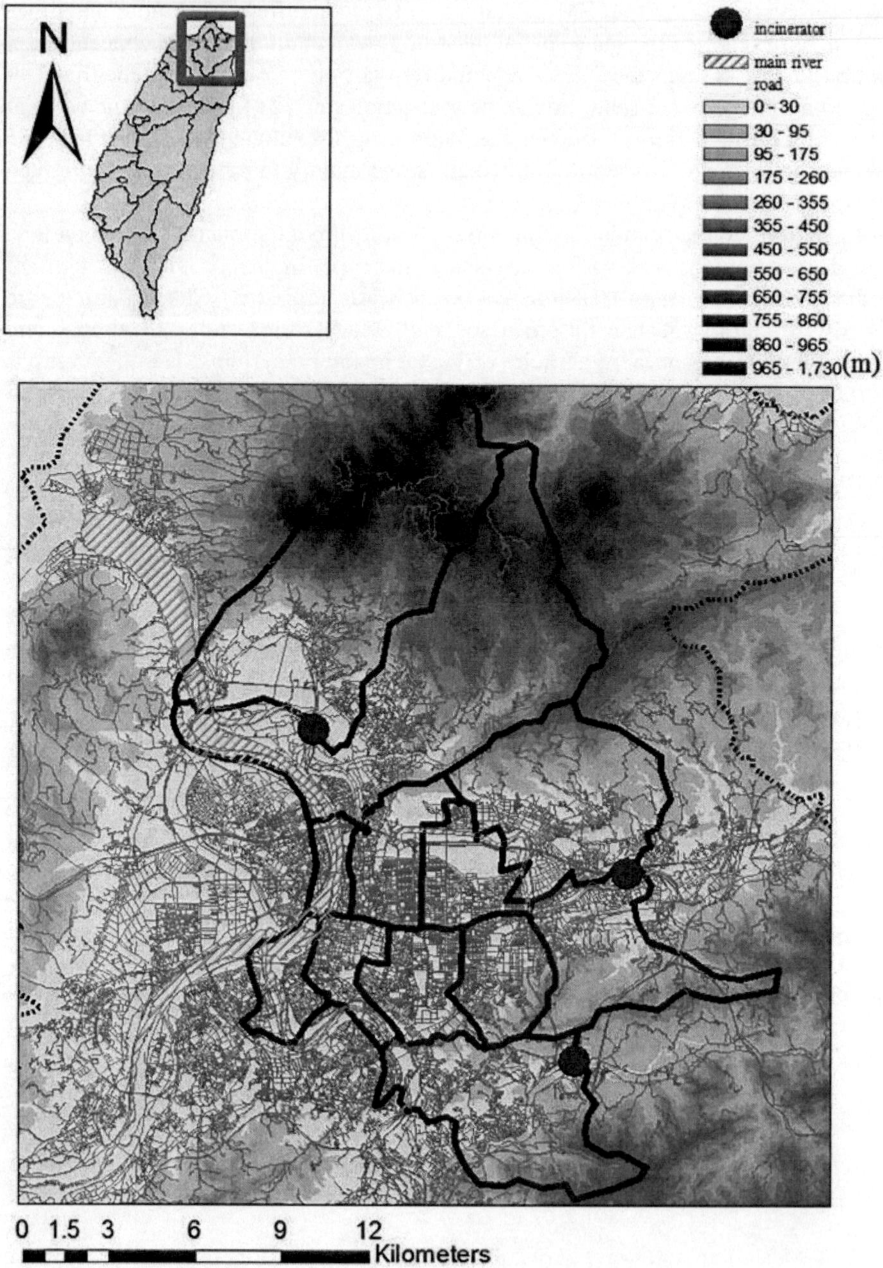

Fig. 24.1 Topographic map of Taipei area/city (*areas bounded by outer/inner solid lines*)

24.2.2 Data

TWEPA has regularly recorded the ambient pollutants and meteorological covariates throughout the Taiwan island since September, 1993. Among them, there are 23 stations located within Taipei city. However, the TWEPA network did not systematically record $PM_{2.5}$ until 2004. Both PM_{10} and $PM_{2.5}$ hourly observations from TWEPA during 2004–2007 are included in this analysis. In addition to TWEPA, Department of Environmental Protection at local governments of Taipei city and Taipei county (TPEDEP) also both independently collect PM data during the study period. Both local governments regularly records TSP data monthly from total of thirty-nine stations. However, PM_{10} were only observed by Taipei city government daily from eight stations. Table 24.1 shows the sampling frequency and instruments of the stations used in this study. As seen in Fig. 24.2, the combination of the PM monitoring networks from central and local governments, i.e. TWEPA and TPEDEP, can provide much better spatial coverage of the study area. To obtain the monthly maps of $PM_{2.5}$, data of $PM_{2.5}$, PM_{10} and TSP from institutes were all re-organized into monthly data following the "three-fourths" criterion that the monthly data can be derived only if the number of daily or hourly measurements should cover over three-fourths of the month; otherwise, the monthly estimation is considered as the missing data (Yu et al., 2009b).

24.3 Method

24.3.1 Brief Review of the BME Method

Since the late 1980s, the implementation of BME theory has allowed the study of attribute distributions in a composite space-time domain, accounting for different kinds of general and site-specific knowledge bases; it does not make any restrictive or unrealistic assumptions (linearity, Normality, independency etc.), and it provides a complete probabilistic characterization at each space-time point instead of a single value. The BME conceptual framework and methods avoid the serious faults of many data-driven techniques of statistical air pollution analysis Dominici et al., (2003a, b; Samet et al., 2000; Smith et al., 2003). Beyond environmental pollution

Table 24.1 Sampling methods of the particulate matter stations

	$PM_{2.5}$	PM_{10}	TSP
Frequency	Continuous automatic monitoring (hourly average value)	Continuous automatic monitoring (hourly average value)	Continuous automatic monitoring (monthly average value)
Instruments	Beta attenuation monitor	Beta attenuation monitor/Tapered element oscillating microbalance (TEOM)	Dusttrak aerosol sampler

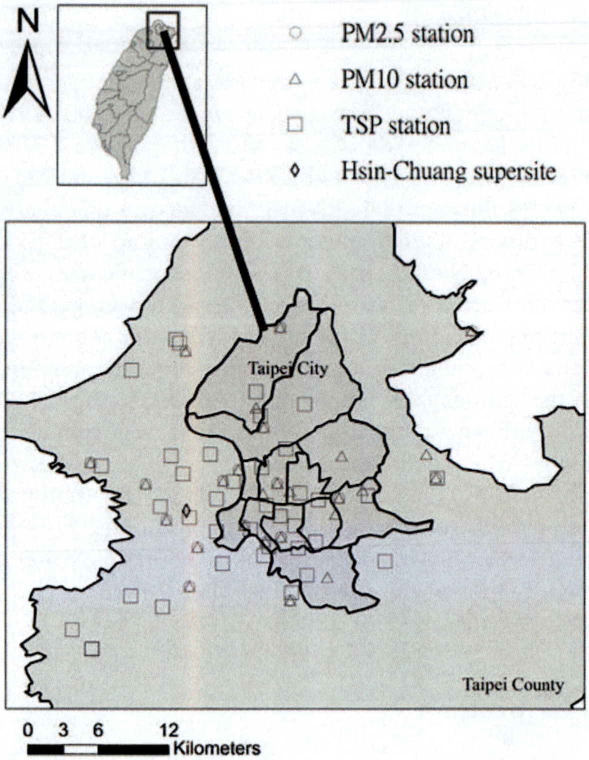

Fig. 24.2 Air quality monitoring stations in Taipei (*circle*: TWEPA $PM_{2.5}$ stations; *triangle*: TWEPA PM_{10} stations; *square*: TPEDEP PM_{10} stations; *solid circle*: TPEDEP TSP stations). The areas of darker and lighter colors are the Taipei City and Taipei county, respectively

and human exposure, in recent times the BME methods have been used in many other scientific and engineering disciplines with considerable success (Bogaert, 2002; Bogaert and D'Or, 2002; Christakos et al., 2005; Douaik et al., 2004; Kolovos et al., 2002; Lee et al., 2008; Orton and Lark, 2007; Serre and Yu, 2003; Serre, 1999; Yu et al., 2007b).

In BME, the air pollution attributes ($PM_{2.5}$, PM_{10}, and TSP) are mathematically represented in terms of spatiotemporal random fields (S/TRF; Christakos 1992). Let $X_{\mathbf{p}} = X_{\mathbf{s},t}$ denote a S/TRF of a pollution attribute; the vector $\boldsymbol{p} = (\boldsymbol{s}, t)$ denotes a spatiotemporal point (s is geographical location and t is time). The model is viewed as the collection of all physically possible realizations of the attribute we seek to represent mathematically under conditions of uncertainty. The S/TRF model is fully characterized by its probability density function (pdf), f_{KB}, where the subscript KB denotes the 'knowledge base' used to construct the pdf. In particular, BME considers a distinction between:

a. General or core KB, denoted as *G*-KB, it includes physical and biological laws, primitive equations, and theoretical models for statistical moments. For the air pollution studies, *G*-KB may include theoretical space-time dependence models (mean, covariance, variogram, generalized covariance, multiple-point statistics, and heterogeneity orders) of the air pollution attribute $X_{\boldsymbol{p}}$ (Kolovos et al., 2002; Porcu et al.,

2008). Among them, the mean and covariance (variogram) are the most commonly used functions in air pollution studies.

b. Site-specific KB, denoted as *S*-KB, which includes, (*b*1) exact numerical values of the attribute across space-time (hard data); and (*b*2) uncertain information, like interval attribute values (there is not a unique value available at a space-time point but, instead, an interval of possible values), and probability functions of the possible attribute values (the datum at the specified space-time location has the form of a probability distribution). The *S*-KB is associated with the vector $\boldsymbol{p}_{\text{data}}$ of space-time points where site-specific information is available. In atmospheric PM modelling studies, the χ_{data} of the *S*-KB can include both hard and soft datasets, i.e., $\boldsymbol{\chi}_{data} = (\boldsymbol{\chi}_{\text{hard}}, \boldsymbol{\chi}_{\text{soft}}) = (\chi_1, \ldots, \chi_m)$ obtained at points $\boldsymbol{p}_i$ $(i = 1, 2, \ldots, m)$ of the specified geographical area. In this case, the $\boldsymbol{\chi}_{\text{hard}} = (\chi_1, \ldots, \chi_{m_h})$ denotes hard data, i.e. exact PM measurements at points $\boldsymbol{p}_i$ $(i = 1, 2, \ldots, m_h)$. In other words, the $\boldsymbol{\chi}_{\text{hard}}$ occurs with probability one. And the vector $\boldsymbol{\chi}_{\text{soft}} = (\chi_{m_h+1}, \ldots, \chi_m)$ denotes soft data at points $\boldsymbol{p}_i$ $(i = m_h + 1, \ldots, m)$ that may include uncertain evidence and other sorts of secondary information. For illustration, assume that 32 exact PM_{10} observations are available at the space-time points $\boldsymbol{p}_{\text{hard}} = (\mathbf{p}_1, \ldots, \mathbf{p}_{32})$, i.e. $X_{\mathbf{p}_1} = 5.1$, ... , $X_{\mathbf{p}_{32}} = 9.3$ (in suitable units); and that 55 uncertain PM_{10} data are available at the points $\boldsymbol{p}_{\text{soft}} = (\mathbf{p}_{33}, \ldots, \mathbf{p}_{87})$, say of the interval form $3.2 < X_{\mathbf{p}_{33}} < 4.1$, ... , $5.2 < X_{\mathbf{p}_{87}} < 6.4$ (in suitable units). This sort of site-specific information is mathematically expressed by $P_S[X_{\mathbf{p}_1} = 5.1\ , \ \ldots\ , \ X_{\mathbf{p}_{32}} = 9.3\ ] = 1$ and $P_S[3.2 < X_{\mathbf{p}_{33}} < 4.1\ , \ \ldots\ , \ 5.2 < X_{\mathbf{p}_{87}} < 6.4\ ] = 1$, respectively. More generally, assume that at point $\mathbf{p}_{24}$ the uncertain datum is expressed by the density function $f_S(\mathbf{p}_{24})$; then, $P_S[X_{\mathbf{p}_{24}} < \chi] = \int_{-\infty}^{\chi} d\chi\, f_S(\mathbf{p}_{24})$. For several examples, see (Wibrin et al., 2006; Yu et al., 2007b).

The total KB is denoted by $K = G \cup S$, i.e. it includes both the general and the site-specific KB. BME method assimilates the KBs by the principle of maximum entropy and operational Bayesian method and generates the spatiotemporal distribution of the attributes in probabilistic forms (Christakos, 2000, 2002). The fundamental BME equations are as follows (for technical details, see (Christakos, 2000; Christakos et al., 2005))

$$\left.\begin{aligned} &\int d\chi\, (\mathbf{g} - \overline{\mathbf{g}})\, e^{\mu^T \mathbf{g}} = 0 \\ &\int d\chi\, \xi_S\, e^{\mu^T \mathbf{g}} - A f_K = 0 \end{aligned}\right\}, \tag{1}$$

where $\boldsymbol{g}$ is a vector of g_α-functions $(\alpha = 1, 2, \ldots)$ that represents stochastically the *G*-KB under consideration (the bar denotes statistical expectation), $\boldsymbol{\mu}$ is a vector of μ_α-coefficients that depend on the space-time coordinates and is associated with $\boldsymbol{g}$ (i.e., the μ_α express the relative significance of each g_α-function), the ξ_S represents the S-KB available, A is a normalization parameter, and f_K is the pollutant or exposure pdf at each space-time point where a space-time prediction (or estimation) is sought – the subscript K means that f_K is based on the blending of the core and site-specific KB. Herein we use the terms “prediction” and “estimation” interchangeably to denote the stochastic approximation of unknown PM concentrations across space-time. The g and ξ_S are the inputs in Eq. (1), whereas the unknowns are the $\boldsymbol{\mu}$ and f_K across space-time. As shown Eq. (1), BME method is a nonlinear estimator without any distributional assumption of the interested attributes. Unlike the common spatiotemporal interpolation methods, e.g. kriging method, they usually assume the linearity and Gaussian-distributed among the space-time attributes.

The pdf f_K in Eq. (1) represents the pdf of the attribute values at each prediction (estimation) point $\boldsymbol{p}_k$ in light of the total K-KB. Given f_K at each p_k, different estimates of pollutant concentrations can be derived at the nodes of the mapping grid (most probable, error minimizing etc. estimates), depending on the objectives of the study. The error minimizing prediction (BMEmean), e.g., is given by $\overline{X_{pk}} = \int_{-\infty}^{\infty} d\chi_k\, \chi_k f_K$ at each grid node $\boldsymbol{p}_k$, and the corresponding BME estimation error variance (BMEvar) is as follows, $\sigma^2_{X_{pk}} = \int_{-\infty}^{\infty} d\chi_k(\chi_k - \overline{X_{pk}})^2 f_K$. Depending on the situation, the BMEmode, BMEmedian and other kind of attribute estimates across space-time can be also calculated from f_K, which does not have a Gaussian shape, in general. The BME method is routinely implemented by means of the several publicly available software libraries, such as SEKS-GUI (Kolovos et al., 2006; Yu et al., 2007b) and QtBME (Ku, 2010), which is a toolbox in the open-sourced Quantum GIS software. In visualization terms, one can generate a series of highly informative space-time maps of the attributes of interest and the associated uncertainties (e.g., see later, Figs. 24.4, 24.5, and 24.6).

24.3.2 BME Spatiotemporal Modelling of $PM_{2.5}$

Let the S/TRF $X_p = X_{s,t}$ denote $PM_{2.5}$ concentration and the $Y_p = Y_{s,t}$ represent PM_{10} or TSP concentrations across space-time (Yu et al., 2009b). The $\frac{PM_{2.5}}{PM_{10}}$ or $\frac{PM_{2.5}}{TSP}$ ratios can be represented as $r_p = X_p\, Y_p^{-1}$, where t is associated with the time period 2004–2007. From existing evidence, the r_p-values change spatially and temporally but the distribution of r_p is yearly-invariant. This assumption implies that the space-time patterns of natural and human pollution sources and the urban and suburban land uses do not change significantly from year to year.

The $r_{s,t}$ can be estimated monthly at every PM monitor station (PM_{10}, $PM_{2.5}$ and TSP stations) based on the recorded or estimated PM data, which can be hard data associated with the existing observations; or soft information (probabilistic or interval types), which is generated by using univariate BME (U-BME) estimation of PM values at missing data locations. In other words, U-BME estimation makes available the three PM measures (in the form of either hard data or soft information) at all monitoring stations, i.e. stations of TWEPA and TPEDEP. Note that in the univariate case, the ratio estimates use only hard data ($PM_{2.5}$, PM_{10}, and TSP observations). Based upon the elaborated PM datasets, soft information of r_p can be obtained in terms of the equations in Table 24.2. Assuming independence between the ratios and the concentrations of PM_{10} or TSP, the spatiotemporal posterior pdf, $f_K(\chi_k)$, of $PM_{2.5}$ at any space-time estimation point $\boldsymbol{p}_k$ can be expressed as

Table 24.2 Deriving soft information of r_p

$PM_{2.5}$	PM_{10} or TSP	Ratio
X_p	Y_p	r_p
$X_p = \chi$	$Y_p = \psi$	$r_p = \chi\,\psi^{-1}$
$X_p = \chi$	$Y_p \sim fs(\psi)$	$r_p \sim fs(\chi\, r_p^{-1})\|-\chi\, r_p^{-2}\|$, where χ is a constant
$X_p \sim fs(\chi)$	$Y_p = \psi$	$r_p \sim fs(\chi\,\psi)\|\psi\|$
$X_p \sim fs_1(\chi)$	$Y_p \sim fs_2(\psi)$	$r_p \sim \int_\psi d\psi\, fs(r_p\,\psi, \psi)\|\psi\|$

Note: $fs(\chi)$ and $fs(\psi)$ can be any pdf representing soft information; in this study, it is the knowledgable f_K's from the BME univariate estimations.

$$f_K(\chi_k) = \int_0^1 dr_k\, r_k^{-1} fs_1(r_k) fs_2(\chi_k r_k^{-1}) \tag{2}$$

where r_k are the estimated ratios at point $\boldsymbol{p}_k$; S_1 and S_2 denote the soft information of ratios and secondary PM data (i.e., PM_{10} and TSP), respectively.

24.4 Results

As shown in Fig. 24.2, the $PM_{2.5}$ and PM_{10} data from TWEPA are mostly collocated, which can provide the required hard information for the $\frac{PM_{2.5}}{PM_{10}}$ ratio. In order to have a more comprehensive spatial coverage of ratio values, spatiotemporal $PM_{2.5}$ estimates were generated by the BME method with space-time $PM_{2.5}$ data at the PM_{10} and TSP stations (obtained from the TPEDEP network). Using the new $PM_{2.5}$ values along with the existing PM_{10} observations, new soft information about the $\frac{PM_{2.5}}{PM_{10}}$ ratios at the PM_{10} stations can complement the spatial coverage of the original $\frac{PM_{2.5}}{PM_{10}}$ dataset. To characterize the spatiotemporal dependence among the $\frac{PM_{2.5}}{PM_{10}}$ ratios, the nested non-separable covariance model below is used as part of the core KB (Fig. 24.3),

$$c(h,\tau) = c_0\, e^{-3\left(\frac{h^2}{a_{r1}^2}+\frac{\tau^2}{a_{\tau 1}^2}\right)} + c_1 e^{-3\left(\frac{h^2}{a_{r2}^2}+\frac{\tau^2}{a_{\tau 2}^2}\right)}, \tag{3}$$

where $I(a_{r2})$ is a 0–1 indicator (it is zero when $h > a_{r2}$); $[c_0,\ c_1] = [0.0067,\ 0.003]$, $[a_{r1},\ a_{r2},\ a_{t1},\ a_{t2}] = [10\ \text{Km}, 6\ \text{Km}, 550\ \text{mo}, 6\ \text{mo}]$ –"Km" means kilometers and "mo" means months. By properly integrating the hard and soft ratio data, the BME method generates the spatiotemporal distribution of the $\frac{PM_{2.5}}{PM_{10}}$ ratios every month during the period 2004–2007. Figure 24.4 shows the spatial distributions of $\frac{PM_{2.5}}{PM_{10}}$ from January to October of 2006, considered in 3-month intervals.

Similarly, in order to obtain the $\frac{PM_{2.5}}{TSP}$ ratios, the BME method was used to estimate $PM_{2.5}$ and TSP concentrations at the monitoring stations of TSP and $PM_{2.5}$, respectively. The soft data of the $\frac{PM_{2.5}}{TSP}$ ratios can be accordingly generated by means of the equations of Table 24.2. A nested non-separable model was used for the spatiotemporal covariance of $\frac{PM_{2.5}}{TSP}$,

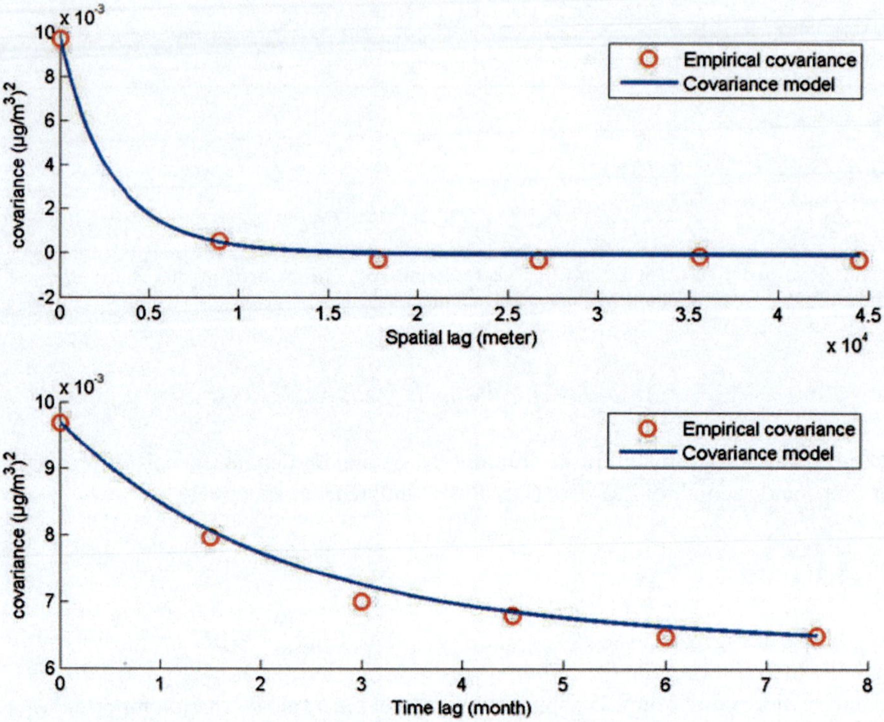

Fig. 24.3 Spatiotemporal covariance of $\frac{PM_{2.5}}{PM_{10}}$ ratios in **a** space ($c(h, \tau = 0)$) and **b** time ($c(h = 0, \tau)$), where *circles* and *lines* denote empirical covariance and theoretical covariance model

$$c(h,\tau) = c_0\, e^{-3\left(\frac{h^2}{a_{r1}^2}+\frac{\tau^2}{a_{\tau 1}^2}\right)} + c_1\left(1-\frac{3}{2}\frac{h}{a_{r2}}+\frac{1}{2}\left(\frac{h}{a_{r2}}\right)^3\right) I(a_{r2}) e^{-\frac{3\tau}{a_{\tau 2}}}, \qquad (4)$$

where $I(a_{r2})$ is a 0–1 indicator (it is zero when $h > a_{r2}$); $[c_0,\ c_1] = [0.004,\ 0.0079]$, $[a_{r1},\ a_{r2}] = [20, 15]$ Km, and $[a_{\tau 1},\ a_{\tau 2}] = [5,\ 110]$ months. Figure 24.5 shows the spatiotemporal $\frac{PM_{2.5}}{TSP}$ ratios in the same months as in Fig. 24.4. Following Eq. (2), soft $PM_{2.5}$ data of a variety of probabilistic types can be generated at every PM_{10} and TSP station over time.

The more informative spatiotemporal distribution of $PM_{2.5}$ can be obtained by the BME method that assimilates the soft and hard $PM_{2.5}$ data (Fig. 24.6). The $PM_{2.5}$ distribution is characterized by the following nested, non-separable spatiotemporal covariance,

$$c(h,\tau) = c_0 e^{-3\left(\frac{3h}{a_{r1}}+\frac{3\tau}{a_{\tau 1}}\right)} + c_1\left(1-\frac{3}{2}\frac{h}{a_{r2}}+\frac{1}{2}\left(\frac{h}{a_{r2}}\right)^3\right) I(a_{r2}) e^{-\frac{3\tau}{a_{\tau 2}}}, \qquad (5)$$

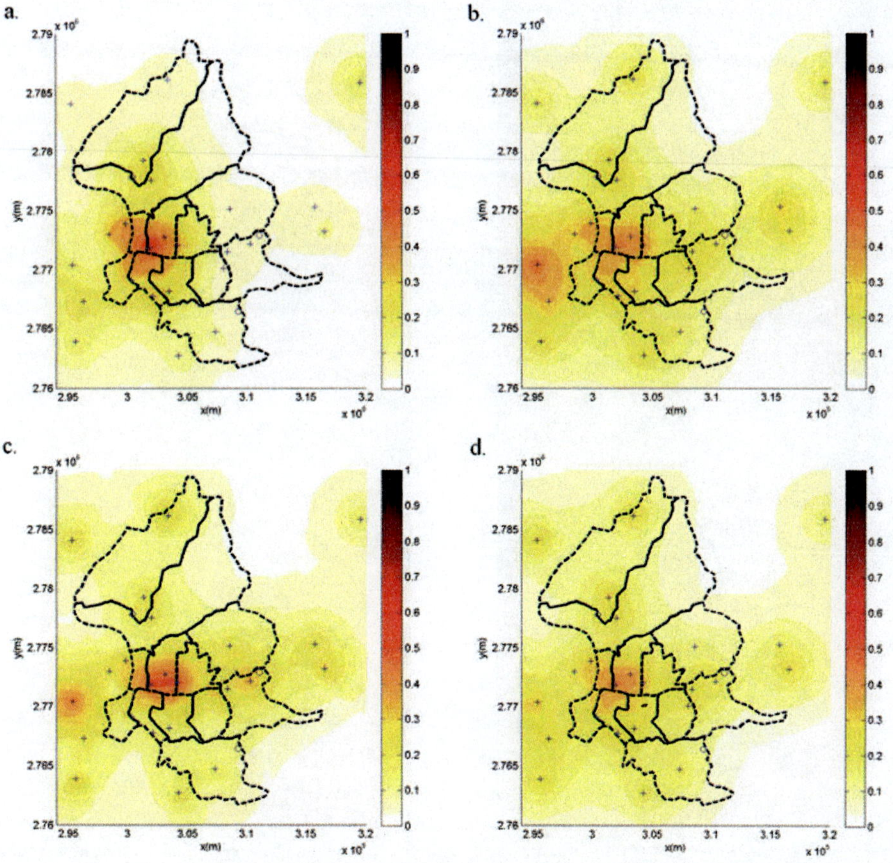

Fig. 24.4 Spatiotemporal distributions of $\frac{PM_{2.5}}{PM_{10}}$ ratios in **a** January, 2006, **b** April, 2006, **c** July, 2006, and **d** October, 2006 (unit: $\mu g/m^3$)

where $I(a_{r2})$ is a 0–1 indicator (it is zero when $h > a_{r2}$);$[c_0, c_1] = [32.9, 57.1]$ and $[a_{r1}, a_{r2}, a_{\tau 1}, a_{\tau 2}] = [80\text{ Km}, 10\text{ Km}, 550\text{ mo}, 8\text{ mo}]$. All the spatiotemporal fitting was performed by an automatic scheme discussed in (Yu et al., 2009a).

Cross-validation was performed at every TWEPA station of $PM_{2.5}$ data monitoring. Figure 24.7 shows the comparison between the $PM_{2.5}$ estimates vs. observations at the stations of Sihlin, Guting, Xizhi, and Cailiao. In addition, cross-validation was performed when only $PM_{2.5}$ data were used for prediction. The comparison of the cross-validation results are shown in Table 24.3.

24.5 Discussion

This study uses the BME approach of spatiotemporal statistics to integrate observations of several PM measures in the prediction (estimation) of fine particulate

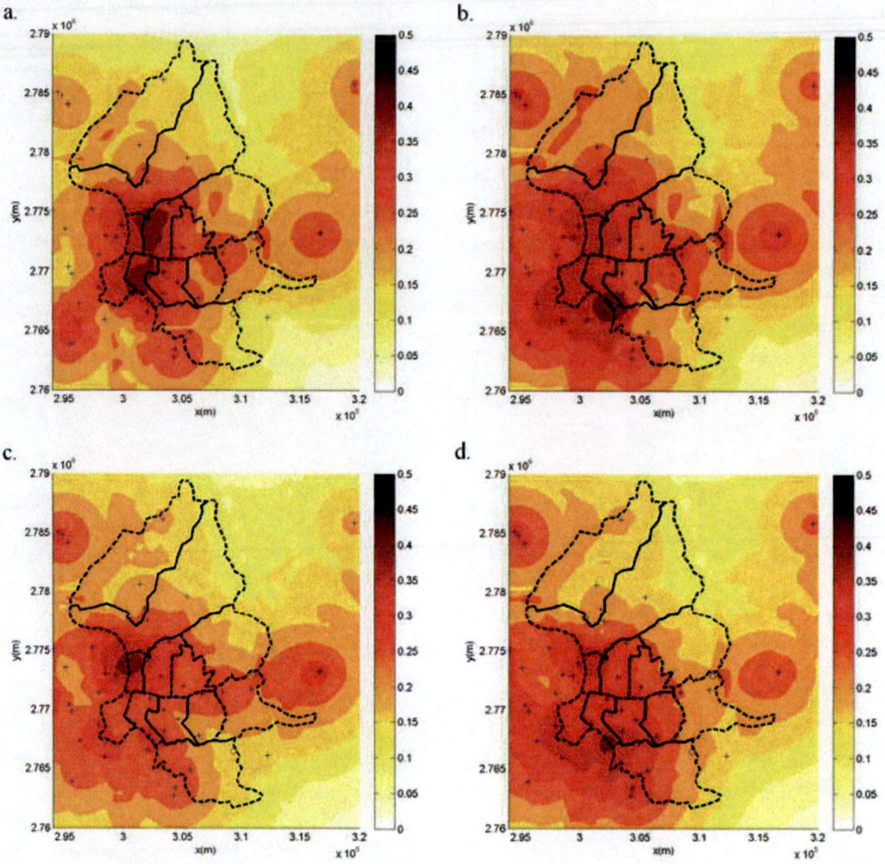

Fig. 24.5 Spatiotemporal distributions of $\frac{PM_{2.5}}{TSP}$ ratios in **a** January, 2006, **b** April, 2006, **c** July, 2006, and **d** October, 2006 (unit: $\mu g/m^3$)

matter concentrations across space-time. This study accounts for different kinds of core and site-specific knowledge bases, without making any restrictive or unrealistic assumptions (linearity, Normality, independency etc.), which are some of the drawbacks that characterize the data-driven statistical models used in other environmental pollution and human exposure studies (Dominici et al., 2003a).

Most of the $PM_{2.5}$ monitoring networks worldwide were established on a systematic basis during this decade when the importance of human exposure to $PM_{2.5}$ and its health effects began to be appreciated. The present study presents an effective way to improve the resolution and $PM_{2.5}$ prediction in composite space-time domain. It involves the original $PM_{2.5}$ monitoring network, and can be useful in the prediction of $PM_{2.5}$ concentration at space-time points where or when the $PM_{2.5}$ observations are absent or limited. In addition to $PM_{2.5}$ prediction and mapping, the modeling of the spatiotemporal ratios $\frac{PM_{2.5}}{PM_{10}}$ and $\frac{PM_{2.5}}{TSP}$ can provide valuable insight

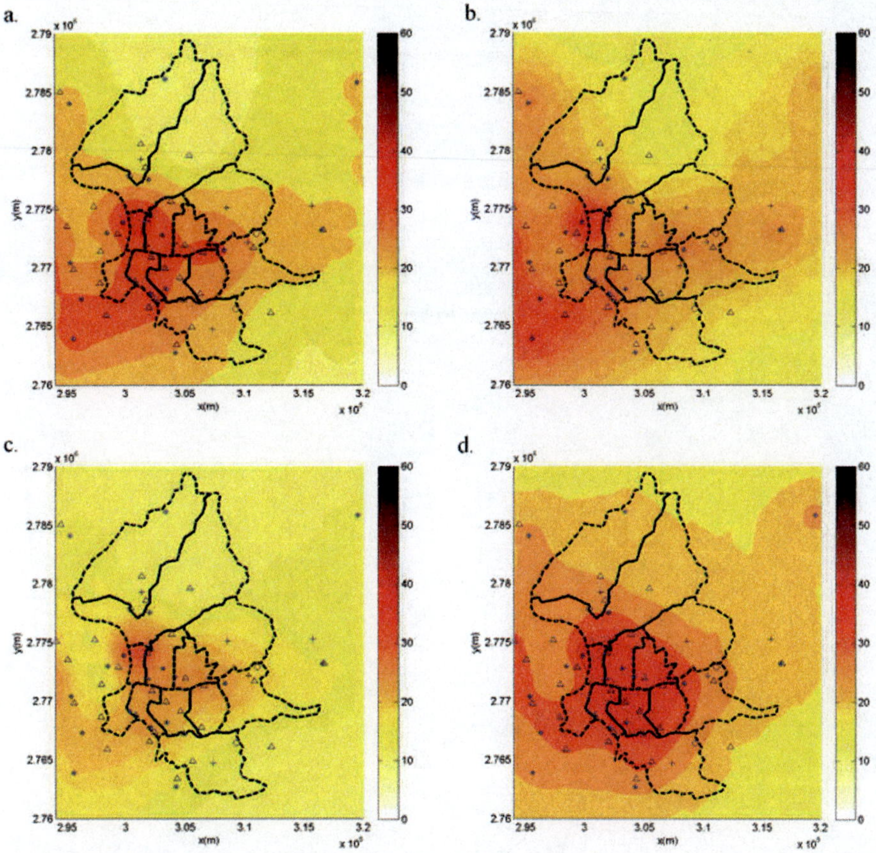

Fig. 24.6 Spatiotemporal distributions of $PM_{2.5}$ concentrations in **a** January, 2006, **b** April, 2006, **c** July, 2006, and **d** October, 2006 (unit: $\mu g/m^3$)

about the underlying PM patterns and mechanisms across space and time. More specifically, the higher ratios of $\frac{PM_{2.5}}{PM_{10}}$ implies that the higher portions of particulate matters are generated by anthropogenic emissions, such as traffic and industrial emissions. The causes of high ratios of $\frac{PM_{2.5}}{TSP}$ can be more complicated which can result from either the high $PM_{2.5}$ concentration due to the anthropogenic emissions or the low TSP concentration owing to certain geographic or atmospheric reasons.

As shown in Eq. (3), the $\frac{PM_{2.5}}{PM_{10}}$ exhibits two processes with different space-time ranges, which characterize the spatiotemporal patterns of the PM size distribution over the Taipei area. The two space-time ranges, [10 km, 155 months] and [6 km, 6 months], show that the $\frac{PM_{2.5}}{PM_{10}}$ can be dominated by the local emissions with the spatial extent about 10 and 6 km wide while distinct temporal ranges which imply the long-term pattern and seasonal variations of the ratios. The spatial ranges of the ratios generally correspond to the ranges of traffic emissions and changes of land use patterns within the city. It implies the spatial distribution of $\frac{PM_{2.5}}{PM_{10}}$ is persistent

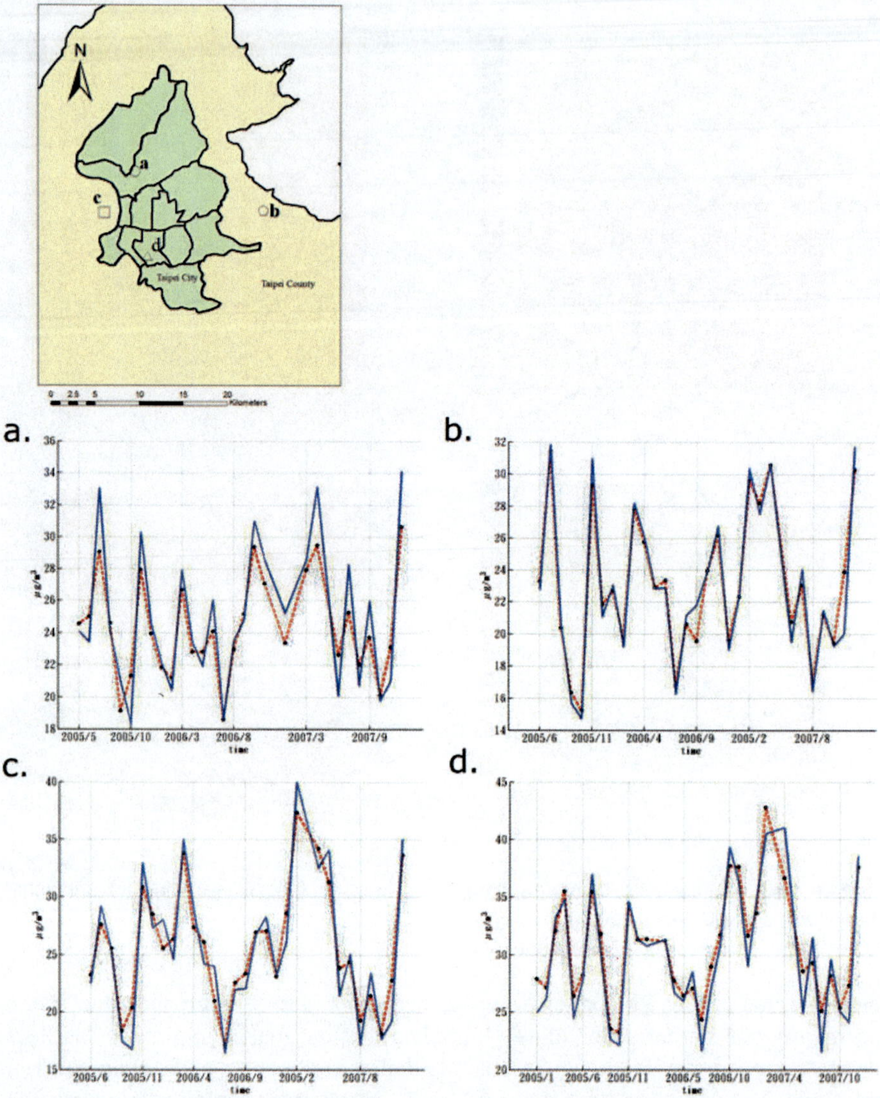

Fig. 24.7 The comparison between $PM_{2.5}$ observations (*solid line*) and estimations (*dash line*) at the four $PM_{2.5}$ stations **a** Shilin station, **b** Xizhi station, **c** Cailiao station, and **d** Guting station

over time with mild seasonal variations. This fact is shown in Fig. 24.4 in which the hotspot of higher $\frac{PM_{2.5}}{PM_{10}}$ values is located at the southwest of Taipei area, where is the area of major transportation hub of the city, i.e. Taipei main station. The shape of the $\frac{PM_{2.5}}{PM_{10}}$ hotspot is generally elliptic with longer East-West axis along the civic boulevard, which connects the major commercial and industrial areas of the city.

Table 24.3 Results of cross-validation (observed-estimated)

Data availability	Mean ($\mu g/m^3$)	Standard deviation ($\mu g/m^3$)	Median ($\mu g/m^3$)	Min value ($\mu g/m^3$)	Max value ($\mu g/m^3$)
$PM_{2.5}$ + PM_{10} and TSP	2.115	1.845	–0.467	–11.516	10.556
$PM_{2.5}$ only	3.242	2.469	–0.456	–11.058	12.917

Similar to the ratios of $\frac{PM_{2.5}}{PM_{10}}$, the spatiotemporal patterns of $\frac{PM_{2.5}}{TSP}$ ratios are also dominated by two local processes with similar spatial ranges, i.e. about 15 and 20 km, and distinct temporal ranges that represent the long-term and seasonal changes of the ratios respectively. It implies the local emissions are the major contributing factor to the spatiotemporal characteristics of size distribution of PM. Figure 24.5 shows that the high $\frac{PM_{2.5}}{TSP}$ values are observed at the areas which are major connections to commute in and out of the city, i.e. the location where is the highway exit from the major freeway (Sun-Yat-Sen freeway) at the northern Taipei, and the area around Fu-He bridge which connects Yong-He city (the major residential area) and Taipei, as well as the areas along the Da-Shui river by which Huan-He and Shui-Yuan highways are surrounded. In general, the high $\frac{PM_{2.5}}{TSP}$ generally exhibits at the major roads which are either at the boundary (i.e. Dan-shui river) of the Taipei downtown or connect the surrounding cities to Taipei by crossing the river. These major road connections generally have relatively high traffic volume with higher $PM_{2.5}$ emissions, yet relatively low TSP values since Taipei city is surrounded by rivers. During the winter, the concentration of high $PM_{2.5}$ plays more important role to $\frac{PM_{2.5}}{TSP}$ patterns that the high $\frac{PM_{2.5}}{TSP}$ follows Jian-Kuo elevated highway, the most important road connecting city north and south.

The space-time maps of Fig. 24.6 present $PM_{2.5}$ concentrations during 2006 in the downtown Taipei area in which higher density of commercial activities are exhibited, and at the west boundary between Taipei city and Taipei county, where the high traffic density over time from the commuters between the two areas. The $PM_{2.5}$ concentrations are generally reduced in summer over the entire study area which contrasts the results of temporal $PM_{2.5}$ distribution in North Carolina (Yu et al., 2007a). This fact can be due to the summer meteorological conditions in Taipei when the higher frequency of precipitation, higher speed of wind, and lower atmospheric pressure contribute to the lower $PM_{2.5}$ concentrations (Tsai et al., 2007). The spatiotemporal mapping of $PM_{2.5}$ as well as the ratios of $\frac{PM_{2.5}}{PM_{10}}$ and $\frac{PM_{2.5}}{TSP}$ show that the areas of high $PM_{2.5}$ closely correspond with the areas of high $\frac{PM_{2.5}}{PM_{10}}$. It implies that the spatial variations of $PM_{2.5}$ are primarily dominated by traffic emissions. The traffic centers have high $PM_{2.5}$ and $\frac{PM_{2.5}}{PM_{10}}$ over the entire study period, even in summer.

For comparison purposes, the cross-validation results in Table 24.3 are used next. The BME estimation of $PM_{2.5}$ by assimilating additional information from PM_{10} and TSP can improve the spatiotemporal prediction of $PM_{2.5}$ concentration over

Taipei during the study period, in the sense that the reduction of the average, standard deviation, maximum, and minimum of the prediction (estimation) errors (i.e., "observed- estimated" $PM_{2.5}$). Figure 24.7, shows a comparison at the four selected stations that represent different regions of Taipei, i.e. the commercial area (Sihlin), the commercial-residential mixture area (Guting and Xizhi), and the commercial-industrial-residential area (Cailiao), show the good agreement of $PM_{2.5}$ predictions over space and time by the BME approach. The spatiotemporal distribution of $PM_{2.5}$ can be used to identify where and when the $PM_{2.5}$ concentrations can be harmful to the health of people living in Taipei. Since the $PM_{2.5}$ standards are still not available in Taipei, in Fig. 24.8, it shows the spatial and temporal distributions of the probability that average monthly $PM_{2.5}$ distribution exceeds the $PM_{2.5}$ standard of daily average designated by USEPA. As shown in Fig. 24.8, the area of higher nonattainment probability is located at southwestern Taipei, especially at the neighborhoods of the Taipei Bridge and Chung-Shan Freeway which are the major connections between Taipei city and Taipei county. BME analysis can not only show the nonattainment areas but show their associated probability (or risk). The results can be a reference for the agencies of public health and environmental protection in Taiwan for the future environmental health policy.

24.6 Conclusion

It is important for environmental health studies to investigate health outcomes resulting from PM exposure by considering its spatiotemporal heterogeneity. This study applied BME method to generate $PM_{2.5}$ maps in a composite space-time domain, which are informative and incorporate secondary information; BME is a non-linear approach that provides the complete PM probability distribution, generally non-Gaussian, at each point across space-time. When core knowledge in the form of epidemiologic laws, scientific theories, physical models etc. is available, BME integrates it with multi-sourced site-specific information at various scales. The results show that the incorporation of multi-sourced soft and hard information by BME analysis and mapping can effectively improve the accuracy of $PM_{2.5}$ estimation across space-time. The $PM_{2.5}$ estimations can be represented in probabilistic form and therefore the nonattainment map of $PM_{2.5}$ can also be expressed as well as its uncertainty. Our results of spatiotemporal mapping can be compared with other geographic data under GIS platform to reveal environmental health information, e.g. causal relationship and high health risk area, which can be important good references for governmental agencies. This analysis also demonstrates that the two dominant space-time mechanisms underlying $PM_{2.5}$ space-time distributions in Taipei are associated with local emissions and seasonal effects. Our modelling results agree with the physical interpretation suggested in relevant substantive studies.

Acknowledgments Support for the research was provided by the National Science Council of Taiwan (NSC98-2625-M-002-012), and California Air Resources Board, USA (55245A).

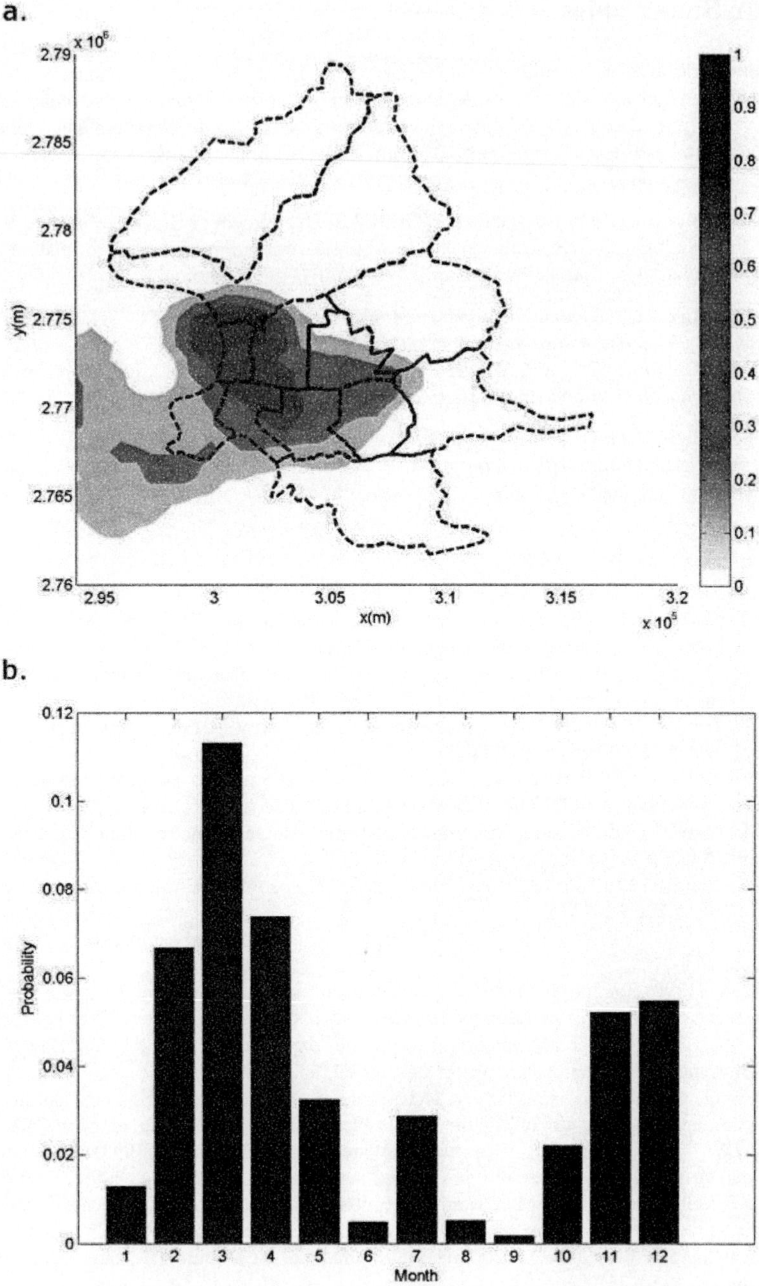

Fig. 24.8 **a** Spatial distribution and **b** temporal distribution of the average probability of monthly $PM_{2.5}$ exceeding the USEPA daily $PM_{2.5}$ standard, i.e. 35 $\mu g/m^3$

Author Biographies

Hwa-Lung Yu is an assistant professor at Department of Bioenvironmental Systems Engineering, National Taiwan University. Dr. Yu works at the laboratory of spatiotemporal environmental modeling (STEMLAB). Currently, Dr. Yu's research efforts focus on the development and applications of spatiotemporal methods in the fields of exposure analysis, climate change, subsurface modeling, as well as disease mapping.

Chih-Hsin Wang is a research assistant at STEMLAB, Department of Bioenvironmental Systems Engineering, National Taiwan University. Mr. Wang is primarily interested in spatiotemporal modeling and mapping of particulate matter by using novel geotechnologies.

Geroge Christakos is the Birch Distinguished Professor at Department of Geography, San Diego State University. He is the author of seven books including "*Spatiotemporal Environmental Health Modelling*", "*Temporal GIS*", and "*Modern Spatiotemporal Geostatistics*". He is the editor of the *Journal of Stochastic Environmental Research and Risk Assessment.*

Yu-Chang Wu is a Ph.D. student at STEMLAB, Department of Bioenvironmental Systems Engineering, National Taiwan University. Mr. Wu is primarily interested in the development of novel spatiotemporal analysis methods.

References

Bogaert P (2002) Spatial prediction of categorical variables: the Bayesian maximum entropy approach. Stoch Environ Res Risk Assess 16:425–448

Bogaert P, D'Or D (2002) Estimating soil properties from thematic soil maps: the Bayesian maximum entropy approach. Soil Sci Soc Am J 66:1492–1500

Chang SC, Lee CT (2007) Evaluation of the trend of air quality in Taipei, Taiwan from 1994 to 2003. Environ Monit Assess 127:87–96

Chang SC, Lee CT (2008) Evaluation of the temporal variations of air quality in Taipei City, Taiwan, from 1994 to 2003. J Environ Manage 86:627–635

Chen M-L, Mao I-F (1998) Spatial variations of airborne particles in metropolitan Taipei. Sci Total Environ 209:225–231

Chen M-L, Mao I-F, Lin I-K (1999) The $PM_{2.5}$ and PM_{10} particles in urban areas of Taiwan. Sci Total Environ 226:227–235

Christakos GA (1990) Bayesian/maximum-entropy view to the spatial estimation problem. Math Geol 22:763–776

Christakos G (1992) Random field models in earth sciences. Academic Press, Inc., San Diego, CA

Christakos G (2000) Modern spatiotemporal geostatistics. Oxford Univ. Press, New York, NY

Christakos G (2002) On the assimilation of uncertain physical knowledge bases: Bayesian and non-Bayesian techniques. Adv in Water Resour 25:1257–1274

Christakos G, Olea RA, Serre ML, Yu H-L, Wang L (2005) Interdisciplinary public health reasoning and epidemic modelling: the case of black death. Springer-Verlag, New York, NY

Dockery DW, Pope CA, Xu XP, Spengler JD, Ware JH, Fay ME et al. (1993) An Association between air-pollution and mortality in 6 united-states cities. N Eng J Med 329:1753–1759

Dominici F, McDermott A, Zeger SL, Samet JM (2003a) National maps of the effects of particulate matter on mortality: exploring geographical variation. Environ Health Perspect 111:39–43

Dominici F, Sheppard L, Clyde M (2003b) Health effects of air pollution: A statistical review. Int Stat Rev 71:243–276

Douaik A, van Meirvenne M, Toth T, Serre M (2004) Space-time mapping of soil salinity using probabilistic bayesian maximum entropy. Stoch Environ Res Risk Assess 18:219–227

Jerrett M, Arain A, Kanaroglou P, Beckerman B, Potoglou D, Sahsuvaroglu T et al. (2005a) A review and evaluation of intraurban air pollution exposure models. J Expo Anal Environ Epidemiol 15:185–204

Jerrett M, Burnett RT, Ma RJ, Pope CA, Krewski D, Newbold KB et al. (2005b) Spatial analysis of air pollution and mortality in Los Angeles. Epidemiology 16:727–736

Kolovos A, Christakos G, Serre ML, Miller CT (2002) Computational Bayesian maximum entropy solution of a stochastic advection-reaction equation in the light of site-specific information. Water Resour Res 38:1318

Kolovos A, Yu H-L, Christakos G (2006) SEKS-GUI v.0.6. User's manual-06 Ed. Deptartment of Geography, San Diego State University, San Diego, CA

Ku S-C (2010) Development of Bayesian maximum entropy method toolbox on quantum GIS – an application of long-term exposure estimation of particulate matter in Taiwan. Department of Bioenvironmental Systems Engineering. Master. National Taiwan University, Taipei, pp. 52

Lee SJ, Balling R, Gober P (2008) Bayesian maximum entropy mapping and the soft data problem in urban climate research. Ann Assoc Am Geogr 98:309–322

Li CS, Lin CH (2002) $PM_1/PM_{2.5}/PM_{10}$ characteristics in the urban atmosphere of Taipei. Aerosol Sci Tech 36:469–473

Liao D, Peuquet DJ, Duan Y, Dou J, Smith RL, Whitsel EA et al. (2005) GIS approaches for estimation of residential-level ambient PM concentrations. Epidemiology 16:S28–S28

Orton TG, Lark RM (2007) Accounting for the uncertainty in the local mean in spatial prediction by Bayesian Maximum Entropy. Stoch Environ Res Risk Assess 21:773–784

Pope CA (2000a) Epidemiology of fine particulate air pollution and human health: biologic mechanisms and who's at risk? Environ Health Perspect 108:713–723

Pope CA (2000b) Review: Epidemiological basis for particulate air pollution health standards. Aerosol Sci Tech 32:4–14

Pope CA, Hansen ML, Long RW, Nielsen KR, Eatough NL, Wilson WE et al. (2004) Ambient particulate air pollution, heart rate variability, and blood markers of inflammation in a panel of elderly subjects. Environ Health Perspect 112:339–345

Porcu E, Mateu J, Saura F (2008) New classes of covariance and spectral density functions for spatio-temporal modelling. Stoch Environ Res Risk Assess 22:S65–S79

Samet JM, Dominici F, Curriero FC, Coursac I, Zeger SL (2000) Fine particulate air pollution and mortality in 20 US Cities, 1987–1994. N Eng J Med 343:1742–1749

Serre ML (1999) Environmental spatiotemporal mapping and ground water flow modelling using the BME and ST methods. University of North Carolina. Deparment. of Environmental Sciences and Engineering, Chapel Hill, NC

Serre M, Yu H-L (2003) Spatiotemporal analysis of particulate matter following the WTC disaster: Initial Results using a geostatistical approach. Center for the Interdisciplinary Study of the Environment 1103. Univ. of North Carolina, Chapel Hill, NC

Smith RL, Kolenikov S, Cox LH (2003) Spatiotemporal modeling of $PM_{2.5}$ data with missing values. J Geophys Res-Atmos 108:STS11.1–STS11.11

Tsai YI, Kuo SC, Lee WJ, Chen CL, Chen PT (2007) Long-term visibility trends in one highly urbanized, one highly industrialized, and two Rural areas of Taiwan. Sci Total Environ 382:324–341

Wibrin MA, Bogaert P, Fasbender D (2006) Combining categorical and continuous spatial information within the Bayesian maximum entropy paradigm. Stoch Environ Res Risk Assess 20:423–433

Wilson JG, Kingham S, Pearce J, Andrew P, Sturmana B (2005) A review of intraurban variations in particulate air pollution: implications for epidemiological research. Atmos Environ 39: 6444–6462

Wilson JG, Zawar-Reza P (2006) Intraurban-scale dispersion modelling of particulate matter concentrations: applications for exposure estimates in cohort studies. Atmos Environ 40:1053–1063

Yang K-L (2002) Spatial and seasonal variation of PM_{10} mass concentrations in Taiwan. Atmos Environ 36:3403–3411

Yu HL, Chen JC, Christakos G, Jerrett M (2009b) BME estimation of residential exposure to ambient PM_{10} and ozone at multiple time scales. Environ Health Perspect 117:537–544

Yu H-L, Christakos G, Chen J-C (2007a) Spatiotemporal air pollution modeling and prediction in epidemiologic research. In: Columbus F (ed) Air pollution research trends. Nova Science Publishers, Inc., Hauppauge, NY, pp. 57–75

Yu H-L, Kolovos A, Christakos G, Chen J-C, Warmerdam S, Dev B (2007b) Interactive spatiotemporal modelling of health systems: the SEKS – GUI framework. Stoch Environ Res Risk Assess 21:555–572

Yu H-L, Wang C-H, Wu Y-Z (2009a) An automatic approach to mean and covariance estimation of spatiotemporal nonstationary processes. In: Dubois G (ed) StatGIS 2009, Milos, Greece

Index

J.A. Maantay, S. McLafferty (eds.), *Geospatial Analysis of Environmental Health*, Geotechnologies and the Environment 4, DOI 10.1007/978-94-007-0329-2,